Invariant Descriptive Set Theory

PURE AND APPLIED MATHEMATICS

A Program of Monographs, Textbooks, and Lecture Notes

MONOGRAPHS AND TEXTBOOKS IN PURE AND APPLIED MATHEMATICS

Recent Titles

Walter Ferrer and Alvaro Rittatore, Actions and Invariants of Algebraic Groups (2005)

Christof Eck, Jiri Jarusek, and Miroslav Krbec, Unilateral Contact Problems: Variational Methods and Existence Theorems (2005)

M. M. Rao, Conditional Measures and Applications, Second Edition (2005)

A. B. Kharazishvili, Strange Functions in Real Analysis, Second Edition (2006)

Vincenzo Ancona and Bernard Gaveau, Differential Forms on Singular Varieties: De Rham and Hodge Theory Simplified (2005)

Santiago Alves Tavares, Generation of Multivariate Hermite Interpolating Polynomials (2005)

Sergio Macías, Topics on Continua (2005)

Mircea Sofonea, Weimin Han, and Meir Shillor, Analysis and Approximation of Contact Problems with Adhesion or Damage (2006)

Marwan Moubachir and Jean-Paul Zolésio, Moving Shape Analysis and Control: Applications to Fluid Structure Interactions (2006)

Alfred Geroldinger and Franz Halter-Koch, Non-Unique Factorizations: Algebraic, Combinatorial and Analytic Theory (2006)

Kevin J. Hastings, Introduction to the Mathematics of Operations Research with *Mathematica®*, Second Edition (2006)

Robert Carlson, A Concrete Introduction to Real Analysis (2006)

John Dauns and Yiqiang Zhou, Classes of Modules (2006)

N. K. Govil, H. N. Mhaskar, Ram N. Mohapatra, Zuhair Nashed, and J. Szabados, Frontiers in Interpolation and Approximation (2006)

Luca Lorenzi and Marcello Bertoldi, Analytical Methods for Markov Semigroups (2006)

M. A. Al-Gwaiz and S. A. Elsanousi, Elements of Real Analysis (2006)

Theodore G. Faticoni, Direct Sum Decompositions of Torsion-Free Finite Rank Groups (2007)

R. Sivaramakrishnan, Certain Number-Theoretic Episodes in Algebra (2006)

Aderemi Kuku, Representation Theory and Higher Algebraic K-Theory (2006)

Robert Piziak and P. L. Odell, Matrix Theory: From Generalized Inverses to Jordan Form (2007)

Norman L. Johnson, Vikram Jha, and Mauro Biliotti, Handbook of Finite Translation Planes (2007)

Lieven Le Bruyn, Noncommutative Geometry and Cayley-smooth Orders (2008)

Fritz Schwarz, Algorithmic Lie Theory for Solving Ordinary Differential Equations (2008)

Jane Cronin, Ordinary Differential Equations: Introduction and Qualitative Theory, Third Edition (2008)

Su Gao, Invariant Descriptive Set Theory (2009)

Invariant Descriptive Set Theory

Su Gao

CRC Press
Taylor & Francis Group
Boca Raton London New York

CRC Press is an imprint of the
Taylor & Francis Group, an **informa** business
A CHAPMAN & HALL BOOK

CRC Press
Taylor & Francis Group
6000 Broken Sound Parkway NW, Suite 300
Boca Raton, FL 33487-2742

First issued in paperback 2019

© 2009 by Taylor & Francis Group, LLC
CRC Press is an imprint of Taylor & Francis Group, an Informa business

No claim to original U.S. Government works

ISBN-13: 978-1-58488-793-5 (hbk)
ISBN-13: 978-0-367-38696-2 (pbk)

Library of Congress Cataloging-in-Publication Data

Gao, Su, 1968-
    Invariant descriptive set theory / Su Gao.
      p. cm. -- (Pure and applied mathematics)
    Includes bibliographical references and index.
    ISBN 978-1-58488-793-5 (alk. paper)
    1. Descriptive set theory. 2. Invariant sets. I. Title.

QA248.G36 2009
511.3'22--dc22   2008031545

Visit the Taylor & Francis Web site at
http://www.taylorandfrancis.com

and the CRC Press Web site at
http://www.crcpress.com

To Shuang, Alvin, and Tony, with love

Contents

Preface

My intention in writing this book is to bring into one place the basics of invariant descriptive set theory, also known as the descriptive set theory of definable equivalence relations. Invariant descriptive set theory has been an active field of research for about 20 years. Many researchers and students are impressed by its fast development and its relevance to other fields of mathematics, and would like to be better acquainted with the theory. I have tried to make this book as self-contained as possible, and at the same time covered what I believe to be the essential concepts, methods, and results.

The book is designed as a graduate text suitable for a year-long course. I have kept the sections short so that they can be used as lecture notes, and most of the sections are followed by a number of exercise problems. Many exercises are propositions and even theorems needed later in the book. So the student is urged to make a serious effort to work them out.

Ideally, the student should have some experience with classical and effective descriptive set theory before reading this book. But since this is most likely not the case, I have only assumed that the student knows some general topology. In the first chapter a review of classical and effective descriptive set theory is given, and throughout the book results are recalled as they become necessary. I can imagine that it is hard, but possible, for a student who has never seen any descriptive set theory to get started on the subject, but I believe that, with patience and diligence, the obstacles will be overcome eventually.

I have to remark, primarily for the experts in the field, that the book is not intended to be a comprehensive account of all aspects of invariant descriptive set theory. A reader who is familiar with the materials of the book and who is interested in further developments should have no problem following the current literature on many topics and applications. The selection of topics contained in this book was greatly influenced by the book of Becker and Kechris [8] and some unpublished notes of Kechris.

I would like to thank Julia Knight for inviting me to give a short course on invariant descriptive set theory at the University of Notre Dame in 2005. The first ideas for this book came from the notes for that short course. I would also like to thank the participants of the short course for typing the notes and for conversations on the topic. I am grateful to Dave Marker and Peter Cholak for the encouragement to write a graduate textbook on the subject. Special thanks are due to Longyun Ding and Vincent Kieftenbeld for comments and suggestions on the manuscript.

I would like to acknowledge the financial support of the National Science Foundation and the University of North Texas for the composition of this book and for related research. It would be impossible for me to write this book without the faculty developmental leave granted by the University of North Texas. Many thanks to all at CRC/Taylor & Francis for their untiring work to make this book a reality.

I am indebted to Greg Hjorth for leading me into the field and to Alekos Kechris for many years of advice and support. It is a privilege for me to be acquainted with many colleagues and experts in the field, too numerous to list (see the References and Index), whose research results shaped this book. I benefited a great deal from communications with them and from their contributions to the literature. I present this book to them with gratitude and pride.

Su Gao
Denton, Texas

Part I

Polish Group Actions

Chapter 1

Preliminaries

This chapter reviews the concepts and results of classical and effective descriptive set theory that will be used in this book. Classical descriptive set theory was founded by Baire, Borel, Lebesgue, Luzin, Suslin, Sierpinski, and others in the first two decades of the twentieth century. The theory studies the *descriptive complexity* of sets of real numbers arising in ordinary mathematics, mostly in topology and analysis. The most striking achievements of this theory are the proofs of regularity properties of low-level definable sets of reals.

Effective descriptive set theory was created later in the century by introducing into the classical theory the new and powerful tools developed from recursion theory (now called computability theory). Computability theory came about from completely different motivations and was invented by another group of great minds such as Gödel, Church, Turing, Kleene, and others. It provided a framework to understand the structural and computational complexity of sets and functions (mostly pertaining to natural numbers). The classical theory is much better understood from the perspective of the effective theory.

In this chapter we review the basic concepts and results of both classical and effective theory that are relevant to the remainder of the book. Readers unfamiliar with these topics should be able to get a working idea about the content and the flavor of the theory by reading this chapter alone, and especially if they are diligent enough to work out the exercise problems in this chapter. Later in the book there will be more specific concepts and results being reviewed as they become necessary tools. In addition, some proofs are given in the appendix. The reader can probably get by with these reviews without ever systematically studying the classical and effective descriptive set theory, but an understanding of the comprehensive theory will be hugely advantageous.

For a complete treatment of these topics the reader can consult the standard references. For classical descriptive set theory the standard textbook is [97]

> A. S. Kechris, *Classical Descriptive Set Theory*. Graduate Texts in Mathematics 156. Springer-Verlag, New York, 1995.

And for effective descriptive set theory our standard source is [126]

> Y. N. Moschovakis, *Descriptive Set Theory*. Studies in Logic and

Foundations of Mathematics 100. North-Holland, Amsterdam, New York, 1980.

In addition, there are more concise treatments of the subject which provide efficient inroads to the theory and can be used as alternatives for or in conjunction with the standard texts: [148] by Srivastava, [125] by Martin and Kechris, [116] by Mansfield and Weitkamp, and [133] by Sacks. Thus in this chapter our review will be pragmatic and sketchy. Many facts, even theorems, are given without proof. Also we have left out important aspects of the theory just to get the reader quickly prepared to deal with the main topics of the book starting in Chapter 2.

Let the story begin.

1.1   Polish spaces

Definition 1.1.1
A topological space is **Polish** if it is separable and completely metrizable.

Some basic properties of Polish spaces are gathered in the following proposition.

Proposition 1.1.2

(a) *Any Polish space is second countable and normal.*

(b) *Any Polish space is Baire. (Recall that a topological space is **Baire** if the intersection of countable many dense open sets is dense.)*

(c) *A finite or countable product of Polish spaces is Polish.*

(d) *A subspace Y of a Polish space X is Polish iff Y is G_δ in X, that is, Y is the intersection of countably many open sets in X.*

(e) *A quotient space of a Polish space is not necessarily Polish.*

Clause (b) in the proposition is a direct consequence of the **Baire category theorem**, which says that a complete metric space is Baire. Examples of Polish spaces are abundant in mathematics. Some of the most familiar examples are listed below.

Example 1.1.3

(1) All countable spaces with the discrete topology are Polish. These include

the following spaces:

$$\mathbb{N} = \omega = \{0, 1, 2, \dots\},$$

$$\mathbb{N}_+ = \mathbb{N} - \{0\} = \{1, 2, \dots\},$$

$$\mathbb{Z} = \{\dots, -2, -1, 0, 1, 2, \dots\}.$$

Throughout this book we use $\mathbb{N}$ and ω interchangeably.

(2) $\mathbb{R}^n$ with the usual topology for $1 \leq n \leq \omega$ are Polish.

(3) The Baire space $\mathcal{N} = \omega^\omega$ is Polish. A complete metric on $\mathcal{N}$ is defined by

$$d(x, y) = \begin{cases} 0, & \text{if } x = y, \\ 2^{-n-1}, & \text{if } n \in \omega \text{ is the least such that } x(n) \neq y(n). \end{cases}$$

(4) The Cantor space 2^ω is a closed subspace of $\mathcal{N}$, hence is Polish.

(5) All separable Banach spaces, such as c_0 and ℓ_p ($1 \leq p < \infty$), are Polish. Note that ℓ_∞ is not separable and therefore not a Polish space.

(6) All compact metrizable spaces are Polish (see Exercise 1.1.1).

Some nontrivial examples of Polish spaces involve hyperspaces of sets or functions. We examine two such examples below.

Let d be a compatible metric on a Polish space X. Then we can define a compatible metric d' on X with the property that $d' \leq 1$ as follows:

$$d'(x, y) = \frac{d(x, y)}{1 + d(x, y)}.$$

Moreover, if d is complete then so is d'. If $x \in X$ and $A \subseteq X$, then we denote

$$d(x, A) = d(A, x) = \inf\{d(x, y) : y \in A\}.$$

Let X be a Polish space. Let $K(X)$ denote the space of all compact subsets of X equipped with the **Vietoris topology** generated by subbasic open sets of the following form:

$$\{K \in K(X) : K \subseteq U\}, \text{ or}$$

$$\{K \in K(X) : K \cap U \neq \emptyset\},$$

for U open in X. Then $K(X)$ is Polish. An explicit compatible metric on $K(X)$ is known as the **Hausdorff metric**. To define it, let d be a compatible metric on X with $d \leq 1$. Then the Hausdorff metric d_H on $K(X)$ is defined by

$$
d_H(K,L) = \begin{cases} 0, & \text{if } K = L = \emptyset, \\ 1, & \text{if exactly one of } K, L \text{ is } \emptyset, \\ \max\{\max\{d(x,L) : x \in K\}, \max\{d(K,y) : y \in L\}\}, \\ & \text{if } K, L \neq \emptyset. \end{cases}
$$

It is routine to check the following facts, which are left as exercises (Exercises 1.1.4 and 1.1.5). d_H is compatible with the Vietoris topology on $K(X)$. Moreover, if d is complete then so is d_H. Let $D \subseteq X$ be a countable dense subset of X. Then the set

$$
\{K \in K(X) : K \subseteq D \text{ is finite}\}
$$

is countable and dense in $K(X)$. These imply that $K(X)$ is a Polish space. Moreover, if X is a compact Polish space then so is $K(X)$.

Another kind of hyperspace comes from Lipschitz functions between Polish metric spaces.

Definition 1.1.4
A **Polish metric space** is a separable complete metric space.

Let $(X, d_X), (Y, d_Y)$ be Polish metric spaces. A map $\varphi : X \to Y$ is **Lipschitz** if for all $x_1, x_2 \in X$,

$$
d_Y(\varphi(x_1), \varphi(x_2)) \leq d_X(x_1, x_2).
$$

Let $L(X,Y)$ denote the space of all Lipschitz maps from X into Y. Equip $L(X,Y)$ with the pointwise convergence topology. Alternatively, let $D \subseteq X$ be a countable dense set and enumerate its elements by $q_1, q_2 \ldots$. Define a metric on $L(X,Y)$ by

$$
d_L(\varphi, \psi) = \sum_{i=1}^{\infty} 2^{-i} d'_Y(\varphi(q_i), \psi(q_i)),
$$

where $d'_Y \leq 1$ is the complete metric on Y compatible with d_Y defined above. Then d_L is a metric on $L(X,Y)$ compatible with the pointwise convergence topology. Moreover, if d_Y is complete so is d_L. Hence $L(X,Y)$ is Polish. However note that the definition of the particular metric d_L depends on the choice of the countable dense set D, thus the metric d_L is not canonical.

An important special kind of Lipschitz maps is the class of distance-preserving functions. A map $\varphi : X \to Y$ is **distance-preserving** (or an **isometric embedding**) if for all $x_1, x_2 \in X$,

$$
d_Y(\varphi(x_1), \varphi(x_2)) = d_X(x_1, x_2).
$$

Let $DP(X,Y)$ denote the space of all distance-preserving maps from X into Y. Equip $DP(X,Y)$ with the pointwise convergence topology. Then $DP(X,Y)$ is a closed subspace of $L(X,Y)$ and hence is Polish.

Exercise 1.1.1 Show that a compact metrizable space is second countable and that any compatible metric is complete. Thus a compact metrizable space is Polish.

Exercise 1.1.2 Show that a locally compact metrizable space is Polish iff its one-point compactification is metrizable.

Exercise 1.1.3 Let d_0 be a compatible metric on a Polish space X. Define a metric d_1 by
$$d_1(x, y) = 1 - e^{-d_0(x,y)}.$$
Show that $d_1 \leq 1$ is a compatible metric on X. Moreover, if d_0 is complete then so is d_1.

Exercise 1.1.4 Let X be a Polish space and $d \leq 1$ a compatible metric on X. Show that the Hausdorff metric d_H is compatible with the Vietoris topology on $K(X)$. Moreover, if d is complete, then so is d_H.

Exercise 1.1.5 Show that if X is compact Polish then so is $K(X)$.

Exercise 1.1.6 Let X be a compact Polish space with a compatible metric d_X. Let Y be a Polish space with a compatible complete metric d_Y. Denote by $C(X, Y)$ the space of all continuous functions from X into Y. Show that the following topologies are equivalent and Polish:

(i) the compact-open topology, that is, the topology given by subbasic open sets of the form
$$\{f \in C(X, Y) : f(K) \subseteq U\}$$
for $K \subseteq X$ compact and $U \subseteq Y$ open;

(ii) the metric topology given by the supnorm metric
$$d_C(f, g) = \sup\{d_Y(f(x), g(x)) : x \in X\}.$$

Exercise 1.1.7 Let $(X, d_X), (Y, d_Y)$ be Polish metric spaces. A map $\varphi : (X, d_X) \to (Y, d_Y)$ is an **isometry** if it is distance-preserving and onto. Denote by $\mathrm{Iso}(X, Y)$ the space of all isometries from X onto Y equipped with the pointwise convergence topology. Show that $\mathrm{Iso}(X, Y)$ is a G_δ subspace of $DP(X, Y)$, and hence is Polish.

Exercise 1.1.8 Let (X, d) be a Polish metric space. For $x \in X$, let $\varphi_x(y) = d(x, y)$. Show that

(1) $\varphi_x \in L(X, \mathbb{R})$ for all $x \in X$.

(2) The map $x \mapsto \varphi_x$ is a Lipschitz map from X into $L(X, \mathbb{R})$, regardless of the choice of the countable set D in the definition of the metric d_L.

Exercise 1.1.9 Let X be a separable metrizable space with a compatible metric $d \leq 1$. Let $L(X, d)$ denote the space $L((X, d), [0, 1])$, that is, the space of all Lipschitz functions from (X, d) into $[0, 1]$ (with the usual metric) with the pointwise convergence topology. Show that $L(X, d)$ is compact metrizable, hence is Polish.

1.2　The universal Urysohn space

In 1927, Urysohn [157] defined a Polish metric space which turned out to contain all other Polish spaces as closed subspaces. Besides the universality, this space also possesses other nice properties. It has attracted a lot of attention in recent years. In this section we give a construction of the universal Urysohn space due to Katětov [88].

Definition 1.2.1
Let (X, d) be a separable metric space. A function $f : X \to \mathbb{R}$ is **admissible** if for all $x, y \in X$,

$$|f(x) - f(y)| \leq d(x, y) \leq f(x) + f(y).$$

Every admissible function on X corresponds to a way to extend X by one point. Let $E(X)$ denote the space of all admissible functions on (X, d). Define a metric on $E(X)$ by

$$d_E(f, g) = \sup\{|f(x) - g(x)| : x \in X\}.$$

We can view $E(X)$ as an extension of X with the following embedding of X into $E(X)$. For $x \in X$ let
$$f_x(y) = d(x, y).$$

Then it is easy to check that $x \mapsto f_x$ is an isometric embedding of X into $E(X)$. For notational simplicity we will identify X with its image in $E(X)$ under this canonical isometric embedding and regard X as a subset of $E(X)$.

Note, however, that $E(X)$ is not necessarily separable (see Exercise 1.2.4). To get separability we need to consider the concept of support for admissible functions.

Definition 1.2.2
Let f be an admissible function on (X, d) and $S \subseteq X$. We say that S is a **support** for f if for all $x \in X$,

$$f(x) = \inf\{f(y) + d(x, y) : y \in S\}.$$

f is called **finitely supported** if there is a finite support for f.

Let now

$$E(X, \omega) = \{f \in E(X) : \ f \text{ is finitely supported}\}.$$

Then $E(X, \omega)$ is a subspace of $E(X)$. Note that the function f_x defined above has a support $\{x\}$, thus in particular it is finitely supported. This shows that $E(X, \omega)$ is also an extension of X. Moreover, $E(X, \omega)$ is separable if X is (see Exercise 1.2.5).

Given any separable metric space (X, d), define by induction a sequence of spaces X_i, $i \in \omega$, as follows:

$$X_0 = X,$$
$$X_{i+1} = E(X_i, \omega).$$

Since each X_{i+1} is a natural extension of X_i as a metric space, it makes sense to take the union of all of X_i. Thus we let

$$X_\omega = \bigcup_{i \geq 0} X_i,$$

and let d_ω be the canonical extension of d on X_ω.

Now the main results of Urysohn imply that, for any separable metric spaces X and Y, the completions of X_ω and of Y_ω are isometric. This unique completion is called the **universal Urysohn space**. It is obvious that it contains every Polish (metric) space as a closed subspace. Officially, we make the following definition.

Notation 1.2.3
The **universal Urysohn space** $\mathbb{U}$ is the completion of $(\mathbb{R}_\omega, d_\omega)$.

In the remainder of this section we prove the results of Urysohn. First we need the following important concept.

Definition 1.2.4
A metric space (X, d) has the **Urysohn property** if for any finite subset F of X and any admissible function f on F, there is $x \in X$ such that $f(p) = d(x, p)$ for any $p \in F$.

Theorem 1.2.5 (Urysohn)
Let (X, d_X) and (Y, d_Y) be Polish metric spaces with the Urysohn property. Then there is an isometry from X onto Y.

Proof. Let D_X be a countable dense subset of X and enumerate its elements as $x_0, x_1, \ldots, x_n, \ldots$. Similarly, let D_Y be a countable dense subset of Y and enumerate it as $y_0, y_1, \ldots, y_n, \ldots$. Using the Urysohn property for both X and Y, we define a partial isometry φ from a subset of X into Y, so that $D_X \subseteq \text{dom}(\varphi)$ and $D_Y \subseteq \text{range}(\varphi)$. This φ is constructed by induction on

n. At stage n we construct a partial isometry φ_n extending φ_{n-1} so that $\mathrm{dom}(\varphi_n)$ is finite, $\{x_0, \ldots, x_n\} \subseteq \mathrm{dom}(\varphi_n)$ and $\{y_0, \ldots, y_n\} \subseteq \mathrm{range}(\varphi_n)$. To begin with we let $\varphi_0(x_0) = y_0$ and φ_0 is otherwise undefined. At stage $n > 0$, suppose that φ_{n-1} has been defined so that $\{x_0, \ldots, x_{n-1}\} \subseteq \mathrm{dom}(\varphi_{n-1})$ and $\{y_0, \ldots, y_{n-1}\} \subseteq \mathrm{range}(\varphi_{n-1})$. If $x_n \in \mathrm{dom}(\varphi_{n-1})$ then we let $\varphi' = \varphi_{n-1}$. Otherwise, let $F = \mathrm{range}(\varphi_{n-1})$ and consider the admissible function on F defined by

$$f(p) = d_X(\varphi_{n-1}^{-1}(p), x_n).$$

By the Urysohn property of Y there is $y \in Y$ so that

$$d_Y(p, y) = f(p) = d_X(\varphi_{n-1}^{-1}(p), x_n).$$

We extend φ_{n-1} to φ' by defining $\varphi'(x_n) = y$. Now if $y_n \in \mathrm{range}(\varphi')$ then we let $\varphi_n = \varphi'$ and go on to the next stage. Otherwise apply the above argument to φ'^{-1} and use the Urysohn property of X to obtain an extension of φ'. Define φ_n to be this extension. We have thus finished the definition of φ_n. Let φ be the union of all φ_n we have defined. Then it has the required properties.

Finally extend φ further to the whole space X by the following definition. For $x \in X - D_X$ let (z_n) be a sequence in D_X with $z_n \to x$ as $n \to \infty$. Then define

$$\varphi(x) = \lim_n \varphi(z_n).$$

Since (z_n) is a d_X-Cauchy sequence, $(\varphi(z_n))$ is a d_Y-Cauchy sequence and therefore the right-hand-side limit exists. By a similar argument we see that $\varphi(x)$ does not depend on the choice of (z_n) and therefore is well defined. It is also easy to see that the extended φ is onto Y and that it is an isometry. ∎

The above proof illustrates a typical use of the Urysohn property in a back-and-forth argument. It is easy to see that, for any separable metric space X the extension space X_ω has the Urysohn property. In the following we show that the completion of a metric space with the Urysohn property retains the Urysohn property. This implies that the universal Urysohn space $\mathbb{U}$ has the Urysohn property as well, and therefore its uniqueness follows from the above theorem.

The rest of the section is devoted to a proof of Theorem 1.2.7, which is elementary but a bit technical. Since the technique is not essential for the rest of this book it is safe for the reader to skip it. We need the following lemma in the proof.

Lemma 1.2.6
Let (Y, d) be a separable complete metric space and $X \subseteq Y$ be dense. Suppose $(X, d \upharpoonright X)$ has the Urysohn property. Then for any finite set $F \subseteq Y$ and admissible function f on F, and for any $\epsilon > 0$, there is $y_\epsilon \in X$ such that

$$|d(p, y_\epsilon) - f(p)| < \epsilon, \text{ for all } p \in F.$$

Proof. Let $F = \{y_0, \ldots, y_n\}$. For definiteness we assume that $f(y_0) \geq f(y_1) \geq \cdots \geq f(y_n) > 0$ and let $\delta = \min\{d(y_i, y_j) : i \neq j \leq n\}$. Choose

$$\epsilon_0 < \min\left\{\frac{\epsilon}{3n + 2}, \frac{\delta}{3n + 2}\right\}.$$

By the density of X we can find $x_0, \ldots, x_n$ such that $d(x_i, y_i) < \epsilon_0$ for all $i \leq n$. Define

$$g(x_i) = f(y_i) + (3i + 1)\epsilon_0, \quad \text{for all } i \leq n.$$

We claim that g is admissible on the set $\{x_0, \ldots, x_n\}$. In fact, for $i < j$,

$$g(x_j) - g(x_i) = f(y_j) - f(y_i) + 3(j - i)\epsilon_0 \leq 3n\epsilon_0$$

$$= (3n + 2)\epsilon_0 - 2\epsilon_0 \leq \delta - (d(x_i, y_i) + d(x_j, y_j))$$

$$\leq d(y_i, y_j) - d(x_i, y_i) - d(x_j, y_j) \leq d(x_i, x_j)$$

and on the other hand,

$$g(x_i) - g(x_j) = f(y_i) - f(y_j) - 3(j - i)\epsilon_0 \leq f(y_i) - f(y_j) - 2\epsilon_0$$

$$\leq d(y_i, y_j) - d(x_i, y_i) - d(x_j, y_j) \leq d(x_i, x_j).$$

For the other inequality we have

$$d(x_i, x_j) \leq d(x_i, y_i) + d(x_j, y_j) + f(y_i) + f(y_j)$$

$$\leq 2\epsilon_0 + f(y_i) + f(y_j) \leq g(x_i) + g(x_j).$$

Thus g is indeed admissible on $\{x_0, \ldots, x_n\} \subseteq X$. By the Urysohn property of X we get $y_\epsilon \in X$ such that $d(x_i, y_\epsilon) = g(x_i)$ for all $i \leq n$. Thus

$$|d(y_i, y_\epsilon) - f(y_i)| \leq |d(x_i, y_\epsilon) - f(y_i)| + |d(x_i, y_\epsilon) - d(y_i, y_\epsilon)| \leq (3i + 1)\epsilon_0 + \epsilon_0 < \epsilon.$$

$\Box$

Theorem 1.2.7 (Urysohn)
Let (X, d) be a separable metric space with the Urysohn property. Then the completion of (X, d) also has the Urysohn property.

Proof. Let Y denote the completion of (X, d) and for simplicity let d denote also the complete metric on Y. Let $F = \{y_0, \ldots, y_n\} \subseteq Y$ and f be an admissible function on F. We need to find $y \in Y$ such that $f(y_i) = d(y_i, y)$ for all $i \leq n$.

Assume again $f(y_0) \geq f(y_1) \geq \cdots \geq f(y_n) > 0$. We define a sequence (z_k) in X by induction on k. Let $z_0 \in X$ be such that $|d(y_i, z_0) - f(y_i)| < f(y_n)$ for all $i \leq n$. Such a z_0 exists by Lemma 1.2.6. In general suppose z_k is already defined with $|d(y_i, z_k) - f(y_i)| \leq 2^{-k} f(y_n)$ for all $i \leq n$. We define z_{k+1} to satisfy $|d(y_i, z_{k+1}) - f(y_i)| \leq 2^{-k-1} f(y_n)$ for all $i \leq n$, and that

$$|d(z_k, z_{k+1})| \leq 2^{-k+1} f(y_n).$$

To define such a z_{k+1} consider the set $F' = \{y_0, \ldots, y_n, z_k\}$ and the function $f' : F' \to \mathbb{R}$ defined by

$$f'(p) = \begin{cases} f(y_i), & \text{if } p = y_i \text{ for } i \leq n, \\ 2^{-k} f(y_n), & \text{if } p = z_k. \end{cases}$$

The inductive hypothesis implies that f' is admissible. Thus by Lemma 1.2.6 again we can find z_{k+1} so that

$$|d(y_i, z_{k+1}) - f(y_i)| \leq 2^{-k-1} f(y_n) \text{ and } |d(z_k, z_{k+1}) - 2^{-k} f(y_n)| \leq 2^{-k-1} f(y_n).$$

This gives the desired z_{k+1}. Finally note that (z_k) is a Cauchy sequence, and thus by letting $y = \lim_k z_k$, we get $d(y_i, y) = f(y_i)$ as desired.   □

Exercise 1.2.1 Show that the map $x \mapsto f_x = d(x, \cdot)$ defined from (X, d) into $(E(X), d_E)$ is an isometric embedding.

Exercise 1.2.2 Let (X, d) be a metric space, $S \subseteq X$, and f an admissible function on S. For all $x \in X$ define

$$g(x) = \inf\{ f(y) + d(x, y) : y \in S\}.$$

Show that g is an admissible function on X, g extends f, and S is a support for g.

Exercise 1.2.3 Show that if $|X| = n < \infty$ then $E(X)$ is homeomorphic to a closed subspace of $\mathbb{R}^n$.

Exercise 1.2.4 Let $X = \omega$ with the trivial metric: $d(x, y) = 1$ iff $x \neq y$. Show that $E(X)$ is not separable.

Exercise 1.2.5 Show that $E(D, \omega)$ is dense in $E(X, \omega)$ if D is dense in X.

Exercise 1.2.6 Show that X_ω has the Urysohn property for any separable metric space X.

A Polish metric space X is **universal** if for any Polish metric space Y there is an isometric embedding from Y into X. A metric space (X, d) is **ultrahomogeneous** if for any finite subsets F_1, F_2 and an isometry φ from F_1 onto F_2 there is an isometry Φ of X onto itself extending φ.

Exercise 1.2.7 Show that $\mathbb{U}$ is a universal Polish metric space.

Exercise 1.2.8 Show that $\mathbb{U}$ is ultrahomogeneous.

Exercise 1.2.9 Show that an ultrahomogeneous, universal Polish metric space has the Urysohn property, hence must be isometric to $\mathbb{U}$.

1.3 Borel sets and Borel functions

The Borel hierarchy is the most important definability notion used to probe the complexity of sets of reals. Let X be a Polish space. Recall that a subset $A \subseteq X$ is **Borel** if it is an element of the smallest σ-algebra on X containing all open sets in X. We define by transfinite induction the collections of $\mathbf{\Pi}_\alpha^0(X)$, $\mathbf{\Sigma}_\alpha^0(X)$, and $\mathbf{\Delta}_\alpha^0(X)$ sets for countable ordinals $\alpha \geq 1$. To begin with, we have

$$\mathbf{\Sigma}_1^0(X) = \text{ the collection of all open sets in } X.$$

Then for $\alpha \geq 1$ and $A \subseteq X$, let

$$A \in \mathbf{\Pi}_\alpha^0(X) \iff X - A \in \mathbf{\Sigma}_\alpha^0(X)$$

and

$$A \in \mathbf{\Delta}_\alpha^0(X) \iff A \in \mathbf{\Sigma}_\alpha^0(X) \text{ and } A \in \mathbf{\Pi}_\alpha^0(X).$$

For $\alpha > 1$ and $A \subseteq X$, let

$$A \in \mathbf{\Sigma}_\alpha^0(X) \iff A = \bigcup_{n \in \omega} B_n, \text{ where } B_n \in \mathbf{\Pi}_{\beta_n}^0 \text{ for } \beta_n < \alpha.$$

Then $A \subseteq X$ is Borel iff $A \in \mathbf{\Sigma}_\alpha^0(X)$ for some $1 \leq \alpha < \omega_1$. A basic fact about the Borel hierarchy on an uncountable Polish space is that for any $1 \leq \alpha < \omega_1$, there are $\mathbf{\Sigma}_\alpha^0$ sets which are not $\mathbf{\Pi}_\alpha^0$. This guarantees that the Borel hierarchy does not collapse.

For low-level Borel sets we will continue to use the well-established terminology: F_σ for $\mathbf{\Sigma}_2^0$ and G_δ for $\mathbf{\Pi}_2^0$.

Let X and Y be Polish spaces. A function $f : X \to Y$ is **continuous** if for any open set U in Y, $f^{-1}(U)$ is open in X. f is **Borel** if for any open set U in Y, $f^{-1}(U)$ is Borel in X. There is a hierarchy for the Borel functions corresponding to the Borel hierarchy of sets. For $\alpha < \omega_1$ a function $f : X \to Y$ is of **Baire class** α if for any open set U in Y, $f^{-1}(U) \in \mathbf{\Sigma}_{\alpha+1}^0(X)$. Thus functions of Baire class 0 are precisely the continuous functions, and Baire class 1 functions pull back open sets to $\mathbf{\Sigma}_2^0$ sets, and so on.

Even a continuous image of a closed set is not necessarily Borel. In contrast we have the following useful fact.

Theorem 1.3.1 (Luzin–Suslin)
Let X, Y be Polish spaces and $f : X \to Y$ be one-to-one and Borel. Then for any Borel set A in X, $f(A)$ is Borel in Y.

The proof is nontrivial and can be found in a standard textbook, for example, Reference [97] Theorem 15.1.

Definition 1.3.2
Let X, Y be Polish spaces. We say that X and Y are **Borel isomorphic** if there is a Borel bijection f between X and Y. A Borel bijection is also called a **Borel isomorphism**.

In view of Theorem 1.3.1 if $f : X \to Y$ is a Borel bijection, then f^{-1} is (well-defined and) Borel; thus being Borel isomorphic is an equivalence notion for Polish spaces. Combining Theorem 1.3.1 with the standard Cantor–Bernstein theorem in set theory, we get a version of the Cantor–Bernstein theorem for Polish spaces as follows.

Theorem 1.3.3 (Cantor–Bernstein)
Let X, Y be Polish spaces. If there exist one-to-one Borel functions $f : X \to Y$ and $g : Y \to X$, then X and Y are Borel isomorphic.

In the rest of this section we give a proof that all uncountable Polish spaces are Borel isomorphic. Our strategy is to show that every uncountable Polish space is Borel isomorphic to the Baire space ω^ω. Let us recall some standard notation related to this space.

Let $\omega^{<\omega}$ denote the set of all finite sequences of natural numbers. This in particular contains the empty sequence, which we denote by $\emptyset$. For $s \in \omega^{<\omega}$, let $\mathrm{lh}(s)$ denote the length of s. For instance, for $s = (s(0), \ldots, s(n-1))$, we have that $\mathrm{lh}(s) = n$. If $m < \mathrm{lh}(s)$ we let

$$s \upharpoonright m = (s(0), \ldots, s(m-1)).$$

For $s, t \in \omega^{<\omega}$, we say that s is an **initial segment** of t, or that t **extends** s, and denote $s \subseteq t$, if $s = t \upharpoonright \mathrm{lh}(s)$. When $t \supseteq s$ and $\mathrm{lh}(t) = \mathrm{lh}(s) + 1$, we write t as $s^\frown n$ if $t(\mathrm{lh}(s)) = n$. We think of the structure of $\omega^{<\omega}$ with the partial order $\subseteq$ as a tree with root $\emptyset$. Then each element of the Baire space ω^ω corresponds to an infinite branch of this tree. Similarly we can define the notation $x \upharpoonright m$ and $s \subseteq x$ for $x \in \omega^\omega$.

Now for $s \in \omega^{<\omega}$ we let

$$N_s = \{x \in \omega^\omega : s \subseteq x\}.$$

Then the collection of all N_s for $s \in \omega^{<\omega}$ is a countable base for the topology of ω^ω, and each N_s is a basic clopen set.

We also note that the same notation and terminology apply to the Cantor space 2^ω.

The Borel hierarchy on ω^ω or 2^ω will be simply denoted as Σ^0_α, Π^0_α, and Δ^0_α.

This finishes the review of notation related to ω^ω. Next we review some basic properties of an uncountable Polish space.

Definition 1.3.4
Let X be a Polish space. A point x in X is **isolated** if $\{x\}$ is open in X. X is **perfect** if there is no isolated point in X. A subset A of X is **perfect** if A is closed in X and perfect as a subspace of X.

Theorem 1.3.5 (Cantor–Bendixson)
Any Polish space X can be uniquely written as $X = P \cup C$, where P is a perfect subset of X and C is countable and open in X.

A proof of this theorem, together with the notion of Cantor–Bendixson rank and derivatives, can be found in any standard reference, for example, Reference [97] Section 6.

Theorem 1.3.6
For every uncountable Polish space X there is a one-to-one continuous function f from the Baire space ω^ω into X.

Proof. Note that there is a one-to-one continuous function from ω^ω into 2^ω (see Exercise 1.3.2). Thus it is enough to show the theorem with ω^ω replaced by 2^ω. By the Cantor–Bendixson theorem X has a nonempty perfect subset. Without loss of generality we assume that X is perfect. Let d be a compatible complete metric on X.

We define a sequence $(U_s)_{s \in 2^{<\omega}}$ of nonempty open sets in X by induction on $\mathrm{lh}(s)$ such that, for any $s \in 2^{<\omega}$,

(i) $\mathrm{diam}(U_s) \le 2^{-\mathrm{lh}(s)}$;

(ii) $\overline{U_{s ^\frown 0}}, \overline{U_{s ^\frown 1}} \subseteq U_s$, where $\overline{U}$ denotes the closure of U in 2^ω;

(iii) $U_{s ^\frown 0} \cap U_{s ^\frown 1} = \emptyset$.

To define $U_\emptyset$ we let $x_\emptyset$ be any element of X and let

$$U_\emptyset = \left\{ x \in X \ : \ d(x, x_\emptyset) < \frac{1}{2} \right\}.$$

Then obviously $U_\emptyset$ is open and $\mathrm{diam}(U_\emptyset) \le 1$.

In general suppose U_s has been defined with $\mathrm{diam}(U_s) \le 2^{-\mathrm{lh}(s)}$. Since X is perfect and U_s is nonempty there exist two distinct elements $x_{s ^\frown 0}, x_{s ^\frown 1} \in U_s$. Let $a > 0$ be sufficiently small so that

$$a < \frac{1}{2} d(x_{s ^\frown 0}, x_{s ^\frown 1}) \le \frac{1}{2^{\mathrm{lh}(s)+1}}$$

and for $i = 0, 1$,

$$\{\, x \in X \, : \, d(x, x_{s^\frown i}) < a \,\} \subseteq U_s.$$

Then for $i = 0, 1$, let

$$U_{s^\frown i} = \left\{ x \in X \, : \, d(x, x_{s^\frown i}) < \frac{a}{2} \right\}.$$

Then it is straightforward to see that (i) through (iii) are satisfied.

Now let $x \in 2^\omega$ and $s_n = x \restriction n$ for all $n \in \omega$. Then by (ii)

$$U_{s_0} = U_\emptyset \supseteq \overline{U_{s_1}} \supseteq U_{s_1} \supseteq \cdots \supseteq U_{s_n} \supseteq \overline{U_{s_{n+1}}} \supseteq U_{s_{n+1}} \supseteq \cdots$$

with $\mathrm{diam}(U_{s_n}) = 2^{-n} \to 0$ as $n \to \infty$. It follows from (i) that $\bigcap_{n \in \omega} U_{s_n} = \bigcap_{n \in \omega} \overline{U_{s_n}}$ is a singleton. Let

$$f(x) = \text{ the unique element of } \bigcap_{n \in \omega} U_{s_n}.$$

This defines a function from 2^ω into X. f is one-to-one because of (iii). That f is continuous is easy to check directly.   ☐

Theorem 1.3.7
For every uncountable Polish space X there is a continuous open surjection $f : \omega^\omega \to X$ and a Baire class 1 injection $g : X \to \omega^\omega$ such that for all $x \in X$, $f(g(x)) = x$.

Proof.   Let $d < 1$ be a compatible complete metric on X.

We define a sequence $(U_s)_{s \in \omega^{<\omega}}$ of nonempty open sets in X by induction on $\mathrm{lh}(s)$ such that, for any $s \in \omega^{<\omega}$,

(i) $\mathrm{diam}(U_s) \leq 2^{-\mathrm{lh}(s)}$;

(ii) for any $n \in \omega$, $\overline{U_{s^\frown n}} \subseteq U_s$;

(iii) $U_s = \bigcup_{n \in \omega} U_{s^\frown n}$.

To begin with, let $U_\emptyset = X$. In general suppose U_s has been defined. For each $x \in U_s$ let

$$V_x = \left\{ z \in U_s \, : \, d(x, z) < \frac{1}{4} \mathrm{diam}(U_s) \text{ and } d(z, x) < d(z, X - U_s) \right\}.$$

Then for every $x \in U_s$, V_x is an open subset of U_s with $\mathrm{diam}(V_x) \leq 2^{-\mathrm{lh}(s)-1}$ and $\overline{V_x} \subseteq U_s$. The collection $\{V_x \, : \, x \in U_s\}$ is an open cover for U_s. In the subspace topology U_s is Polish, and therefore second countable. This implies that $\{V_x \, : \, x \in U_s\}$ has a countable subcover. We let the sequence $(U_{s^\frown n})_{n \in \omega}$ enumerate this countable subcover (possibly with repetitions). Then (i) through (iii) are satisfied.

Similar to the preceding proof, we define a function $f : \omega^\omega \to X$ by letting, for each $x \in \omega^\omega$,

$$f(x) = \text{ the unique element of } \bigcap_{k \in \omega} U_{x \restriction k}.$$

It is then straightforward to check that f is continuous. To see that f is open, note that for any $s \in \omega^{<\omega}$, $f(N_s) = U_s$.

Now for any $z \in X$ we obtain $g(z) \in \omega^\omega$ by defining $g(z)(k)$ by induction on $k \in \omega$. Let $g(z)(0)$ be the least $n \in \omega$ such that $z \in U_{\langle n \rangle}$. In general suppose $g(z)(j)$ have been defined for $j \le k$. Let $s = (g(z)(0), \dots, g(z)(k))$. Then define $g(z)(k+1)$ to be the least $n \in \omega$ such that $z \in U_{s^\frown n}$. This defines a function g as well as confirms that f is onto. The only nontrivial claim left to check is that g is Baire class 1. For this let $s \in \omega^{<\omega}$. It is enough to see that

$$g^{-1}(N_s) = \{x \in X : g(x) \in N_s\} \in \Sigma_2^0(X).$$

Note

$$x \in g^{-1}(N_s) \iff \forall j < k \, (\, x \in U_{s \restriction j} \text{ and } \forall i < s(j-1) \, x \notin U_{s \restriction (j-1)^\frown i}).$$

This shows that $g^{-1}(N_s)$ is a finite Boolean combination of open sets and therefore $g^{-1}(N_s) \in \Sigma_2^0(X)$. ☐

We refer to the following corollary as the **Borel isomorphism theorem**.

Corollary 1.3.8
If X and Y are both uncountable Polish spaces, then there is a Borel isomorphism from X onto Y.

Proof. Let X be uncountable Polish. Theorem 1.3.6 gives a one-to-one continuous function from ω^ω into X. The Baire class 1 function from X into ω^ω given by Theorem 1.3.7 is also one-to-one. Thus by the Cantor–Bernstein theorem X is Borel isomorphic to ω^ω. ☐

Exercise 1.3.1 Show that for every Polish space X and $1 \le \alpha < \omega_1$, $\Sigma_\alpha^0(X)$ is closed under countable unions, $\Pi_\alpha^0(X)$ is closed under countable intersections, and $\Delta_\alpha^0(X)$ is closed under complementations.

Fix a bijection $\langle \cdot, \cdot \rangle : \omega \times \omega \to \omega$ and for each $n \in \omega$, let $(n)_0, (n)_1$ be the unique numbers such that

$$\langle (n)_0, (n)_1 \rangle = n.$$

Exercise 1.3.2 Define a function from ω^ω into 2^ω as follows: for each $x \in \omega^\omega$, let $f(x) \in 2^\omega$ be given by

$$f(x)(n) = \text{ the } (n)_1\text{-th least significant digit in the binary expansion}$$
$$\text{of } x((n)_0).$$

Show that f is one-to-one and continuous.

Exercise 1.3.3 Show that the range of the above function f is $\mathbf{\Pi}^0_3$.

Exercise 1.3.4 Let X be a Polish space and $A \subseteq X$. Show that A contains a perfect subset iff there is a continuous injection from 2^ω into A.

Exercise 1.3.5 Prove Theorem 1.3.3.

Exercise 1.3.6 Show that the Baire space ω^ω is homeomorphic to the subspace $\mathbb{R} - \mathbb{Q}$ of irrationals in $\mathbb{R}$.

Exercise 1.3.7 Show that for any Polish space X there is a continuous surjection φ from the Baire space ω^ω onto X. Moreover, if $d < 1$ is a compatible complete metric on X, then there is a Lipschitz map φ from ω^ω onto X.

Exercise 1.3.8 Show that

(a) there is a Borel isomorphism $h : [0, 1] \to (0, 1)$ such that both h and h^{-1} are of Baire class 1;

(b) there is a Borel isomorphism $h : \omega^\omega \to \mathbb{R}$ such that both h and h^{-1} are of Baire class 2.

Exercise 1.3.9 Show that ℓ_p $(1 \le p \le \infty)$ is a $\mathbf{\Sigma}^0_2$ subset of $\mathbb{R}^\omega$ and c_0 is a $\mathbf{\Pi}^0_3$ subset of $\mathbb{R}^\omega$.

1.4 Standard Borel spaces

The Borel isomorphism theorem (Corollary 1.3.8) shows that the Borel structures on all uncountable Polish spaces are essentially the same. In this section we study this Borel structure.

First we note that the Borel sets can be coded by single elements of some Polish spaces. Since we do not lose any generality by focusing on one Polish space, we will choose to focus on the Baire space ω^ω, where this coding is easy to describe. First recall that elements of ω^ω can be viewed as the branches of the tree $\omega^{<\omega}$. In general a **tree** T on ω is a subset of $\omega^{<\omega}$ which is closed under taking initial segments, that is, if $s \subseteq t$ and $t \in T$, then $s \in T$. A node s in T is **terminal** if s has no proper extension in T; otherwise it is **nonterminal** or **intermediate**. If T is a tree on ω, the set of **branches** of T is defined as

$$[T] = \{\, x \in \omega^\omega : \forall n \in \omega \; x{\restriction}n \in T \,\}$$

and is a closed subset of ω^ω (see Exercise 1.4.1). T is **well-founded** if $[T] = \emptyset$, and **ill-founded** otherwise.

Recall also that $(N_s)_{s\in\omega^{<\omega}}$ is a countable base for the topology of ω^ω consisting of basic clopen sets. We fix once and for all a bijection between the countable set $\omega^{<\omega}$ and ω and denote it by $\langle\cdot\rangle : \omega^{<\omega} \to \omega$. We can now give the formal definition of a Borel code.

Definition 1.4.1
A **Borel code** c is a labeled well-founded tree on ω, formally a pair (T, λ), where T is a nonempty well-founded tree on ω and $\lambda : T \to \omega$, such that

(1) if $s \in T$ is nonterminal, then $\lambda(s) \in \{0, 1\}$;

(2) if $s \in T$ is nonterminal and $\lambda(s) = 0$, then there is a unique $t \in T$ with $s \subseteq t$ and $\mathrm{lh}(t) = \mathrm{lh}(s) + 1$.

It is easy to see that a Borel code is essentially an element of the Baire space. In fact, to each Borel code $c = (T, \lambda)$ we may associate an element $x_c \in \omega^\omega$ by letting, for each $s \in \omega^{<\omega}$,

$$x_c(\langle s \rangle) = \begin{cases} 0, & \text{if } s \notin T, \\ \lambda(s) + 1, & \text{if } s \in T. \end{cases}$$

Next we interpret each Borel code by a Borel subset of ω^ω. To do this we need some more notation. For $s, t \in \omega^{<\omega}$ we define $s^\frown t \in \omega^{<\omega}$ by induction on $\mathrm{lh}(t)$:

$$s^\frown\emptyset = s;$$
$$s^\frown(t^\frown n) = (s^\frown t)^\frown n.$$

Recall that for a well-founded tree T on ω we can define the **rank** of T as follows. For $s \in T$, define by induction

$$\mathrm{rk}(s) = \begin{cases} 0, & \text{if } s \text{ is terminal,} \\ \sup\{\, \mathrm{rk}(s^\frown n) + 1 \; : \; s^\frown n \in T\}, & \text{otherwise;} \end{cases}$$

and let $\mathrm{rk}(T) = \mathrm{rk}(\emptyset)$. If T is a tree on ω and $s \in T$, we define the subtree T_s by

$$t \in T_s \iff s^\frown t \in T.$$

Similarly, for a labeled tree (T, λ) and $s \in T$, the labeled subtree (T_s, λ_s) can be defined, where $\lambda_s(t) = \lambda(s^\frown t)$.

Given a Borel code $c = (T, \lambda)$ we now define the Borel set B_c coded by c. The definition is by induction on the rank of the well-founded tree T.

(a) If $\emptyset$ is the only node of T then $B_c = N_s$, where s is the unique element of $\omega^{<\omega}$ such that $\langle s \rangle = \lambda(\emptyset)$.

(b) If $\emptyset$ is a nonterminal node of T and $\lambda(\emptyset) = 0$, then by Definition 1.4.1 (ii) T contains a unique element s of length 1, and we let $B_c = \omega^\omega - B_d$, where $d = (T_s, \lambda_s)$.

(c) If $\emptyset$ is a nonterminal node of T and $\lambda(\emptyset) = 1$, then enumerating all nodes of T with length 1 as $s_0, s_1, \ldots$, we let $B_c = \bigcup_n B_{d_n}$, where $d_n = (T_{s_n}, \lambda_{s_n})$.

Note that we do not have intersection in the codes. Intersections are treated as combinations of complements and unions following de Morgan's laws. It is obvious that every Borel subset of ω^ω is coded by some Borel code c.

Now back to the Borel structure of general Polish spaces. A **Borel space** is a pair $(X, \mathcal{B})$ where X is a set and $\mathcal{B}$ is a σ-algebra over X. In a Borel space $(X, \mathcal{B})$ a set $B \subseteq X$ is **Borel** if $B \in \mathcal{B}$. We will often omit the σ-algebra when we denote the Borel space, if there is no danger of confusion. If X and Y are Borel spaces, a function $f : X \to Y$ is **Borel** if for any Borel set B in Y, $f^{-1}(B)$ is Borel in X. f is a **Borel isomorphism** if f is a bijection between X and Y and both f and f^{-1} are Borel. The standard example of a Borel space is when X is a Polish space and $\mathcal{B}$ the σ-algebra of Borel sets in X. Thus we have the following concept.

Definition 1.4.2
A Borel space $(X, \mathcal{B})$ is **standard** if there is a Polish topology on X so that $\mathcal{B}$ is the σ-algebra of Borel sets in this topology.

It follows from the Luzin–Suslin Theorem 1.3.1 that any Borel bijection between two standard Borel spaces is a Borel isomorphism, and from the Borel isomorphism theorem (Corollary 1.3.8) that there is only one uncountable standard Borel space up to Borel isomorphism. This immediate corollary is also referred to as the **Borel isomorphism theorem** for standard Borel spaces.

Corollary 1.4.3
Any two uncountable standard Borel spaces are Borel isomorphic.

It also follows quickly from the properties of Polish spaces that a countable product of standard Borel spaces (with the product Borel structure) is still standard Borel. The following nontrivial fact is another nice closure property for the class of standard Borel spaces.

Theorem 1.4.4
If X be a standard Borel space and $B \subseteq X$ is Borel, then the subspace B with the inherited Borel structure is standard Borel.

The proof is nontrivial and can be found in a standard reference. In fact we will give the proof later in the book (see Theorem 4.2.3 and Corollary 4.2.4) because it is related to an important technique known as change of topology.

One of the most important examples of standard Borel spaces comes from Effros Borel spaces.

Definition 1.4.5
Let X be a Polish space and $F(X)$ denote the space of all closed subsets of X. The **Effros Borel structure** on $F(X)$ is the σ-algebra generated by sets of the form
$$\{F \in F(X) : F \cap U \neq \emptyset\}$$
for U open in X.

The space $F(X)$ with the Effros Borel structure is a standard Borel space (see Exercise 1.4.5 or Reference [97] Theorem 12.6). Note that if X is compact Polish, then $F(X)$ coincides with the space $K(X)$, and in fact the Effros Borel structure coincides with the Borel structure given by the Vietoris topology (see Exercise 1.4.4).

Let $F^*(X)$ be the space $F(X) - \{\emptyset\}$ with the Borel structure inherited from $F(X)$. Then $F^*(X)$ is also a standard Borel space. By the Borel isomorphism theorem $F^*(X)$ is Borel isomorphic to $F(X)$. The following theorem is known as the **Borel selection theorem**. The function in the theorem is called a **Borel selector** for $F^*(X)$.

Theorem 1.4.6 (Kuratowski–Ryll-Nardzewski)
There is a Borel function $s : F^*(X) \to X$ such that for any $F \in F^*(X)$, $s(F) \in F$.

Proof. A proof of the theorem when $X = \omega^\omega$ is outlined in Exercise 1.4.6. We deduce the general case here. Let $f : \omega^\omega \to X$ be a continuous open surjection given by Theorem 1.3.7. For each $F \in F(X)$, $f^{-1}(F)$ is a closed subset of ω^ω. Let s_0 be the Borel selector for $F^*(\omega^\omega)$ given by Exercise 1.4.6. Then define $s(F) = f(s_0(f^{-1}(F)))$ for any $F \in F^*(X)$. It is clear that $s(F) \in F$. To see that s is Borel, it suffices to verify that $F \mapsto f^{-1}(F)$ is a Borel map from $F^*(X)$ to $F^*(\omega^\omega)$. For this let $U \subseteq \omega^\omega$ be open. Then $f^{-1}(F) \cap U \neq \emptyset$ iff $F \cap f(U) \neq \emptyset$. Since f is an open map $f(U)$ is open in X, and the condition is Borel by the definition of the Effros Borel structure. ∎

A useful application of the Effros Borel space is that it allows us to talk about the space of all Polish metric spaces. Recall that the universal Urysohn space contains an isometric copy of every Polish metric space. Thus $F(\mathbb{U})$ naturally represents the space of all Polish metric spaces. We will return to this topic in Chapter 14.

The Borelness of the operations in $F(X)$ is not always intuitive. For instance, the intersection operator $(F_1, F_2) \mapsto F_1 \cap F_2$ is not necessarily Borel (see Reference [97] Exercise 27.7).

Exercise 1.4.1 Show that $C \subseteq \omega^\omega$ is closed iff there is a tree T on ω such that $C = [T]$.

A Borel code $c = (T, \lambda)$ is a Σ_1^0-**code** if $\lambda(\emptyset) = 1$ and all nodes of length 1 in T are terminal. For $1 \leq \alpha < \omega_1$ a Borel code $c = (T, \lambda)$ is a Π_α^0-**code** if $\lambda(\emptyset) = 0$ and, letting s be the unique node of length 1 in T, (T_s, λ_s) is a Σ_α^0-code. For $1 < \alpha < \omega_1$ a Borel code $c = (T, \lambda)$ is a Σ_α^0-**code** if $\lambda(\emptyset) = 1$ and for each node s of length 1 in T, (T_s, λ_s) is a Π_β^0-code for some $\beta < \alpha$.

Exercise 1.4.2 Show that for any $\alpha < \omega_1$ and $A \subseteq \omega^\omega$, $A \in \Sigma_\alpha^0$ iff $A = B_c$ for a Σ_α^0-code c, and $A \in \Pi_\alpha^0$ iff $A = B_c$ for a Π_α^0-code c.

Exercise 1.4.3 Show that if Y is a Polish space and X a Polish subspace of Y then there is a one-to-one Borel map from $F(X)$ into $F(Y)$.

Exercise 1.4.4 Let X be compact Polish, $\mathcal{S}$ a countable base for the topology of X, and U open in X. Let $\mathcal{D}$ be the collection of finite covers of $X - U$ by elements of $\mathcal{S}$, that is, finite sets $\{O_1, \ldots, O_n\} \subseteq \mathcal{S}$ such that $X - U \subseteq O_1 \cup \cdots \cup O_n$. Show that for any compact set K in X,

$$K \subseteq U \iff \exists \{O_1, \ldots, O_n\} \in \mathcal{D} \; \forall 1 \leq i \leq n \; K \cap O_i = \emptyset.$$

Exercise 1.4.5 Let X be a Polish space and $\overline{X}$ a compactification of X. For each $F \in F(X)$ let $\overline{F}$ denote its closure in $\overline{X}$. Note that $\overline{F} \in K(\overline{X}) = F(\overline{X})$.

(a) Show that the set $Y = \{\overline{F} : F \in F(X)\}$ is G_δ in $K(\overline{X})$, and hence is Polish.

(b) Show that the Polish topology on $F(X)$ obtained by pulling back the Polish topology on Y via the map $F \mapsto \overline{F}$ generates the Effros Borel structure on $F(X)$.

Recall that the **lexicographic order** $\leq_{\text{lex}}$ on $\omega^{\leq\omega} = \omega^\omega \cup \omega^{<\omega}$ is defined by

$$s \leq_{\text{lex}} t \iff s \subseteq t \text{ or } \exists i < \text{lh}(s) \, (\, s(i) < t(i) \text{ and } \forall j < i \; s(j) = t(j) \,).$$

Exercise 1.4.6 For any tree T on ω with $[T] \neq \emptyset$ we can define the **leftmost branch** b_T of T to be the $\leq_{\text{lex}}$-least element of $[T]$, that is, for any $b \in [T]$, $b_T \leq_{\text{lex}} b$.

(a) Show that if $[T] \neq \emptyset$ then there exists a leftmost branch of T.

(b) Show that the map $[T] \mapsto b_T$ is a Borel selector for $F^*(\omega^\omega)$.

Exercise 1.4.7 Let X be a Polish space. Show that the union operator

$$\cup : F(X) \times F(X) \to F(X)$$
$$(F_1, F_2) \quad \mapsto F_1 \cup F_2$$

is Borel.

1.5   The effective hierarchy

The introduction of the notion of computability into descriptive set theory profoundly changed the subject. The resulting theory is often referred to as effective descriptive set theory, and concepts and results obtained in the pre-computability theory era is now called classical descriptive set theory. The effective theory is more fundamental since it can be developed independently and it covers the classical theory completely. All of the classical results have effective proofs; and for some the effective proofs are the only ones known. Classical proofs are still desirable and sometimes preferred since they are easier to communicate to nonlogicians, but more and more new results are proved using effective methods and it becomes harder to find classical proofs.

In the remaining sections of this chapter we give a review of the basic results of the effective theory. A complete account of the theory can be found in [126].

We need to assume that the reader is familiar with the notion of a computable function from ω^n to ω. A discussion of the equivalent definitions of computability and the Church–Turing Thesis can be found in most textbooks on computability theory (or recursion theory, as it was called before the 1990s) or mathematical logic, for example, see References [24], [12], [137], [11]. Here, for the sake of completeness we recall one of several equivalent formal definitions below. By convention, we use $f : \omega^n \rightharpoonup \omega$ to indicate that $\mathrm{dom}(f) \subseteq \omega^n$.

Definition 1.5.1
The class of all **partial computable functions** is the smallest class of functions containing the zero function $Z : \omega \to \omega$, the successor function $S : \omega \to \omega$, and the projection functions $P_i^n : \omega^n \to \omega$ $(1 \le i \le n)$, where

$$Z(x) = 0,$$
$$S(x) = x + 1,$$
$$P_i^n(x_1, \ldots, x_n) = x_i,$$

and is closed under the following operations on partial functions:

(1) Substitution: if $f : \omega^n \rightharpoonup \omega$ and $g_1, \ldots, g_n : \omega^m \rightharpoonup \omega$ are partial computable, then so is $h : \omega^m \rightharpoonup \omega$, where

$$h(x_1, \ldots, x_m) = f(g_1(x_1, \ldots, x_m), \ldots, g_n(x_1, \ldots, x_m));$$

(2) Recursion: if $f : \omega^n \rightharpoonup \omega$ and $g : \omega^{n+2} \rightharpoonup \omega$ are partial computable, then so is $h : \omega^{n+1} \rightharpoonup \omega$, where

$$h(x_1, \ldots, x_n, 0) = f(x_1, \ldots, x_n),$$
$$h(x_1, \ldots, x_n, m + 1) = g(h(x_1, \ldots, x_n, m), x_1, \ldots, x_n, m);$$

(3) Minimization: if $f : \omega^{n+1} \rightharpoonup \omega$ is partial computable, then so is $g : \omega^n \rightharpoonup \omega$, where

$$g(x_1, \ldots, x_n) = \mu m[f(x_1, \ldots, x_n, m) = 0]$$
$$= \text{the least } m \text{ such that}$$

(i) $f(x_1, \ldots, x_n, k)$ is defined for all $k \leq m$,
(ii) $f(x_1, \ldots, x_n, k) \neq 0$ for all $k < m$, and
(iii) $f(x_1, \ldots, x_n, m) = 0$.

All formal definitions of computability have been shown to be equivalent to the above one. In fact, when working with computable functions we will not deal with the details of the above definition, but will rather adopt the **Church–Turing Thesis**, which states that any of the formal definitions of computable functions captures the intuitive notion of computability. Thus if a function is intuitively computable by an informal algorithm then by the Church–Turing Thesis we may conclude that it is formally computable without checking the details of the formal definitions.

A function $f : \omega^n \to \omega$ is **computable** if it is partial computable and total. Recall that a subset of ω^n is **computable** if its characteristic function is computable. We also fix various coding functions, namely computable bijections between various sets and ω, and denote all of them by $\langle \cdot \rangle$ (see Exercise 1.5.1). With these codings every partial computable function has an index. We will use the standard notation $\{e\}$ to denote the partial computable function with index e (if there is no confusion about the arity of the function), and call it the **e-th partial computable function**. Another standard notation is W_e, which denotes the domain of the e-th partial computable unary function.

Moreover, we will often use relative computations in some given oracle, usually an element of the Baire space. The oracle is used to provide extra information not otherwise available from an algorithm. Given $x \in \omega^\omega$, the class of all **partial functions computable in the oracle x** is the smallest class of functions containing x and all partial computable functions and is closed under substitutions, recursions, and minimizations. Most results in computability theory can be relativized in any oracle, and the proofs of relativized results are always repetitions of those of unrelativized results. Throughout this book relativization of notions will be used freely without elaboration. For $x \in \omega^\omega$ the e-th partial computable function with oracle x will be denoted by $\{e\}^x$ and its domain by W_e^x.

For the rest of this section we define the effective Borel hierarchy, also known as the hyperarithmetical hierarchy.

First we need to recall the notion of computable ordinals. Let

$$\mathrm{LO} = \{ < \, \subseteq \omega \times \omega : \; < \text{ is a linear ordering of } \omega \},$$

that is,

$$< \in \mathrm{LO} \iff \forall n, m \in \omega \, (n < m \text{ or } m < n \text{ or } m = n) \text{ and}$$
$$\forall n, m, k \in \omega \, (n < m \text{ and } m < k \to n < k) \text{ and}$$
$$\forall n, m \in \omega \, (n < m \to m \not< n).$$

Then let

$$\mathrm{WO} = \{ < \in \mathrm{LO} : \ < \text{ is a well-ordering of } \omega \},$$

that is,

$$< \in \mathrm{WO} \iff \ < \in \mathrm{LO} \text{ and any subset of } \omega \text{ has a } <\text{-least element.}$$

For $< \in \mathrm{WO}$ let $\mathrm{ot}(<)$ be the order type of $<$. Note that $\mathrm{ot}(<)$ is a countable ordinal, and $\sup\{ \mathrm{ot}(<) : < \in \mathrm{WO} \} = \omega_1$. Let

$$\omega_1^{\mathrm{CK}} = \sup\{ \mathrm{ot}(<) : < \in \mathrm{WO} \text{ and } < \text{ is computable} \}.$$

Since there are only countably many computable well-orderings of ω, ω_1^{CK} is a countable ordinal. The CK in the superscript stands for Church–Kleene. Ordinals $\alpha < \omega_1^{\mathrm{CK}}$ are called **computable ordinals** since each of them is the order type of a computable well-ordering of ω (Exercise 1.5.3).

It is useful to note that ω_1^{CK} is also the supremum of the ranks of all well-founded trees which are computable subsets of $\omega^{<\omega}$ (see Exercise 1.5.9).

Just as the Borel hierarchy is indexed by all countable ordinals $\alpha < \omega_1$, the effective Borel hierarchy is indexed by ordinals $\alpha < \omega_1^{\mathrm{CK}}$. Furthermore, all notions can be relativized in a straightforward way. Thus given any element $x \in \omega^\omega$, we also speak of the relativized notions $\omega_1^{\mathrm{CK}(x)}$ and the effective Borel hierarchy relativized to x, which is indexed by ordinals $\alpha < \omega_1^{\mathrm{CK}(x)}$. Note that

$$\omega_1 = \sup\{ \omega_1^{\mathrm{CK}(x)} : x \in \omega^\omega \}.$$

Our objective is to define the effective Borel hierarchy for subsets of ω^ω. Nevertheless we will need to consider more general product spaces of the form

$$X = \omega^k \times (\omega^\omega)^l,$$

where $k, l \in \omega$. We fix some notation related to this space. Given a tuple

$$t = (n_1, \ldots, n_k, s_1, \ldots, s_l)$$

where $n_1, \ldots, n_k \in \omega$ and $s_1, \ldots, s_l \in \omega^{<\omega}$, define

$$N_t = \{n_1\} \times \cdots \times \{n_k\} \times N_{s_1} \times \cdots \times N_{s_l}.$$

The set of N_t for $t \in \omega^k \times (\omega^{<\omega})^l$ is a base for the topology on the product space X. We can fix a computable bijection between ω and the index set

$$T = \omega^k \times (\omega^{<\omega})^l,$$

and this makes the notion of computable functions and sets pertaining to T clearly fixed.

Now we are ready to define the first level of the effective Borel hierarchy, the **effectively open sets**: for $A \subseteq X$, let

$$A \in \Sigma_1^0 \iff A = \bigcup_{n \in \omega} N_{f(n)}, \text{ for some computable } f : \omega \to T.$$

Note that an index e for the function f codes all the information needed to recover the set A. For this reason we regard e as an effective Borel index for the set A.

To define the other levels of the effective hierarchy we first fix this notion of effective Borel index for sets in the effective Borel hierarchy. Since the sets in this hierarchy are usually called hyperarithmetical sets, we will call the indices hyperarithmetical indices. Hyperarithmetical indices are similar to Borel codes as they contain information about labeled well-founded trees, with rank $< \omega_1^{\mathrm{CK}}$; however, literally they will be natural numbers.

Definition 1.5.2
The set of all **hyperarithmetical indices** is defined by induction:

(1) If $e \in \omega$, then $\langle 0, e \rangle$ is a hyperarithmetical index;

(2) If e is a hyperarithmetical index, then $\langle 1, e \rangle$ is a hyperarithmetical index;

(3) If every element of W_e is a hyperarithmetical index, then $\langle 2, e \rangle$ is a hyperarithmetical index.

The sets coded by the hyperarithmetical indices are given by the following definition.

Definition 1.5.3
For each hyperarithmetical index $c \in \omega$ we define the **hyperarithmetical set** coded by c, denoted H_c, by induction:

(1) If $c = \langle 0, e \rangle$, then $H_c = \bigcup_{n \in W_e} N_{\langle n \rangle}$ where $\langle \cdot \rangle : \omega \to T$ is a fixed coding function;

(2) If $c = \langle 1, e \rangle$, where e is a hyperarithmetical index, then $H_c = X - H_e$;

(3) If $c = \langle 2, e \rangle$, where W_e is a set of hyperarithmetical indices, then $H_c = \bigcup_{d \in W_e} H_d$.

By clause (1) of this definition, an official hyperarithmetical index for an effectively open set is of the form $\langle 0, e \rangle$. Clause (2) describes the coding of complementation and clause (3) describes the coding of the operation of effective countable union. In general for $\alpha < \omega_1^{CK}$ we let

$$A \in \Pi_\alpha^0 \iff X - A \in \Sigma_\alpha^0$$

and

$$A \in \Delta_\alpha^0 \iff A \in \Sigma_\alpha^0 \text{ and } A \in \Pi_\alpha^0.$$

For $1 < \alpha < \omega_1^{CK}$, define $A \in \Sigma_\alpha^0$ iff $A = H_c$, where $c = \langle 2, e \rangle$ is a hyperarithmetical index and for all $d \in W_e$, $H_d \in \Pi_\beta^0$ for some $\beta < \alpha$.

An equivalent but less formal definition is as follows. For any $\alpha < \omega_1^{CK}$,

$$A \in \Sigma_{\alpha+1}^0 \iff A = \{x \in X : \exists n \in \omega(n, x) \in B\} \text{ for some } B \in \Pi_\alpha^0 \text{ in } \omega \times X.$$

If $1 < \alpha < \omega_1^{CK}$ is a limit ordinal, then

$$A \in \Sigma_\alpha^0 \iff A \text{ is an effective union of sets in the collection } \bigcup_{\beta < \alpha} \Pi_\beta^0.$$

Note that there is still ambiguity in this last characterization, since we need to fix an effective enumeration of $\bigcup_{\beta < \alpha} \Pi_\beta^0$ in order to deal with effective unions of sets in the collection. We leave it to the reader to fill up the details.

Given any oracle $x \in \omega^\omega$ we can define in the same fashion the relativized effective hierarchy $\Sigma_\alpha^0(x)$, $\Pi_\alpha^0(x)$, $\Delta_\alpha^0(x)$ for $\alpha < \omega_1^{CK(x)}$. Sets in this hierarchy are called **hyperarithmetical in** x. The following fact is simple yet unravels a beautiful connection between the classical and effective notions.

Theorem 1.5.4
Let $\alpha < \omega_1$ and $A \subseteq \omega^\omega$. Then the following are equivalent:

(i) *$A \in \Sigma_\alpha^0$.*

(ii) *There is $x \in \omega^\omega$ with $\alpha < \omega_1^{CK(x)}$ such that $A \in \Sigma_\alpha^0(x)$.*

The key idea for the nontrivial direction of the theorem is the observation that a Borel set can be effectively recovered from a Borel code. The theorem is also true for the Π-side and Δ-side of the hierarchies. The proof is left to the reader (Exercise 1.5.5).

A basic fact for the effective Borel hierarchy is that for each $1 \leq \alpha < \omega_1^{CK}$ there are Σ_α^0 sets which are not Π_α^0. This also relativizes.

Exercise 1.5.1

(a) Show that

$$\langle m_1, m_2 \rangle_2 = \frac{(m_1 + m_2)(m_1 + m_2 + 1)}{2} + m_2$$

is a computable bijection from ω^2 onto ω.

(b) Define $\langle \cdots \rangle_n$ for $n > 2$ by induction:

$$\langle m_1, \ldots, m_n \rangle_n = \langle \langle m_1, \ldots, m_{n-1} \rangle_{n-1}, m_n \rangle_2.$$

Show that $\langle \cdots \rangle_n$ is a computable bijection from ω^n onto ω.

(c) Define $\langle \cdot \rangle_{<\omega}$ from $\omega^{<\omega}$ by

$$\langle s \rangle = \begin{cases} 0, & \text{if } s = \emptyset, \\ \langle n, \langle m_1, \ldots, m_n \rangle_n \rangle_2, & \text{if } s = (m_1, \ldots, m_n) \text{ for } n \geq 1. \end{cases}$$

Show that $\langle \cdot \rangle_{<\omega}$ is a bijection from $\omega^{<\omega}$ onto ω.

Exercise 1.5.2 Show that $C \subseteq \omega^\omega$ is Π_1^0 iff there is a computable tree $T \subseteq \omega^{<\omega}$ such that $C = [T]$.

Exercise 1.5.3 Show that any ordinal $< \omega_1^{\mathrm{CK}}$ is the order type of a computable well-ordering of ω.

Exercise 1.5.4 Show that $\omega_1^{\mathrm{CK}(x)}$ is a countable limit ordinal for any $x \in \omega^\omega$.

Exercise 1.5.5 Give the definition of the relativized effective hierarchy and prove Theorem 1.5.4.

Exercise 1.5.6 Let $1 \leq \alpha < \omega_1$. Show that there is a Σ_α^0 subset U of $\omega^\omega \times \omega^\omega$ such that for every $x \in \omega^\omega$, $U_x = \{y : (x, y) \in U\}$ is Σ_α^0, and for every Σ_α^0 subset A of ω^ω there is $x \in \omega^\omega$ such that $U_x = A$. (Such a set is said to be **universal** for Σ_α^0 subsets of ω^ω.)

For any tree T on ω define the **Kleene–Brouwer ordering** of T by

$$s <_{\mathrm{KB}}^T t \iff t \subsetneq s \vee \exists i \in \omega \, (\forall j < i \, s(j) = t(j) \wedge s(i) < t(i)).$$

Exercise 1.5.7 Show that for any tree T on ω the Kleene–Brouwer ordering $<_{\mathrm{KB}}^T$ is a linear order, and that it is a well-order iff T is well-founded.

Exercise 1.5.8 Show that for any tree T on ω the Kleene–Brouwer ordering $<_{\mathrm{KB}}^T$ is computable in T.

Exercise 1.5.9 Let $T \subseteq \omega^{<\omega}$ be a computable well-founded tree on ω. Show that $\mathrm{rk}(T) < \omega_1^{\mathrm{CK}}$.

1.6 Analytic sets and Σ_1^1 sets

In this section we continue to consider subsets of product spaces $X = \omega^k \times (\omega^\omega)^l$. Recall that a subset $A \subseteq X$ is **analytic** (or $\mathbf{\Sigma}_1^1$) if there is a closed subset C of $X \times \omega^\omega$ such that, for all $x \in X$,

$$x \in A \iff \exists y \in \omega^\omega \ (x, y) \in C.$$

Similarly, we define that $A \subseteq X$ is Σ_1^1 if there is a Π_1^0 set C of $X \times \omega^\omega$ such that

$$x \in A \iff \exists y \in \omega^\omega \ (x, y) \in C.$$

In addition, $A \subseteq X$ is **coanalytic** (or $\mathbf{\Pi}_1^1$) if $X - A$ is analytic. A is $\mathbf{\Delta}_1^1$ if it is both analytic and coanalytic. Similarly, A is Π_1^1 if $X - A$ is Σ_1^1, and is Δ_1^1 if it is both Σ_1^1 and Π_1^1. In addition, the lightface classes can all be relativized, thus it makes sense to talk about pointclasses $\Sigma_1^1(x)$, $\Pi_1^1(x)$, and $\Delta_1^1(x)$.

The following fundamental theorems connect together the classes of sets we will focus on in the rest of this book.

Theorem 1.6.1 (Luzin)
Let A and B be disjoint analytic sets in X. Then there is a Borel set C separating A from B, that is, $A \subseteq C$ and $B \cap C = \emptyset$.

Theorem 1.6.2 (Luzin–Addison)
Let A and B be disjoint Σ_1^1 sets in X. Then there is a hyperarithmetical set C separating A from B.

Theorem 1.6.3 (Suslin)
Let $A \subseteq X$. Then A is $\mathbf{\Delta}_1^1$ iff it is Borel.

Theorem 1.6.4 (Suslin–Kleene)
Let $A \subseteq X$. Then A is Δ_1^1 iff it is hyperarithmetical.

The first of these theorems is called the **Luzin separation theorem**. The third, **Suslin's theorem**, is a direct corollary of it. The **Suslin–Kleene theorem** was proved by Kleene and can be similarly derived from the **Luzin–Addison separation theorem**. Each of the lightface theorems can be relativized. Besides the analogy between the boldface classes and lightface classes, the two types of classes are also directly connected (see Exercise 1.6.4). Using these connections one can easily derive the boldface theorems from their lightface counterparts.

In the classical study of analytic sets the Suslin operation $\mathcal{A}$ plays an important role. We briefly recall its definition and basic properties.

Definition 1.6.5

Let Y be a set and $\{A_s\}_{s \in \omega^{<\omega}}$ be a family of subsets of Y. Then the set

$$A = \bigcup_{x \in \omega^\omega} \bigcap_{n \in \omega} A_{x \restriction n}$$

is said to be obtained from $\{A_s\}$ by the **Suslin operation** or **operation** $\mathcal{A}$. If in addition Y is a topological space and every A_s is closed, then A is called a **Suslin set**.

Note that the concept of Suslin sets is defined for general Polish spaces other than the product spaces. In the product space X the Suslin sets coincide with the analytic sets (Exercise 1.6.6). In general, every Borel set is Suslin (see Exercise 1.6.7) and the result of a Suslin operation on a sequence of Borel sets is Suslin (see Exercise 1.6.8). Thus the concept of Suslin sets can also be used as the definition of analytic sets on a standard Borel space.

We will use the following basic theorem known as the **perfect set theorem** for analytic sets.

Theorem 1.6.6 (Suslin)

Any uncountable analytic set in a Polish space contains a perfect subset.

In the rest of this section we review the boundedness principles for various sets of well-orders. These fundamental results will be very useful in later chapters.

Recall that LO is the set of all linear orderings of ω. When $x \in$ LO we denote it as $<_x$ to emphasize that it is a linear order and speak of its order type $\mathrm{ot}(<_x)$. It is straightforward to check that LO is a Borel subset of $2^{\omega \times \omega}$. In contrast, the set WO of all well-orderings of ω is Π_1^1, and in particular $\mathbf{\Pi}_1^1$. We note below that it is not $\mathbf{\Sigma}_1^1$.

Lemma 1.6.7

For any Π_1^1 set $A \subseteq \omega^\omega$ there is a continuous function $f : \omega^\omega \to$ LO such that $A = f^{-1}(\mathrm{WO})$.

Proof. Let $C \subseteq \omega^\omega \times \omega^\omega$ be a closed subset such that

$$x \notin A \iff \exists y \in \omega^\omega \ (x, y) \in C.$$

By Exercise 1.4.1 there is a tree T on $\omega \times \omega$ such that $C = [T]$. For each $x \in \omega^\omega$, define a tree T_x on ω by

$$s \in T_x \iff (x \restriction \mathrm{lh}(s), s) \in T.$$

It follows that

$$x \in A \iff \forall y \in \omega^\omega (x, y) \notin [T] \iff [T_x] = \emptyset \iff T_x \text{ is well-founded.}$$

Let $<_{\text{KB}}^{T_x}$ be the Kleene–Brouwer ordering of T_x (defined for Exercise 1.5.7). Then we have further

$$x \in A \iff <_{\text{KB}}^{T_x} \text{ is a well-order.}$$

Thus if we let $f(x) = <_{\text{KB}}^{T_x}$, we have that $f : \omega^\omega \to \text{LO}$ and $A = f^{-1}(\text{WO})$. It is easy to see that f is a continuous function. $\square$

It follows immediately from Lemma 1.6.7 and Suslin's theorem that WO is not Σ_1^1, since otherwise it would be Borel and in fact Σ_α^0 for some $\alpha < \omega_1$, which implies by Lemma 1.6.7 that any Borel set would be Σ_α^0, and the Borel hierarchy collapses, a contradiction.

For each $\alpha < \omega_1$, let

$$\text{WO}_\alpha = \{x \in \text{WO} : \text{ot}(<_x) \le \alpha\}.$$

It is easy to show by transfinite induction that each WO_α is a Borel set.

We also consider the effective classes of well-orders. For each $x \in \omega^\omega$, we let

$$\text{WO}^x = \{y \in \text{WO} : y \text{ is computable in } x \},$$

and

$$\text{WO}^0 = \{y \in \text{WO} : y \text{ is computable}\}.$$

Then similar to Lemma 1.6.7 we have that WO^0 is Π_1^1 but not Σ_1^1. Likewise each WO^x is $\Pi_1^1(x)$ but not $\Sigma_1^1(x)$. However, each of these sets is countable, thus trivially Borel.

We have that $\text{WO} = \bigcup_{x \in \omega^\omega} \text{WO}^x$, and for any computable $x \in \omega^\omega$, $\text{WO}^x = \text{WO}^0$. Recall that

$$\omega_1^{\text{CK}} = \sup\{\text{ot}(<_y) : y \in \text{WO}^0\}$$

and

$$\omega_1^{\text{CK}(x)} = \sup\{\text{ot}(<_y) : y \in \text{WO}^x\}.$$

Every $\omega_1^{\text{CK}(x)}$ is a limit ordinal (Exercise 1.5.4). For $\alpha < \omega_1$ we also let

$$\text{WO}_\alpha^0 = \text{WO}^0 \cap \text{WO}_\alpha$$

and

$$\text{WO}_\alpha^x = \text{WO}^x \cap \text{WO}_\alpha.$$

When $\alpha < \omega_1^{\text{CK}}$, WO_α^0 is Δ_1^1. This also relativizes.

The following theorem is known as the **boundedness principle** for Σ_1^1 sets of computable well-orders.

Theorem 1.6.8
If $A \subseteq \text{WO}^0$ is Σ_1^1 then there is $\alpha < \omega_1^{\text{CK}}$ such that $A \subseteq \text{WO}_\alpha^0$.

Proof. Otherwise we have that for all $y \in \mathrm{LO}$, $y \in \mathrm{WO}^0$ iff

$$y \text{ is computable } \wedge \; \exists z \in A \; \exists f \; (\; f : \omega \to \omega \text{ is an order-preserving injection from } (\omega, <_y) \text{ into } (\omega, <_z) \;).$$

This shows that WO^0 is Σ_1^1, a contradiction. $\quad\square$

A similar proof gives the relativization of the above theorem, stated below.

Theorem 1.6.9
Let $x \in \omega^\omega$. If $A \subseteq \mathrm{WO}^x$ is $\Sigma_1^1(x)$ then there is $\alpha < \omega_1^{\mathrm{CK}(x)}$ such that $A \subseteq \mathrm{WO}_\alpha^x$.

Also a similar proof gives the classical **boundedness principle** for Σ_1^1 sets of well-orders.

Theorem 1.6.10
If $A \subseteq \mathrm{WO}$ is Σ_1^1 then there is $\alpha < \omega_1$ such that $A \subseteq \mathrm{WO}_\alpha$.

We will also use the following boundedness principle for Σ_1^1 sets of well-founded trees.

Theorem 1.6.11
If A is a Σ_1^1 set of well-founded trees then there is $\alpha < \omega_1$ such that $\mathrm{rk}(T) < \alpha$ for all $T \in A$.

The proof is another application of the Kleene–Brouwer ordering (see Exercise 1.5.9) and is left as an exercise (Exercise 1.6.5).

Finally we note the following theorem of Spector about Δ_1^1 well-orders.

Theorem 1.6.12 (Spector)
For any $x \in \omega^\omega$, $\omega_1^{\mathrm{CK}(x)} = \sup\{\mathrm{ot}(<_y) : y \in \Delta_1^1(x), \; y \in \mathrm{WO}\}$.

Proof. We show the unrelativized version only, and for this it suffices to show that if $y \in \Delta_1^1$ and $y \in \mathrm{WO}$, then $\mathrm{ot}(<_y) < \omega_1^{\mathrm{CK}}$. Assume not, and fix $y \in \Delta_1^1$ with $\mathrm{ot}(<_y) \geq \omega_1^{\mathrm{CK}}$. Then for all $z \in \mathrm{LO}$, $z \in \mathrm{WO}^0$ iff

$$z \text{ is computable } \wedge \; \exists f \; (\; f : \omega \to \omega \text{ is an order-preserving injection from } (\omega, <_z) \text{ into } (\omega, <_y) \;).$$

This shows that WO^0 is Σ_1^1, a contradiction. The theorem follows from a relativization of this argument. $\quad\square$

Exercise 1.6.1 Show that $A \subseteq \omega^\omega$ is Σ_1^1 iff there is a computable tree $T \subseteq (\omega \times \omega)^{<\omega}$ such that $A = p[T] =_{\mathrm{def}} \{x : \exists y \; (x, y) \in [T]\}$.

Exercise 1.6.2 Show that the collection of all analytic sets is closed under countable unions and countable intersections.

Exercise 1.6.3 Suppose $B \subseteq X \times \omega^\omega$ is Borel and $A \subseteq X$ is defined by

$$x \in A \iff \exists y \in \omega^\omega \ (x, y) \in B.$$

Show that A is analytic.

Exercise 1.6.4 Show that

(i) $\Sigma_1^1 = \bigcup_{x \in \omega^\omega} \Sigma_1^1(x)$,

(ii) $\Pi_1^1 = \bigcup_{x \in \omega^\omega} \Pi_1^1(x)$, and

(iii) $\Delta_1^1 = \bigcup_{x \in \omega^\omega} \Delta_1^1(x)$.

Exercise 1.6.5 Prove Theorem 1.6.11.

Exercise 1.6.6 Show that in the product space X the class of all Suslin sets is exactly the class of all analytic sets.

Exercise 1.6.7 Show that every Borel set in a Polish space is Suslin.

Exercise 1.6.8 Let Y be a Polish space and $\{A_s\}_{s \in \omega^{<\omega}}$ be a sequence of Borel subsets of Y. Let A be obtained from $\{A_s\}$ by the Suslin operation. Show that A is Suslin.

Exercise 1.6.9 Let Y be a set, $\{A_s\}_{s \in \omega^{<\omega}}$ subsets of Y, and A obtained from $\{A_s\}$ by the Suslin operation. Define by transfinite induction on $\alpha < \omega_1$ the following sequences:

$$A_s^0 = A_s, \quad A_s^{\alpha+1} = A_s^\alpha \cap \bigcup_{n \in \omega} A_{s^\frown n}^\alpha, \quad \text{and} \quad A_s^\lambda = \bigcap_{\alpha < \lambda} A_s^\alpha \ (\lambda \text{ limit});$$

$$B_\alpha = A_\emptyset^\alpha \quad \text{and} \quad D_\alpha = \bigcup_s (A_s^\alpha - A_s^{\alpha+1}).$$

Show that

$$A = \bigcup_{\alpha < \omega_1} B_\alpha = \bigcup_{\alpha < \omega_1} (B_\alpha - D_\alpha).$$

1.7 Coanalytic sets and Π_1^1 sets

We have seen that the Borel hierarchy is connected to the effective Borel hierarchy by relativization, and that the Suslin–Kleene theorem connects the structure of Σ_1^1 or Π_1^1 sets with Δ_1^1 or hyperarithmetical sets. This connection makes Σ_1^1 and Π_1^1 sets a vital tool in the study of Borel sets. Many results about Borel sets proved by exploring this connection have not been proved by classical methods. In this section we continue our review of the effective theory for Π_1^1 sets; we will encounter some theorems without straightforward classical analogs.

Theorem 1.7.1
*The class of all Π_1^1 sets on X has the **reduction property**, that is, for any $A, B \in \Pi_1^1$, there are $A', B' \in \Pi_1^1$ with $A' \subseteq A$, $B' \subseteq B$, $A' \cup B' = A \cup B$, and $A' \cap B' = \emptyset$.*

The reduction property for Π_1^1 is a very basic property since it implies immediately that the class of Σ_1^1 sets has the **separation property**, that is, for any disjoint $A, B \in \Sigma_1^1$ there is a Δ_1^1 set C separating A from B. Furthermore, from this separation property for Σ_1^1 sets and the Luzin–Addison separation theorem one can derive the Suslin–Kleene theorem as a corollary. The reduction property relativizes and implies the same property for $\mathbf{\Pi}_1^1$ sets.

We have seen that WO is Π_1^1 but not Σ_1^1. Recall the following definition of completeness in a boldface pointclass.

Definition 1.7.2
A set $S \subseteq \omega^\omega$ is $\mathbf{\Pi}_1^1$**-complete** or **complete coanalytic** if S is $\mathbf{\Pi}_1^1$ and for all $\mathbf{\Pi}_1^1$ subset A of X there is a continuous function $f : X \to X$ such that $A = f^{-1}(S)$.

The function f in the definition is called a **reduction function**. Note that WO is a subset of $2^{\omega \times \omega}$, and by a standard coding can be viewed as a subset of ω^ω. By Lemma 1.6.7 we have that WO is $\mathbf{\Pi}_1^1$-complete. Similarly one can define $\mathbf{\Sigma}_1^1$-completeness (or complete analyticity), and it is easy to derive that LO − WO is $\mathbf{\Sigma}_1^1$-complete. In fact for all $1 \leq \alpha < \omega_1$ similar definitions can be employed to define $\mathbf{\Sigma}_\alpha^0$- or $\mathbf{\Pi}_\alpha^0$-completeness. It is a basic result of descriptive set theory that for each $1 \leq \alpha < \omega_1$, $\mathbf{\Sigma}_\alpha^0$-complete sets exist (and the existence of $\mathbf{\Pi}_\alpha^0$-complete sets follows by taking complements) (see, for example, Reference [97] Exercise 22.12).

A closely related basic result is the existence of universal Π_1^1 sets.

Theorem 1.7.3
There is a Π_1^1 set $P \subseteq \omega \times X$ such that for any Π_1^1 set $A \subseteq X$ there is $n \in \omega$ such that
$$A = P_n = \{x \in X : (n, x) \in P\}.$$

The proof of this theorem is simple and is left as an exercise (Exercise 1.7.2). The set in the theorem is said to be **universal** for all Π_1^1 subsets of X. It is immediate that the complement of P is universal for all Σ_1^1 subsets of X. Note that the theorem gives an enumeration of all Π_1^1 subsets of X (and therefore their complements as well), which leads to the following theorem on a Π_1^1 coding of all Δ_1^1 sets. This is in contrast with the hyperarithmetic indices we discussed in Section 1.5.

Theorem 1.7.4
There are Π_1^1 subsets $P^+, P^- \subseteq \omega \times X$ and $D \subseteq \omega$ such that

(i) for any $n \in D$, P_n^+ and P_n^- are complements of each other, and

(ii) for any Δ_1^1 set A there is $n \in D$ such that $A = P_n^+$.

Proof. Fix a computable bijection $\langle \cdot, \cdot \rangle : \omega \times \omega \to \omega$. Let $P \subseteq \omega \times X$ be universal for all Π_1^1 subsets of X. Define

$$Q^+(\langle m, k \rangle, x) \iff P(m, x),$$
$$Q^-(\langle m, k \rangle, x) \iff P(k, x).$$

Then Q^+, Q^- are both Π_1^1 subsets of $\omega \times X$. By the reduction property for Π_1^1 sets let P^+, P^- be Π_1^1 subsets of $\omega \times X$ such that $P^+ \cup P^- = Q^+ \cup Q^-$ and $P^+ \cap P^- = \emptyset$. Define

$$n \in D \iff Q_n^+ \cup Q_n^- = X \iff P_n^+ \cup P_n^- = X$$
$$\iff \forall x \in X \ (Q^+(n, x) \lor Q^-(n, x)).$$

Then $D \in \Pi_1^1$. It is easy to see that properties (i) and (ii) are satisfied. ∎

In the computation of descriptive complexity involving Σ_1^1 and Π_1^1 sets the following theorems are very useful. They give another canonical representation of Π_1^1 sets and show that the pointclass Π_1^1 is closed under the existential Δ_1^1 quantification.

Theorem 1.7.5 (Kleene)
If $A \subseteq X \times \omega^\omega$ is Π_1^1 and

$$x \in B \iff \exists y \in \Delta_1^1(x) \ (x, y) \in A,$$

then B is also Π_1^1.

Theorem 1.7.6 (Spector–Gandy)
Let $A \subseteq X$. Then A is Π_1^1 iff there is Π_1^0 set $C \subseteq X \times \omega^\omega$ such that

$$x \in A \iff \exists y \in \Delta_1^1(x) \ (x, y) \in C.$$

In Section 8.3 we will also use some facts about Π_1^1 singletons. A real $x \in \omega^\omega$ is called a Π_1^1 **singleton** if $\{x\}$ is Π_1^1. It is easy to find examples of Π_1^1 singletons. For instance, if $x \subseteq \omega$ is hyperarithmetical, then x, as an element of 2^ω, is a Π_1^1 singleton, since

$$y \in \{x\} \iff \forall n \ (n \in y \leftrightarrow n \in x).$$

The usefulness of this concept comes from the following basis theorem ([126] 4E.2 or [133] Corollary 9.4).

Theorem 1.7.7 (Addison–Kondo)
Every nonempty Π_1^1 subset of ω^ω contains a Π_1^1 singleton.

The following fact can be derived as an easy corollary to the above basis theorem.

Corollary 1.7.8
For any Π_1^1 singleton x there exists a Π_1^1 singleton y such that y is not computable in x.

Proof. Let x be a Π_1^1 singleton. Consider the set P of all $y \in \omega^\omega$ not computable in x, that is,

$$y \in P \iff \forall e \in \omega \exists n \in \omega \ y(n) \neq \{e\}^x(n).$$

P is obviously nonempty Δ_1^1. Thus by Theorem 1.7.7 P contains a Π_1^1 singleton y as required. ☐

Exercise 1.7.1 Show that a subset A of ω^ω is $\mathbf{\Pi}_1^1$ iff there is a Borel function $f : \omega^\omega \to \mathrm{LO}$ such that $A = f^{-1}(\mathrm{WO})$.

Exercise 1.7.2 Prove Theorem 1.7.3. (*Hint*: Use the definition of Π_1^1 sets and the hyperarithmetical indices for the effectively open sets in the definition.)

Exercise 1.7.3 A real $x \in \omega^\omega$ is Δ_1^1 if its graph $\{(n, m) : x(n) = m\} \subseteq \omega^2$ is Δ_1^1. Show that the following are equivalent:

(1) x is Δ_1^1;

(2) $\{x\}$ is Δ_1^1;

(3) $\{x\}$ is Σ_1^1.

Moreover, the statement can be relativized.

1.8 The Gandy–Harrington topology

In this section we introduce the Gandy–Harrington topology and review some facts relevant to this topic. This topic is more technical than the results reviewed in the previous sections, but is necessary in the study of equivalence relations in Part II of this book.

To keep the logic clear we only give a quick introduction of the concepts and main results in this section and postpone the technical proofs to Appendix

A. All the nontrivial claims, including theorems and some exercises, of this section are proved there.

The techniques developed so far are sufficient for all the proofs in Appendix A except that for Theorem 1.8.5, which will have to wait until we investigate strong Choquet spaces in Chapter 4. By the end of Part I of this book all the results of this section will have been established and will be ready for applications.

To keep the notation simple we continue to work on the product space X in this section. Also we remark that every statement about the Gandy–Harrington topology has a relativized version given an arbitrary real oracle.

Definition 1.8.1
The **Gandy–Harrington topology** on X is the topology generated by all Σ_1^1 sets.

Since there are only countably many Σ_1^1 sets, the Gandy–Harrington topology is second countable. The collection of all Σ_1^1 sets is closed under finite unions and intersections. This implies that the Σ_1^1 sets form a base for the Gandy–Harrington topology. Also, any open set in the Gandy–Harrington topology is the union of countably many Σ_1^1 sets, thus it is Σ_1^1 by Exercise 1.6.2.

If the underlying space is countable (that is, of the form ω^k), then the Gandy–Harrington topology on it is discrete, since all singletons are Σ_1^1. In contrast, the collection of all Σ_1^1 sets in ω^k is nontrivial.

For general X, note that for every $t \in T = \omega^k \times (\omega^{<\omega})^l$, the basic clopen set N_t in the usual topology is Σ_1^1, hence also open in the Gandy–Harrington topology. This implies that the Gandy–Harrington topology is finer than the usual topology on X, and that it is Hausdorff. But the Gandy–Harrington topology is not regular (see Exercise 1.8.1).

A rather surprising fact is that there is a canonical subspace on which the Gandy–Harrington topology is Polish.

Definition 1.8.2
An element $x \in X$ is **low** if $\omega_1^{\mathrm{CK}(x)} = \omega_1^{\mathrm{CK}}$. We denote

$$X_{\mathrm{low}} = \{x \in X : \omega_1^{\mathrm{CK}(x)} = \omega_1^{\mathrm{CK}}\}.$$

The following is a useful characterization of low elements.

Theorem 1.8.3
Let $x \in X$. Then x is low iff for any Σ_1^1 set A either $x \in A$ or else there is a Σ_1^1 set B with $x \in B$ and $A \cap B = \emptyset$.

This in particular implies that every Σ_1^1 subset of X_{low} is relatively clopen in X_{low}, and that the subspace X_{low} is regular. The most important properties of X_{low} lie in the following theorems. The first one of these is also called the **Gandy basis theorem** for Σ_1^1 sets.

Theorem 1.8.4 (Gandy)
X_{low} is dense open in the Gandy–Harrington topology.

Theorem 1.8.5
X_{low} (with the Gandy–Harrington topology) is a Polish space.

Here is an important corollary.

Corollary 1.8.6
The product space X with the Gandy–Harrington topology is a Baire space. That is, the intersection of countably many dense open sets is dense.

This theorem allows category arguments to be used (for instance the concept of meagerness and the Kuratowski–Ulam Theorem 3.2.1). The Gandy–Harrington topology is an important tool in the study of Borel equivalence relations. We will come back to this topic several times later in the book.

Exercise 1.8.1 Show that the Gandy–Harrington topology on ω^ω is not regular. (*Hint*: Use the fact that there is a Π^1_1 set that is not Σ^1_1.)

Exercise 1.8.2 Show that X_{low} is Σ^1_1.

Exercise 1.8.3 (Louveau) Let τ_1 be the Gandy–Harrington topology on X and τ_2 be the Gandy–Harrington topology on $X \times X$. (Note that τ_2 is finer than $\tau_1 \times \tau_1$.) Let $C, D \subseteq X$ be comeager sets in τ_1. Show that $C \times D$ is comeager in τ_2. Similarly, if V is τ_1-open and C, D are both τ_1-comeager in V, then $C \times D$ is τ_2-comeager in $V \times V$.

Chapter 2

Polish Groups

Invariant descriptive set theory is a new branch of descriptive set theory that deals with complexity of equivalence relations. On any Polish space there is always the identity equivalence relation (or the equality equivalence relation, that is, every element is equivalent to nothing but itself), and any subset of a Polish space is an invariant set for this equivalence relation. Therefore on a rather fundamental level invariant descriptive set theory encompasses and strengthens the classical and effective theories.

The focus on equivalence relations, as opposed to other possible objects and structures, originated from connections with other areas of mathematics. The connection with model theory was first recognized and developed by Vaught in the early 1970s. By the end of the 1980s, through the work of many, but especially Becker, Harrington, Kechris, and Louveau, it had become clear that invariant descriptive set theory provides a good framework for the classification problems in many fields of mathematics.

In the process of this development it was also recognized that orbit equivalence relations are the most important class of equivalence relations to be studied. These are equivalence relations induced by actions of Polish groups. What made the subject more exciting was that there was already a vast theory about Polish group actions from dynamical systems and operator algebras, and invariant descriptive set theory is a perfect place to mingle the techniques of logic, topology, algebra, and some other fields.

Starting from this chapter we will present this exciting theory and its applications. Our emphasis will be on orbit equivalence relations, but general Borel, analytic, or even coanalytic equivalence relations will also be investigated. In order to study the complexity of orbit equivalence relations we need to have a good understanding of the actions that induce them. Thus in the rest of Part I we will develop a theory of Polish group actions.

In this chapter, though, we will set the modest goal to review the basic facts about Polish groups. Much of the material in this chapter is of topological nature, but the reader can already find a good dose of logic, algebra, and geometry.

One thing that might be good for the reader to keep in mind is that there are still many open questions about Polish groups. Thus there is really a lot more to do in the invariant descriptive set theory, or even just the theory of Polish groups.

2.1   Metrics on topological groups

A **topological group** is a group $(G, \cdot, 1_G)$ with a topology on G such that the map $(x, y) \mapsto xy^{-1}$ from G^2 to G is continuous.

A topological group G is **metrizable** if there is a metric d on G which induces the topology on G. If d induces the topology on a topological group G we say that d is a **compatible** metric.

A metric d on G is **left-invariant** if $d(gh, gk) = d(h, k)$ for all $g, h, k \in G$.

Recall that a topological space X is **first countable** if every element of X has a countable nbhd base. A topological group G is first countable iff its identity element 1_G has a countable nbhd base.

It is easy to see that a metrizable topological space is Hausdorff and first countable. The following classical theorem shows that the converse is also true for topological groups.

Theorem 2.1.1 (Birkhoff–Kakutani)
Let G be a topological group. Then G is metrizable iff G is Hausdorff and first countable. Moreover, if G is metrizable, then G admits a compatible left-invariant metric.

Proof.   We only need to show the ($\Leftarrow$) direction. Let $\{U_n\}$ be a countable nbhd base for 1_G. Without loss of generality assume $U_0 = G$ and that all U_n are open. Since G is Hausdorff, $\bigcap_n U_n = \{1_G\}$. We define a sequence $\{V_n\}$ of open nbhds of 1_G with the following properties:

(i) $V_0 = G$; $V_{n+1} \subseteq V_n$;

(ii) $V_n = V_n^{-1} =_{\mathrm{def}} \{g^{-1} \mid g \in V_n\}$;

(iii) $V_{n+1}^3 \subseteq V_n$;

(iv) $V_n \subseteq U_n$

for all $n \in \mathbb{N}$. By induction, suppose $V_n \subseteq U_n \cap V_{n-1}$ has been defined with $V_n = V_n^{-1}$. By the continuity of the multiplication operation there are open nbhds W_1, W_2, W_3 of 1_G such that $W_1 \cdot W_2 \cdot W_3 \subseteq V_n$. In particular $W_1, W_2, W_3 \subseteq V_n$. By the continuity of the inverse operation $W_1^{-1}, W_2^{-1}, W_3^{-1}$ are also open nbhds of 1_G. Define

$$V_{n+1} = U_{n+1} \cap U_{n+1}^{-1} \cap W_1 \cap W_1^{-1} \cap W_2 \cap W_2^{-1} \cap W_3 \cap W_3^{-1}.$$

Then V_{n+1} is still an open nbhd of 1_G. It is clear that the properties (i) through (iv) are maintained. By property (iv) we know that $\bigcap_n V_n = \{1_G\}$.

For $h, k \in G$ let

$$\rho(h, k) = \inf \left\{ 2^{-n} \mid h^{-1}k \in V_n \right\}.$$

Then $\rho(h, k) \geq 0$, $\rho(h, k) = 0$ iff $h = k$, and $\rho(h, k) = \rho(k, h)$ by (ii). By (iii) ρ also has the following property (*):

for any $\epsilon > 0$, if $\rho(g_0, g_1), \rho(g_1, g_2), \rho(g_2, g_3) \leq \epsilon$, then $\rho(g_0, g_3) \leq 2\epsilon$.

Finally define

$$d(h, k) = \inf \left\{ \sum_{i=0}^{l} \rho(g_i, g_{i+1}) \mid g_0 = h, g_{l+1} = k, g_1, \ldots, g_l \in G, l \in \mathbb{N} \right\}.$$

We check that d is a compatible left-invariant metric on G.

It is obvious that $d(h, k) \geq 0$ and that $d(h, h) = 0$. Also d is symmetric since ρ is symmetric. The triangle inequality $d(h, k) \leq d(h, g) + d(g, k)$ is also easy to check. Thus to show that d is a metric, it suffices to prove that $d(h, k) \neq 0$ for $h \neq k$. For this we claim that $d(h, k) \geq \frac{1}{2}\rho(h, k)$ for $h \neq k$. In fact, by induction on $l \in \mathbb{N}$ we argue that

$$\sum_{i=0}^{l} \rho(g_i, g_{i+1}) \geq \frac{1}{2}\rho(g_0, g_{l+1}) \tag{2.1}$$

for any $g_0, \ldots, g_{l+1} \in G$. When $l \leq 2$ inequality (2.1) follows from property (*) above. For a general inductive step $l \geq 3$ we assume that inequality (2.1) holds for all $l' < l$. Let S be the sum on the left-hand side. If $\rho(g_0, g_1) \geq \frac{1}{2}S$, then by the inductive hypothesis

$$\rho(g_1, g_{l+1}) \leq 2(S - \rho(g_0, g_1)) \leq 2S - 2\rho(g_0, g_1) \leq S,$$

and hence by property (*) we have that $\rho(g_0, g_{l+1}) \leq 2S$. A symmetric argument also finishes the proof when $\rho(g_l, g_{l+1}) \geq \frac{1}{2}S$. Hence in the remainder we only consider the case that $\rho(g_0, g_1), \rho(g_l, g_{l+1}) < \frac{1}{2}S$. Let m be the largest such that $\sum_{i=0}^{m-1} \rho(g_i, g_{i+1}) \leq \frac{1}{2}S$. Then $1 \leq m < l$. By the inductive hypothesis

$$\rho(g_0, g_m) \leq 2 \sum_{i=0}^{m-1} \rho(g_i, g_{i+1}) \leq S.$$

Since $\sum_{i=0}^{m} \rho(g_i, g_{i+1}) > \frac{1}{2}S$, we have that $\sum_{i=m+1}^{l} \rho(g_i, g_{i+1}) \leq \frac{1}{2}S$ and by the inductive hypothesis again

$$\rho(g_{m+1}, g_{l+1}) \leq 2 \sum_{i=m+1}^{l} \rho(g_i, g_{i+1}) \leq S.$$

Trivially $\rho(g_m, g_{m+1}) \leq S$. It follows from (*) that $\rho(g_0, g_{l+1}) \leq 2S$. This finishes the proof of the claim and that d is a metric.

It is clear that ρ is left-invariant and consequently so is d.

Finally we verify that d is compatible with the topology of G. Let U be open in G and $g \in U$. Then for some $n \in \mathbb{N}$, $gV_n \subseteq U$. We check that $B_d(g, 2^{-n-1}) = \{h \in G \mid d(g, h) < 2^{-n-1}\} \subseteq U$. Let $h \in B_d(g, 2^{-n-1})$. Then $d(h, g) \in 2^{-n-1}$. By the claim in the proof above $\rho(g, h) \leq 2d(g, h) < 2^{-n}$. From the definition of ρ, $g^{-1}h \in V_n$. Thus $h \in gV_n \subseteq U$. Conversely, let U be open in the topology given by d and let $g \in U$. For some $n \in \mathbb{N}$ we have that $B_d(g, 2^{-n}) \subseteq U$. We check that $gV_{n+1} \subseteq U$. Let $h \in gV_{n+1}$. Then $\rho(g, h) \leq 2^{-n-1}$ and by the definition of d, $d(g, h) \leq \rho(g, h) \leq 2^{-n-1} < 2^{-n}$. Therefore $h \in B_d(g, 2^{-n})$ and hence $h \in U$.   □

For any metric space there is the notion of a metric completion. Next we investigate the similar process on a metric group. Let G be a topological group with compatible left-invariant metric d. Define

$$D(h, k) = d(h, k) + d(h^{-1}, k^{-1}).$$

Then D is a also a compatible metric on G (see Exercise 2.1.1). Let $\overline{G}$ be the completion of (G, D).

Lemma 2.1.2
If $(h_n), (k_n)$ are D-Cauchy sequences in G, then both $(h_n^{-1}k_n)$ and $(k_n^{-1}h_n)$ are d-Cauchy.

Proof.   Let $\epsilon > 0$ and let $N_0 \in \mathbb{N}$ be such that for all $n, m \geq N_0$, $D(k_n, k_m) < \frac{1}{3}\epsilon$. By continuity of the multiplication operation there is $\delta > 0$ such that for all $g \in G$ with $d(g, 1_G) < \delta$, $d(k_{N_0}^{-1} g k_{N_0}) < \frac{1}{3}\epsilon$. Let $N \geq N_0$ be such that for all $n, m \geq N$, $D(h_n, h_m) < \delta$. Then for $n, m \geq N$,

$$d(h_m h_n^{-1}, 1_G) = d(h_n^{-1}, h_m^{-1}) \leq D(h_n, h_m) < \delta$$

and hence

$$d(k_{N_0}^{-1} h_m h_n^{-1} k_{N_0}, 1_G) < \frac{1}{3}\epsilon.$$

Then for $n, m \geq N$, by Exercise 2.1.2 below,

$$d(h_n^{-1}k_n, h_m^{-1}k_m) = d(k_m^{-1} h_m h_n^{-1} k_n, 1_G)$$

$$\leq d(k_m^{-1}k_{N_0}, 1_G) + d(k_{N_0}^{-1} h_m h_n^{-1} k_{N_0}, 1_G) + d(k_{N_0}^{-1}k_n, 1_G)$$

$$< \tfrac{1}{3}\epsilon + \tfrac{1}{3}\epsilon + \tfrac{1}{3}\epsilon = \epsilon.$$

This shows that $(h_n^{-1}k_n)$ is d-Cauchy. The sequence $(k_n^{-1}h_n)$ is also d-Cauchy by symmetry.   □

Thus if (h_n) and (k_n) are D-Cauchy it follows immediately that $(h_n^{-1}k_n)$ is D-Cauchy. Therefore we can define the extended multiplication operation on

$\overline{G}$ by

$$(h_n) \cdot (k_n) = (h_n k_n)$$

for $(h_n), (k_n) \in \overline{G}$. Moreover, we can define an extension of d to $\overline{G}$ by

$$\overline{d}((h_n), (k_n)) = \lim_n d(h_n, k_n).$$

By Lemma 2.1.2 $\overline{d}$ is well-defined. It also follows routinely from Lemma 2.1.2 that $\overline{G}$ becomes a topological group with the extended multiplication operation and that $\overline{d}$ is a compatible left-invariant metric on $\overline{G}$. Thus we have obtained the following theorem.

Theorem 2.1.3
Let G be a topological group with a compatible left-invariant metric d. Let D be defined as above, and let $\overline{G}$ be the completion of (G, D). Then the multiplication operation of G extends uniquely onto $\overline{G}$, making $\overline{G}$ a topological group. Moreover, there is a unique extension of d onto $\overline{G}$ which is compatible left-invariant.

The details of the proof are left to the reader.

Example 2.1.4
(1) All groups with the discrete topology are metrizable groups. Moreover the trivial metric

$$d(g, h) = \begin{cases} 0, & \text{if } g = h, \\ 1, & \text{otherwise} \end{cases}$$

is a compatible metric that is left-invariant and complete.

(2) The groups $(\mathbb{R}, +), (\mathbb{Q}, +), (\mathbb{Z}, +)$ with their usual topologies are all metrizable topological groups.

(3) All Banach spaces are topological groups under the addition operation. The norm metric is a compatible metric that is both left-invariant and complete.

Exercise 2.1.1 A metric d on a group G is **right-invariant** if $d(hg, kg) = d(h, k)$ for all $h, g, k \in G$. For a metric d on G define $d^{-1}(h, k) = d(h^{-1}, k^{-1})$. Show that

(a) d^{-1} is a metric on G.

(b) d and d^{-1} induce the same topology on G.

(c) d is left-invariant iff d^{-1} is right-invariant.

Exercise 2.1.2 Let G be a group and d a left-invariant metric on G. Show that $d(hk, 1_G) \le d(h, 1_G) + d(k, 1_G)$ for all $h, k \in G$.

Exercise 2.1.3 Prove Theorem 2.1.3.

A metric d on a group G is **two-sided invariant** if

$$d(g_1 h g_2, g_1 k g_2) = d(h, k)$$

for all $g_1, g_2, h, k \in G$.

Exercise 2.1.4 Show that a topological group G admits a compatible two-sided invariant metric iff G is Hausdorff and 1_G has a countable nbhd base $\{U_n\}$ such that $g U_n g^{-1} = U_n$ for all $g \in G$ and $n \in \mathbb{N}$.

Exercise 2.1.5 Show that any compact metrizable group admits a compatible two-sided invariant metric. (*Hint*: For any open set U containing 1_G and $g \notin U$, there are open sets V, W with $g \in V$ and $1_G \in W$ such that $V \cdot W \cdot V^{-1} \subseteq U$.)

Exercise 2.1.6 Let G be a group and d a metric on G. Show that d is two-sided invariant iff $d(h_1 h_2, k_1 k_2) \leq d(h_1, k_1) + d(h_2, k_2)$ for all $h_1, h_2, k_1, k_2 \in G$.

Exercise 2.1.7 Let G be a group and d a two-sided invariant metric on G. Show that G is a topological group in the topology induced by d.

Exercise 2.1.8 Let G be a topological group with compatible two-sided invariant metric d. Suppose $(h_n), (k_n)$ are sequences in G. Show that $\lim_n h_n k_n = 1_G$ iff $\lim_n k_n h_n = 1_G$.

Exercise 2.1.9 Let $SL(2, \mathbb{R})$ be the topological group of all 2×2 real matrices A with $\det(A) = 1$, equipped with the subspace topology it inherits from $\mathbb{R}^4$. Show that $SL(2, \mathbb{R})$ does not admit any compatible two-sided invariant metric. (*Hint*: Consider

$$A_n = \begin{pmatrix} \dfrac{1}{n} & \dfrac{1}{n} \\ 0 & n \end{pmatrix} \text{ and } B_n = \begin{pmatrix} n & \dfrac{1}{n} \\ 0 & \dfrac{1}{n} \end{pmatrix}.$$

Use Exercise 2.1.8.)

2.2   Polish groups

A topological group G is **Polish** if its underlying topology is Polish, that is, separable and completely metrizable. It follows from Theorem 2.1.3 that any separable metrizable group is densely embedded in a Polish group. For Polish subgroups of a Polish group, we have the following characterization.

Proposition 2.2.1
Let G be a Polish group and H a subgroup of G with the subspace topology. Then the following are equivalent:

(i) *H is Polish.*

(ii) *H is G_δ in G.*

(iii) *H is closed in G.*

Proof. The implications (iii) $\Rightarrow$ (ii) $\Rightarrow$ (i) $\Rightarrow$ (ii) are standard consequences of Proposition 1.1.2 (d). To see that (ii) $\Rightarrow$ (iii), let H be a G_δ subgroup of G. Then the closure of H, $\overline{H}$, is a closed subgroup of H, and H is a dense G_δ in $\overline{H}$. If $\overline{H} \neq H$ there is at least one coset $gH \neq H$ in $\overline{H}$. In $\overline{H}$ the coset gH is also a dense G_δ. It follows from Proposition 1.1.2 (b) that $H \cap gH \neq \emptyset$, a contradiction. Hence $H = \overline{H}$ and so H is closed. □

Corollary 2.2.2
Let G be a Polish group with compatible left-invariant metric d. Then $D(h, k) = d(h, k) + d(h^{-1}, k^{-1})$ is a compatible complete metric for G.

Proof. By Theorem 2.1.3 the completion $\overline{G}$ of (G, D) is a Polish group in which G is a dense G_δ, thus G is closed in $\overline{G}$ by Proposition 2.2.1 and therefore $G = \overline{G}$. This shows that D is a compatible complete metric on G. □

The following proposition is straightforward.

Proposition 2.2.3
A countable product of Polish groups is a Polish group.

Proof. Let G_n, $n \in \omega$, be Polish groups with compatible complete metrics d_n. The metric $d'_n \leq 1$ defined in Section 1.1 is still a compatible complete metric on G_n. For $g = (g_n), h = (h_n) \in G = \prod G_n$, define

$$d(g, h) = \sum_{n=0}^{\infty} 2^{-n} d'_n(g_n, h_n).$$

Then it is routine to check that d is a compatible complete metric on G. □

Example 2.2.4
(1) All countable groups with the discrete topology are Polish groups.

(2) The additive groups of $\mathbb{R}^n$, $\mathbb{C}^n$ $(1 \leq n \leq \omega)$ are Polish groups. The additive group of $\mathbb{Q}$ with the usual topology is not Polish by Proposition 2.2.1.

(3) The multiplicative group of $\mathbb{R}_+ = \{x \in \mathbb{R} : x > 0\}$ is a Polish group.

(4) The additive group of any separable Banach space is a Polish group.

(5) The ω-power of the two element group $\mathbb{Z}_2$, denoted $\mathbb{Z}_2^\omega$, is a Polish group. Note that $\mathbb{Z}_2^\omega$ as a topological space is homeomorphic to the Cantor space 2^ω.

(6) If X is a compact Polish space with a compatible metric d, let $H(X)$ be the space of all homeomorphisms of X onto itself, equipped with the compact-open topology (see Exercise 1.1.6). With composition as the group operation $H(X)$ is a Polish group. The supnorm metric

$$d_{H(X)}(f, g) = \sup\{d(f(x), g(x)) : x \in X\}$$

is a compatible complete metric on $H(X)$ (see Exercise 2.2.2).

(7) If (X, d) is a Polish metric space, let $\mathrm{Iso}(X, d)$ be the space of all isometries of X onto itself, equipped with the pointwise convergence topology (see Exercise 1.1.7). With composition as the group operation $\mathrm{Iso}(X, d)$ is a Polish group (see Exercise 2.2.3).

Every Polish group admits a compatible left-invariant metric and a compatible complete metric. However, not every Polish group admits a compatible complete left-invariant metric. For brevity we make the following definition.

Definition 2.2.5

A Polish group is **cli** if it admits a compatible complete left-invariant metric.

Proposition 2.2.6

Let G be a Polish group with compatible left-invariant metric d. Then G is cli iff d is complete.

Proof. We only prove the nontrivial direction ($\Rightarrow$). Let δ be a compatible complete left-invariant metric on G. Let $\epsilon > 0$. By compatibility of δ and d there is $\lambda > 0$ such that

$$B_d(1_G, \lambda) =_{\mathrm{def}} \{g \in G : d(g, 1_G) < \lambda\} \subseteq B_\delta(1_G, \epsilon).$$

Now let (g_n) be a d-Cauchy sequence in G. Then there exists $N \in \mathbb{N}$ such that for all $n, m \geq N$, $d(g_n, g_m) < \lambda$. Thus for all $n, m \geq N$, $d(g_m^{-1}g_n, 1_G) < \lambda$ and hence $\delta(g_m^{-1}g_n, 1_G) < \epsilon$ or $\delta(g_n, g_m) < \epsilon$. This shows that (x_n) is also δ-Cauchy and hence it converges. We have thus proved that d is complete. $\square$

What we essentially showed in the above proof is that if d and δ are both compatible left-invariant metrics for a Polish group, then a sequence is d-Cauchy iff it is δ-Cauchy.

It follows from the proof of Proposition 2.2.3 that a countable product of cli Polish groups is still cli.

Example 2.2.7
S_∞ is the group of all permutations of ω, with the topology inherited from the Baire space ω^ω. The usual metric

$$d(x,y) = \begin{cases} 0, & \text{if } x = y, \\ 2^{-n}, & \text{if } x \neq y \text{ and } n \text{ is the least such that } x(n) \neq y(n). \end{cases}$$

is a compatible left-invariant metric on S_∞. Since d is not complete (why?), S_∞ is not cli.

Next we study quotients of Polish groups. Let G be a Polish group and H a closed subgroup of G. We denote by G/H the set of right cosets of H and by $\pi_H : G \to G/H$ the canonical projection map $\pi_H(g) = Hg$. G/H is equipped with the quotient topology, that is, $U \subseteq G/H$ is open iff $\pi_H^{-1}(U)$ is open in G. Note that π_H is continuous and open. We next show that G/H is metrizable.

Lemma 2.2.8
Let G be a Polish group with compatible left-invariant metric d. Let H be a closed subgroup of G. Let

$$d^*(Hg_1, Hg_2) = \inf\{\, d(k_1, k_2) \,|\, k_1 \in Hg_1, k_2 \in Hg_2\}.$$

Then d^* is a compatible metric for G/H.

Proof. We first verify that d^* is a metric. For this the only nontrivial part is to show that $d^*(Hg_1, Hg_2) > 0$ for $Hg_1 \neq Hg_2$. Assume $d^*(Hg_1, Hg_2) = 0$. Then by the definition of d^* there are sequences $(h_{1,n}), (h_{2,n})$ of elements of H such that $d(h_{1,n}g_1, h_{2,n}g_2) \to 0$ as $n \to \infty$. By left-invariance of d, we have that $d(g_1^{-1}h_{1,n}^{-1}h_{2,n}g_2, 1_G) \to 0$ as $n \to \infty$. Let $h_n = h_{1,n}^{-1}h_{2,n}$. Then $g_1^{-1}h_ng_2 \to 1_G$ as $n \to \infty$. And by the continuity of the multiplication operation $h_n \to g_1g_2^{-1}$ as $n \to \infty$. Since H is closed, this implies that $g_1g_2^{-1} \in H$, or $g_1 \in Hg_2$. Thus $Hg_1 = Hg_2$.

Before showing that d^* is compatible with the quotient topology on G/H we note that G/H is first countable. In fact, if $Hg \in G/H$ and $\{V_n\}$ is a countable open nbhd base for $g \in G$, then $\{\pi_H(V_n)\}$ is an open nbhd base for Hg in G/H. With this it is enough to show that a subset $A \subseteq G/H$ is closed iff for any sequence (Hg_n) in A and Hg_∞ in G/H with $d^*(Hg_n, Hg_\infty) \to 0$ as $n \to \infty$, $Hg_\infty \in A$.

We first assume that $A \subseteq G/H$ is closed, $Hg_n \in A$, and $d^*(Hg_n, Hg_\infty) \to 0$ as $n \to \infty$. In particular for any $k \in \mathbb{N}$ there is n_k such that $d^*(Hg_{n_k}, Hg_\infty) < 2^{-k}$. It follows from the definition of d^* and the left-invariance of d that for any k there are n_k and $h_k \in H$ such that $d(h_kg_{n_k}, g_\infty) < 2^{-k}$. This means that g_∞ is an accumulation point of $\pi_H^{-1}(A)$. Since A is closed in G/H, $\pi_H^{-1}(A)$ is closed in G, and thus $g_\infty \in \pi_H^{-1}(A)$, or $\pi_H(g_\infty) = Hg_\infty \in A$.

Conversely, let A be closed in the d^*-topology. We show that $\pi_H^{-1}(A)$ is closed in G. For this let (g_n) be a sequence in $\pi_H^{-1}(A)$ and assume that

$g_n \to g_\infty$ as $n \to \infty$. Then $d^*(Hg_n, Hg_\infty) \to 0$ as $n \to \infty$ and $Hg_n \in A$. It follows that $Hg_\infty \in A$ and thus $g_\infty \in \pi_H^{-1}(A)$, as required.   □

Thus G/H is separable metrizable. G/H is indeed a Polish space, but for this we need the following theorem of Sierpinski.

Theorem 2.2.9 (Sierpinski)
Let X be a Polish space, Y a metrizable space, and $\pi : X \to Y$ a continuous and open surjection. Then Y is Polish.

Proof. Let d_X be a compatible complete metric on X. Let $(U_s)_{s \in \omega^{<\omega}}$ be a sequence of nonempty open sets in X such that, for any $s \in \omega^{<\omega}$,

(i) $\mathrm{diam}(U_s) \leq 2^{-\mathrm{lh}(s)}$;

(ii) for any $n \in \omega$, $\overline{U_{s^\frown n}} \subseteq U_s$;

(iii) $U_\emptyset = X$; $U_s = \bigcup_{n \in \omega} U_{s^\frown n}$.

Such a sequence was constructed in the proof of Theorem 1.3.7. Let $V_s = \pi(U_s)$. Since π is open, each V_s is open in Y.

Now let d_Y be a compatible metric on Y. Let $\hat{Y}$ be the d_Y-completion of Y. Let N_s be the interior of the closure of V_s in $\hat{Y}$. Then it is easy to see that $N_s \cap Y = V_s$. Also note that $V_\emptyset = Y$ and $N_\emptyset = \hat{Y}$.

We define a sequence (M_s) of open sets in $\hat{Y}$ such that, for any $s \in \omega^{<\omega}$,

(a) $M_s \subseteq N_s$;

(b) for any $n \in \omega$, $M_{s^\frown n} \subseteq M_s$;

(c) for any $y \in \hat{Y}$ there are only finitely many n with $y \in M_{s^\frown n}$;

(d) $M_s \cap \bigcup_n M_{s^\frown n} = M_s \cap \bigcup_n N_{s^\frown n}$.

The sequence (M_s) is defined by induction on $\mathrm{lh}(s)$. To begin with let $M_\emptyset = N_\emptyset = \hat{Y}$. In general suppose M_s has been defined. We define $M_{s^\frown n}$ for $n \in \omega$. For each n by normality we can find an increasing sequence $(F_{n,m})_{m \in \omega}$ of closed subsets of $N_{s^\frown n}$ such that $N_{s^\frown n} = \bigcup_m F_{n,m}$. We let

$$M_{s^\frown n} = M_s \cap N_{s^\frown n} - \bigcup_{m<n} F_{m,n}.$$

Then conditions (a) and (b) are immediate. To see (c) and (d), let n be the least such that $y \in M_s \cap N_{s^\frown n}$ and let m be the least such that $y \in F_{n,m}$. Then $y \in M_{s^\frown n}$ and for $k > \max\{m, n\}$, $y \notin M_{s^\frown k}$.

We claim that if $y \in \hat{Y} - Y$ then there are at most finitely many s with $y \in M_s$. For this assume $y \in \hat{Y} - Y$ and consider the set $T = \{s \in \omega^{<\omega} : y \in M_s\}$. By (b) and (c) T is a finite splitting tree. Assume that T is infinite.

Then by König's lemma there is $z \in \omega^\omega$ such that $y \in M_{z\restriction n}$ for all $n \in \omega$. Consider instead the element $x \in X$ with $\{x\} = \bigcap_n U_{z\restriction n}$. Since $y \neq \pi(x)$ there is an open set V in $\hat{Y}$ such that $\pi(x) \in V$ and $y \notin \overline{V}$. Let n be such that $x \in U_{z\restriction n} \subseteq \pi^{-1}(V)$. Then $V_{z\restriction n} \subseteq V$ and $y \notin \overline{V_{z\restriction n}}$. But this is a contradiction since $y \in M_{z\restriction n} \subseteq N_{z\restriction n} \subseteq \overline{V_{z\restriction n}}$ by our construction.

The claim implies that $\hat{Y} - Y$ is an F_σ. Thus Y is G_δ in the Polish space $\hat{Y}$, and therefore is Polish by Proposition 1.1.2 (d). □

Suppose in addition that H is a normal subgroup of G, then the above discussion gives that G/H is a Polish group. Moreover, the metric d^* is a compatible left-invariant metric on G/H (see Exercise 2.2.7). These results are summarized in the following theorem.

Theorem 2.2.10
Let G be a Polish group and H be a closed subgroup of G. Then G/H is a Polish space. If in addition H is a normal subgroup of G, then G/H is a Polish group.

We close this section with a theorem on extensions of cli Polish groups.

Theorem 2.2.11
Let G be a Polish group and H a closed normal subgroup of G. Then the following are equivalent:

(i) *G is cli;*

(ii) *Both H and G/H are cli.*

Proof. Let d be a compatible left-invariant metric on G. Then $d \restriction H$ and d^* are compatible left-invariant metrics on H and G/H, respectively. By Proposition 2.2.6 it suffices to show that d is complete iff both $d \restriction H$ and d^* are complete.

First suppose d is complete. Then $d \restriction H$ is obviously complete since H is closed. For the completeness of d^*, let (Hg_n) be a d^*-Cauchy sequence in G/H. By passing to a subsequence we may assume that $d^*(Hg_n, Hg_m) < 2^{-n}$ for all $m > n$. Then define a sequence (k_n) with $k_n \in Hg_n$ by induction. Let $k_0 = g_0$. In general let $k_{n+1} \in Hg_{n+1}$ be such that $d(k_{n+1}, k_n) < 2^{-n}$. Such a k_{n+1} exists by the definition of d^* and left-invariance of d. Now it is clear that (k_n) is d-Cauchy in G, and by the completeness of d there is k_∞ such that $k_n \to k_\infty$ as $n \to \infty$. It follows that $Hg_k \to Hg_\infty$ as required.

Conversely, assume both $d \restriction H$ and d^* are complete. Let (g_n) be a d-Cauchy sequence. By passing to a subsequence we may assume that $d(g_n, g_m) < 2^{-n}$ for $n < m$. It follows that (Hg_n) is d^*-Cauchy in G/H, and therefore there is g_∞ such that $Hg_n \to Hg_\infty$ as $n \to \infty$. And by the definition of d^* and left-invariance of d we obtain a sequence (h_n) in H such that $h_n g_n \to g_\infty$ as $n \to$

∞. Again by passing to a subsequence we may assume that $d(h_n g_n, g_\infty) < 2^{-n}$. By left-invariance of d, $d(g_n, h_n^{-1} g_\infty) < 2^{-n}$. Since H is a normal subgroup there is a sequence (h'_n) in H such that $h_n^{-1} g_\infty = g_\infty h'_n$. Thus $d(g_n, g_\infty h'_n) < 2^{-n}$. It follows by the triangle inequality from our assumption that $d(g_n, g_m) < 2^{-n}$ for $n < m$ that $(g_\infty h'_n)$ is d-Cauchy. By left-invariance of d we have that (h'_n) is d-Cauchy. Now since (h'_n) is in H there exists h_∞ such that $h'_n \to h_\infty$ as $n \to \infty$. We thus get that $g_n \to g_\infty h_\infty$ as $n \to \infty$. □

Exercise 2.2.1 Show that any locally compact, second countable, Hausdorff group is Polish.

Exercise 2.2.2 Let (X, d) be a compact Polish metric space. Show that $H(X)$ is a Polish group and that the supnorm metric is a compatible complete metric for $H(X)$.

Exercise 2.2.3 Let (X, d) be a Polish metric space and $D \subseteq X$ be a countable dense subset. Show that $\mathrm{Iso}(X, d)$ is a Polish group and that the following metric defined in Section 1.1 is a compatible left-invariant metric for $\mathrm{Iso}(X, d)$:

$$d_L(\varphi, \psi) = \sum_{i=1}^{\infty} 2^{-i} d'(\varphi(q_i), \psi(q_i)),$$

where (q_i) is an enumeration of D.

Exercise 2.2.4 Let G be a Polish group with compatible two-sided invariant metric d. Show that d is complete.

Exercise 2.2.5 Show that any locally compact Polish group is cli.

Exercise 2.2.6 Let G be a Polish group and H a closed subgroup of G. Show that G/H is at most countable iff H is clopen.

Exercise 2.2.7 Let G be a Polish group with compatible left-invariant metric d, and H a closed normal subgroup of G. Prove that d^* is a compatible left-invariant metric on G/H.

Exercise 2.2.8 Let G be a Polish group with compatible two-sided invariant metric d, and H a closed normal subgroup of G. Show that G/H admits a compatible two-sided invariant metric.

Exercise 2.2.9 Show that any solvable Polish group admits a compatible complete left-invariant metric.

2.3 Continuity of homomorphisms

The automatic continuity of homomorphisms among Polish groups we will discuss in this section is one of many special and nice properties of Polish group topologies. Its proof is also significant because it marks the beginning of the use of Baire category methods in invariant descriptive set theory.

Recall that a topological group is **Baire** if it is Baire as a topological space, that is, the intersection of countably many open dense sets is dense. In any topological space a subset is **meager** if it is a subset of a union of countably many nowhere dense sets, **nonmeager** if it is not meager, and **comeager** if its complement is meager. A topological space is Baire iff every nonempty open subset of it is nonmeager, thus in particular a Baire space is nonmeager (in itself).

We also recall the following concept. Let X be a Baire space. A subset $A \subseteq X$ has the **Baire property** if there is an open set $U \subseteq X$ such that $A \triangle U =_{\text{def}} (A - U) \cup (U - A)$ is meager. It is easy to see that the subsets of X with the Baire property form a σ-algebra and it is generated by all open sets and meager sets. A nontrivial theorem of Nikodym (see Reference [97] Corollary 29.14) states that the collection of sets with the Baire property in any topological space is closed under the Suslin operation $\mathcal{A}$. It follows that every analytic set (or coanalytic set) has the Baire property.

The next proposition gives equivalent conditions for a topological group to be Baire.

Proposition 2.3.1
Let G be a topological group. Then the following are equivalent:

(i) *G is Baire;*

(ii) *G is nonmeager;*

(iii) *Every nonempty open subset of G is nonmeager;*

(iv) *There exists a nonmeager open subset of G;*

(v) *There exists a nonmeager subset of G with the Baire property.*

Proof. We noted that (i)$\Leftrightarrow$(iii)$\Rightarrow$(ii). It is also trivial that (ii)$\Leftrightarrow$(iv)$\Leftrightarrow$(v). The only nontrivial implication is (ii)$\Rightarrow$(i). Assume that G is nonmeager but some open subset U of G is meager. Then note that $G = \bigcup_{g \in G} gU$ and each gU is meager. Consider the collection of all families $\mathcal{F}$ of disjoint open subsets of G such that each $V \in \mathcal{F}$ is contained in some gU, $g \in G$. Partially order the collection by inclusion. Then it is clear that the Zorn's lemma applies and thus we obtain a maximal family $\mathcal{F}$ in this collection. Let $\mathcal{F} = \{V_i\}_{i \in I}$ and $V = \bigcup_{i \in I} V_i$. Then the maximality of $\mathcal{F}$ implies that V is dense open in G, and thus $G - V$ is nowhere dense and thus meager. Each V_i, being a subset of gU for some $g \in G$, is meager. Let $F_{i,n}$ be nowhere dense sets such that $V_i \subseteq \bigcup_n F_{i,n}$. Let $F_n = \bigcup_{i \in I} F_{i,n}$. Then each F_n is again nowhere dense. To see this, assume F_n is dense in an open set W. Since $G - V$ is nowhere dense, $W \cap V_i \neq \emptyset$ for some $i \in I$, and without loss of generality we may assume that $W \subseteq V_i$ for some $i \in I$. However, $F_n \cap W \subseteq F_n \cap V_i \subseteq F_{i,n}$, the latter being nowhere dense, a contradiction. We have thus seen that $V \subseteq \bigcup_n F_n$ is meager, and hence $G = (G - V) \cup V$ is meager. □

Moreover, each of (iii) to (v) is equivalent to a strengthened condition in which we require that the set contains 1_G.

Theorem 2.3.2 (Pettis)
Let G be a topological group and $A \subseteq G$ a nonmeager subset with the Baire property. Then $A^{-1}A$ contains an open nbhd of 1_G.

Proof. Let U be open such that $A \triangle U$ is meager. Fix $g \in U$ and let V be open with $1_G \in V$ such that $VV^{-1} \subseteq g^{-1}U$. We claim that $V \subseteq A^{-1}A$. To see this, take $h \in V$ and we show that $A \cap Ah \neq \emptyset$. For this, first note that $g \in U \cap Uh$, and hence $U \cap Uh \neq \emptyset$. Next note that $(A \cap Ah) \triangle (U \cap Uh) \subseteq (A \triangle U) \cup (A \triangle U)h$, and therefore $(A \cap Ah) \triangle (U \cap Uh)$ is meager. Thus $A \cap Ah \neq \emptyset$ since otherwise the nonempty open set $U \cap Uh$ would be meager in the space G which is Baire. □

If in the above theorem A is in addition a subgroup, then A is open, and therefore clopen (see Exercise 2.3.3).

The phenomenon revealed in the following theorem is often referred to as automatic continuity. Recall that a function $f : X \to Y$ is **Baire measurable** if for every open set $U \subseteq Y$, $f^{-1}(U)$ has the Baire property.

Theorem 2.3.3
Let G, H be Polish groups and $\varphi : G \to H$ be a group homomorphism. If φ is Baire measurable, then φ is continuous. If moreover $\varphi(G)$ is nonmeager, then φ is also open.

Proof. Let $U \subseteq Y$ be open and $g \in \varphi^{-1}(U)$. Let $V \subseteq Y$ be open such that $1_H \in V$ and $V^{-1}V \subseteq \varphi(g)^{-1}U$. Let $D \subseteq H$ be a countable dense subset such that $D = D^{-1}$. Note that $\bigcup_{h \in D} hV = H$. It follows that $\bigcup_{h \in D} \varphi^{-1}(hV) = G$. Since G is Baire, there is some $h \in D$ such that $A = \varphi^{-1}(hV)$ is nonmeager. Since φ is Baire measurable, A has the Baire property. Thus by Theorem 2.3.2 $A^{-1}A$ contains an open nbhd N of 1_G. Then

$$\varphi(gN) \subseteq \varphi(gA^{-1}A) \subseteq \varphi(g)\varphi(A)^{-1}\varphi(A)$$
$$\subseteq \varphi(g)(hV)^{-1}(hV) \subseteq \varphi(g)V^{-1}V \subseteq U.$$

This shows that φ is continuous.

We note that $\varphi(G)$ is Borel in H. In fact, let $G_0 = \varphi^{-1}(1_H) = \ker(\varphi)$. Then G_0 is a closed normal subgroup of G and G/G_0 is a Polish group by Theorem 2.2.10. Let $\tilde{\varphi} : G/G_0 \to H$ be defined by

$$\tilde{\varphi}(gG_0) = \varphi(g).$$

Then $\tilde{\varphi}$ is a continuous homomorphism from G/G_0 into H which is moreover an injection. Thus by the Luzin–Suslin Theorem 1.3.1 $\tilde{\varphi}(G/G_0) = \varphi(G)$ is Borel in H. This in particular implies that $\varphi(G)$ has the Baire property.

Now suppose in addition that $\varphi(G)$ is nonmeager, then $\varphi(G)$ is clopen by Exercise 2.3.3. In particular, $\varphi(G)$ is a Polish group. If $W \subseteq G/G_0$ is open, then $\tilde{\varphi}(W)$ being a continuous injective image of an open set is Borel and has the Baire property. This shows that the inverse homomorphism $\tilde{\varphi}^{-1} : \varphi(G) \to G/G_0$ is Baire measurable, thus by the first part of the proof $\tilde{\varphi}^{-1}$ is continuous. Therefore $\varphi : G \to \varphi(G)$ is open, since for any open $E \subseteq G$, $\varphi(E) = \varphi(EG_0) = \tilde{\varphi}(\pi(E))$, where $\pi : G \to G/G_0$ is the projection map $g \mapsto gG_0$. Finally, since $\varphi(G)$ is indeed open in H, φ is an open map from G into H. ☐

The proof of the second part of the above theorem gives the following useful corollary.

Corollary 2.3.4
Let G, H be Polish groups and $\varphi : G \to H$ be a surjective continuous homomorphism. Then the natural group isomorphism $\varphi^* : G/\ker(\varphi) \to H$ induced by φ is a topological group isomorphism.

Exercise 2.3.1 Let X be a Baire space and $A \subseteq X$ has the Baire property. Suppose for every nonempty open set $U \subseteq X$, $A \cap U$ is nonmeager. Show that A is comeager.

Exercise 2.3.2 Let X be a Baire space, Y a second countable space, and $f : X \to Y$. Show that if f is Baire measurable then there is a dense G_δ set $A \subseteq X$ such that $f \restriction A$ is continuous.

Exercise 2.3.3 Let G be a Baire group and $A \leq G$ a nonmeager subgroup with the Baire property. Show that A is a clopen subgroup of G.

Exercise 2.3.4 Let G, H be topological groups and $\varphi : G \to H$ a group homomorphism. Show that if G is Baire, H is separable and φ is Baire measurable, then φ is continuous.

Exercise 2.3.5 Show that any Baire measurable group isomorphism between Polish groups is a topological group isomorphism.

Exercise 2.3.6 Let G be a Polish group and $H \leq G$ a closed subgroup. Show that the Borel structure generated by the quotient topology on G/H is the same as the Effros Borel structure.

Exercise 2.3.7 Let G be a group and τ_1 and τ_2 be both Polish topologies on G making G topological groups. Show that if τ_1 and τ_2 have the same Borel structure then $\tau_1 = \tau_2$.

Exercise 2.3.8 Let G be a group, d a left-invariant metric on G, and consider the topology induced by d. Show that if the right-translation $g \mapsto gh$ is continuous for each $h \in G$, then the inverse $g \mapsto g^{-1}$ is continuous. Deduce that if multiplication is continuous, then G is a topological group.

Exercise 2.3.9 (Kallman) Show that the usual topology on S_∞ is the unique Polish group topology on S_∞. (*Hint:* For any $n \in \omega$ the partial identity $s_n : \{0,1,\ldots,n\} \to \{0,1,\ldots,n\}$ with $s_n(k) = k$ gives rise to a nbhd of the identity N_{s_n}. Let B_n be the subgroup $\{g \in S_\infty : g(l) = l \text{ for all } l > n \}$. Show that N_{s_n} consists of exactly all group elements h commuting with all elements of B_n. Thus in any Polish group topology N_{s_n} is closed.)

2.4   The permutation group S_∞

One of the most important examples of Polish groups is the permutation group S_∞. Many of the results in invariant descriptive set theory were first obtained for S_∞ and later generalized to arbitrary Polish groups. S_∞ and its closed subgroups are closely related to countable model theory, and because of this natural but profound connection with logic the study of S_∞ and its actions has been the most fruitful part of invariant descriptive set theory. In this section we characterize S_∞ and its closed subgroups as Polish groups and study their basic properties.

Recall from Example 2.2.7 that S_∞ consists of all permutations (bijections) of ω and is a Polish subspace of the Baire space ω^ω. A compatible left-invariant metric on S_∞ is

$$d(x, y) = \begin{cases} 0, & \text{if } x = y, \\ 2^{-n}, & \text{if } x \neq y \text{ and } n \text{ is the least such that } x(n) \neq y(n). \end{cases}$$

S_∞ does not admit any compatible complete left-invariant metric. A compatible complete metric on S_∞ is

$$D(x, y) = d(x, y) + d(x^{-1}, y^{-1})$$

by Corollary 2.2.2.

Note that the metric d is in fact an ultrametric. Recall that an **ultrametric** d is a metric satisfying

$$d(x, z) \leq \max\{d(x, y), d(y, z)\}$$

instead of the usual triangle inequality.

The following theorem characterizes Polish groups that are essentially closed subgroups of S_∞.

Theorem 2.4.1
Let G be a Polish group. Then the following are equivalent:

(i) *G is isomorphic to a closed subgroup of S_∞.*

(ii) *G admits a compatible left-invariant ultrametric.*

(iii) *G admits a countable nbhd base of 1_G consisting of open subgroups.*

(iv) *G admits a countable base closed under left (or right) multiplication, that is, there is a countable base $\mathcal{B}$ of G so that for any $U \in \mathcal{B}$ and $g \in G$, $gU \in \mathcal{B}$.*

Proof. (i)$\Rightarrow$(ii) is obvious. For (ii)$\Rightarrow$(iii), let d_G be a compatible left-invariant ultrametric on G. For each $n \in \omega$ let

$$U_n = \{x \in G : d_G(x, 1_G) < 2^{-n}\}.$$

Then $\{U_n\}$ is a countable nbhd base for 1_G. Since d_G is a left-invariant ultrametric, for $x, y \in G$ we have

$$d_G(x^{-1}y, 1_G) = d_G(y, x) \leq \max\{d_G(x, 1_G), d_G(y, 1_G)\}$$

and thus U_n is a subgroup. For (iii)$\Rightarrow$(iv), let $\{U_n\}$ be a countable nbhd base of 1_G so that each U_n is an open subgroup. Note that each U_n has only countably many left (or right) cosets. Let $\mathcal{B}$ be the collection of all left cosets of U_n for all $n \in \omega$. Then $\mathcal{B}$ is a countable base for G.

It remains to show the implication (iv)$\Rightarrow$(i). Let $\mathcal{B}$ be a countable base closed under left multiplication. Let $U_0, U_1, \ldots$ enumerate $\mathcal{B}$ without repetition. We define an isomorphic embedding $\varphi : G \to S_\infty$ by letting

$$\varphi(g)(n) = m \iff gU_n = U_m.$$

It is straightforward to see that φ is an injective continuous homomorphism. To finish the proof it suffices to show that φ is an open map from G onto $\varphi(G)$, since then $\varphi(G)$ is homeomorphic to G and is Polish and by Proposition 2.2.1 $\varphi(G)$ is closed in S_∞.

For this let U be open in G and let $g \in U$. We find a basic open V in $\varphi(G)$ so that $\varphi(g) \in V \subseteq \varphi(U)$. Since $g^2 \cdot g^{-1} = g$, there are open subsets W_0, W_1 of G with $g^2 \in W_0$, $g \in W_1$ such that $W_0 W_1^{-1} \subseteq U$. Let $U_n \in \mathcal{B}$ be such that $g \in U_n \subseteq g^{-1} W_0 \cap W_1$. Let $U_m = g U_n \in \mathcal{B}$ and let $V = \{\pi \in S_\infty : \pi(n) = m\} \cap \varphi(G)$. Then V is a basic open set in $\varphi(G)$ and $\varphi(g) \in V$. To see that $V \subseteq \varphi(U)$, let $h \in G$ with $\varphi(h) \in V$. Then $h U_n = U_m$. It follows that $h \in U_m U_n^{-1} = g U_n U_n^{-1} \subseteq W_0 W_1^{-1} \subseteq U$ and hence $\varphi(h) \in \varphi(U)$. $\qquad\Box$

The following proposition is immediate either as a corollary of the above theorem or from a direct proof.

Proposition 2.4.2
A countable product of closed subgroups of S_∞ is isomorphic to a closed subgroup of S_∞.

Since S_∞ is totally disconnected, no nontrivial connected group can be isomorphic to a closed subgroup of S_∞.

Example 2.4.3
(1) Any countable group G with the discrete topology is isomorphic to a closed subgroup of S_∞, because $\{1_G\}$ is a nbhd base of itself which is an open subgroup.

(2) Each of the product groups $\mathbb{Z}_2^\omega$, $\mathbb{Z}^\omega$ is isomorphic to a closed subgroup of S_∞. In fact, any countable product of countable groups is isomorphic to a closed subgroup of S_∞ by (1) and Proposition 2.4.2.

(3) None of the groups $\mathbb{R}^n$, $\mathbb{C}^n$ is isomorphic to a closed subgroup of S_∞.

Next we explore the role S_∞ plays in countable model theory. We first review some basics of mathematical logic. A **relational language** is a set L of (relation) symbols together with a function $a : L \to \omega$. An **interpretation** of a relational language L consists of a set A and an assignment which, to each relation symbol R in L, associates a relation R^A of arity $a(R)$, that is, $R^A \subseteq A^{a(R)}$. Under these circumstances we call the tuple

$$M = (A, \{R^A : R \in L\})$$

an L-**structure** or a **model** of L. The set A is called the **universe** of the model M and is also denoted $|M|$. The interpretation of a relation R in a model M is also denoted as R^M (instead of R^A as above). Models of the same language L maintain a formal resemblance but can be of an entirely different nature. For instance, let $L = \{R\}$ where R is a binary relation symbol. Any directed graph $G = (V, E)$ is an L-structure with $E = R^G$. Any partial order $P = (A, <)$ is also an L-structure with $<= R^P$.

A general language can also contain function symbols, that is, symbols to be interpreted as functions (again of various arity) on the universe of the model. For instance, a group can be regarded as a model of the language $L = \{F_1, F_2\}$, where F_1 is a unary function (to be interpreted as the inverse) and F_2 is a binary function (to be interpreted as the multiplication). Note, however, that functions in models can be alternatively interpreted as relations anyway. For instance the inverse function in a group can be completely determined by a binary relation $R(x, y)$ whose intended interpretation is $x^{-1} = y$. Thus when describing models it suffices to consider relational languages.

For an L-structure M, an **automorphism** is a bijection $f : |M| \to |M|$ such that, for any n-ary relation symbol R in L and any $x_1, \ldots, x_n \in |M|$,

$$R^M(x_1, \ldots, x_n) \iff R^M(f(x_1), \ldots, f(x_n)).$$

Let $\mathrm{Aut}(M)$ be the group of all automorphisms of M. Then $\mathrm{Aut}(M)$ is a subset of the product space $|M|^{|M|}$. Equip $|M|$ with the discrete topology and $|M|^{|M|}$ the product topology, then $\mathrm{Aut}(M)$ becomes a topological group. When $|M|$ is countable, $\mathrm{Aut}(M)$ becomes a Polish group (Exercise 2.4.7). The following theorem shows that every closed subgroup of S_∞ is isomorphic to the automorphism group of some countable model.

Theorem 2.4.4
Let G be a Polish group. Then the following are equivalent:

(i) *G is isomorphic to a closed subgroup of S_∞.*

(ii) *There is a relational language L and a countable L-structure M such that $G \cong \mathrm{Aut}(M)$.*

(iii) *There is a countable relational language L and a countable L-structure M such that $G \cong \mathrm{Aut}(M)$.*

Proof. We show that (ii)$\Rightarrow$(i)$\Rightarrow$(iii). For (ii)$\Rightarrow$(i), assume L is a relational language and M a countable L-structure. If $|M|$ is finite then it is isomorphic to a closed subgroup of S_∞. We assume that $|M|$ is infinite, and without loss of generality assume that $|M| = \omega$. Then $\mathrm{Aut}(M)$ is a subgroup of S_∞. It suffices to note that it is closed in S_∞. Indeed, let (f_k) be a sequence in $\mathrm{Aut}(M)$ and suppose $f_k \to f$ as $k \to \infty$ for $f \in S_\infty$. We verify that $f \in \mathrm{Aut}(M)$. For this let $R \in L$ be an n-ary relation symbol and $x_1, \ldots, x_n \in |M|$. For each $1 \leq i \leq n$, there is K_i such that for all $k \geq K_i$, $f_k(x_i) = f(x_i)$. Let $K = \max\{K_i : 1 \leq i \leq n\}$. Then for $k \geq K$,

$$R^M(x_1, \ldots, x_n) \iff R^M(f_k(x_1), \ldots, f_k(x_n)) \iff R^M(f(x_1), \ldots, f(x_n)).$$

This shows that f is an automorphism of M. For (i)$\Rightarrow$(iii), let G be a closed subgroup of S_∞. For each $n \in \omega$, define an equivalence relation $\sim_n$ on ω^n by

$$(a_1, \ldots, a_n) \sim_n (b_1, \ldots, b_n) \iff \exists f \in G \ \forall 1 \leq i \leq n \ f(a_i) = b_i.$$

Let C_n be the set of all $\sim_n$-equivalence classes. Let $C = \bigcup_n C_n$. Let a language L consist of an n-ary relation symbol R_c associated with each n-ary element c of C. Note that L is countable. Define an L-structure M with the universe ω as follows: for each n-ary $c \in C$, let $R_c^M = c$. We claim that $\mathrm{Aut}(M) = G$. That $G \subseteq \mathrm{Aut}(M)$ is obvious from our definition. For the other inclusion, we take $f \in \mathrm{Aut}(M)$. Since G is closed, it suffices to show that, given any $n \in \omega$, there is $f_n \in G$ so that $f_n(i) = f(i)$ for $i < n$. Let $n \in \omega$ be fixed. Since $f \in \mathrm{Aut}(M)$ we have that for all n-ary relation $R_c \in L$,

$$R_c^M(0, \ldots, n-1) \iff R_c^M(f(0), \ldots, f(n-1)).$$

In particular, $(0, \ldots, n-1) \sim_n (f(0), \ldots, f(n-1))$. Thus by the definition of $\sim_n$, there is $f_n \in G$ so that $f_n(i) = f(i)$ for all $i < n$. $\Box$

Note that S_∞ can also be viewed as the isometry group for the discrete space ω with the trivial metric (Exercise 2.4.4). It turns out that many properties of S_∞ can be generalized to arbitrary isometry groups via this trivial connection. Here we give just one example. Let (X, d) be a Polish metric space and $\{R_i\}_{i \in I}$ be a set of closed relations over X. Let $\mathrm{Iso}(X, d, \{R_i\}_{i \in I})$ denote the set of elements φ of $\mathrm{Iso}(X, d)$ such that for every $i \in I$, if R_i is an n-ary closed relation over X (that is, a closed subset of X^n), then for any $x_1, \ldots, x_n \in X$,

$$R_i(x_1, \ldots, x_n) \iff R_i(\varphi(x_1), \ldots, \varphi(x_n)).$$

This is still a Polish group (Exercise 2.4.8). We have the following straightforward generalization of the above theorem to arbitrary isometry groups. The proof is similar to that of Theorem 2.4.4 and is left as an exercise.

Theorem 2.4.5
Let (X, d) be a Polish metric space and G a Polish group. Then the following are equivalent:

(i) *G is isomorphic to a closed subgroup of $\mathrm{Iso}(X, d)$.*

(ii) *There is a set of closed relations $\{R_i\}_{i \in I}$ over X such that*

$$G \cong \mathrm{Iso}(X, d, \{R_i\}_{i \in I}).$$

(iii) *There are countably many closed relations $R_n \subseteq X^n$ such that $G \cong \mathrm{Iso}(X, d, \{R_n\}_{n \in \omega})$.*

Exercise 2.4.1 Let G be a Polish group and H a closed normal subgroup of G. Show that G is isomorphic to a closed subgroup of S_∞ iff both H and G/H are isomorphic to closed subgroups of S_∞.

Exercise 2.4.2 Show that a closed subgroup G of S_∞ admits a compatible complete left-invariant metric iff G is closed in the Baire space ω^ω.

Exercise 2.4.3 Show that a closed subgroup G of S_∞ is compact iff for every $n \in \omega$ the set $\{g(n) : g \in G\}$ is finite.

Exercise 2.4.4 Let (X, d) be a countable discrete metric space. Show that $\mathrm{Iso}(X, d)$ is isomorphic to a closed subgroup of S_∞.

Exercise 2.4.5 Let (X, d) be a compact Polish ultrametric space. Show that $H(X)$ is isomorphic to a closed subgroup of S_∞.

Exercise 2.4.6 Let X be a countable compact Polish space. Show that $H(X)$ is isomorphic to a closed subgroup of S_∞.

Exercise 2.4.7 Let L be a relational language and M a countable L-structure. Show that $\mathrm{Aut}(M)$ is a Polish group.

Exercise 2.4.8 Let (X, d) be a Polish metric space and $\{R_i\}_{i \in I}$ a set of closed relations over X. Show that $\mathrm{Iso}(X, d, \{R_i\}_{i \in I})$ is a closed subgroup of $\mathrm{Iso}(X, d)$, and hence is Polish.

Exercise 2.4.9 Prove Theorem 2.4.5.

2.5 Universal Polish groups

One of the prominent properties of the class of Polish groups is the existence of universal objects. In Section 1.2 we studied the universal Urysohn space, which is a universal Polish metric space. In this section we consider topological group isomorphic embeddings among Polish groups and universal objects in this category. Thus a Polish group is called **universal** if every Polish group is isomorphic to a closed subgroup of it. Based on the properties of the universal Urysohn space $\mathbb{U}$, we will show that $\mathrm{Iso}(\mathbb{U})$ is a universal Polish group.

Lemma 2.5.1
For every Polish group G there is a Polish metric space (X, d) such that G is isomorphic to a closed subgroup of $\mathrm{Iso}(X, d)$.

Proof. Let d_G be a compatible left-invariant metric on G. Let (X, d) be the completion of (G, d_G). For any $g \in G$ and $x \in X$, letting (h_n) be a d_G-Cauchy sequence in G with $d(h_n, x) \to 0$ as $n \to \infty$, define $\varphi_g(x) = \lim_n gh_n$. The limit exists since (gh_n) is also d_G-Cauchy by left-invariance of d_G. It is easy to see that the definition does not depend on the choice of (h_n) and that φ_g is an isometry of (X, d). Moreover the map $g \mapsto \varphi_g$ is a continuous homomorphism from G into $\mathrm{Iso}(X,d)$. To finish the proof it suffices to show that $\{\varphi_g : g \in G\}$ is closed in $\mathrm{Iso}(X,d)$, since then $\{\varphi_g : g \in G\}$ is a closed subgroup of $\mathrm{Iso}(X,d)$, hence is Polish, and by Theorem 2.3.3 the map $g \mapsto \varphi_g$ is a topological group isomorphism.

Suppose (g_n) is a sequence in G such that $\varphi_{g_n} \to \psi$ as $n \to \infty$ for some $\psi \in \mathrm{Iso}(X,d)$. In particular, $\varphi_{g_n}(1_G) = g_n \to \psi(1_G)$ and (g_n) is d_G-Cauchy. On the other hand, $\varphi_{g_n}^{-1} \to \psi^{-1}$ by the continuity of the inverse in $\mathrm{Iso}(X,d)$. But $\varphi_{g_n}^{-1} = \varphi_{g_n^{-1}}$, hence similarly (g_n^{-1}) is also d_G-Cauchy. By Corollary 2.2.2 (g_n) is D-Cauchy for the complete metric D on G, and there is $h \in G$ such that $g_n \to h$ as $n \to \infty$. It follows that $\varphi_{g_n} \to \varphi_h$ and hence $\varphi_h = \psi$, as required.   □

Recall that in Section 1.2 we defined extensions $E(X)$ and $E(X,\omega)$ for arbitrary separable metric spaces X. We also noted that X can be isometrically embedded into $E(X,\omega)$. Here we make some more observations so as to study their isometry groups. For notational simplicity we suppress the metrics when writing the isometry groups, since there is no danger of confusion. If $\varphi \in \mathrm{Iso}(X)$, we let $E(\varphi) \in \mathrm{Iso}(E(X))$ be given by

$$E(\varphi)(f)(x) = f(\varphi^{-1}(x)).$$

When $f \in E(X,\omega)$, then $E(\varphi)(f) \in E(X,\omega)$ (see Exercise 2.5.2). Thus $E(\varphi)$ can be regarded as an isometry of $E(X,\omega)$. Moreover, for each $x \in X$, $E(\varphi)(f_x)(y) = d(x, \varphi^{-1}(y)) = d(\varphi(x), y) = f_{\varphi(x)}(y)$. Thus $E(\varphi) \upharpoonright X = \varphi$. It is also easy to see that $E(\varphi \circ \psi) = E(\varphi) \circ E(\psi)$ when $\varphi, \psi \in \mathrm{Iso}(X)$.

Recall also from Section 1.2 the general procedure to construct an isometric copy of the universal Urysohn space. Given an arbitrary Polish metric space X define inductively $X_0 = X$, $X_{n+1} = E(X_n,\omega)$ and $X_\omega = \bigcup_n X_n$. Then $\mathbb{U}$ is isometric to $\overline{X_\omega}$, the completion of X_ω. We are now ready to prove the universality of $\mathrm{Iso}(\mathbb{U})$ in the class of all Polish groups.

Theorem 2.5.2 (Uspenskij)
Iso($\mathbb{U}$) is a universal Polish group.

Proof. In view of the preceding lemma it suffices to show that for any Polish metric space X the isometry group $\mathrm{Iso}(X)$ is isomorphic to a closed subgroup of $\mathrm{Iso}(\overline{X_\omega})$. For any $\varphi \in \mathrm{Iso}(X)$, inductively define a sequence (φ_n) by $\varphi_0 = \varphi, \varphi_{n+1} = E(\varphi_n)$. Then each φ_n is an isometry of X_n. Moreover, each φ_{n+1} is an extension of φ_n. Thus we may define $\varphi_\omega = \bigcup_n \varphi_n$ and φ_ω is then an isometry of X_ω. Finally φ_ω has a unique extension to $\overline{X_\omega}$ similar to the proof of the preceding lemma. Let φ^* be this unique extension. It is then straightforward to see that $\varphi \mapsto \varphi^*$ is a continuous homomorphism from $\mathrm{Iso}(X)$ into $\mathrm{Iso}(\overline{X_\omega})$.

To finish the proof it suffices to show that $\varphi \mapsto \varphi^*$ is an open map. But this follows immediately from the fact that φ^* is an extension of φ for any

$\varphi \in \text{Iso}(X)$. To see this, let $(\varphi_k), \psi \in \text{Iso}(X)$ and assume $(\varphi_k)^* \to \psi^*$ as $k \to \infty$. Then $\varphi_k = (\varphi_k)^* \restriction X \to \psi^* \restriction X = \psi$. □

In the rest of this section we mention without proof some further facts about universal Polish groups. First we should point out the following theorem of Uspenskij parallel to the previous theorem.

Theorem 2.5.3 (Uspenskij)
The Polish group $H([0,1]^\omega)$ of homeomorphisms of the Hilbert cube $[0,1]^\omega$ is a universal Polish group.

Note that every Polish space is homeomorphic to a subspace of the Hilbert cube $[0,1]^\omega$ (see Exercise 2.5.1). Thus in this theorem the Hilbert cube is playing the role of the universal Urysohn space in Theorem 2.5.2. A proof of this theorem can be found in Reference [97] (Theorem 9.18).

In general suppose $\mathcal{G}$ is a class of Polish groups so that for any $G \in \mathcal{G}$ and closed subgroup $H \leq G$, $H \in \mathcal{G}$. One can ask whether $\mathcal{G}$ has a universal object, that is, whether there is $G \in \mathcal{G}$ such that every group $H \in \mathcal{G}$ is a closed subgroup of G. Some natural properties that define such classes of Polish groups are commutativity, compactness, local compactness, admitting two-sided invariant metric, and admitting complete left-invariant metric. It is known that there are universal abelian Polish groups [136] and there are no universal locally compact Polish groups (folklore). For the other classes the answers are not known.

We introduce some related concepts.

Definition 2.5.4
Let G and H be Polish groups. We say that H is **involved** in G or G **involves** H, denoted $H \leq_w G$, if H is isomorphic to a closed subgroup of a quotient group of G.

It is not immediate, but easy to show that $\leq_w$ is a transitive relation (Exercise 2.5.5).

Definition 2.5.5
A Polish group G is **surjectively universal** if every Polish group is isomorphic to a quotient group of G. A Polish group G is **weakly universal** if every Polish group is involved in G.

Similarly one can define the same concepts for a class of Polish groups closed under taking quotients. All the concepts mentioned above give rise to such classes. Thus it is of interest to ask whether there are surjectively universal or weakly universal objects in these classes. It is unknown whether there is a surjectively universal Polish group. On the positive side, we will show in the next section that a construction of Graev gives a surjectively universal Polish group with two-sided invariant metric, which implies the existence of a surjectively universal abelian Polish group. In fact, some of the familiar groups turn out to be weakly universal for abelian Polish groups, such as ℓ_1 and $C(0,1)$ under addition [59]. However, nothing seems to be known for cli Polish groups.

Exercise 2.5.1 Let X be a Polish space, $d \leq 1$ a compatible complete metric, and $D = \{x_n : n \in \omega\}$ a countable dense subset of X. For any $x \in X$ define $f(x) = (d(x, x_n))_{n \in \omega}$. Show that f is a homeomorphic embedding of X into $[0,1]^\omega$.

Exercise 2.5.2 Show that if $\varphi \in \mathrm{Iso}(X)$ and $f \in E(X, \omega)$ then $E(\varphi)(f) \in E(X, \omega)$.

Exercise 2.5.3 Show that $E(\varphi)$ is the unique extension of φ to an isometry of $E(X, \omega)$ so that $E(\varphi) \upharpoonright X = \varphi$.

Exercise 2.5.4 Let G and H be Polish groups. Show that if H is a quotient group of a closed subgroup of G then $H \leq_w G$.

Exercise 2.5.5 Show that if G, H, K are Polish groups and $K \leq_w H \leq_w G$, then $K \leq_w G$.

Exercise 2.5.6 Show that there is a universal compact subgroup of S_∞. (*Hint*: Use Exercise 2.4.3 to show that every compact subgroup of S_∞ is homeomorphic to a subgroup of $\prod_n S_n$, where S_n is the symmetric group for an n-element set.)

2.6   The Graev metric groups

The Graev metrics are constructed on algebraically free groups with continuum many generators (due to Graev [62] based on earlier work of Markov [119]). These are two-sided invariant metrics that make the completions of the free groups Polish. The construction gives rise to examples of surjectively universal groups in the classes of abelian Polish groups and of Polish groups admitting two-sided invariant metrics [135]. In this section we follow the approach of Ding and Gao [27] and give a construction of Graev metrics more general than the original one.

For a nonempty set X, let $X^{-1} = \{x^{-1} : x \in X\}$ be a disjoint copy of X and let $e \notin X \cup X^{-1}$. Put $\overline{X} = X \cup X^{-1} \cup \{e\}$. We use the notational convention that $(x^{-1})^{-1} = x$ for $x \in X$ and $e^{-1} = e$. Since we will mostly discuss elements of $\overline{X}$ we also make the convention that the lowercase letters

x, y, x_0, x_1, and so on. will denote elements of $\overline{X}$ if we do not explicitly specify otherwise.

Let $W(X)$ be the set of words over the alphabet $\overline{X}$. A word $w \in W(X)$ is **irreducible** if $w = e$ or else $w = x_0 \cdots x_n$, where for any i, $x_i \neq e$ and $x_{i+1} \neq x_i^{-1}$. Let $\mathbb{F}(X)$ be the set of irreducible words. For $w \in W(X)$, we denote by $\mathrm{lh}(w)$ the length of w. Then note that an irreducible word has positive length. For each $w \in W(X)$ **the reduced word** for w, denoted w', is the unique irreducible word obtained by successively replacing any occurrence of xx^{-1} in w by e and eliminating e from any occurrence of the form $w_1 e w_2$, where at least one of w_1 and w_2 is nonempty. We can turn $\mathbb{F}(X)$ into a group, which is called the **free group** on X, by defining $w \cdot u = (wu)'$ where wu is the concatenation of words w and u. Note that this is similar to but slightly different from the usual way free groups are obtained. In this treatment the identity element of $\mathbb{F}(X)$ is e but not the empty word.

Assume now (X, d) is a metric space. Assume that a metric $\overline{d}$ on $\overline{X}$ satisfies the following conditions for all $x, y \in X$:

(i) $\overline{d}(x, y) = \overline{d}(x^{-1}, y^{-1}) = d(x, y)$;

(ii) $\overline{d}(x, e) = \overline{d}(x^{-1}, e)$;

(iii) $\overline{d}(x, y^{-1}) = \overline{d}(x^{-1}, y)$.

If $w = x_0 \cdots x_n$ and $u = y_0 \cdots y_n$ are two words in $W(X)$ of the same length, then put

$$\rho(w, u) = \sum_{i=0}^{n} \overline{d}(x_i, y_i).$$

Next call a word $w \in W(X)$ **trivial** if $w' = e$. A trivial word is also called a **trivial extension** of e. For $w \in W(X)$ with $\mathrm{lh}(w) > 0$ but $w \neq e$, a **trivial extension** of $w = x_0 \cdots x_n$ is a word of the form $u_0 x_0 u_1 x_1 \cdots u_n x_n u_{n+1}$, where each of $u_0, \ldots, u_{n+1}$ is either trivial or empty. In particular, if $w \in \mathbb{F}(X)$ and $w^* \in W(X)$, then w^* is a trivial extension of w if and only if $(w^*)' = w$.

Definition 2.6.1
The **Graev metric** on $\mathbb{F}(X)$ is defined as

$$\delta(u, v) = \inf\{\rho(u^*, v^*) : u^*, v^* \in W(X), \mathrm{lh}(u^*) = \mathrm{lh}(v^*), (u^*)' = u, (v^*)' = v\}.$$

We will verify that δ is a two-sided invariant metric on $\mathbb{F}(X)$ extending d, that the group $\mathbb{F}(X)$ equipped with the topology given by δ becomes a topological group, and that if X is separable then so is $\mathbb{F}(X)$. To prove these results we need first to develop a method of computation for Graev metrics.

Definition 2.6.2
Let $m, n \in \mathbb{N}$ and $m \leq n$. A bijection θ on $\{m, \ldots, n\}$ is a **match** if

(1) $\theta \circ \theta = \mathrm{id}$; and

(2) there is no $m \leq i, j \leq n$ such that $i < j < \theta(i) < \theta(j)$.

Lemma 2.6.3

Let θ be a bijection on $\{m, \ldots, n\}$ such that $\theta \circ \theta = \mathrm{id}$. Then θ is a match iff for any nonempty set X, if $w = x_m \cdots x_n \in W(X)$ is such that $x_{\theta(i)} = x_i^{-1}$ for any $m \leq i \leq n$, then w is trivial.

Proof. ($\Rightarrow$) We argue by induction on $\mathrm{lh}(w)$. Without loss of generality we may assume $m = 0$. If $\theta(0) = 0$, then $w = x_0 w_1$ and $x_0 = x_0^{-1} = e$. Since $\theta \restriction \{1, \ldots, n\}$ is a match on $\{1, \ldots, n\}$, the inductive hypothesis implies that w_1 is trivial. Then $w = e w_1$ is trivial. Otherwise, $\theta(0) > 0$ and $w = x_0 w_1 x_{\theta(0)} w_2$ with $x_{\theta(0)} = x_0^{-1}$. Thus if $0 < i < \theta(0)$ then $0 < \theta(i) < \theta(0)$, and if $\theta(0) < i$ then $\theta(0) < \theta(i)$. By inductive hypothesis, both w_1 and w_2 are trivial. Therefore $w = x_0 w_1 x_0^{-1} w_2$ is trivial.

($\Leftarrow$) Assume there are $m \leq i, j \leq n$ so that $i < j < \theta(i) < \theta(j)$. Let X contain at least two distinct elements x, y and let $w = x_m \ldots x_n \in W(X)$ be the word so that $x_i = x$, $x_j = y$, $x_{\theta(i)} = x^{-1}$, $x_{\theta(j)} = y^{-1}$, and $x_k = e$ for any other index k. Then $w' = xyx^{-1}y^{-1} \neq e$. So w is not trivial. □

Lemma 2.6.4

For any trivial word $w = x_0 \cdots x_n$ there is a match θ such that for any $i \leq n$,
$$x_{\theta(i)} = x_i^{-1}.$$

Proof. We prove by induction on $\mathrm{lh}(w)$. If $\mathrm{lh}(w) = 1$ then $w = e$; let $\theta(0) = 0$ and θ is a match. For $\mathrm{lh}(w) > 1$ or $n > 0$ we consider two cases. Case 1: $x_n \neq x_0^{-1}$. Then it follows that $w = w_1 w_2$, while $w_1 = x_0 \cdots x_m$ and $w_2 = x_{m+1} \cdots x_n$ are both nonempty and trivial. By inductive hypothesis there are two matches θ_1 on $\{0, \ldots, m\}$ and θ_2 on $\{m + 1, \ldots, n\}$ such that $x_{\theta_k(i)} = x_i^{-1}$, $k = 1, 2$. Let $\theta = \theta_1 \cup \theta_2$. Then θ is as required. Case 2: $x_n = x_0^{-1}$. Then $w = x_0 w_1 x_0^{-1}$ and $w_1 = x_1 \ldots x_{n-1}$ is trivial. By inductive hypothesis there is a match θ_1 on $\{1, \ldots, n - 1\}$ such that $x_{\theta_1(i)} = x_i^{-1}$. Then the required θ can be defined as $\theta = \theta_1 \cup \{(0, n), (n, 0)\}$. □

For any match θ on $\{0, \ldots, n\}$ and $w = x_0 \cdots x_n \in W(X)$, we define

$$x_i^\theta = \begin{cases} x_i, & \text{if } \theta(i) > i, \\ e, & \text{if } \theta(i) = i, \\ x_{\theta(i)}^{-1}, & \text{if } \theta(i) < i. \end{cases}$$

Let $w^\theta = x_0^\theta \ldots x_n^\theta$. Then $x_{\theta(i)}^\theta = (x_i^\theta)^{-1}$ for any $i \leq n$, so w^θ is trivial by Lemma 2.6.3. Moreover,

$$\rho(w, w^\theta) = \sum_{i=0}^n \bar{d}(x_i, x_i^\theta) = \sum_{\theta(i)=i} \bar{d}(x_i, e) + \sum_{\theta(i)>i} \bar{d}(x_i^{-1}, x_{\theta(i)}).$$

Note that for each $w \in W(X)$ there are only finitely many matches θ on $\{0, \ldots, \mathrm{lh}(w) - 1\}$ and thus only finitely many trivial words of the form w^θ. The following theorem shows how matches are used in the computation of Graev metrics.

Theorem 2.6.5 (Ding–Gao)
For any $w \in \mathbb{F}(X)$, $\delta(w, e) = \min\{\rho(w, w^\theta) : \theta$ is a match$\}$.

Proof. We need to show that for any trivial extension w^* of w and any trivial word u with $\mathrm{lh}(w^*) = \mathrm{lh}(u)$, there is a match θ on $\{0, \ldots, \mathrm{lh}(w) - 1\}$ so that $\rho(w^*, u) \geq \rho(w, w^\theta)$. For this let $w = x_0 \cdots x_n$, $w^* = y_0 \cdots y_l$ be a trivial extension of w, and $u = z_0 \cdots z_l$ a trivial word. We can assume that

$$w^* = w_0 x_0 w_1 \cdots w_n x_n w_{n+1},$$

where each w_i is either trivial or empty, and

$$x_i = y_{k_i},$$
$$w_p = y_{k_{p-1}+1} \cdots y_{k_p-1},$$

for $i \leq n$, $p \leq n+1$ and a sequence $k_{-1} = -1 < k_0 < \cdots < k_n < k_{n+1} = l+1$. For each $p \leq n + 1$, let φ_p be a match on $\{k_{p-1} + 1, \ldots, k_p - 1\}$ given by Lemma 2.6.4 such that $y_{\varphi_p(j)} = y_j^{-1}$. Let $\varphi = \bigcup_{p \leq n+1} \varphi_p$. Then $\varphi \circ \varphi = \mathrm{id}$. Similarly since u is trivial we can let μ be a match on $\{0, \ldots, l\}$ such that $z_{\mu(j)} = z_j^{-1}$. We now define a match θ on $\{0, \ldots, n\}$.

First we consider the following graph defined on the set $\{0, \ldots, l\}$. For each pair j and j', we put an edge between j and j' if either $\varphi(j) = j'$ or $\mu(j) = j'$. Let Γ be the resulting graph. Then note that every element of the form k_i for $i \leq n$ is incident with exactly one edge since it is not in the domain of φ. However, for any other $j \leq l$ there are exactly two edges incident with j, one from $\varphi(j)$ and one from $\mu(j)$. Note that it is possible to have loops in Γ, that is, elements j with an edge to itself. Now consider the connected components of Γ. For each $i \leq n$, the connected component containing k_i must be a simple path with k_i as one of its endpoints. If j is the other endpoint of this path, then either j has only one edge incident with it, in which case j must be of the form $k_{i'}$ for $i' \neq i$, or else j has two edges incident with it, one of which must be a loop. In the first case, we define $\theta(i) = i'$; in the second case, we let $\theta(i) = i$. It is clear that θ is a bijection of $\{0, \ldots, n\}$ with $\theta \circ \theta = \mathrm{id}$.

We check that θ is a match. By Lemma 2.6.3 it suffices to show that if X' is any nonempty set and $v = x_0' \cdots x_n' \in W(X')$ satisfies $x_{\theta(i)}' = x_i'^{-1}$, then v is trivial. To do this we first define a word $\bar{w} = \bar{y}_0 \cdots \bar{y}_l$. For any $i \leq n$, consider the connected component of k_i in the graph Γ. As noted above it must be a path with k_i as one of its endpoints. In fact the other elements of this path are successively $\mu(k_i)$, $\varphi\mu(k_i)$, $\mu\varphi\mu(k_i)$, $\varphi\mu\varphi\mu(k_i)$, and so on. Moreover, if the other endpoint is some $k_{i'}$ for $i' \neq i$, then the number of elements in the path is even. In view of this, we also successively let $\bar{y}_{k_i} = x_i'$,

$\bar{y}_{\mu(k_i)} = x_i'^{-1}$, $\bar{y}_{\varphi\mu(k_i)} = x_i'$, $\bar{y}_{\mu\varphi\mu(k_i)} = x_i'^{-1}$, and so on, until all indices in the connected component of k_i are exhausted. If $\theta(i) \neq i$ the construction guarantees that $\bar{y}_{k_{\theta(i)}} = x_i'^{-1} = x_{\theta(i)}'$. Thus $\bar{y}_j$ is well defined for any j in a connected component of some k_i. Finally for any $j \leq l$ such that $\bar{y}_j$ has not been defined above, let $\bar{y}_j = e$. Now let $\bar{w}_i = \bar{y}_{k_{p-1}+1} \cdots \bar{y}_{k_p-1}$. Since $\bar{y}_{\varphi_p(j)} = \bar{y}_j^{-1}$ whenever $k_{p-1} + 1 \leq j \leq k_p - 1$, we have that $\bar{w}_i$ is trivial. So $\bar{w}$ is a trivial extension of v. On the other hand $\bar{y}_{\mu(j)} = \bar{y}_j^{-1}$ for any $j \leq l$, and thus $\bar{w}$ is trivial. This implies that $e = (\bar{w})' = (v)'$, so v is trivial too.

For any $j \leq l$, we have

$$\bar{d}(y_j, z_j) + \bar{d}(y_{\mu(j)}, z_{\mu(j)}) = \bar{d}(y_j, z_j) + \bar{d}(y_{\mu(j)}, z_j^{-1})$$

$$= \bar{d}(y_j, z_j) + \bar{d}(y_{\mu(j)}^{-1}, z_j) \geq \bar{d}(y_j, y_{\mu(j)}^{-1}).$$

Thus, if $\theta(i) > i$, then the connected component of Γ containing k_i can be enumerated as

$$k_i, \mu(k_i), \varphi\mu(k_i), \cdots, \mu(\varphi\mu)^m(k_i) = k_{\theta(i)}$$

for some $m \geq 0$. Since φ is a match, we have that

$$\bar{d}(y_{k_i}, z_{k_i}) + \bar{d}(y_{\mu(k_i)}, z_{\mu(k_i)})$$
$$+ \bar{d}(y_{\varphi\mu(k_i)}, z_{\varphi\mu(k_i)}) + \bar{d}(y_{\mu(\varphi\mu)(k_i)}, z_{\mu(\varphi\mu)(k_i)})$$
$$+ \cdots \cdots$$
$$+ \bar{d}(y_{(\varphi\mu)^m(k_i)}, z_{(\varphi\mu)^m(k_i)}) + \bar{d}(y_{\mu(\varphi\mu)^m(k_i)}, z_{\mu(\varphi\mu)^m(k_i)})$$

$$\geq \bar{d}(x_i, y_{\mu(k_i)}^{-1}) + \bar{d}(y_{\varphi\mu(k_i)}, y_{\mu\varphi\mu(k_i)}^{-1}) + \cdots + \bar{d}(y_{\mu(\varphi\mu)^m(k_i)}, x_{\theta(i)}^{-1})$$

$$= \bar{d}(x_i, y_{\mu(k_i)}^{-1}) + \bar{d}(y_{\mu(k_i)}^{-1}, y_{\mu\varphi\mu(k_i)}^{-1}) + \cdots + \bar{d}(y_{(\varphi\mu)^m(k_i)}^{-1}, x_{\theta(i)}^{-1})$$

$$\geq \bar{d}(x_i, x_{\theta(i)}^{-1}) = \bar{d}(x_i^{-1}, x_{\theta(i)}).$$

In the case of $\theta(i) = i$, the connected component of Γ containing k_i can be enumerated as either

$$k_i, \mu(k_i), \varphi\mu(k_i), \ldots, \mu(\varphi\mu)^m(k_i) = (\varphi\mu)^{m+1}(k_i)$$

or

$$k_i, \mu(k_i), \varphi\mu(k_i), \ldots, (\varphi\mu)^m(k_i) = \mu(\varphi\mu)^m(k_i).$$

In the first case if j is the last index in the enumeration then $\varphi(j) = j$ and therefore $y_j = e$. Then a similar computation as the above gives that

$$\bar{d}(y_{k_i}, z_{k_i}) + \bar{d}(y_{\mu(k_i)}, z_{\mu(k_i)}) + \cdots + \bar{d}(y_j, z_j) \geq \bar{d}(x_i, e).$$

In the second case if j is the last index in the enumeration then $\mu(j) = j$ and therefore $z_j = e$. Again a similar computation as above, but with one less term, gives that

$$\overline{d}(y_{k_i}, z_{k_i}) + \overline{d}(y_{\mu(k_i)}, z_{\mu(k_i)}) + \cdots + \overline{d}(y_j, z_j)$$

$$\geq \overline{d}(x_i, y_{\mu(k_i)}^{-1}) + \overline{d}(y_{\mu(k_i)}^{-1}, y_{\mu\varphi\mu(k_i)}^{-1}) + \cdots$$
$$+ \overline{d}(y_{(\varphi\mu)^{m-1}(k_i)}^{-1}, y_{\mu(\varphi\mu)^{m-1}(k_i)}^{-1}) + \overline{d}(y_{(\varphi\mu)^m(k_i)}, z_{(\varphi\mu)^m(k_i)})$$

$$= \overline{d}(x_i, y_{\mu(k_i)}^{-1}) + \overline{d}(y_{\mu(k_i)}^{-1}, y_{\mu\varphi\mu(k_i)}^{-1}) + \cdots$$
$$+ \overline{d}(y_{(\varphi\mu)^{m-1}(k_i)}^{-1}, y_{\mu(\varphi\mu)^{m-1}(k_i)}^{-1}) + \overline{d}(y_{\mu(\varphi\mu)^{m-1}(k_i)}^{-1}, e)$$

$$\geq \overline{d}(x_i, e).$$

In summary,

$$\rho(w^*, u) = \sum_{j=0}^{l} \overline{d}(y_j, z_j) \geq \sum_{\theta(i)=i} \overline{d}(x_i, e) + \sum_{\theta(i)>i} \overline{d}(x_i^{-1}, x_{\theta(i)}) = \rho(w, w^\theta).$$

☐

With Theorem 2.6.5 the fundamental results of Graev follow easily.

Theorem 2.6.6 (Graev)
Let (X, d) be a metric space. Then the Graev metric δ is a two-sided invariant metric on $\mathbb{F}(X)$ extending d. Furthermore, $\mathbb{F}(X)$ is a topological group in the topology induced by δ. If X is separable, so is $\mathbb{F}(X)$.

Proof. It is immediate from Definition 2.6.1 that δ is two-sided invariant and for any $D \subseteq X$ dense in X, $\mathbb{F}(D)$ is dense in $\mathbb{F}(X)$. That $\delta \geq 0$ and its symmetry are also clear.

For the triangle inequality, by two-sided invariance it suffices to check that for any $w, u \in \mathbb{F}(X)$, $\delta(w \cdot u, e) \leq \delta(w, e) + \delta(u, e)$. By Theorem 2.6.5 let θ, φ be matches so that $\delta(w, e) = \rho(w, w^\theta)$ and $\delta(u, e) = \rho(u, u^\varphi)$. Then $\rho(wu, w^\theta u^\varphi) = \rho(w, w^\theta) + \rho(u, u^\varphi)$. Since wu is a trivial extension of $w \cdot u$ and $w^\theta u^\varphi$ is trivial, $\delta(w \cdot u, e) \leq \rho(wu, w^\theta u^\varphi) = \delta(w, e) + \delta(u, e)$.

To see that $\delta(w, u) = 0$ implies $w = u$ for $w, u \in \mathbb{F}(X)$, by two-sided invariance it suffices to show that $\delta(w, e) = 0$ implies $w = e$. Suppose $\delta(w, e) = 0$ and let θ be a match so that $0 = \delta(w, e) = \rho(w, w^\theta)$. Then the definition of ρ gives that $w = w^\theta$, so w is trivial after all. But the only trivial element of $\mathbb{F}(X)$ is e, hence $w = e$.

Now if $x, y \in X$ are distinct, then $\delta(x, y) = \delta(y^{-1}x, e)$ and $(y^{-1}x)^\theta$ can only result in $y^{-1}y$ and ee. Then $y^{-1}y$ achieves the smaller value $\overline{d}(x, y) = d(x, y)$. This shows that δ extends d.

The continuity of the group operations follows from the two-sided invariance and the triangle inequality as usual. $\qquad$ ☐

Let $(\overline{\mathbb{F}}(X), \overline{\delta})$ denote the completion of $(\mathbb{F}(X), \delta)$. Then by Theorem 2.1.3 $\overline{\mathbb{F}}(X)$ becomes a Polish group with a two-sided invariant compatible metric $\overline{\delta}$.

Definition 2.6.7
The **Graev metric group** $\mathcal{G}$ is $\overline{\mathbb{F}}(\omega^\omega)$ with the usual metric

$$d(x, y) = \begin{cases} 0, & \text{if } x = y, \\ 2^{-\min\{n \,:\, x(n) \neq y(n)\}}, & \text{if } x \neq y, \end{cases}$$

and the metric $\overline{d}$ on $\overline{\omega^\omega}$ extending d is determined by letting

$$\overline{d}(x^{-1}, y^{-1}) = d(x, y), \quad \overline{d}(x, e) = \overline{d}(x^{-1}, e) = \overline{d}(x, y^{-1}) = 1$$

for $x, y \in \omega^\omega$.

Theorem 2.6.8
The Graev metric group $\mathcal{G}$ is surjectively universal for the class of all Polish groups admitting compatible two-sided invariant metrics.

Proof. Let G be a Polish group with a compatible two-sided invariant metric d_G. Then by Exercise 2.2.4 d_G is complete. Without loss of generality we may assume $d_G < 1$. And then by Exercise 1.3.7 there is a Lipschitz map φ from ω^ω onto G, that is, $d_G(\varphi(x), \varphi(y)) \leq d(x, y)$ for all $x, y \in \omega^\omega$. Let $\hat{\varphi} : \mathbb{F}(\omega^\omega) \to G$ be the canonical group homomorphism extending φ. We claim that $\hat{\varphi}$ is also Lipschitz, that is, $d_G(\hat{\varphi}(u), \hat{\varphi}(v)) \leq \delta(u, v)$ for $u, v \in \mathbb{F}(\omega^\omega)$. By invariance and homomorphism properties it suffices to show that $d_G(\hat{\varphi}(u), 1_G) \leq \delta(u, e)$ for all $u \in \mathbb{F}(\omega^\omega)$. Let v be a trivial word with $\mathrm{lh}(v) = \mathrm{lh}(u)$ and $\rho(u, v) = \delta(u, e)$. Assume $u = x_1 \ldots x_n$ and $v = y_1 \ldots y_n$. Then by Exercise 2.1.6 we have that

$$\delta(u, e) = \rho(u, v) = \sum_{i=1}^{n} d(x_i, y_i) \geq \sum_{i=1}^{n} d_G(\hat{\varphi}(x_i), \hat{\varphi}(y_i))$$

$$\geq d_G(\hat{\varphi}(u), \hat{\varphi}(v)) = d_G(\hat{\varphi}(u), 1_G).$$

Finally we define $\overline{\varphi} : \overline{\mathbb{F}}(\omega^\omega) \to G$. Let $\sigma \in \overline{\mathbb{F}}(\omega^\omega)$ and (u_n) be a δ-Cauchy sequence in $\mathbb{F}(\omega^\omega)$ with $u_n \to \sigma$ as $n \to \infty$. We let $\overline{\varphi}(\sigma) = \lim_n \hat{\varphi}(u_n)$. The right-hand side makes sense since $(\hat{\varphi}(u_n))$ is d_G-Cauchy by the Lipschitz condition. It is routine to check that the definition of $\overline{\varphi}$ does not depend on the choice of the Cauchy sequence. Moreover $\overline{\varphi}$ is still Lipschitz, hence in particular it is a continuous homomorphism from $\mathcal{G}$ onto G. ▯

Exercise 2.6.1 Consider $X = \{x_n \,:\, n \in \omega\}$ with metric $d(x_n, x_m) = 2^{-\min(n,m)}$. Extend d to $\overline{d}$ on $\overline{X}$ by letting $\overline{d}(x_n, e) = 2^{-n}$ and $\overline{d}(x_n, x_m^{-1}) = 2^{-\min(n,m)}$. Compute $\rho(x_1 x_2 x_3, x_2 x_3 x_1)$ and $\delta(x_1 x_2 x_3, x_2 x_3 x_1)$.

Exercise 2.6.2 Show that for any $w, u \in \mathbb{F}(X)$ there are trivial extensions w^* of w and u^* of u such that $\delta(w, u) = \rho(w^*, u^*)$.

Exercise 2.6.3 Let (X, d) be a compact metric space and for $n \in \omega$ let $\mathbb{F}(X, n) = \{u \in \mathbb{F}(X) : \mathrm{lh}(u) \leq n\}$. Show that $\mathbb{F}(X, n)$ is a compact subset of $(\mathbb{F}(X), \delta)$.

Exercise 2.6.4 Let (X, d) be a metric space and $Y \subseteq X$. Let δ_Y be the Graev metric on $\mathbb{F}(Y)$ induced from the subspace $(Y, d \restriction Y)$ and let δ_X be the Graev metric on $\mathbb{F}(X)$. Show that $\delta_Y = \delta_X \restriction \mathbb{F}(Y)$. In particular, if Y is dense in X, then $\overline{F}(Y) = \overline{F}(X)$.

Exercise 2.6.5 (Shakhmatov–Pelant–Watson) Let $\mathcal{G}$ be the Graev metric group. Let G be the commutator subgroup of $\mathcal{G}$, that is, G is the group generated by elements of the form $g^{-1}h^{-1}gh$ for $g, h \in \mathcal{G}$. Let N be the closure of G. Show that N is a closed normal subgroup of $\mathcal{G}$ and that the quotient group $\mathcal{G}/N$ is surjectively universal for all abelian Polish groups. (*Hint:* Recall from group theory that if H is a normal subgroup of G then G/H is abelian iff H contains the commutator subgroup of G.)

Chapter 3

Polish Group Actions

In this chapter we turn to an overview of the basic concepts and results about Borel actions of Polish groups. Many of the results in this chapter originated in the study of operator algebras, representation theory, and dynamical systems at least for special classes of Polish groups.

As explained at the beginning of Chapter 2, our ultimate goal is to study the complexity of equivalence relations, among which orbit equivalence relations are important examples.

3.1 Polish G-spaces

Definition 3.1.1
Let G be a group and X a set. An **action** of G on X is a map

$$a : G \times X \to X$$

such that, for all $x \in X$ and $g, h \in G$,

(i) $a(1_G, x) = x$;

(ii) $a(g, a(h, x)) = a(gh, x)$.

When a is an action of G on X, the triple (X, G, a) is called a **transformation group**, or alternatively the pair (X, a) is called a G-**space**.

Some informal notation is often used to avoid the awkward mention of the action a, especially when composition is involved. When the action a is understood, we write $g \cdot x$ for $a(g, x)$, (X, G) as the transformation group, and X as a G-space.

Definition 3.1.2
Let G be a topological group, X a topological space, and a an action of G on X. If a is continuous, then (X, G, a) is called a **topological transformation group** and (X, a) a **topological G-space**. Furthermore, if G is a Polish group, X is a Polish space and a is continuous, then (X, G, a) is called a **Polish transformation group** and (X, a) a **Polish G-space**.

We first give a useful criterion for the continuity of an action of a Polish group on a Polish space. For this we need the following definition.

Definition 3.1.3
Let G be a topological group, X a topological space, and a an action of G on X. The action a is **separately continuous** if for any $g \in G$ the map $a_g : X \to X$ defined by $a_g(x) = a(g, x)$ is continuous and for any $x \in X$ the map $a_x : G \to X$ defined by $a_x(g) = a(g, x)$ is continuous.

For a separately continuous action a, each a_g, $g \in G$, is a homeomorphism of the space X, with $a_{g^{-1}}$ as its inverse. We next show that separate continuity is sufficient to guarantee continuity of an action of a Polish group.

Theorem 3.1.4
Let G be a Polish group, X a metrizable space, and a an action of G on X. Then a is continuous iff it is separately continuous.

Proof. The ($\Rightarrow$) direction is obvious. We only need to show ($\Leftarrow$). Assume that a is separately continuous. We show that a is continuous at every point (g_0, x_0). First we claim that there is some $g_1 \in G$ such that a is continuous at (g_1, x_0). For this let d be a compatible metric on X. For $n, k \in \omega$ let

$$F_{n,m} = \{g \in G : \forall x (d(x, x_0) < 2^{-n} \Rightarrow d(a(g, x), a(g, x_0)) \leq 2^{-m})\}.$$

Then it follows from separate continuity that $F_{n,m}$ is a closed subset of G and that $G = \bigcap_m \bigcup_n F_{n,m}$. Let $D = \bigcup_m \bigcup_n (F_{n,m} - \text{Int}(F_{n,m}))$. Then D is meager in G and thus there is $g_1 \in G - D$. We check that a is continuous at (g_1, x_0). For this let $\{(h_k, y_k)\}$ be a sequence converging to (g_1, x_0). Let $\epsilon > 0$. Let m be large enough such that $2^{-m+1} < \epsilon$. Then for some n we have $g_1 \in F_{n,m}$. Fix such an n and we note that $g_1 \in \text{Int}(F_{n,m})$ since $g_1 \notin D$. Since $h_k \to g_1$ as $k \to \infty$ there is K_0 such that for all $k \geq K_0$, $h_k \in F_{n,m}$. Also since $y_k \to x_0$ as $k \to \infty$ there is K_1 such that for all $k \geq K_1$, $d(y_k, x_0) < 2^{-n}$. Thus for all $k \geq \max\{K_0, K_1\}$, we have both $h_k \in F_{n,m}$ and $d(y_k, x_0) < 2^{-n}$, and therefore $d(a(h_k, y_k), a(h_k, x_0)) \leq 2^{-m}$. From left continuity of a we also have K_2 such that for all $k \geq K_2$, $d(a(h_k, x_0), a(g_1, x_0)) < 2^{-m}$. Thus for all $k \geq \max\{K_0, K_1, K_2\}$, we have $d(a(h_k, y_k), a(g_1, x_0)) < 2^{-m+1} < \epsilon$. We have thus shown that a is continuous at (g_1, x_0). Finally we note that $a(g_0, x_0) = a(g_0 g_1^{-1}, a(g_1, x_0))$, thus by the left continuity it follows that a is continuous at (g_0, x_0). $\square$

Example 3.1.5
(1) Let G be any topological group. Then G acts on itself by left multiplication:
$$a_l : G \times G \to G$$
$$a_l(g, h) = gh.$$

Then a_l is continuous. Similarly G also acts continuously on itself by right multiplication, where $a_r(g,h) = hg^{-1}$, and by conjugacy, where $a_c(g,h) = ghg^{-1}$.

(2) Let (X,d) be a Polish metric space and G a closed subgroup of $\text{Iso}(X)$. Then there is a natural **application action** of G on X given by $g \cdot x = g(x)$. This action is continuous since it is separately continuous.

(3) Let X be a compact Polish space and G a closed subgroup of $H(X)$. Then the application action of G on X is also continuous.

We recall some algebraic notion associated with group actions.

Definition 3.1.6
Let a group G act on a set X. For each $x \in X$, the G-**orbit** of x, denoted $[x]_G$ or $G \cdot x$, is the set $\{g \cdot x : g \in G\}$. A subset A of X is G-**invariant** if $G \cdot x \subseteq A$ for every $x \in A$.

Any orbit is an invariant set. Any invariant set is the union of a collection of orbits.

Definition 3.1.7
Let a group G act on a set X. The action is said to be **free** if for every $g \in G$ with $g \neq 1_G$ and every $x \in X$, $g \cdot x \neq x$. For any G-space X, the **free part** of X (or of the action) is the subset $F = \{x \in X : \forall g \in G \, (g \neq 1_G \Rightarrow g \cdot x \neq x)\}$. The subset $X - F$ is the **nonfree part** of the action.

The free part of an action on X is an invariant subset of X.

Definition 3.1.8
Let a group G act on a set X. For every $x \in X$, the **stabilizer** of x, denoted G_x, is $\{g \in G : g \cdot x = x\}$.

The stabilizer of any element is a subgroup of G. For $g \in G$ and $x \in X$, $G_{g \cdot x} = gG_x g^{-1}$. For any $x \in X$, $G_x = \{1_G\}$ iff x is in the free part of the action. Also note that there is a canonical map from G/G_x onto $G \cdot x$ given by $gG_x \mapsto g \cdot x$. This map is a bijection. (See Exercise 3.1.4.)

Definition 3.1.9
Let G be a group, X, Y G-spaces, and $f : X \to Y$. f is called a G-**map** if for any $g \in G$ and $x \in X$,

$$f(g \cdot x) = g \cdot f(x).$$

An injective G-map is called a G-**embedding**.

Under a G-map the orbit structures are preserved: if C is a G-orbit then $f(C)$ is a G-orbit; if A is G-invariant then so is $f(A)$. If f is a G-embedding all the algebraic structures are preserved: for any $x \in X$, $G_x = G_{f(x)}$; thus if F is the free part of X then $f(F)$ is contained in the free part of Y.

In the context of a continuous Polish group action, the orbits are demonstrating nice properties. First note that every stabilizer group is closed in G. And thus by Theorem 2.2.10 G/G_x becomes a Polish space. For each $x \in X$, the canonical map from G/G_x onto $G \cdot x$ is continuous.

Proposition 3.1.10 (Ryll-Nardzewski)
Let G be a Polish group and X a Polish G-space. Then every orbit is Borel.

Proof. The canonical map is a continuous injection from the Polish space G/G_x onto $G \cdot x$. By the Luzin–Suslin Theorem 1.3.1 $G \cdot x$ is Borel.   ⬚

Exercise 3.1.1 Let G be a group and X a set. Consider the space X^G of all functions from G into X. Define $a_l : G \times X^G \to X^G$ by

$$a_l(g, p)(h) = p(g^{-1}h).$$

Show that a_l is an action of G on X^G. This is called the left action of G on X^G. Similarly define a right action and a conjugacy action of G on X^G.

Exercise 3.1.2 Let G be a countable group with the discrete topology, X a topological space, and X^G equipped with the product topology. Show that the left, right, and conjugacy actions of G on X^G defined above are all continuous.

Exercise 3.1.3 Let G be a Polish group and X a compact Polish G-space. For each $g \in G$ let $h_g : X \to X$ be defined by $h_g(x) = y$ iff $g \cdot x = y$. Show that h_g is a homeomorphism of X and the map $g \mapsto h_g$ is a topological group isomorphic embedding from G into $H(X)$.

Exercise 3.1.4 Let a group G act on a set X. Let $g \in G$ and $x \in X$. Prove the following statements:

(a) $G_{g \cdot x} = g G_x g^{-1}$;

(b) $G_x = \{1_G\}$ iff x is in the free part of the action;

(c) The canonical map from G/G_x to $G \cdot x$ is a bijection.

Exercise 3.1.5 Let G be a group, X, Y G-spaces, and $f : X \to Y$. Let $x \in X$. If f is a G-map show that $f(G \cdot x) = G \cdot f(x)$. If f is a G-embedding show that $G_{f(x)} = G_x$.

Exercise 3.1.6 Let G be a countable group and X a Polish space. Let X_l^G be the Polish G-space with the left action (see Exercises 3.1.1 and 3.1.2 above) and X_r^G the Polish G-space with the right action. Show that there is a homeomorphic G-map (G-homeomorphism) between X_l^G and X_r^G. Hence we may refer to either action as X^G with the shift action.

Exercise 3.1.7 Let G be a countable group and $([0,1]^\omega)^G$ be the Polish G-space with the shift action. Show that for any Polish G-space X there is a continuous G-embedding from X into $([0,1]^\omega)^G$. (*Hint*: Use Exercise 2.5.1.)

Exercise 3.1.8 Let G be a Polish group, d a compatible complete metric on G, and Y a Polish metric space. Let $L(G,Y)$ be the Polish space of all Lipschitz functions from (G,d) into Y. Consider the action of G on $L(G,Y)$ defined by

$$(g \cdot f)(h) = f(g^{-1}h).$$

Show that under this action $L(G,Y)$ is a Polish G-space.

Exercise 3.1.9 Let G be a Polish group and $d \leq 1$ a compatible metric on G. Recall from Exercise 1.1.9 that $L(G,d)$ denotes the space of all Lipschitz functions from (G,d) into $[0,1]$. Consider the action of G on $L(G,d)$ defined by

$$(g \cdot f)(h) = f(hg).$$

Show that under this action $L(G,d)$ is a compact Polish G-space.

3.2   The Vaught transforms

   Baire category method plays the central role in many results in invariant descriptive set theory.

   It is very useful to use the following logical notation of category quantifiers in the study of Polish group actions. Let X be a topological space, $A \subseteq X$, and $x \in X$, then we write $A(x)$ for the statement $x \in A$, and $\neg A(x)$ for $x \notin A$. In this fashion a subset is identified with a property. We also write $\forall^* x \, A(x)$ for the statement that A is comeager in X, and read the quantifier $\forall^* x$ as "for comeager many x." Similarly, we write $\exists^* x \, A(x)$ for the statement that A is nonmeager in X, and read the quantifier $\exists^* x$ as "for nonmeager many x." Sometimes we specify the domain of the quantifiers by writing $\forall^* x \in X$ and $\exists^* x \in X$, especially when there is more than one space involved.

   We will use the following basic theorem throughout the book, whose proof can be found in Reference [97] Theorem 8.41.

Theorem 3.2.1 (Kuratowski–Ulam)
Let X, Y be second countable topological spaces and $A \subseteq X \times Y$ with the Baire property. Then

$$\forall^* (x,y) \, A(x,y) \iff \forall^* x \, \forall^* y \, A(x,y) \iff \forall^* y \, \forall^* x A(x,y).$$

Definition 3.2.2
Let G be a topological group and X a G-space. For $A \subseteq X$ and nonempty open set $V \subseteq G$, the **Vaught transforms** are the sets

$$A^{\triangle V} = \{x \in X : \exists^* g \in V \ g \cdot x \in A\},$$
$$A^{*V} = \{x \in X : \forall^* g \in V \ g \cdot x \in A\}.$$

The following proposition is a basic algebraic property of the Vaught transforms.

Proposition 3.2.3
Let G be a topological group and X a G-space. Then for any $A \subseteq X$, both $A^{\triangle G}$ and A^{*G} are G-invariant subsets of X. Conversely, if G is a Baire group and $A \subseteq X$ is G-invariant then $A = A^{\triangle G} = A^{*G}$.

Proof. To see that $A^{\triangle G}$ is G-invariant, let $x \in A^{\triangle G}$ and $g \in G$. The set $\{h \in G : h \cdot x \in A\}$ is nonmeager in G. Note, however, that $\{h \in G : h \cdot (g \cdot x) \in A\} = \{h \in G : h \cdot x \in A\} g^{-1}$. It follows that $\{h \in G : h \cdot (g \cdot x) \in A\}$ is nonmeager in G. Hence $g \cdot x \in A^{\triangle G}$. A similar argument shows that A^{*G} is invariant as well. The converse is easy. ◻

The following theorem of Effros is fundamental for the study of orbits in a Polish G-space.

Theorem 3.2.4 (Effros)
Let G be a Polish group and X a Polish G-space. Then the following are equivalent for each $x \in X$:

(i) *The canonical map from G/G_x onto $G \cdot x$ is a homeomorphism;*

(ii) *$G \cdot x$ is nonmeager in its relative topology;*

(iii) *$G \cdot x$ is G_δ in X.*

Proof. The directions (i)⇒(iii)⇒(ii) are obvious. It suffices to show (ii)⇒(i). For this we fix $x \in X$ and assume $G \cdot x$ is nonmeager in its relative topology. Let $\varphi : G \cdot x \to G/G_x$ be the inverse of the canonical map, that is, $\varphi(g \cdot x) = gG_x$. Then φ is Borel. This is because, given any open set U in G/G_x, since φ^{-1} is continuous and injective, $\varphi^{-1}(U)$ is Borel by the Luzin–Suslin Theorem 1.3.1. Thus there is a dense G_δ subset C of $G \cdot x$ such that $\varphi \restriction C$ is continuous (see Exercise 2.3.2). It follows that for every $h \in G$, the set $h^{-1} \cdot C$ is also dense G_δ, hence comeager in $G \cdot x$. This means that $\forall h \in G \ \forall^* y \in G \cdot x \ (h \cdot y \in C)$. By the Kuratowski–Ulam Theorem 3.2.1, it follows that $\forall^* y \in G \cdot x \ \forall^* h \ (h \cdot y \in C)$ or $\forall^* y \in G \cdot x \ (y \in C^{*G})$. Since C^{*G} is G-invariant, we must have that $G \cdot x \subseteq C^{*G}$. Now we are ready to deduce that φ is continuous. For this let $(g_n), g_\infty \in G$ be such that

$g_n \cdot x \to g_\infty \cdot x$ as $n \to \infty$. For each $n \in \omega$ let $D_n = \{h \in G : h \cdot (g_n \cdot x) \in C\}$. Also let $D_\infty = \{h \in G : h \cdot (g_\infty \cdot x) \in C\}$. Since every D_n is comeager, their intersection is comeager. Therefore there is $h \in \bigcap_n D_n \cap D_\infty$. Thus $h \cdot (g_n \cdot x) = (hg_n) \cdot x \in C$ and $h \cdot (g_\infty \cdot x) = (hg_\infty) \cdot x \in C$. By the continuity of the action we have that $(hg_n) \cdot x \to (hg_\infty) \cdot x$ as $n \to \infty$. Thus by the continuity of φ on C we have that $hg_n G_x \to hg_\infty G_x$, which implies that $g_n G_x \to g_\infty G_x$. This shows that φ is continuous. □

In the next proposition we list some other basic properties of the Vaught transforms.

Proposition 3.2.5
Let G be a Polish group and X a G-space. Then the following hold:

(i) *(Monotonicity) For $A \subseteq B \subseteq X$ and nonempty open $V \subseteq U \subseteq G$,*

$$A^{*V} \subseteq A^{\triangle V}; \quad A^{\triangle V} \subseteq A^{\triangle U}; \quad A^{*U} \subseteq A^{*V}; \quad A^{\triangle V} \subseteq B^{\triangle V}; \quad A^{*V} \subseteq B^{*V}.$$

(ii) *(De Morgan's laws) For $A \subseteq X$ and nonempty open $V \subseteq G$,*

$$X - A^{\triangle V} = (X - A)^{*V}; \quad X - A^{*V} = (X - A)^{\triangle V}.$$

(iii) *(Countable union rule) For $A = \bigcup_{n \in \omega} A_n$, where $A_n \subseteq X$, and $V = \bigcup_{m \in \omega} V_m$, where $V_m \subseteq G$ nonempty open,*

$$A^{\triangle V} = \bigcup_{n \in \omega} A_n^{\triangle V} = \bigcup_{m \in \omega} A^{\triangle V_m} = \bigcup_n \bigcup_m A_n^{\triangle V_m}.$$

(iv) *(Countable intersection rule) For $A = \bigcap_{n \in \omega} A_n$, where $A_n \subseteq X$, and $V = \bigcap_{m \in \omega} V_m$, where $V_m \subseteq G$ nonempty open,*

$$A^{*V} = \bigcap_{n \in \omega} A_n^{*V} = \bigcap_{m \in \omega} A^{*V_m} = \bigcap_n \bigcap_m A_n^{*V_m}.$$

(v) *(G-invariance) For $A \subseteq X$, nonempty open $V \subseteq G$ and $g \in G$,*

$$g \cdot A^{\triangle V} = A^{\triangle(Vg^{-1})}; \quad g \cdot A^{*V} = A^{*(Vg^{-1})}.$$

The proof of this proposition is straightforward and is left as an exercise (Exercise 3.2.1). In the context of Polish transformation groups the two Vaught transforms can characterize each other.

Proposition 3.2.6
Let G be a Polish group, X a Polish G-space, $A \subseteq X$ a Borel subset, and $V \subseteq G$ nonempty open. Then

$$A^{\triangle V} = \bigcup \{A^{*U} : \ U \subseteq V \text{ nonempty open}\},$$

$$A^{*V} = \bigcap \{A^{\triangle U} : \ U \subseteq V \text{ nonempty open}\}.$$

Proof. We only show the first equality. The second follows from the De Morgan's laws. Let $x \in A^{\triangle V}$ and consider the set $C = \{g \in V : g \cdot x \in A\}$. Then C is Borel in G, hence has the Baire property. Since C is nonmeager, there is nonempty open $U \subseteq V$ such that C is comeager in U (that is, $U - C$ is meager). Thus we have that $x \in A^{*U}$. The converse is obvious. □

The union and the intersection in this proposition can be made countable if we fix a countable base for G (see Exercise 3.2.2). This has the following immediate corollary.

Theorem 3.2.7 (Vaught)
Let G be a Polish group and X a Polish G-space. Let $A \subseteq X$ and $V \subseteq G$ nonempty open. Let $1 \le \alpha < \omega_1$. Then $A^{\triangle V} \in \mathbf{\Sigma}^0_\alpha$ if $A \in \mathbf{\Sigma}^0_\alpha$, and $A^{*V} \in \mathbf{\Pi}^0_\alpha$ if $A \in \mathbf{\Pi}^0_\alpha$.

Proof. We first check that if A is open then $A^{\triangle V}$ is open. For this let $x \in A^{\triangle V}$. Since $\exists^* g \in V$ $(g \cdot x \in A)$ we can fix a $g \in V$ with $g \cdot x \in A$. By the continuity of the action there are open sets $U \subseteq V$ and $W \subseteq X$ such that $g \in U$, $x \in W$, and $U \cdot W \subseteq A$. Then $W \subseteq A^{\triangle V}$. This shows that $A^{\triangle V}$ is open. Using Exercise 3.2.2 and the countable union and intersection rules in Proposition 3.2.5 the required statements now can be easily shown by a transfinite induction on α. □

Thus Vaught transforms of Borel sets are Borel. In his seminal paper [161] in which the Vaught transforms were first introduced, Vaught also showed that the Vaught transforms of analytic sets are analytic. In the rest of this section we give a proof of this theorem.

Recall from Section 1.6 that analytic sets in general Polish spaces are defined as the Suslin sets. To study the Vaught transforms on analytic sets we will consider the effect of category quantifiers on the Suslin operation. We will use the basic facts about the Suslin operation reviewed in Section 1.6, including the exercises in that section, and in particular Exercise 1.6.9. Also, we will introduce a game in the proof. This leads to another important method of descriptive set theory we will come back to several times later. However, we remark that the result itself will not be used throughout this book, so it is safe for the reader to skip the rest of this section.

Let G be a Polish group and X a Polish G-space. Let $\mathcal{B} = \{V_n\}_{n \in \omega}$ be a countable base for G. Let $\{A_s\}_{s \in \omega^{<\omega}}$ be a family of closed subsets of X. Given $x \in X$, we define a two-player game $\mathsf{G}(\{A_s\}, x)$ as follows. Players I and II take turns to play natural numbers as indicated below:

$$
\begin{array}{llllll}
\text{I} & n_0 & & n_1 & \cdots & n_k & \cdots \\[2ex]
\text{II} & & (m_0, p_0) & (m_1, p_1) & \cdots & (m_k, p_k) & \cdots
\end{array}
$$

The rules of the play are: $V_{m_k} \subseteq V_{n_k}$ and $V_{n_{k+1}} \subseteq V_{m_k}$. The first player who breaks the rules loses. When an infinite play is finished, Player II wins the play if $x \in A^{*V_{m_k}}_{(p_0,\cdots,p_k)}$ for all $k \in \omega$.

Theorem 3.2.8
Let A be the set obtained from $\{A_s\}$ by the Suslin operation. Then $x \in A^{*G}$ iff Player II has a winning strategy in the game $\mathsf{G}(\{A_s\},x)$.

Proof. Let W be the set of all $x \in X$ such that Player II has a winning strategy in $\mathsf{G}(\{A_s\},x)$. Let $A_s^\alpha, B_\alpha, D_\alpha$ $(\alpha < \omega_1)$ be the sets defined in Exercise 1.6.9. Note that $A = \bigcap_\alpha B_\alpha = \bigcup_\alpha(B_\alpha - D_\alpha)$. We show that

$$A^{*G} = \bigcap_\alpha B_\alpha^{*G} = \bigcup_\alpha (B_\alpha - D_\alpha)^{*G} = W.$$

We first show that $A^{*G} \subseteq W$. For this let $x \in A^{*G}$. Then

$$\forall^* g \in G \; (g \cdot x \in \bigcup_{z \in \omega^\omega} \bigcap_{n \in \omega} A_{z \restriction n}).$$

For each $s \in \omega^{<\omega}$ let $E_s = \bigcup_z \bigcap_n A_{s \frown z \restriction n}$. Then $A = E_\emptyset$ and for each $s \in \omega^{<\omega}$ we have $E_s = \bigcup_{p \in \omega} E_{s \frown p}$. We describe a winning strategy for Player II. Let Player I play n_0 in his first move. Since there are comeager many $g \in V_{n_0}$ such that $g \cdot x \in A = E_\emptyset = \bigcup_p E_{(p)}$, there is $p_0 \in \omega$ such that there are nonmeager many $g \in V_{n_0}$ such that $g \cdot x \in E_{(p_0)}$. It follows that there is a basic open $V_{m_0} \subseteq V_{n_0}$ such that there are comeager many $g \in V_{m_0}$ such that $g \cdot x \in E_{(p_0)}$. This finishes our description of the winning strategy for the first round of the play. It is apparent that the strategy can be repeated in general. We verify that it is winning for Player II. For this just note that $E_s \subseteq A_s$ for any $s \in \omega^{<\omega}$. Thus if m_k, p_k are played in the k-th round then we have that $\forall^* g \in V_{m_k} \; (g \cdot x \in E_{(p_0,...,p_k)})$, and therefore $\forall^* g \in V_{m_k} \; (g \cdot x \in A_{(p_0,...,p_k)})$, as required.

Next we show that $W \subseteq \bigcap_\alpha B_\alpha^{*G}$. Given $V_{n_0} \supseteq V_{m_0} \supseteq \cdots \supseteq V_{n_k} \supseteq V_{m_k}$ and $s = (p_0,\ldots,p_k)$, we let $W_s(n_0,m_0,\ldots,n_k,m_k)$ be the set of all $x \in X$ such that in the game $\mathsf{G}(\{A_s\},x)$ if $n_0,(m_0,p_0),\ldots,n_k,(m_k,p_k)$ have been played in the first k rounds then Player II still has a winning strategy in the remaining game. We show that for any $\alpha < \omega_1$,

$$W_s(n_0,m_0,\ldots,n_k,m_k) \subseteq (A_s^\alpha)^{*V_{m_k}}.$$

This includes the case $\mathrm{lh}(s) = 0$. We proceed by induction on $\alpha < \omega_1$. The case $\alpha = 0$ is immediate. If λ is a limit, $A_s^\lambda = \bigcap_{\alpha<\lambda} A_s^\alpha$. Since this is a countable intersection, we have that $(A_s^\lambda)^{*V_{m_k}} = \bigcap_{\alpha<\lambda}(A_s^\alpha)^{*V_{m_k}}$. For the successor case, note that

$$(A_s^{\alpha+1})^{*V_{m_k}} = (A_s^\alpha)^{*V_{m_k}} \cap \left(\bigcup_p A_{s \frown p}^\alpha\right)^{*V_{m_k}}.$$

But by the inductive hypothesis

$$\left(\bigcup_p A^\alpha_{s^\frown p}\right)^{*V_{m_k}} = \bigcap_{V_{n_{k+1}} \subseteq V_{m_k}} \bigcup_{V_{m_{k+1}} \subseteq V_{n_{k+1}}} \bigcup_p (A^\alpha_{s^\frown p})^{*V_{m_{k+1}}}$$

$$\supseteq \bigcap_{V_{n_{k+1}} \subseteq V_{m_k}} \bigcup_{V_{m_{k+1}} \subseteq V_{n_{k+1}}} \bigcup_p W_{s^\frown p}(n_0, m_0, \ldots, n_{k+1}, m_{k+1})$$

$$= W_s(n_0, m_0, \ldots, n_k, m_k).$$

Hence by the inductive hypothesis again

$$W_s(n_0, m_0, \ldots, n_k, m_k) \subseteq (A^{\alpha+1}_s)^{*V_{m_k}}.$$

This finishes the induction, and for $\mathrm{lh}(s) = 0$ we obtain $W \subseteq B^{*G}_\alpha$ for all $\alpha < \omega_1$.

Finally we show that $\bigcap_\alpha B^{*G}_\alpha \subseteq \bigcup_\alpha (B_\alpha - D_\alpha)^{*G} \subseteq A^{*G}$. For this let $x \in \bigcap_\alpha B^{*G}_\alpha$. Then for all $\alpha < \omega_1$, $\forall^* g \in G$ $(g \cdot x \in B_\alpha)$. We claim that there is $\beta < \omega_1$ such that $\forall^* g \in G$ $(g \cdot x \notin D_\alpha)$, and thus $x \in \bigcup_\alpha (B_\alpha - D_\alpha)^{*G}$. Assume that for every $\alpha < \omega_1$ there are nonmeager many $g \in G$ with $g \cdot x \in D_\alpha$. Since $\{D_\alpha\}_{\alpha < \omega_1}$ is a pairwise disjoint collection of Borel sets, the collection of sets $\{g \in G : g \cdot x \in D_\alpha\}$ is thus an uncountable collection of pairwise disjoint nonmeager Borel subsets of G. This contradicts the second countability of G. Now the last inclusion follows immediately from the equality $\bigcup_\alpha (B_\alpha - D_\alpha) = A$. □

Theorem 3.2.9 (Vaught)
Let G be a Polish group and X a Polish G-space. For any open $V \subseteq G$ and analytic $A \subseteq X$, both $A^{\triangle V}$ and A^{*V} are analytic.

Proof. Fix a sequence $\{A_s\}$ of closed sets so that A is obtained from $\{A_s\}$ by the Suslin operation. Let $\mathcal{B} = \{V_n\}_{n \in \omega}$ be a countable base for G.

To deal with A^{*V} consider an enhanced game of $\mathsf{G}(\{A_s\}, x)$ in which Player I must play his first move n_0 so that $V_{n_0} \subseteq V$. Then the same proof of the preceding theorem gives that $x \in A^{*V}$ iff Player II has a winning strategy in this enhanced game. Since a winning strategy in this game of natural numbers is essentially an element of ω^ω, it follows from a straightforward computation that having a winning strategy in this game is an analytic condition. Thus A^{*V} is analytic.

Now $A^{\triangle V} = \bigcup_{V_n \subseteq V} A^{*V_n}$. Since analytic sets are closed under countable unions, $A^{\triangle V}$ is also analytic. □

Exercise 3.2.1 Prove Proposition 3.2.5.

Exercise 3.2.2 Let G be a Polish group and $\{U_n\}$ be a countable base for G. Let X be a Polish G-space and $A \subseteq X$ a Borel subset. Show that for any nonempty open set $V \subseteq G$,

$$A^{\triangle V} = \bigcup \{A^{*U_n} \; : \; U_n \subseteq V\} \quad \text{and} \quad A^{*V} = \bigcap \{A^{\triangle U_n} \; : \; U_n \subseteq V\}.$$

Exercise 3.2.3 Let G be a Polish group and X be a Polish G-space. For nonempty open $U \subseteq G$ let $G_U = \{g \in G \; : \; Ug = U\}$. Show that G_U is a closed subgroup of G and that for any set $A \subseteq X$, A^{*U} is invariant under the induced action of G_U.

Exercise 3.2.4 Let G be a Polish group, X a Polish G-space, and $A \subseteq X$ an invariant meager set. Show that A is included in a countable union of closed, nowhere dense sets, each of which is G_U-invariant (see above) for some nonempty open $U \subseteq G$.

Exercise 3.2.5 Let G be a Polish group, X a Polish G-space, and $A \subseteq X$ an invariant analytic set. Show that A is the union of ω_1 many invariant Borel sets.

Exercise 3.2.6 Let X, Y be Polish spaces and $B \subseteq X \times Y$ Borel. Show that the sets

$$B_\triangle = \{x \in X \; : \; \exists^* y \in Y \; (x, y) \in B\},$$
$$B_* = \{x \in X \; : \; \forall^* y \in Y \; (x, y) \in B\}$$

are both Borel.

3.3 Borel G-spaces

Definition 3.3.1
Let G be a Polish group, X a standard Borel space, and $a : G \times X \to X$ an action of G on X. If a is a Borel function then we say that X is a **(standard) Borel G-space**.

Theorem 3.3.2 (Miller)
Let G be a Polish group and X a Borel G-space. Then for any $x \in X$, G_x is closed and $G \cdot x$ is Borel.

Proof. Fix $x \in X$ and consider $H = \overline{G_x}$. H is a closed subgroup of G and the action of G on X restricted to H is Borel. Thus without loss of generality we may assume $H = G$ and therefore G_x is dense in G. Note that G_x is

Borel, and therefore has the Baire property. It thus suffices to show that G_x is nonmeager, since then it is clopen by Exercise 2.3.3.

We fix a Polish topology on X and let $\{U_n\}$ be a countable base. Let $A_n = \{g \in G : g \cdot x \in U_n\}$. Again each A_n is Borel, and therefore has the Baire property. We claim that each A_n is either meager or comeager. Toward a contradiction, assume some A_n is neither meager nor comeager. By Exercise 2.3.1 there are nonempty open sets V_1 and V_2 in G such that A_n is comeager in V_1 and meager in V_2. Since G_x is dense there is $h \in G_x \cap V_1^{-1}V_2$, and then $V_1 h \cap V_2 \neq \emptyset$. However, note that $A_n h = A_n$, and therefore A_n is comeager in $V_1 h$ as well as in $V_1 h \cap V_2$. This means that A_n is nonmeager in V_2, a contradiction.

Now note that for any $g \in G$ we have that $gG_x = \bigcap\{A_n : g \in A_n\}$. Thus if G_x is meager then for every $g \in G$ there is some n with $g \in A_n$ and A_n meager. This implies that $G = \bigcup\{A_n : A_n \text{ meager}\}$, and thus G is meager, a contradiction. This finishes the proof of the first part.

Now we have shown that G_x is closed, it follows that G/G_x is Polish. We note that the canonical map from G/G_x onto $G \cdot x$ is a Borel injection. To see this, note that the Borel structure generated by the quotient topology on G/G_x is exactly the Effros Borel structure (Exercise 2.3.6). By the Borel selection Theorem 1.4.6 there is a Borel function $s : G/G_x \to G$ such that $s(gG_x) \in gG_x$ for any $g \in G$. By the Luzin–Suslin Theorem 1.3.1 the range of s, $T = s(G/G_x)$, is a Borel subset of G. Furthermore the natural map $e : g \mapsto g \cdot x$ from T into $G \cdot x$ is injective and Borel since the action is Borel. Hence the canonical map from G/G_x into $G \cdot x$, being $e \circ s$, is a Borel injection. Thus $G \cdot x$ is Borel again by the Luzin–Suslin Theorem 1.3.1. $\square$

The first part of the proof uses a category argument which is a special case of a more general principle called the first topological 0-1 law (see Exercise 3.3.2).

The Vaught transforms continue to make sense for Borel G-spaces. In addition, the basic algebraic properties Propositions 3.2.3 and 3.2.5 continue to hold for Borel G-spaces. Next we show that Vaught transforms preserve Borel sets even in Borel G-spaces.

Theorem 3.3.3
Let G be a Polish group and X be a Borel G-space. Let $A \subseteq X$ be Borel and $V \subseteq G$ be nonempty open. Then both $A^{\triangle V}$ and A^{*V} are Borel.

Proof. Since the action is Borel for a Borel set A the set $\{(g,x) : g \cdot x \in A\}$ is a Borel subset of $G \times X$. Thus to prove the theorem it suffices to show that for any Borel $B \subseteq G \times X$ and nonempty open $V \subseteq G$, the sets $B_{\triangle V} = \{x \in X : \exists^* g \in V \ (g,x) \in B\}$ and $B_{*V} = \{x \in X : \forall^* g \in V \ (g,x) \in B\}$ are both Borel. Since $B_{*V} = X - (X - B)_{\triangle V}$, we only need to show the Borelness of $B_{\triangle V}$. For this consider the collection of sets $B \subseteq G \times X$ with the property that for any nonempty open $V \subseteq G$ $B_{\triangle V}$ is Borel. We show that the collection contains all sets of the form $U \times A$ for $A \subseteq X$ Borel and $U \subseteq G$ open, and is closed under complement and countable union. Consider first $B = U \times A$. Then $B_{\triangle V} = A$ if $U \cap V \neq \emptyset$ and $B_{\triangle V} = \emptyset$ if $U \cap V = \emptyset$. In either case $B_{\triangle V}$ is Borel. Next fix a countable base $\{U_n\}$ for G. Note that for any $B \subseteq G \times X$,

$$(X - B)_{\triangle V} = X - B_{*V}$$

$$= X - \bigcap \{B_{\triangle U_n} : U_n \subseteq V\} = \bigcup \{X - B_{\triangle U_n} : U_n \subseteq V\}.$$

Thus the collection is closed under complement. Finally let $B = \bigcup_m B_m$. Then $B_{\triangle V} = \bigcup_m (B_m)_{\triangle V}$. Thus the collection is closed under countable union, and we are done. ☐

Let $F(G)$ be the Effros Borel space with the right action of G:

$$g \cdot F = Fg^{-1}.$$

Then $F(G)$ becomes a Borel G-space. Let $F(G)^\omega$ be the product of infinitely many copies of $F(G)$. Then with the coordinatewise right action $F(G)^\omega$ is still a Borel G-space. The following theorem shows that it is a **universal** Borel G-space.

Theorem 3.3.4 (Becker–Kechris)
Let G be a Polish group and X a Borel G-space. Then there is a Borel G-embedding from X into $F(G)^\omega$.

Proof. Fix a Polish topology on X and let $\{A_n\}$ be a countable base for X in this topology. Let $\{V_m\}$ be a countable base for G. Define a map $\theta : X \to F(G)^\omega$ by

$$\theta(x)(n) = \{h \in G : \forall m \ (h \in V_m \Rightarrow x \in A_n^{\triangle V_m})\}.$$

It is easy to see that the set defined is closed in G. We check that θ is a Borel G-embedding. To see that $\theta(g \cdot x) = g \cdot \theta(x)$, we fix $n \in \omega$ and note that

$$h \in \theta(g \cdot x)(n) \iff \forall m \ (h \in V_m \Rightarrow g \cdot x \in A_n^{\triangle V_m})$$

$$\iff \forall m \ (h \in V_m \Rightarrow x \in A_n^{\triangle V_m g})$$

$$\iff \forall m \ (hg \in V_m g \Rightarrow x \in A_n^{\triangle V_m g})$$

$$\iff hg \in \theta(x)(n)$$

$$\iff h \in \theta(x)(n)g^{-1} = (g \cdot \theta(x))(n)$$

To see that θ is Borel, note that for any nonempty open $V \subseteq G$,

$$\theta(x)(n) \cap V \neq \emptyset \iff x \in A_n^{\triangle V},$$

and $A_n^{\triangle V}$ is Borel by Theorem 3.3.3. Finally to see that θ is injective, we assume $\theta(x) = \theta(y)$. Then it follows that for every n and m, $x \in A_n^{\triangle V_m}$ iff $y \in A_n^{\triangle V_m}$. Thus for comeager many $g \in G$, $g \cdot x \in A_n \iff g \cdot y \in A_n$. In particular, there is some $g \in G$ with $g \cdot x \in A_n \iff g \cdot y \in A_n$. Since $\{A_n\}$ is a base we get that $g \cdot x = g \cdot y$, and therefore $x = y$. □

The space $F(G)$ also admits a left action of G: $g \cdot F = gF$. Under this action $F(G)$ is also a Borel G-space. It is obvious that there is a Borel isomorphic G-map (Borel G-isomorphism) between the two spaces with the left and the right actions, namely $F \mapsto F^{-1}$. And thus we may regard the two G-spaces as the same.

It is also worthwhile to state the following version of the Cantor–Bernstein theorem for Borel G-embeddings, although its proof is classical and is left as an exercise (Exercise 3.3.3).

Theorem 3.3.5
Let G be a Polish group and X, Y be Borel G-spaces. Suppose there is a Borel G-embedding from X into Y and also a Borel G-embedding from Y into X. Then there is a Borel G-isomorphism between X and Y.

Exercise 3.3.1 Let G be a Polish group and X a Borel G-space. For any $x, y \in X$ define $G_{x,y} = \{g \in G : g \cdot x = y\}$. Show that $G_{x,y}$ is closed for any $x, y \in X$.

Exercise 3.3.2 Let X be a Baire space and G a group of homeomorphisms on X. Suppose for any nonempty open sets U, V in X there is a $g \in G$ with $g \cdot U \cap V \neq \emptyset$. Let $A \subseteq X$ be a G-invariant subset of X with the Baire property. Show that A is either meager or comeager.

Exercise 3.3.3 Prove Theorem 3.3.5.

Exercise 3.3.4 Let G be a countable group. Let $(2^\omega)^G$ be the Polish G-space with the (left or right) shift action. Show that for any Borel G-space X there is a Borel G-embedding from X into $(2^\omega)^G$. Moreover, if X is a Polish G-space then the Borel G-embedding can be a function of Baire class 1. Furthermore, if X is a 0-dimensional Polish G-space then the G-embedding can be continuous.

Exercise 3.3.5 Let G be a Polish group with a compatible complete left-invariant metric d. Let $L(G, d)$ denote the Polish G-space of all Lipschitz functions from G into $\mathbb{R}$ (see Exercise 3.1.8). Define $\varphi : F(G) \to L(G, d)$ by

$$\varphi(F)(x) = d(x, F).$$

Show that φ is a Borel G-embedding from $F(G)$ into $L(G, d)$. Deduce that $L(G, d)^\omega$ is a universal Borel G-space.

Exercise 3.3.6 Let G be a Polish group and d a compatible left-invariant metric on G.

(a) Let $(\overline{G}, d)$ denote the d-completion of G, with the extended metric still denoted by d. Consider the action of G on $\overline{G}$ defined by

$$g \cdot \{h_n\} = \{gh_n\}.$$

Show that under this action $\overline{G}$ becomes a Polish G-space.

(b) Let $L(\overline{G}, d)$ denote the Polish space of all Lipschitz functions from $\overline{G}$ to $\mathbb{R}$. Consider the action of G on $L(\overline{G}, d)$ defined by

$$(g \cdot f)(x) = f(g^{-1} \cdot x).$$

Show that under this action $L(\overline{G}, d)$ becomes a Polish G-space.

(c) Define $\overline{\varphi} : F(G) \to L(\overline{G}, d)$ by

$$\overline{\varphi}(F)(x) = d(x, \overline{F}),$$

where $\overline{F}$ is the closure of F in $\overline{G}$. Show that $\overline{\varphi}$ is a Borel G-embedding. Deduce that $L(\overline{G}, d)^\omega$ is a universal Borel G-space.

3.4 Orbit equivalence relations

Equivalence relations are the main objects studied in invariant descriptive set theory. In this section we review some generalities about equivalence relations and prove some basic results about orbit equivalence relations induced by actions of Polish groups.

Definition 3.4.1
For an arbitrary equivalence relation E on a set X, we let $[x]_E$ denote the $(E-)$**equivalence class** of $x \in X$, that is, $[x]_E = \{y \in X : xEy\}$. More generally, if $A \subseteq X$ we denote by $[A]_E$ the $(E-)$**saturation** of A, that is, $[A]_E = \{y \in X : \exists x \in A \, (xEy)\}$. We call A $(E-)$**invariant** if $[A]_E = A$. A is called a **complete section** or **full** if $[A]_E = X$. We denote by X/E the **quotient space** of E, that is, the set of all E-equivalence classes $\{[x]_E : x \in X\}$.

Definition 3.4.2
Let G be a group and X be a G-space. Then the **orbit equivalence relation**, denoted E_G^X (or simply E_G), is given by

$$x E_G^X y \iff \exists g \in G \, (g \cdot x = y).$$

For an orbit equivalence relation E_G, the quotient space X/E is also denoted X/G.

Definition 3.4.3
Let E be an equivalence relation on a set X. A **transversal** for E is a set $T \subseteq X$ meeting each E-equivalence class at exactly one point, that is, $|T \cap [x]_E| = 1$ for all $x \in X$. A **selector** for E is a map $s : X \to X$ such that $s(x) E x$ and $s(x) = s(y)$ for all $x E y$.

An equivalence relation E on X is a subset of $X \times X$. When X is Polish or standard Borel, the descriptive complexity of the equivalence relation E is closely related to the descriptive complexity of its orbits and to the topological properties of the quotient space. First we note that if G is a Polish group and X a Borel G-space, then the orbit equivalence relation E_G^X is $\boldsymbol{\Sigma}_1^1$. However, not all $\boldsymbol{\Sigma}_1^1$ equivalence relations on a standard Borel space are orbit equivalence relations. This is because, by Theorem 3.3.2, every orbit of an E_G^X is Borel.

On the other hand, if E is a Borel equivalence relation on a standard Borel space X, then every E-equivalence class is Borel. We will see that the converse is not true in general. The following theorem is one of a few positive results connecting the complexity of an orbit equivalence relation and that of the orbits.

Theorem 3.4.4 (Effros)
Let G be a Polish group and X a Polish G-space. Then the following are equivalent:

(i) *E_G is G_δ.*

(ii) *Each orbit $G \cdot x$ is G_δ.*

(iii) *X/G is T_0, that is, for any $p, q \in X/G$ with $p \neq q$, $\overline{\{p\}} \neq \overline{\{q\}}$.*

Proof. The implication (i)$\Rightarrow$(ii) is obvious. We show that (ii)$\Rightarrow$(iii)$\Rightarrow$(i). Let $\pi : X \to X/G$ be the quotient map. We first note that π is open. This is because for any open $U \subseteq X$, $\pi^{-1}(\pi(U)) = [U]_G = \bigcup_{g \in G} g \cdot U$ is open since each $g \cdot U$ is open. Since X is second countable, it follows that X/G is also second countable. Also note that for any $p \in X/G$,

$$\overline{\pi^{-1}(p)} = \pi^{-1}\left(\overline{\{p\}}\right).$$

The inclusion $\overline{\pi^{-1}(p)} \subseteq \pi^{-1}\left(\overline{\{p\}}\right)$ follows from continuity of π. For the other inclusion, note that $F = \overline{\pi^{-1}(p)}$ is G-invariant by the continuity of the action, and thus $\pi(F)$ is closed. As $p \in \pi(F)$, $\pi^{-1}\left(\overline{\{p\}}\right) \subseteq \pi^{-1}(\pi(F)) = F$.

Now for the proof of the direction (ii)$\Rightarrow$(iii), suppose each orbit is G_δ. Assume $p, q \in X/G$ with $p \neq q$ but $\overline{\{p\}} = \overline{\{q\}}$. Thus

$$\overline{\pi^{-1}(p)} = \pi^{-1}\left(\overline{\{p\}}\right) = \pi^{-1}\left(\overline{\{q\}}\right) = \overline{\pi^{-1}(q)} = C.$$

Now both $\pi^{-1}(p)$ and $\pi^{-1}(q)$ are G_δ orbits which are dense in C, hence they are both comeager in C and therefore $\pi^{-1}(p) \cap \pi^{-1}(q) \neq \emptyset$. But this contradicts the assumption that $p \neq q$.

For the implication (iii)$\Rightarrow$(i), we fix a countable base $\{U_n\}$ for the quotient space X/G. Then $\{\pi^{-1}(U_n)\}$ is a countable family of open sets separating the orbits, that is,

$$x E_G y \iff \forall n \ (x \in \pi^{-1}(U_n) \leftrightarrow y \in \pi^{-1}(U_n)).$$

This demonstrates that E_G is G_δ. □

It is straightforward to see that X/E is T_1 iff every E-equivalence class is closed.

The following proposition characterizes when X/G is Hausdorff. The proof is left as an exercise (Exercise 3.4.1).

Proposition 3.4.5
Let G be a Polish group and X a Polish G-space. Then X/G is Hausdorff iff E_G is closed.

In the next proposition some condition between T_1 and Hausdorffness is considered. It gives a sufficient condition for the existence of Borel selectors. Note that if $s : X \to X$ is a Borel selector for E then $\{x : s(x) = x\}$ is a Borel transversal for E. Conversely, if T is a Borel transversal for E then the function defined by

$$s(x) = y \iff y \in T \text{ and } x E y$$

is a selector for E, and s is Borel if E is Borel.

Proposition 3.4.6
Let X be a Polish space and E an equivalence relation on X. If every E-equivalence class is closed and, for any open $A \subseteq X$, the saturation of A is Borel, then there is a Borel selector for E. In particular, if G is a Polish group, X is a Polish G-space and every orbit is closed, then there is a Borel selector for E_G.

Proof. Let $\pi : X \to X/E$ be the quotient map. Then X/E is a subset of the Effros Borel space $F(X)$, and as a map from X into $F(X)$, π is Borel. This is because, for any $x \in X$ and open set $U \subseteq X$,

$$[x]_E \cap U \neq \emptyset \iff x \in [U]_E.$$

Now by Theorem 1.4.6 there is a Borel function $\sigma : F(X) \to X$ with $\sigma(F) \in F$ for every nonempty closed $F \subseteq X$. Let $s = \sigma \circ \pi$. Then s is a Borel selector for E. In case of a continuous G-action, the saturation of any open set is open, and thus the conclusion holds.   □

Finally the following proposition gives a sufficient condition for X/G to be Polish.

Proposition 3.4.7
Let G be a Polish group and X a Polish G-space. Suppose for any closed subset $A \subseteq X$, the saturation of A is closed. Then X/G is Polish.

Proof. The assumption implies that the quotient map is closed. Thus it is closed, open, continuous, and onto. Since a closed continuous image of a normal space is normal, we conclude that X/G is normal. Since a continuous open image of a second countable space is second countable, we conclude that X/G is second countable. Now it follows from the Urysohn metrization theorem that X/G is metrizable. Thus by Theorem 2.2.9 X/G is Polish.   □

We will see that there are orbit equivalence relations which are non-Borel. In general, it is of great interest to decide if a particular orbit equivalence relation is Borel. We will give some characterizations in Section 8.2. The following proposition gives a sufficient condition.

Proposition 3.4.8
Let G be a Polish group and X a standard Borel space. If the action of G on X is free, then E_G is Borel.

Proof. This follows from the Luzin–Suslin Theorem 1.3.1. To see this, let

$$A = \{(g, x, y) : g \cdot x = y\}.$$

Then A is a Borel subset of $G \times X \times X$, and hence is a standard Borel space. Let $\pi : A \to X \times X$ be the projection map $\pi(g, x, y) = (x, y)$. Then π is injective since the action is free. Since $\pi(A) = E_G$, it follows that E_G is Borel.   □

Exercise 3.4.1 Prove Proposition 3.4.5.

Exercise 3.4.2 Let G be a compact Polish group and X a Polish G-space. Show that E_G is closed.

Exercise 3.4.3 Let (X, d) be a Polish metric space and G a compact subgroup of $\mathrm{Iso}(X)$. Show that X/G is Polish.

Exercise 3.4.4 Let G be a compact Polish group and X a Borel G-space. Show that E_G is Borel.

Exercise 3.4.5 Let G be a locally compact Polish group and X a Polish G-space. Show that E_G is F_σ.

Exercise 3.4.6 Let G be a locally compact Polish group and X a Borel G-space. Show that E_G is Borel.

Exercise 3.4.7 Let G be a Polish group and X a Polish G-space. Show that if E_G^X is Borel then there is $\alpha < \omega_1$ such that all orbits are $\mathbf{\Pi}_\alpha^0$.

3.5   Extensions of Polish group actions

In Section 3.3 we saw that every Polish group G admits a universal Borel G-space. In this section we consider extensions of Borel (or Polish) G-spaces to H-spaces when H involves G. First note that if G is a quotient of H then any G-space is trivially an H-space, and thus any Borel (or Polish) G-space is also a Borel (respectively Polish) H-space. Next we deal with the case G is a closed subgroup of H.

Theorem 3.5.1 (Hjorth)
Let H be a Polish group, G a closed subgroup, and X a Polish G-space. Consider the continuous action of G on $X \times H$ by

$$g \cdot (x, h) = (g \cdot x, gh).$$

Then $(X \times H)/G$ is Polish.

Proof.   Let π be the quotient map from $X \times H$ onto $(X \times H)/G$. Then π is continuous and open. Since $X \times H$ is second countable, so is $(X \times H)/G$. Let E_G be the orbit equivalence relation on $X \times H$. Then note that E_G is closed, since

$$(x, h)E_G(y, k) \iff kh^{-1} \in G \text{ and } kh^{-1} \cdot x = y.$$

Thus $(X \times H)/G$ is Hausdorff by Proposition 3.4.5. By Sierpinski's Theorem 2.2.9 it suffices to show that $(X \times H)/G$ is metrizable, and by the Urysohn metrization theorem it suffices to show that $(X \times H)/G$ is regular. The regularity of $(X \times H)/G$ is equivalent to the following property of the space $X \times H$:

> Let $F \subseteq X \times H$ be an invariant closed set and $(x, h) \notin F$. There are open sets $U \subseteq X$ and $N \subseteq H$ with $x \in U$ and $1_H \in N$ such that for any $(y, k) \in F$ there is an open set $W \subseteq X \times H$ with $(y, k) \in W$ so that $G \cdot (U \times hN) \cap W = \emptyset$.

For this fix such F and (x, h). First let $U_0 \subseteq X$ and $N_0 \subseteq H$ be open sets with $x \in U_0$, $1_H \in N_0$, and $(U_0 \times N_0 h) \cap F = \emptyset$. Let $N_1 \subseteq N_0$ be open with $N_1^{-1} = N_1$ and $U \subseteq U_0$ be open with $x \in U$ such that $\overline{N_1^2 \cdot U} \subseteq U_0$. Let $N \subseteq N_1$ be open with $N^{-1} = N$ and $hN^2 h^{-1} \subseteq N_1$. We verify that U and N are as required.

For this let $(y, k) \in F$. Consider two cases. Case 1: $Gk \cap N_1 h = \emptyset$. In this case let $W = X \times kN$. Then W is open and $(y, k) \in W$. Assume $W \cap G \cdot (U \times hN) \neq \emptyset$. Then $kN \cap GhN \neq \emptyset$. This implies that $Gk \cap hN^2 \neq \emptyset$. But $hN^2 \subseteq N_1 h$, contradicting our case assumption. Case 2: $Gk \cap N_1 h \neq \emptyset$. Let $g_0 \in G$ be such that $g_0 k \in N_1 h$. Note that $g_0 \cdot (y, k) \notin U_0 \times N_0 h$ but $g_0 k \in N_1 h \subseteq N_0 h$. It follows that $g_0 \cdot y \notin U_0$, and thus $g_0 \cdot y \notin \overline{N_1^2 \cdot U}$. Let $V \subseteq X$ be open such that $g_0 \cdot y \in V$ and $V \cap N_1^2 \cdot U = \emptyset$. Let $W = g_0^{-1} \cdot (V \times N_1 h)$. Then W is open and $(y, k) \in W$. Assume $W \cap G \cdot (U \times hN) \neq \emptyset$. Then $(V \times N_1 h) \cap G \cdot (U \times hN) \neq \emptyset$. Let $g \in G$ be such that $(V \times N_1 h) \cap g \cdot (U \times hN) \neq \emptyset$. Then $g \notin N_1^2$ since $V \cap g \cdot U \neq \emptyset$. However, $N_1 h \cap ghN \neq \emptyset$. And thus $g \in N_1 h N^{-1} h^{-1} \subseteq N_1^2$, a contradiction. $\qquad \Box$

Theorem 3.5.2 (Mackey–Hjorth)

Let H be a Polish group and G a closed subgroup of H. Let X be a Polish G-space with action a. Then there is a Polish H-space Y with action b such that:

(i) *X is a closed subset of Y;*

(ii) *$a(g, x) = b(g, x)$ for all $g \in G$ and $x \in X$;*

(iii) *Every H-orbit in Y contains exactly one G-orbit in X.*

Proof. Let $Y = (X \times H)/G$ be as in the preceding theorem. Let π be the quotient map from $X \times H$ onto Y. For $(x, h) \in X \times H$, denote $\pi(x, h) = [(x, h)]_G$ by $[x, h]$ for simplicity. Define $j : X \to Y$ by $j(x) = [x, e]$. Then it is easy to see that j is a continuous embedding. We claim that j is in fact a homeomorphic embedding and that $j(X)$ is closed in Y. In fact $(x, h) \in \pi^{-1}(j(X))$ iff $h \in G$. Thus $\pi^{-1}(j(X))$ is closed and therefore so is $j(X)$. To see that j is open onto $j(X)$, let $U \subseteq X$ be open. Then $(x, g) \in \pi^{-1}(j(U))$ iff $g^{-1} \cdot x \in U$. Thus there are open sets $N \subseteq G$ and $V \subseteq X$ with $g^{-1} \in N$, $x \in V$, and $N \cdot V \subseteq U$, and thus $V \times N^{-1} \subseteq \pi^{-1}(j(U))$. This shows that $\pi^{-1}(j(U))$ is open in $X \times H$, and so is $j(U)$ in $j(X)$. Therefore we have essentially verified (i).

Let H act on Y by

$$k \cdot [x, h] = [x, hk^{-1}].$$

To verify (ii) let $g \in G$ and $x \in X$. It suffices to note that $[g \cdot x, e] = [x, g^{-1}]$. To verify (iii) note that the H-orbit of $[x, h]$ contains the unique G-orbit of $[x, e]$.

It remains to check that the H-action on Y is continuous. For this we check that it is separately continuous. First fix $[x, h] \in Y$ and let $W \subseteq Y$ be open. Then $k \cdot [x, h] \in W$ iff $(x, hk^{-1}) \in \pi^{-1}(W)$, and thus there is an open $N \subseteq G$ with $k \in N$ such that $N \cdot [x, h] \subseteq W$. Next fix $k \in H$ and let $W \subseteq Y$ be open. Then $(x, h) \in \pi^{-1}(k^{-1} \cdot W)$ iff $(x, hk^{-1}) \in \pi^{-1}(W)$, and thus there are open sets $U \subseteq X$ and $N \subseteq G$ with $x \in U$, $h \in N$, and $U \times Nk^{-1} \subseteq \pi^{-1}(W)$. This shows that $k^{-1} \cdot W$ is open. ◻

Theorem 3.5.3 (Mackey)
Let H be a Polish group and G a closed subgroup of H. Let X be a Borel G-space with action a. Then there is a Borel H-space Y with action b such that:

(i) *X is a Borel subset of Y;*

(ii) *$a(g, x) = b(g, x)$ for all $g \in G$ and $x \in X$;*

(iii) *Every H-orbit in Y contains exactly one G-orbit in X.*

Proof. Define the action of G on $X \times H$, the space Y, and the action of H on Y the same as in the preceding theorem. We first verify that Y with the quotient Borel structure is standard Borel. For this let T be a Borel transversal of the right G-cosets in H. Then the quotient map π restricted on $X \times T$ is a Borel isomorphism onto Y. Since $X \times T$ is standard Borel, so is Y.

The injection $j : X \to Y$ is now a Borel embedding, and therefore $j(X)$ is a Borel subset of Y. This essentially verified (i). Clauses (ii) and (iii) are algebraic properties with the same proof as in the preceding theorem. Finally it follows from also the same proof that the action of H on Y is Borel. ◻

As immediate corollaries in both Theorems 3.5.2 and 3.5.3 we may assume $G \leq_w H$.

Exercise 3.5.1 Let E_G^X and E_H^Y denote the orbit equivalence relations in Theorems 3.5.2 and 3.5.3. Show that E_H^Y is Borel iff E_G^X is Borel.

Exercise 3.5.2 Let actions a and b be given by Theorems 3.5.2 and 3.5.3. Show that if a is a free action then so is b.

3.6   The logic actions

Recall from Section 2.4 that the infinite permutation group S_∞ and its closed subgroups can be viewed as automorphism groups of countable models. In this section we relate general actions of S_∞ and its closed subgroups to certain special actions on classes of countable models. These special actions are called logic actions.

Let L be a countable relational language, that is, $L = \{R_i\}_{i \in I}$, where I is a countable set of indices and each R_i is an n_i-ary relation symbol. Let $\mathrm{Mod}(L)$ denote the space of all countable L-models with the underlying universe ω. Each element of $\mathrm{Mod}(L)$ can be viewed as an element of the product space

$$X_L = \prod_{i \in I} 2^{\omega^{n_i}}.$$

Thus $\mathrm{Mod}(L)$ becomes a compact Polish space (that is homeomorphic to the Cantor space) with the product topology on X_L.

More formally for each $x \in X_L$ let $M_x \in \mathrm{Mod}(L)$ be the countable model coded by x. Then for any $i \in I$ and $(k_1, \ldots, k_{n_i}) \in \omega^{n_i}$,

$$R_i^{M_x}(k_1, \ldots, k_{n_i}) \iff x_i(k_1, \ldots, k_{n_i}) = 1.$$

With this correspondence $\mathrm{Mod}(L)$ and X_L can be used interchangeably.

The **logic action** of S_∞ on $\mathrm{Mod}(L)$ is defined by letting $g \cdot M = N$ iff

$$R_i^N(k_1, \ldots, k_{n_i}) \iff R_i^M(g^{-1}(k_1), \ldots, g^{-1}(k_{n_i}))$$

for all $i \in I$ and $(k_1, \ldots, k_{n_i}) \in \omega^{n_i}$. Thus $g \cdot M = N$ iff g is an isomorphism from M onto N. This is clearly a continuous action, and the orbit equivalence relation is exactly the isomorphism relation on $\mathrm{Mod}(L)$, which we denote by $\cong_L$.

The following theorem shows that $\mathrm{Mod}(L)$ is a universal Borel S_∞-space if L is sufficiently rich.

Theorem 3.6.1
Let $L = \{R_i\}_{i \in I}$ be a countable relational language with each R_i an n_i-ary relation symbol. If $\{n_i : i \in I\}$ is unbounded in ω, then the Polish S_∞-space $\mathrm{Mod}(L)$ is a universal Borel S_∞-space, that is, for any Borel S_∞-space X there is a Borel S_∞-embedding from X into $\mathrm{Mod}(L)$.

Proof. By Theorem 3.3.4 $F(S_\infty)^\omega$ is a universal Borel S_∞-space. It suffices to construct a Borel S_∞-embedding from $F(S_\infty)^\omega$ into $\mathrm{Mod}(L)$. This is done in several steps.

First consider the action of S_∞ on ω^ω by $g \cdot x = g \circ x$. This action is continuous. Moreover, it extends to an action of S_∞ on $F(\omega^\omega)$: $g \cdot F = \{g \cdot x : x \in F\}$. It is easy to see that this action is Borel. Now let S_∞ act on $F(\omega^\omega)^\omega$ coordinatewise. Then $F(\omega^\omega)^\omega$ is a Borel S_∞-space. We claim that there is a Borel S_∞-embedding from $F(S_\infty)$ into $F(\omega^\omega)$. This will in an obvious way give a Borel S_∞-embedding of $F(S_\infty)^\omega$ into $F(\omega^\omega)^\omega$. In fact the desired embedding from $F(S_\infty)$ into $F(\omega^\omega)$ is just $F \mapsto \overline{F}$, where $\overline{F}$ denotes the closure of F in ω^ω. It is injective since $\overline{F} \cap S_\infty = F$. It is easy to verify that this is indeed a Borel S_∞-map.

Next we consider a language L_0 with infinitely many relation symbols of each arity. Thus $L_0 = \{R_{m,l}\}_{(m,l) \in \omega^2}$, where each $R_{m,l}$ is an l-ary relation symbol. We claim that there is a Borel S_∞-embedding from $F(\omega^\omega)^\omega$ into $\mathrm{Mod}(L_0)$. To see this note that each nonempty closed set $F \in F(\omega^\omega)$ is correspondent to a unique pruned tree $T_F \subseteq \omega^{<\omega}$ so that $[T_F] = F$. More explicitly, $s \in T_F$ iff there is $x \in F$ with $s \subseteq x$. Now for $\vec{F} = (F_0, F_1, \dots) \in F(\omega^\omega)^\omega$, define $M_{\vec{F}}$ so that

$$R^{M_{\vec{F}}}_{m,l}(k_1, \dots, k_l) \iff (k_1, \dots, k_l) \in T_{F_m}.$$

Then $\vec{F} \mapsto M_{\vec{F}}$ is the desired Borel S_∞-embedding.

Finally we verify that there is a Borel S_∞-embedding from $\mathrm{Mod}(L_0)$ into $\mathrm{Mod}(L)$. Since $\{n_i : i \in I\}$ is unbounded, we can find an injection $j : \omega^2 \to I$ such that $n_{j(m,l)} \geq l$ for all $m, n \in \omega$. Let $J \subseteq I$ be the range of j. Then we define $M_0 \mapsto M$ from $\mathrm{Mod}(L_0)$ to $\mathrm{Mod}(L)$ by letting

$$R^M_{j(m,l)}(k_1, \dots, k_{n_{j(m,l)}}) \iff R^{M_0}_{m,l}(k_1, \dots, k_l)$$

for any $m, l, k_1, \dots, k_{n_{j(m,l)}}$ and letting R^M_i always hold if $i \notin J$. It is straightforward to check that this is a Borel S_∞-embedding. □

Next we show that the hypothesis of unbounded arity for the language L in the above theorem is necessary. For this we need to introduce a new concept. First we fix the following notation. For $n \in \omega$, let D_n denote the set of all bijections $s : A \to B$, where A and B are n-element subsets of ω. Let $D = \bigcup_n D_n$. For $s \in D$ and $g \in S_\infty$, we write $s \subseteq g$ if g is an extension of s. For $s \in D$, let $N_s = \{g \in S_\infty : s \subseteq g\}$. Then the collection $\{N_s : s \in D\}$ is a clopen base for S_∞.

Definition 3.6.2

Let X be an S_∞-space and $n \in \omega$. An element $x \in X$ is called n-**based** if for any $g \in S_\infty$, if $g \notin (S_\infty)_x$ then there is $s \in D_n$ such that $s \subseteq g$ and $N_s \cap (S_\infty)_x = \emptyset$. The action of S_∞ on X is n-**based** if every element $x \in X$ is n-based.

If L is a relational language consisting of relation symbols of arity $\leq n$, then $\mathrm{Mod}(L)$ is n-based. If X and Y are both S_∞-spaces, Y is n-based, and there is an S_∞-embedding from X into Y, then X is also n-based.

Theorem 3.6.3 (Dougherty)

Let $n > 0$ and L be a countable relational language containing an n-ary relation symbol. Then there is $M \in \mathrm{Mod}(L)$ which is not $(n-1)$-based.

Proof. Let R be an n-ary relation symbol in L. Without loss of generality we assume $L = \{R\}$. Let I be the set of all n-tuples $(a_1, \ldots, a_n)$ so that $a_i \neq a_j$ for $i \neq j$. We will construct $M = (\omega, R^M)$ so that $R^M \neq \emptyset$, $R^M \neq I$, and for any $s \in D_{n-1}$ there is $g \in N_s \cap \mathrm{Aut}(M)$. These imply that M is not $(n-1)$-based.

For any $t \in D$, we let $\mathcal{A}_t$ be the set

$$\{S \subseteq I : \forall (a_1, \ldots, a_n) \in \mathrm{dom}(t)^n \; S(a_1, \ldots, a_n) \iff S(t(a_1), \ldots, t(a_n))\}.$$

Intuitively, $\mathcal{A}_t$ is the collection of all n-ary relations S so that t is a partial automorphism of the model (ω, S). Thus if $t \in D_m$ for $m < n$, then $\mathcal{A}_t = \mathcal{P}(I)$. If g is an automorphism for (ω, S), then $S \in \bigcap_{t \subseteq g} \mathcal{A}_t$. Conversely, if $S \in \bigcap t \subseteq g \mathcal{A}_t$ then g is an automorphism for (ω, S). For $t \in D_m$ and $k \in \omega$, we also let

$$\mathcal{E}_{t,k}^+ = \{S \subseteq I : S \in \mathcal{A}_t \Rightarrow \exists u \supseteq t \; (u \in D_{m+1} \wedge k \in \mathrm{dom}(u) \wedge S \in \mathcal{A}_u)\},$$

and

$$\mathcal{E}_{t,k}^- = \{S \subseteq I : S \in \mathcal{A}_t \Rightarrow \exists u \supseteq t \; (u \in D_{m+1} \wedge k \in \mathrm{range}(u) \wedge S \in \mathcal{A}_u)\}.$$

Now the sets $\mathcal{A}_t$, $\mathcal{E}_{t,k}^+$, and $\mathcal{E}_{t,k}^-$ can all be viewed as subsets of 2^I, and as such they are in fact open subsets. We claim that both $E_{t,k}^+$ and $E_{t,k}^-$ are dense in 2^I.

We only verify the claim for $E_{t,k}^+$, since the argument for $E_{t,k}^-$ is similar. Let $i_1, \ldots, i_l \in I$ be finitely many elements in I, with $i_j = (a_1^j, \ldots, a_n^j)$. Let $A = \{a_1^j, \ldots, a_n^j : 1 \leq j \leq l\}$. Let also $S_0 \in 2^{i_1, \ldots, i_l}$. We need to define $S \in 2^I$ so that $S_0 \subseteq S$ and $S \in \mathcal{E}_{t,k}^+$. If there is an $S \subseteq S_0$ so that $S \notin \mathcal{A}_t$ then we are done. Otherwise we have that any $S \subseteq S_0$ is an element of $\mathcal{A}_t$. We define $k_0 = k$ if $k \notin \mathrm{dom}(t)$ and k_0 to be any element not in $\mathrm{dom}(t)$ if $k \in \mathrm{dom}(t)$. To define u it is enough to define $u(k_0)$. For this we let $u(k_0)$ to be arbitrarily chosen from $\omega - A$. Then we can extend S_0 to S_1 with a finite domain so that any extension S of S_1 is an element of $\mathcal{A}_u$. Now let $S \supseteq S_1$ be arbitrary, and we have obtained $S \in \mathcal{E}_{t,k}^+$.

It follows that $\bigcap_{t,k} \mathcal{E}_{t,k}^+ \cap \mathcal{E}_{t,k}^-$ is comeager in 2^I, and thus there is $S \in 2^I$ so that $S \in \mathcal{E}_{t,k}^+$ and $S \in \mathcal{E}_{t,k}^-$ for all $t \in D$, $k \in \omega$, and so that $S \neq \emptyset$ and $S \neq I$. It remains to check that for any $s \in D_{n-1}$ there is $g \in N_s \cap \mathrm{Aut}(M)$. For this we define a sequence $t_0 \subseteq t_1 \subseteq \ldots$ and let $g = \bigcup_m t_m$. To begin with, let $t_0 = s$. Then since $S \in \mathcal{E}^+_{t_0,0}$ there is $u_0 \supseteq t_0$ such that $S \in \mathcal{A}_{u_0}$, and since $S \in \mathcal{E}^-_{u_0,0}$ there is $t_1 \supseteq u_0$ such that $S \in \mathcal{A}_{t_1}$. We now have that $0 \in \mathrm{dom}(t_1) \cap \mathrm{range}(t_1)$. In general, suppose t_m is defined. Then since $S \in \mathcal{E}^+_{t_m,m}$ there is $u_m \supseteq t_m$ such that $S \in \mathcal{A}_{u_m}$, and since $S \in \mathcal{E}^-_{u_m,m}$ there is $t_{m+1} \supseteq u_m$ such that $S \in \mathcal{A}_{t_{m+1}}$. Since $S \in \bigcup_m \mathcal{A}_{t_m}$, we have that $g \in \mathrm{Aut}(M)$. □

Corollary 3.6.4
Let $n > 0$ and L be a countable relational language consisting of relation symbols of arity $\leq n$. If L' is a countable relational language containing an $(n + 1)$-ary relation symbol, then there is no S_∞-embedding from $\mathrm{Mod}(L')$ into $\mathrm{Mod}(L)$. In particular, $\mathrm{Mod}(L)$ is not a universal Borel S_∞-space.

In Part III of this book we will give a more thorough treatment of countable model theory and logic actions.

Exercise 3.6.1 Show that being n-based is an invariant property. That is, if X is an S_∞-space, $x \in X$, $y \in [x]$, and x is n-based, then y is n-based.

Exercise 3.6.2 Show that any Borel S_∞-space X can be decomposed into disjoint Borel subsets $\{X_n\}$ so that each X_n is n-based.

Exercise 3.6.3 For each $n > 0$ let L_n be the relational language with exactly n unary relation symbols. Let also L_ω be the relational language with countably many unary relation symbols. Show that for $n, m \leq \omega$, there is a Borel S_∞-embedding from $\mathrm{Mod}(L_n)$ into $\mathrm{Mod}(L_m)$ iff $n \leq m$.

Exercise 3.6.4 Let L be a countable relational language with only unary relation symbols. Show that $\cong_L$ is Borel.

Chapter 4

Finer Polish Topologies

In this chapter we introduce the very useful technique of change of topology on Polish G-spaces. With the technique and results developed in this chapter we will be able to unify the study of Polish G-spaces and Borel G-spaces. We will also encounter a technical topological concept of strong Choquet space, which will be used several times in the proofs of important theorems later on.

4.1 Strong Choquet spaces

Given a nonempty topological space X, the **strong Choquet game** G_X is a two-player game defined as follows:

$$
\begin{array}{llll}
\text{I} & x_0, U_0 & & x_1, U_1 \\
& & & & \cdots\cdots \\
\text{II} & & V_0 & & V_1
\end{array}
$$

Players I and II take turns to play open sets U_i, V_i and elements $x_i \in X$, $i \in \omega$. The rules of the game are $x_i \in V_i \subseteq U_i$ and $U_{i+1} \subseteq V_i$, for all $i \in \omega$. The first player who violates these rules loses. If an infinite sequence of objects are played following the rules then Player II wins the play if $\bigcap_n U_n = \bigcap_n V_n \neq \emptyset$. Also in the case that X is empty, we regard that Player I has no legitimate move and therefore he loses.

Definition 4.1.1
A topological space X is a **strong Choquet space** if Player II has a winning strategy in the strong Choquet game G_X.

It is easy to show that any completely metrizable space or any locally compact Hausdorff space is strong Choquet (Exercises 4.1.2 and 4.1.3). It is also clear that any strong Choquet space is a Baire space (Exercise 4.1.4). We prove some closure properties for strong Choquet spaces.

Theorem 4.1.2
(a) If X is strong Choquet and $Y \subseteq X$ is G_δ, then Y is strong Choquet.

(b) If $f : X \rightarrow Y$ is continuous, open, and surjective and X is strong Choquet, then Y is strong Choquet.

(c) If $X = \prod_{i \in I} X_i$, then X is strong Choquet iff each X_i, $i \in I$, is strong Choquet.

Proof. (a) Let X be strong Choquet and $Y = \bigcap_n W_n \subseteq X$, where each W_n is open. We need to describe a winning strategy for Player II in the game G_Y. For this suppose $x_0, U_0, x_1, U_1, \ldots$ are Player I's moves in a play of the game G_Y. We inductively define Player II's responses in this play. To begin with let U_0' be an open set in X with $U_0' \cap Y = U_0$. Then in the game G_X let Player I play x_0 and $U_0^* = U_0' \cap W_0$ as his first move. Let V_0^* be Player II's response in G_X according to her winning strategy. Then in the game G_Y we let $V_0 = V_0^* \cap Y$ be Player II's first response. Note that $x_0 \in V_0$ since $x_0 \in V_0^* \cap Y \subseteq V_0^* \cap W_0$. In general suppose $x_0, U_0, V_0, \ldots, x_n, U_n, V_n$ and U_n', U_n^*, V_n^* have been defined so that $U_n' \cap Y = U_n$, $U_n^* = U_n' \cap W_n$, and $V_n = V_n^* \cap Y$. Also suppose that in the $(n+1)$-th round Player I has played x_{n+1} and U_{n+1} in the game G_Y. Then without loss of generality assume $U_{n+1} \subseteq V_n$. Let U_{n+1}' be an open set in X with $U_{n+1}' \cap Y = U_{n+1}$. Again without loss of generality we may choose $U_{n+1}' \subseteq V_n^*$ since $U_{n+1} \subseteq V_n$. Then in the game G_X let Player I play the moves x_n and $U_{n+1}^* = U_{n+1}' \cap W_{n+1}$ in the $(n+1)$-th round. Suppose V_{n+1}^* is Player II's response in the $(n+1)$-th round in the game G_X according to her winning strategy. Then in the game G_Y let Player II's response in the $(n+1)$-th round be $V_{n+1} = V_{n+1}^* \cap Y$. It is then easy to see that the rules of the game are followed, that is, $x_{n+1} \subseteq V_{n+1} \subseteq U_{n+1}$. Moreover, from the play of the auxiliary game G_X we get that $\bigcap_n U_n^* \neq \emptyset$. Thus $\bigcap_n U_n = \bigcap_n (U_n' \cap Y) = \bigcap_n U_n' \cap \bigcap_n W_n = \bigcap_n (U_n' \cap W_n) = \bigcap_n U_n^* \neq \emptyset$. This shows that the strategy we just described is winning for Player II.

(b) The auxiliary game is again G_X. If $y_0, U_0, y_1, U_1, \ldots$ are Player I's moves in G_Y then let $x_0, f^{-1}(U_0), x_1, f^{-1}(U_1), \ldots$ be the moves in the auxiliary game, where x_n is an arbitrary element of X with $f(x_n) = y_n$. If Player II responds with $W_0, W_1, \ldots$ in the auxiliary game according to her winning strategy, then let $V_n = f(W_n)$ be Player II's responses in the game G_Y. We thus have described a winning strategy of Player II in the game G_Y.

(c) For each $i \in I$ let π_i be the projection map from X onto X_i. Then π_i is continuous, open, and surjective. If X is strong Choquet then each X_i is strong Choquet by part (b). Conversely, suppose for each $i \in I$, X_i is strong Choquet. We use a system of auxiliary games G_{X_i}, $i \in I$. Now we describe a winning strategy for Player II in the game G_X. To begin with, let x_0, U_0 be Player I's first move in the game G_X. Let W_0 be a basic open set with $x_0 \in W_0 \subseteq U_0$. Then there are $i_1, \ldots, i_n \in I$ such that $W_0 = \pi_{i_1}^{-1}(U_{i_1}) \cap \cdots \cap \pi_{i_n}^{-1}(U_{i_n})$, where $U_{i_k} \subseteq X_{i_k}$ is open. Then in each of the auxiliary game $\mathsf{G}_{X_{i_k}}$ let $\pi_{i_k}(x_0)$ and $\pi_{i_k}(W_0) = U_{i_k}$ be Player I's first move, and using Player II's winning strategies get responses V_{i_k}. Back in the game G_X we then let Player II's first response to be $V_0 = \pi_{i_1}^{-1}(V_{i_1}) \cap \cdots \cap \pi_{i_n}^{-1}(V_{i_n})$. The general inductive definition of n-th round moves is similar. Note that at the end of the play at most countably many auxiliary games are played, and each of them is won by Player II. This gives a nonempty intersection of the moves made in the game G_X, and therefore Player II has won. $\quad\Box$

Theorem 4.1.3
Let X be a Polish space and $Y \subseteq X$. Then Y is strong Choquet iff Y is Polish.

Proof. By Exercise 4.1.2 if Y is completely metrizable then it is strong Choquet. Conversely, suppose Y is strong Choquet and without loss of generality assume that Y is dense in X. We show that Y is a G_δ subset of X.

We need to make use of the following claim about collections of open sets in X: any collection $\mathcal{U}$ of open sets in X has a point-finite refinement, that is, a collection $\mathcal{U}'$ of open sets so that

(i) $\bigcup \mathcal{U} = \bigcup \mathcal{U}'$;

(ii) for any $W \in \mathcal{U}'$ there is some $U \in \mathcal{U}$ such that $W \subseteq U$; and

(iii) for any $x \in X$ there are at most finitely many elements $W \in \mathcal{U}'$ with $x \in W$.

Let d be a compatible complete metric on X. We may also require that

(iv) for any $\epsilon > 0$, $\text{diam}(W) < \epsilon$ for all $W \in \mathcal{U}'$.

For a proof of the claim, let $\mathcal{U}$ be given. By refining we may assume that $\text{diam}(U) < \epsilon$ for all $U \in \mathcal{U}$. Since X is second countable we may also get a countable subcover of $\mathcal{U}$, and hence we may assume that $\mathcal{U}$ is countable. Let $U_0, U_1, \ldots, U_n, \ldots$ enumerate $\mathcal{U}$. For each $n \in \omega$ let $(U_{n,m})_{m \in \omega}$ be a sequence of open sets such that $U_{n,m} \subseteq U_{n,m+1}$, $\overline{U_{n,m}} \subseteq U_n$, and $\bigcup_m U_{n,m} = U_n$. Then define $U'_n = U_n - \bigcup_{m<n} \overline{U_{m,n}}$. Now $\mathcal{U}' = \{U'_n : n \in \omega\}$ is a refinement of $\mathcal{U}$ with $\bigcup \mathcal{U} = \bigcup \mathcal{U}'$, since for any $x \in X$, if n is the least such that $x \in U_n$, then $x \in U'_n$. To see that $\mathcal{U}'$ is point-finite, suppose $x \in X$ and again n is the least such that $x \in U_n$. Let $m \geq n$ be the least such that $x \in U_{n,m}$. Then for all $k \geq m$, $x \notin U'_k$. This finishes the proof of the claim.

Using the claim we now define a tree T consisting of sequences of the form

$$(U'_0, x_0, V_0, U'_1, x_1, V_1, \ldots, U'_{n-1}, x_{n-1}, V_{n-1}, U'_n)$$

for $n \geq 1$. To define T we fix a winning strategy σ for Player II in the game G_Y. First consider the set S_1 of all sequences of the form (U'_0, x_0, V_0, U'_1) where $U'_0 = X$, V_0 is obtained by applying σ to the play of the game G_Y in which Player I's first move is $x_0, U'_0 \cap Y = Y$, and $U'_1 \cap Y \subseteq V_0$ (thus $U'_1 \cap Y$ is a legitimate partial move of Player I for the next round). Let $\mathcal{U}_1 = \{U'_1 : (X, x_0, V_0, U'_1) \in S_1$ for some $x_0 \in Y\}$ and $W_1 = \bigcup \mathcal{U}_1$. Then it is clear that W_1 is open and $Y \subseteq W_1$. Now by the above claim we can get a point-finite refinement of $\mathcal{U}_1$, which we denote by $\mathcal{U}_1'$, with $\mathrm{diam}(U) < 2^{-1}$ for every $U \in \mathcal{U}_1'$. Now let $T_1 \subseteq S_1$ so that for any $U_1' \in \mathcal{U}_1'$ there is a unique $(X, x_0, V_0, U_1') \in T_1$. In general we consider the set S_n of all sequences of the form $(U_0', x_0, V_0, \dots, U_{n-1}', x_n, V_{n-1}, U_n')$ where $(U_0', x_0, V_0, \dots, U_{n-1}') \in T_{n-1}$, V_n is obtained by applying σ to the play of the game G_Y in which the following moves have been made:

I x_0, Y $x_1, U_1 = U_1' \cap Y$ $x_{n-1}, U_{n-1} = U_{n-1}' \cap Y$

$\cdots\cdots$

II V_0 V_1 V_{n-1}

and finally $U_n' \cap Y \subseteq V_{n-1}$. Let $\mathcal{U}_n = \{U_n' : (X, x_0, V_0, \dots, U_n') \in S_n\}$ and $W_n = \bigcup \mathcal{U}_n$. Still $Y \subseteq W_n$. Then for each $p = (X, x_0, V_0, \dots, U_{n-1}') \in T_{n-1}$, the collection $\mathcal{U}_p = \{U_n' \in \mathcal{U}_n : (X, x_0, V_0, \dots, U_n')$ extends $p\}$ has a point-finite refinement $\mathcal{U}_p'$ with elements of diameter $< 2^{-n}$. We let $\mathcal{U}_n' = \bigcup_{p \in T_{n-1}} \mathcal{U}_p'$. Then in particular $\mathcal{U}_n'$ is a point-finite refinement of $\mathcal{U}_n$. We then define $T_n \subseteq S_n$ so that for any $U_n' \in \mathcal{U}_n'$ there is a unique $(X, x_0, V_0, \dots, U_n') \in T_n$. Finally let $T = \bigcup_n T_n$.

To finish the proof of the theorem we check that $Y = \bigcap_n W_n$. We have seen that $Y \subseteq \bigcap_n W_n$. It suffices to check that $\bigcap_n W_n \subseteq Y$. For this suppose $x \in \bigcap_n W_n$. Thus for every $n \in \omega$ there is $(X, x_0, V_0, \dots, U_n') \in T$ with $x \in U_n'$. Consider $T_x = \{(X, x_0, V_0, \dots, U_n') \in T : x \in U_n'\}$. By the construction of T the subtree T_x is finite-splitting, and therefore by König's lemma there is an infinite sequence $(X, x_0, V_0, \dots, U_n', x_n, V_n, \dots)$ such that $x \in \bigcap_n U_n'$ and the sequence $(x_0, Y, V_0, \dots, x_n, U_n = U_n' \cap Y, V_n, \dots)$ is a play in the game G_Y obtained by applying σ. Since σ is a winning strategy for Player II, $\bigcap_n U_n \neq \emptyset$. However, it follows that $\bigcap_n U_n' \neq \emptyset$; but since $\mathrm{diam}(U_n') < 2^{-n}$, $\bigcap_n U_n'$ contains a unique element, and thus $\bigcap_n U_n = \bigcap_n U_n' = \{x\}$. This shows that $x \in Y$. $\quad\square$

Theorem 4.1.4 (Choquet)
A topological space is Polish iff it is second countable, T_3, and strong Choquet.

Proof. Suppose X is second countable, T_3, and strong Choquet. By the Urysohn metrization theorem X is metrizable. Let d be a compatible metric on X. Let $(\hat{X}, \hat{d})$ be the completion of (X, d). X is now a strong Choquet subspace of the Polish space $\hat{X}$, and by the above theorem X is Polish. $\quad\square$

Recall from Section 1.8 that the Gandy–Harrington topology on ω^ω is generated by Σ_1^1 subsets. This turns out to be an important example of strong Choquet space.

Theorem 4.1.5
The space ω^ω with the Gandy–Harrington topology is strong Choquet.

Proof. Recall from Exercise 1.6.1 that every Σ_1^1 set can be represented as $p[T]$ for some computable tree T on $\omega \times \omega$. We describe a winning strategy for Player II in the strong Choquet game on ω^ω.

Let Player I start by playing x_0 and U_0. Let $A_0 \in \Sigma_1^1$ be such that $x_0 \in A_0 \subseteq U_0$ and let T_0 be a computable tree on $\omega \times \omega$ with $A_0 = p[T_0]$. Since $x_0 \in p[T_0]$ there is y_0 such that $(x_0, y_0) \in [T_0]$. To define Player II's response we use the following notation. For any tree T on $\omega \times \omega$ and $(s, t) \in T$, let $T_{(s,t)} = \{(s', t') \in T : (s' \subseteq s \text{ or } s' \supseteq s) \text{ and } (t' \subseteq s \text{ or } t' \supseteq t)\}$. Then $T_{(s,t)}$ is a subtree of T and if T is computable so is $T_{(s,t)}$. Let $s_0 = x_0 \upharpoonright 1$ and $t_0^0 = y_0 \upharpoonright 1$. We let Player II play $V_0 = p[(T_0)_{(s_0, t_0^0)}]$. It is easy to see that $x_0 \in V_0 \subseteq A_0 \subseteq U_0$, and thus the rules are obeyed.

Let Player I next play x_1 and U_1, with $x_1 \in U_1 \subseteq V_0$. Then since $x_1 \in V_0$ there is y_0' with $(x_1, y_0') \in [(T_0)_{(s_0, t_0^0)}]$. Let $s_1 = x_1 \upharpoonright 2$ and $t_1^0 = y_0' \upharpoonright 2$. Then $s_0 \subseteq s_1$ and $t_0^0 \subseteq t_1^0$. Also let $A_1 \in \Sigma_1^1$ with $x_1 \in A_1 \subseteq U_1$, T_1 be a computable tree on $\omega \times \omega$ with $A_1 = p[T_1]$, and y_1 be such that $(x_1, y_1) \in [T_1]$. Note that $s_0 \subseteq x_1$. So we just let $t_0^1 = y_1 \upharpoonright 1$, and we get $x_1 \in p[(T_1)_{(s_0, t_0^1)}]$. Thus if we let Player II play $V_1 = p[(T_0)_{(s_1, t_1^0)}] \cap p[(T_1)_{(s_0, t_0^1)}]$, then we have that $x_1 \in V_1 \subseteq A_1 \subseteq U_1$.

Proceeding this way, when Player I plays $x_0, U_0, x_1, U_1, \dots$ Player II responds with $V_0, V_1, \dots$ with $x_n \in V_n \subseteq U_n$. In the process Player II also defines computable trees T_n on $\omega \times \omega$ and sequences $s_0 \subseteq s_1 \subseteq \dots, t_0^n \subseteq t_1^n \subseteq \dots$ with the following properties:

(i) $x_n \in A_n = p[T_n] \subseteq U_n$;

(ii) $\mathrm{lh}(s_n) = n + 1$ and $s_n \subseteq x_n$;

(iii) $(s_k, t_k^n) \in T_n$.

Moreover, $V_n = p[(T_0)_{(s_n, t_n^0)}] \cap p[(T_1)_{(s_{n-1}, t_{n-1}^1)}] \cap \dots \cap p[(T_n)_{(s_0, t_0^n)}]$. Now let $x = \bigcup_n s_n$. We check that $x \in \bigcap_n A_n \subseteq_n V_n$. For each n, let $y_n = \bigcup_k t_k^n$. Then $(s_k, t_k^n) \in T_n$ and hence $(x, y_n) \in p[T_n] = A_n$ as required. □

Theorem 4.1.5 was used in the proof of Theorem 1.8.5 (see the proof of Theorem A.2.2).

Exercise 4.1.1 Let X be a topological space and $\mathcal{B}$ be a base for the topology of X. Let $\mathsf{G}_X(\mathcal{B})$ be the game G_X in which Players I and II must play elements of $\mathcal{B}$ for the open sets. Show that X is strong Choquet iff Player II has a winning strategy in $\mathsf{G}_X(\mathcal{B})$.

Exercise 4.1.2 Show that any completely metrizable space is strong Choquet.

Exercise 4.1.3 Show that any locally compact Hausdorff space is strong Choquet.

Exercise 4.1.4 Show that any strong Choquet space is Baire.

Exercise 4.1.5 Use Choquet's Theorem 4.1.4 to give another proof of Sierpinski's Theorem 2.2.9.

Exercise 4.1.6 Let X be a Polish space and $Y \subseteq X$. Show that if Y is meager then Player I has a winning strategy in the game G_Y.

4.2   Change of topology

The technique of change of topology is very useful in the study of equivalence relations. We shall employ the technique to turn Borel actions into continuous actions and equivalence relations with certain characteristics into canonical equivalence relations.

Lemma 4.2.1
Let X be a space, τ a Polish topology on X, and $F \subseteq X$ a closed subset. Let σ be the topology generated by $\tau \cup \{F\}$. Then σ is a Polish topology on X.

Proof. Since both F and $X - F$ are Polish subspaces of X, there are compatible complete metrics $d_F < 1$ and $d_{X-F} < 1$ on F and $X - F$ respectively. It is easy to see that the topology σ is compatible with the following complete metric d defined on X:

$$
d(x, y) = \begin{cases} d_F(x, y), & \text{if } x, y \in F, \\ d_{X-F}(x, y), & \text{if } x, y \in X - F, \\ 1, & \text{otherwise.} \end{cases}
$$

The topology σ is obviously second countable, hence separable.   ☐

Lemma 4.2.2
Let X be a space and $(\tau_n)_{n \in \omega}$ be a sequence of Polish topologies on X with $\bigcap_n \tau_n$ Hausdorff. Let τ be the topology generated by $\bigcup_n \tau_n$. Then τ is a Polish topology on X.

Proof. Let $X_n = (X, \tau_n)$ and $Y = \prod_n X_n$ be the product space. Consider the diagonal $D \subseteq Y$ defined by

$$
D = \{(x_n) \in Y : x_n = x_{n+1} \text{ for all } n \in \omega\}.
$$

Then D is closed in Y since $\bigcap_n \tau_n$ is Hausdorff. Therefore D is Polish. Consider the embedding $e : X \to Y$ given by $e(x)_n = x$ for all $n \in \omega$. Then e

is in fact a homeomorphism from (X, τ) onto the subspace D, and so (X, τ) is Polish. In fact e is obviously a bijection. The continuity of e follows from the fact that $\tau_n \subseteq \tau$ for all $n \in \omega$. To see that e^{-1} is open let U be subbasic open in (X, τ). Thus $U \in \bigcup_n \tau_n$ and $U \in \tau_{n_0}$ for some $n_0 \in \omega$. Now for any $(x_n) \in D$, $(x_n) \in e(U)$ iff $x_{n_0} \in U$. Therefore $e(U) \cap D$ is open in D. □

Theorem 4.2.3
Let X be a Polish space, $1 \le \alpha < \omega_1$, and $\mathcal{A}$ a countable collection of Σ^0_α subsets of X. Then there is a countable collection $\mathcal{B}$ of Σ^0_α sets such that $\mathcal{A} \subseteq \mathcal{B}$ and the topology generated by $\mathcal{B}$ is Polish and is finer than the original topology on X.

Proof. We prove it by a transfinite induction on $1 \le \alpha < \omega_1$. The base case $\alpha = 1$ is trivial. For the successor case, assume the theorem is true for $\alpha \ge 1$ and consider a countable collection $\mathcal{A} = \{A_0, A_1, \dots\}$ of $\Sigma^0_{\alpha+1}$ sets. For each $n \in \omega$, let $A_n = \bigcup_{m \in \omega} B_{n,m}$ where each $B_{n,m}$ is Π^0_α. Then the collection $\mathcal{C} = \{X - B_{n,m} : n, m \in \omega\}$ is a countable collection of Σ^0_α sets, and by the inductive hypothesis there is a countable collection $\mathcal{D} \supseteq \mathcal{C}$ of Σ^0_α sets so that the topology generated by $\mathcal{D}$ is Polish and is finer than the original topology on X. Now consider X with this new Polish topology σ generated by $\mathcal{D}$. Each $B_{n,m}$ is σ-closed, thus by Lemma 4.2.1 the topology $\tau_{n,m}$ generated by $\sigma \cup \{B_{n,m}\}$ is Polish, whereas $\bigcap_{n,m} \tau_{n,m} \supseteq \sigma$ is Hausdorff. Thus by Lemma 4.2.2 the topology τ generated by $\sigma \cup \{B_{n,m} : n, m \in \omega\}$ is again Polish, and in addition each A_n is open in τ. Now if we let $\mathcal{B} = \mathcal{B}_0 \cup \{B_{n,m} : n, m \in \omega\} \cup \mathcal{A}$ where $\mathcal{B}_0$ is a countable base for the topology σ, then $\mathcal{B}$ generates the topology τ, as required.

Suppose now α is a limit ordinal and $\mathcal{A} = \{A_0, A_1, \dots\}$ is a countable collection of Σ^0_α sets. Again assume $A_n = \bigcup_{m \in \omega} B_{n,m}$ where each $B_{n,m}$ is in $\Pi^0_{\beta(n,m)}$ for some $\beta(n, m) < \alpha$. Let $\mathcal{C} = \{B_{n,m} : n, m \in \omega\}$. Fix a cofinal sequence in α, $\gamma_0 < \gamma_1 < \cdots < \alpha$, and for each $k \in \omega$ let $\mathcal{C}_k = \{B \in \mathcal{C} : B \in \Pi^0_{\gamma_k}\}$. Then each $\mathcal{C}_k$ is a countable collection of $\Sigma^0_{\gamma_k+1}$ sets. Since $\gamma_k + 1 < \alpha$ for all $k \in \omega$, by the inductive hypothesis there is $\mathcal{D}_k \supseteq \mathcal{C}_k$ so that the topology generated by $\mathcal{D}_k$ is Polish and is finer than the original topology on X. By Lemma 4.2.2 the topology τ generated by $\bigcup_k \mathcal{D}_k$ is Polish. It is clear that each A_n is open in τ. Thus if we let $\mathcal{B} = \bigcup_k \mathcal{D}_k \cup \mathcal{A}$ then $\mathcal{B}$ generates τ as well. □

Corollary 4.2.4
Let X be a Polish space with topology τ and $\mathcal{A}$ a countable collection of Borel sets of X. Then there is a Polish topology σ finer than τ such that all elements of $\mathcal{A}$ are σ-open.

In fact the property of being contained in a finer Polish topology characterizes Borelness (see Exercise 4.2.1). Next we show a nice analog for analytic sets.

Theorem 4.2.5

Let X be a Polish space with topology τ and $\mathcal{A}$ a countable collection of $\boldsymbol{\Sigma}_1^1$ sets of X. Then there is a second countable strong Choquet topology σ on X finer than τ such that all elements of $\mathcal{A}$ are σ-open.

Proof. Without loss of generality we may assume X is uncountable and thus is Borel isomorphic to ω^ω. It then suffices to prove the theorem for ω^ω. In fact, let $\varphi : (X, \tau) \to \omega^\omega$ be a Borel isomorphism and $\mathcal{B}$ be a countable base for τ. Then $\{\varphi(A) : A \in \mathcal{A} \cup \mathcal{B}\}$ is a countable collection of $\boldsymbol{\Sigma}_1^1$ sets of ω^ω. If σ' is a second countable strong Choquet topology on ω^ω finer than the usual topology such that all elements of $\{\varphi(A) : A \in \mathcal{A} \cup \mathcal{B}\}$ are σ'-open, then the topology σ generated by $\{\varphi^{-1}(B) : B \in \sigma'\}$ on X is a second countable strong Choquet topology finer than τ such that all elements of $\mathcal{A}$ are σ-open.

Thus we assume $X = \omega^\omega$. Then there is some $x \in X$ such that $\mathcal{A} \subseteq \boldsymbol{\Sigma}_1^1(x)$. Now the relativized Gandy–Harrington topology σ generated by all $\boldsymbol{\Sigma}_1^1(x)$ sets is second countable and strong Choquet by Theorem 4.1.5. Thus σ is as required. $\qquad\square$

Theorem 4.2.6 (Becker)

Let τ be a Polish topology on X and σ a second countable strong Choquet topology finer than τ. Then every σ-open set is $\boldsymbol{\Sigma}_1^1$ in τ.

Proof. Let $A \in \sigma$. We show that A is Suslin in τ. Since A is open in the strong Choquet space (X, σ), the subspace $(A, \sigma \restriction A)$ is a strong Choquet space. We fix a winning strategy for Player II in the game G_A. Similar to the proof of Theorem 4.1.3 we can construct a tree T of sequences of the form $(U_0, x_0, V_0, U_1, x_1, V_1, \ldots, U_n, x_n, V_n, U_{n+1})$ where $x_0, U_1, \ldots, x_n, U_n$ are Player I's moves in G_A, $V_0, V_1, \ldots, V_n$ are Player II's responses according to her winning strategy, and U_{n+1} is a partial move for Player I in the $(n+1)$-th round. Then we define a Suslin system $(P_s)_{s \in \omega^{<\omega}}$ as follows. Let $P_\emptyset = \overline{A}^\tau$. To define P_s for $\mathrm{lh}(s) = 1$ consider elements $(U_0, x_0, V_0, U_1) \in T$ where $U_0 = A$ and $\mathrm{diam}(U_1) < 2^{-1}$. Note that such elements exist since σ is finer than τ. In fact in the topology σ the collection of all possible U_1 components from such elements form an open cover of A (since $x_0 \in A$ can be arbitrarily chosen, the collection of V_0 components form an open cover). Now that σ is second countable, there exists a countable subcover $\mathcal{U}_1$ consisting of such U_1 components. We then let the collection $\{P_s : \mathrm{lh}(s) = 1\}$ be an enumeration of $\{\overline{U_1}^\tau : U_1 \in \mathcal{U}_1\}$. In general if P_s has been defined for each s with $\mathrm{lh}(s) = n$ and is correspondent to some element $(U_0, x_0, V_0, U_1, \ldots, U_n) \in T$ where $U_0 = A$ and $\mathrm{diam}(U_n) < 2^{-n}$, the collection $\{P_{s^\frown k} : k \in \omega\}$ is an enumeration of the τ-closures of some elements U_{n+1} with $\mathrm{diam}(U_{n+1}) < 2^{n+1}$ where the collection of all of them is a σ-open cover of U_n and $(U_0, x_0, V_0, U_1, \ldots, U_n, x_n, V_n, U_{n+1}) \in T$ for some x_n and V_n.

We claim that A is obtained from $(P_s)_{s \in \omega^{<\omega}}$ by the Suslin operation, that is, $A = \bigcup_z \bigcap_n P_{z \restriction n}$. To see first $A \subseteq \bigcup_z \bigcap_n P_{z \restriction n}$, let $x \in A$. By our construction one can find $U_1 \supseteq U_2 \supseteq \ldots$ and $s_1 \subseteq s_2 \subseteq \ldots$ with $\mathrm{lh}(s_n) = n$ such that $x \in U_n \subseteq P_{s_n}$. Let $z = \bigcup_n s_n$. Then $x \in \bigcap_n P_{z \restriction n}$. For the converse suppose $x \in \bigcap_n P_{z \restriction n}$ for some $z \in \omega^\omega$. Let $s_n = z \restriction n$ and U_n from the construction such that $P_{s_n} = \overline{U_n}^\tau$. Since the construction followed a winning strategy of Player II it follows that $\bigcap_n U_n \neq \emptyset$. On the other hand since $\mathrm{diam}(U_n) < 2^{-n}$ it follows that $\mathrm{diam}(P_{s_n}) \leq 2^{-n}$, and in particular $\bigcap_n U_n \subseteq \bigcap_n P_{s_n}$ can have at most one element. This implies that $x \in \bigcap_n U_n$ and thus $x \in A$. ☐

Corollary 4.2.7
Let X be a Polish space with topology τ and $A \subseteq X$. Then A is Σ^1_1 iff there is a second countable strong Choquet topology σ on X finer than τ such that A is σ-open.

Exercise 4.2.1 Let X be a Polish space with topology τ and $A \subseteq X$. Show that A is Borel iff there is a Polish topology σ on X finer than τ such that A is σ-open.

Exercise 4.2.2 Show that the topology on ω^ω generated by all Δ^1_1 sets is Polish.

Exercise 4.2.3 Let τ be a Polish topology on X and σ a finer second countable strong Choquet topology. Show that there is a second countable strong Choquet topology σ' finer than σ and a dense open subset Y of X with respect to σ' such that $(Y, \sigma' \restriction Y)$ is Polish. (*Hint*: Use Theorem 1.8.5.)

Exercise 4.2.4 Prove the following uniform version of Theorem 4.2.3: Let X, Y be Polish spaces, $1 \leq \alpha < \omega_1$, and $A \subseteq X \times Y$ be Σ^0_α. Then there is a Σ^0_α subset B of $X \times Y \times \omega$ such that $A(x, y) \iff B(x, y, 0)$, and for every $y \in Y$, a Polish topology on X finer than the original topology is generated by the countable collection $\{B_{y,n} : n \in \omega\}$, where $B_{y,n} = \{x \in X : (x, y, n) \in B\}$.

4.3   Finer topologies on Polish G-spaces

In this section we extend the technique of change of topology to Polish G-spaces. We would like to maintain that the action is still continuous with respect to the finer topology. However, there are simple obstacles for a full generalization of the results in the preceding section.

Consider a nontrivial continuous action of a connected Polish group G on a Polish space X. By the continuity of the action every G-orbit is connected. Let $x \in X$ with more than one element in $[x]_G$. Then there is no finer Polish topology on X so that $\{x\}$ is open and the action is still continuous, since otherwise $[x]_G$ would be discrete.

It turns out that the modification needed in this context is to consider the Vaught transforms.

Lemma 4.3.1
Let G be a Polish group, (X, τ) a Polish G-space, and $F \subseteq X$ closed. Then there is a Polish topology $\sigma \subseteq \Sigma^0_2(X, \tau)$ finer than τ such that the action is continuous with respect to σ and $F^{\triangle W}$ is σ-open for all nonempty open $W \subseteq G$.

Proof. We are going to make use of a special Polish G-space $L(G, d)$ introduced in Exercise 3.1.9. Let $d \le 1$ be a compatible right-invariant metric on G. Define, for each $x \in X$, $f_x \in L(G, d)$ by

$$f_x(g) = d(g, \{h \in G : h \cdot x \notin F\}) = \inf\{d(g, h) : h \cdot x \notin F\}.$$

We claim that $x \mapsto f_x$ is a G-map. In fact, let $k \in G$ and $x \in X$, we verify that $k \cdot f_x = f_{k \cdot x}$ by the following computation using the right-invariance of d:

$$
\begin{aligned}
(k \cdot f_x)(g) = f_x(gk) &= \inf\{d(gk, h) : h \cdot x \notin F\} \\
&= \inf\{d(gk, h'k) : h'k \cdot x \notin F\} \\
&= \inf\{d(g, h') : h' \cdot (k \cdot x) \notin F\} = f_{k \cdot x}(g).
\end{aligned}
$$

Let G act on $X \times L(G, d)$ diagonally: $g \cdot (x, f) = (g \cdot x, g \cdot f)$. Let

$$\varphi : X \to X \times L(G, d)$$

be given by $\varphi(x) = (x, f_x)$. It is clear that φ is a G-embedding.

We next claim that $\varphi(X)$ is a G_δ subset of $X \times L(G, d)$. To see this let G_0 be a countable dense subset of G. Then for $(x, f) \in X \times L(G, d)$,

$$
\begin{aligned}
(x, f) \in \varphi(X) &\iff f = f_x \iff \forall g \in G_0 \; f(g) = f_x(g) \\
&\iff \forall q \in \mathbb{Q} \; \forall g \in G_0 \; (f(g) < q \leftrightarrow f_x(g) < q) \\
&\iff \forall q \in \mathbb{Q} \; \forall g \in G_0 \; (f(g) < q \leftrightarrow \\
&\qquad\qquad \exists h \in G_0 \; (d(g, h) < q \wedge h \cdot x \notin F)).
\end{aligned}
$$

The computation shows that for fixed q and g, the sets $\{x \in X : f_x(g) < q\}$ and $\{f \in L(G, d) : f(g) < q\}$ are both open. It follows that $\varphi(X)$ is a countable intersection of Boolean combinations of open sets in $X \times L(G, d)$, and therefore is G_δ in $X \times L(G, d)$.

Now $\varphi(X)$ is an invariant G_δ subset of $X \times L(G, d)$, hence it is a Polish G-space with the diagonal action. Let σ be the topology on X pulled back from $\varphi(X)$ along φ, that is, $A \subseteq X$ is σ-open iff $\varphi(A)$ is open in $X \times L(G, d)$. It is clear that (X, σ) is a Polish G-space since it is G-homeomorphic to the Polish G-space $\varphi(X)$. We claim that σ has the other required properties.

First note that φ is an open mapping, since $\varphi^{-1} : \varphi(X) \to X$ is just the projection of the product space onto its first factor, and is thus continuous. It follows that σ is a Polish topology finer than τ. Next we check that $\sigma \subseteq \Sigma_2^0(X, \tau)$. This is equivalent to checking that φ is a Baire class 1 function from (X, τ) into $X \times L(G, d)$. Again let G_0 be a countable dense subset of G. Then a subbase for $X \times L(G, d)$ consists of sets of the form $O \times \{f \in L(G, d) : f(g) < q\}$ or $O \times \{f \in L(G, d) : f(g) > q\}$ for $O \subseteq X$ open, $g \in G_0$, and $q \in \mathbb{Q}$. Now

$$\varphi(x) \in O \times \{f \in L(G, d) : f(g) < q\}$$
$$\iff x \in O \wedge f_x(g) < q$$
$$\iff x \in O \wedge \exists h \in G_0 \ (d(g, h) < q \wedge h \cdot x \notin F),$$

implying that $\varphi^{-1}(O \times \{f \in L(G, d) : f(g) < q\})$ is open, and

$$\varphi(x) \in O \times \{f \in L(G, d) : f(g) > q\}$$
$$\iff x \in O \wedge \exists q' \in \mathbb{Q}_+ \ \forall h \in G_0 \ (d(g, h) < q + q' \to h \cdot x \in F),$$

implying that $\varphi^{-1}(O \times \{f \in L(G, d) : f(g) > q\})$ is Σ_2^0.

Finally we check that $F^{\triangle W}$ is σ-open for any $W \subseteq G$ nonempty open. For this consider

$$A = \{(x, f_x) : f_x(1_G) > 0\}.$$

A is open in $\varphi(X)$, and also

$$\varphi(x) \in A^{\triangle W} \iff \exists^* g \in W \ (g \cdot f_x)(1_G) = f_x(g) > 0$$
$$\iff \exists^* g \in W \ g \cdot x \in F \iff x \in F^{\triangle W}.$$

In the computation we used the fact that $f_x(g) > 0 \Rightarrow g \cdot x \in F$ and the following version of the converse. If $U \subseteq W$ is an open set such that for densely many $g \in U$, $g \cdot x \in F$, then by continuity of the action we have $U \cdot x \subseteq F$ and thus for all $g \in U$, $f_x(g) > 0$. □

The next lemma is an analog of Lemma 4.2.2 with a similar proof. The proof is left as an exercise (Exercise 4.3.4).

Lemma 4.3.2
Let G be a Polish group, X a G-space, and $(\tau_n)_{n \in \omega}$ a sequence of Polish topologies on X with $\bigcap_n \tau_n$ Hausdorff. Suppose that the action is continuous with respect to τ_n for all $n \in \omega$. Let τ be the topology generated by $\bigcup_n \tau_n$. Then (X, τ) is a Polish G-space.

We use the following notation. Let G be a topological group and X a G-space. For a collection $\mathcal{A}$ of subsets of X and a collection $\mathcal{U}$ of nonempty open subsets of G, let

$$\mathcal{A}^{\triangle \mathcal{U}} = \{A^{\triangle U} : A \in \mathcal{A}, \ U \in \mathcal{U}\}.$$

Theorem 4.3.3 (Hjorth)

Let G be a Polish group, (X, τ) a Polish G-space, $\mathcal{U} \subseteq G$ a countable collection of nonempty open subsets of G, $1 \leq \alpha < \omega_1$, and $\mathcal{A}$ a countable collection of $\mathbf{\Sigma}^0_\alpha$ subsets of (X, τ). Then there is a Polish topology $\sigma \subseteq \mathbf{\Sigma}^0_\alpha(X, \tau)$ finer than τ such that (X, σ) is a Polish G-space and $\mathcal{A}^{\triangle \mathcal{U}} \subseteq \sigma$.

Proof. Without loss of generality we may assume that $\mathcal{U}$ is a countable base for the topology of G. We prove the theorem by a transfinite induction on α. Note that there is nothing to prove when $\alpha = 1$. For the successor case, suppose $\alpha = \beta + 1$ with $\beta \geq 1$. Enumerate $\mathcal{A}$ as $\{A_i\}_{i \in \omega}$ and $\mathcal{U}$ as $\{U_j\}_{j \in \omega}$. Also let $\{B_{i,k}\}_{i,k \in \omega}$ be $\mathbf{\Pi}^0_\beta$ sets with $A_i = \bigcup_k B_{i,k}$ for all $i \in \omega$. Let $\mathcal{C} = \{X - B_{i,k} : i, k \in \omega\}$. Then $\mathcal{C} \subseteq \mathbf{\Sigma}^0_\beta(X, \tau)$. By the inductive hypothesis there is a Polish topology $\tau' \subseteq \mathbf{\Sigma}^0_\beta(X, \tau)$ finer than τ such that (X, τ') is a Polish G-space and $\mathcal{C}^{\triangle \mathcal{U}} \subseteq \tau'$. Each $B_{i,k}$ is τ'-closed, hence by Lemma 4.3.1 for each $i, k \in \omega$ there is a Polish topology $\tau_{i,k} \subseteq \mathbf{\Sigma}^0_2(X, \tau')$ finer than τ' such that $(X, \tau_{i,k})$ is a Polish G-space and $B^{\triangle W}_{i,k}$ is $\tau_{i,k}$-open for all nonempty open $W \subseteq G$. Let σ be the topology generated by $\bigcup_{i,k} \tau_{i,k}$. Then by Lemma 4.3.2 (X, σ) is a Polish G-space. It is clear that σ consists of $\mathbf{\Sigma}^0_\alpha(X, \tau)$ sets since each $\tau_{i,k}$ consists of such sets. To see that $\mathcal{A}^{\triangle \mathcal{U}} \subseteq \sigma$, note that

$$A_i^{\triangle U_j} = \left(\bigcup_k B_{i,k} \right)^{\triangle U_j} = \bigcup_k B^{\triangle U_j}_{i,k},$$

which is σ-open.

Finally consider the case α is a limit. Again let $\mathcal{A} = \{A_i\}_{i \in \omega}$, $\mathcal{U} = \{U_j\}_{j \in \omega}$ and for each $i \in \omega$, $A_i = \bigcup_k B_{i,k}$ with $B_{i,k} \in \mathbf{\Pi}^0_{\beta(i,k)}$ with $\beta(i, k) < \alpha$. Then each $B_{i,k} \in \mathbf{\Sigma}^0_{\beta(i,k)+1}$ and $\beta(i, k) + 1 < \alpha$. By the inductive hypothesis for each $i, k \in \omega$ there is a Polish topology $\tau_{i,k} \subseteq \mathbf{\Sigma}^0_{\beta(i,k)+1}$ finer than τ such that $(X, \tau_{i,k})$ is a Polish G-space and $B^{\triangle U_j}_{i,k}$ is $\tau_{i,k}$-open for all $j \in \omega$. Let σ be the topology generated by $\bigcup_{i,k} \tau_{i,k}$. Then by Lemma 4.3.2 (X, σ) is a Polish G-space. Again $A_i^{\triangle U_j} = \bigcup_k B^{\triangle U_j}_{i,k}$ is σ-open. □

Corollary 4.3.4 (Becker–Kechris)

Let G be a Polish group, (X, τ) a Polish G-space, and $A \subseteq X$ invariant Borel. Then there is a Polish topology σ finer than τ such that (X, σ) is a Polish G-space and A is σ-open.

Proof. This follows immediately from the preceding theorem since $A = A^{\triangle G}$. □

The Becker–Kechris theorem was first proved by a different method which yielded a slightly weaker result than Theorem 4.3.3. In the next section we give this method and use it to prove another result on topological realization of Borel G-spaces.

Exercise 4.3.1 Let G be a Polish group with a compatible right invariant metric d and X a Polish G-space. For each $x \in X$ and open set $O \subseteq X$ define

$$f_x^O(g) = d(g, \{h \in G : h \cdot x \in O\}).$$

Show that if $\mathcal{O} = \{O_i\}_{i \in \omega}$ is a countable base for X then $x \mapsto (f_x^{O_i})_{i \in \omega}$ is a G-embedding from X into $L(G, d)^\omega$.

Exercise 4.3.2 Show that the G-embedding in Exercise 4.3.1 is of Baire class 1.

Exercise 4.3.3 In Exercise 4.3.1 let σ be the weak topology on X given by the G-embedding into $L(G, d)^\omega$. Show that (X, σ) is a Polish G-space.

Exercise 4.3.4 Prove Lemma 4.3.2.

Exercise 4.3.5 Prove the following uniform version of Corollary 4.3.4: Let G be a Polish group, X a Polish G-space, Y a Polish space, $1 < \alpha < \omega_1$, and $A \subseteq X \times Y$ a Σ_α^0 subset such that for all $y \in Y$, $A_y = \{x \in X : (x, y) \in A\}$ is invariant. Then there is a Borel set $B \subseteq X \times Y \times \omega$ such that $(x, y) \in A \iff (x, y, 0) \in B$, and for all $y \in Y$, the topology σ_y generated by the countable collection $\{B_{y,n} : n \in \omega\}$, where $B_{y,n} = \{x \in X : (x, y, n) \in B\}$, is finer than the original topology on X, and such that (X, σ_y) is a Polish G-space. (*Hint*: See Exercise 4.2.4.)

4.4   Topological realization of Borel G-spaces

In this section we give the method used by Becker and Kechris in Reference [8] to generate finer topologies on Polish G-spaces. A benefit of the method is the useful result that any Borel G-space can be turned into a Polish G-space. The method is based on a very simple construction. Let G be a Polish group, $\mathcal{U}$ a base for the topology of G, X a Borel G-space, and $\mathcal{A}$ a collection of analytic subsets of X. We will show that, under suitable assumptions, the topology generated by $\mathcal{A}^{\triangle\mathcal{U}}$ already has the desired nice properties.

We prove these properties in a series of lemmas below. For brevity, in these lemmas we make the same assumptions about $G, \mathcal{U}, X$, and $\mathcal{A}$ as above unless otherwise specified.

Lemma 4.4.1

Let $g \in G$, $x \in X$, $A \in \Sigma_1^1(X)$, and $W \subseteq G$ nonempty open. Suppose $g \cdot x \in A^{\triangle W}$. Then there are nonempty $U, V \in \mathcal{U}$ such that $x \in A^{\triangle U}$, $g \in V$, and $UV^{-1} \subseteq W$.

Proof. It follows from $g \cdot x \in A^{\triangle W}$ that $x \in A^{\triangle(Wg)}$. Thus the set $H = \{h \in Wg : h \cdot x \in A\}$ is nonmeager. Since H is Σ_1^1 in G, it has the Baire property. Thus for some $U_0 \in \mathcal{U}$ with $U_0 \subseteq Wg$, H is comeager in U_0; that is, $x \in A^{*U_0}$. Now $U_0 g^{-1} \subseteq W$, so there are $U, V \in \mathcal{U}$ with $U \subseteq U_0$, $g \in V$ and $UV^{-1} \subseteq W$. Then $x \in A^{*U_0} \subseteq A^{*U} \subseteq A^{\triangle U}$. $\quad\square$

Lemma 4.4.2

The action of G on X is continuous with respect to the topology on X generated by $\mathcal{A}^{\triangle \mathcal{U}}$.

Proof. Let τ be the topology on X generated by $\mathcal{A}^{\triangle \mathcal{U}}$. Fix $g_0 \in G$, $x_0 \in X$, $A \in \mathcal{A}$, $W \in \mathcal{U}$ with $g_0 \cdot x_0 \in A^{\triangle W}$. By Lemma 4.4.1 there are $U, V \in \mathcal{U}$ such that $x_0 \in A^{\triangle U}$, $g_0 \in V$, and $UV^{-1} \subseteq W$. Let $N = A^{\triangle U}$. Note that $N \in \mathcal{A}^{\triangle \mathcal{U}}$ and thus is τ-open. We check that for all $g \in V$ and $x \in N$, $g \cdot x \in A^{\triangle W}$. Since $UV^{-1} \subseteq W$, $Ug^{-1} \subseteq W$. Thus $x \in N = A^{\triangle U} \subseteq A^{\triangle(Wg)}$. Hence $g \cdot x \in A^{\triangle W}$. $\quad\square$

Lemma 4.4.3

Let $\mathcal{B}$ be a base for a Polish topology on X generating the Borel structure of X and $\mathcal{B} \subseteq \mathcal{A}$. Then the topology generated by $\mathcal{A}^{\triangle \mathcal{U}}$ is T_1.

Proof. Let $x, y \in X$ with $x \neq y$. It suffices to find $B \in \mathcal{B}$ and $U \in \mathcal{U}$ such that $x \in B^{\triangle U} \in \mathcal{A}^{\triangle \mathcal{U}}$ and $y \notin B^{\triangle U}$. Let τ be the topology on X generated by $\mathcal{B}$. Consider the function $f : G \to X \times X$ defined by $f(g) = (g \cdot x, g \cdot y)$. Since X is a Borel G-space and τ generates the Borel structure on X, f is a Borel function from G into $(X, \tau) \times (X, \tau)$. Thus there is a dense G_δ subset H of G such that $f \restriction H$ is continuous. Let $g_0 \in H$. Since $g_0 \cdot x \neq g_0 \cdot y$, there are $B, C \in \mathcal{B}$ with $B \cap C = \emptyset$ such that $(g_0 \cdot x, g_0 \cdot y) \in B \times C$. By the continuity of $f \restriction H$, there is an open set $U \subseteq G$ with $g_0 \in U \cap H$ such that $f(U \cap H) \subseteq B \times C$. Since $U \cap H$ is comeager in U, we have that $x \in B^{*U}$ and $y \in C^{*U}$. Thus in particular $x \in B^{\triangle U}$ and $y \notin B^{\triangle U}$ since $B^{*U} \subseteq B^{\triangle U}$ and $B^{\triangle U} \subseteq (X - C)^{\triangle U} = X - C^{*U}$. $\quad\square$

Lemma 4.4.4

Suppose that for any $A \in \mathcal{A}$ and $U \in \mathcal{U}$, $X - A^{\triangle U} \in \mathcal{A}$. Then the topology generated by $\mathcal{A}^{\triangle \mathcal{U}}$ is regular.

Proof. Let $x \in A^{\triangle W}$ for $A \in \mathcal{A}$ and $W \in \mathcal{U}$. It suffices to find $U \in \mathcal{U}$ such that $x \in A^{\triangle U} \subseteq \overline{A^{\triangle U}} \subseteq A^{\triangle W}$. Since $1_G \cdot x = x$, by Lemma 4.4.1 we obtain nonempty $U_1, V_1 \in \mathcal{U}$ such that $x \in A^{\triangle U_1}$, $1_G \in V_1$, and $U_1 V_1^{-1} \subseteq W$. Now by Lemma 4.4.1 again we obtain $U_2, V_2 \in \mathcal{U}$ such that $x \in A^{\triangle U}$, $1_G \in V_2$, and $U_2 V_2^{-1} \subseteq U_1$. Let $U = U_2$ and $N = A^{\triangle U}$. Let $V \in \mathcal{U}$ be such that $V \subseteq V_1^{-1} \cap V_2$ and $F = (A^{\triangle U_1})^{*V}$. We claim that F is closed and $x \in N \subseteq F \subseteq A^{\triangle W}$.

To see that F is closed, note that $X - F = (X - A^{\triangle U_1})^{\triangle V}$ is open by our assumption on the closure property of $\mathcal{A}$.

For $N \subseteq F$, let $y \in N$. Note that for any $g \in V \subseteq V_2$, $U = U_2 \subseteq U_1 g$ since $U_2 g^{-1} \subseteq U_2 V_2^{-1} \subseteq U_1$. Thus for any $g \in V$, $y \in N = A^{\triangle U} \subseteq A^{\triangle(U_1 g)}$; it follows that $g \cdot y \in A^{\triangle U_1}$. Thus we have that $\forall^* g \in V (g \cdot y \in A^{\triangle U_1})$ and thus $y \in (A^{\triangle U_1})^{*V} = F$.

For $F \subseteq A^{\triangle W}$, let $y \in F = (A^{\triangle U_1})^{*V}$. Let $g_0 \in V$ be such that $g_0 \cdot y \in A^{\triangle U_1}$. Then $y \in A^{\triangle(U_1 g_0)}$. But since $U_1 g_0 \subseteq U_1 V_1^{-1} \subseteq W$, we get that $y \in A^{\triangle W}$. □

It is easy to construct collections $\mathcal{A}$ with the closure properties specified in the preceding lemmas. For instance, one can start by choosing a countable base $\mathcal{B}$ for a Polish topology τ on X generating the given Borel structure. Then inductively define a sequence of collections by letting $\mathcal{A}_0 = \mathcal{B}$ and $\mathcal{A}_{n+1} = \{X - A^{\triangle U} : A \in \mathcal{A}_n, U \in \mathcal{U}\}$. Finally let $\mathcal{A} = \bigcup_n \mathcal{A}_n$. Such a collection $\mathcal{A}$ satisfies the assumptions of both Lemmas 4.4.3 and 4.4.4, and if $\mathcal{U}$ is a countable base for G then the topology generated by $\mathcal{A}^{\triangle \mathcal{U}}$ is second countable (since both $\mathcal{A}$ and $\mathcal{U}$ are countable) and T_3, and the action of G on X is continuous with respect to it. Also note that every element of this $\mathcal{A}$ is $\Sigma_n^0(X, \tau)$ for some $n < \omega$, which is not as economical as Lemma 4.3.1. However, the advantage here is that the action is always continuous.

Next in order to guarantee that the topology is Polish, we seek closure conditions to guarantee that the topology is strong Choquet. One condition desired by the following lemma is the existence of a Polish topology finer than the one generated by $\mathcal{A}^{\triangle \mathcal{U}}$. To do this we combine the construction of Theorem 4.2.3 into the above iterative closure construction of $\mathcal{A}$. Thus a minimal collection $\mathcal{A}$ with all desired closure properties can be given as follows. Assume $\mathcal{U}$ is a countable base for G. Let $\mathcal{A}_0$ be a countable base for a Polish topology τ on X generating the given Borel structure. For each $n \in \omega$, let $\mathcal{A}_{n+1} \supseteq \{X - A^{\triangle U} : A \in \mathcal{A}_n, U \in \mathcal{U}\} \cup \mathcal{A}^{\triangle \mathcal{U}}$ be given by Theorem 4.2.3 so that $\mathcal{A}_{n+1}$ generates a Polish topology on X. Let $\mathcal{A} = \bigcup_n \mathcal{A}_n$. Then $\mathcal{A}$ generates a Polish topology by Lemma 4.2.2, and $\mathcal{A}^{\triangle \mathcal{U}} \subseteq \mathcal{A}$. Note that every element of this $\mathcal{A}$ is still $\Sigma_n^0(X, \tau)$ for some $n \in \omega$.

The presentation of the construction can be made a bit more succinct by requiring that $\mathcal{A}$ is a countable Boolean algebra. The details are left as an exercise (see Exercise 4.4.3).

Lemma 4.4.5
Let $\mathcal{U}$ be a countable base for G and $\mathcal{A}$ a countable Boolean algebra of Borel

subsets of X such that $\mathcal{A}^{\triangle \mathcal{U}} \subseteq \mathcal{A}$ and the topology on X generated by $\mathcal{A}$ is Polish. Then the topology generated by $\mathcal{A}^{\triangle \mathcal{U}}$ is strong Choquet.

Proof. Let τ be the topology generated by $\mathcal{A}^{\triangle \mathcal{U}}$ and σ be that by $\mathcal{A}$. Let $\mathcal{B}$ be the collection of nonempty intersections of finitely many elements of $\mathcal{A}^{\triangle \mathcal{U}}$. Then $\mathcal{B}$ is a countable base for τ. Since σ is Polish, we can fix a complete metric d on X giving rise to σ. We also fix a compatible complete metric on G.

Now to show that τ is strong Choquet, we consider the strong Choquet game $\mathsf{G}_{(X,\tau)}$:

$$\text{I} \quad x_0, M_0 \qquad x_1, M_1$$
$$\cdots\cdots$$
$$\text{II} \qquad N_0 \qquad\qquad N_1$$

We informally describe a winning strategy for Player II as follows.

Suppose $x_0, M_0, x_1, M_1, \ldots$ are played by Player I in a run of the game. We inductively define nonempty τ-open N_i, and in addition, $A_i \in \mathcal{A}$ and $U_i \in \mathcal{U}$, so that the following are satisfied:

(i) $x_i \in N_i \subseteq \overline{N_i}^\tau \subseteq M_i$;

(ii) $N_i \subseteq A_i^{\triangle U_i}$;

(iii) $\mathrm{diam}(A_i) < 2^{-i}$ and $\overline{A_i}^\sigma \subseteq A_{i-1}$;

(iv) $\mathrm{diam}(U_i) < 2^{-i}$ and $\overline{U_i}^\sigma \subseteq U_{i-1}$.

To see that such N_i, A_i, U_i exist, assume $x_i \in M_i \subseteq N_{i-1} \subseteq A_{i-1}^{\triangle U_{i-1}}$ by the inductive hypothesis. Since

$$A_{i-1} = \bigcup \{ A \in \mathcal{A} : \overline{A}^\sigma \subseteq A_{i-1}, \mathrm{diam}(A) < 2^{-i} \},$$

$$U_{i-1} = \bigcup \{ U \in \mathcal{U} : \overline{U}^\sigma \subseteq U_{i-1}, \mathrm{diam}(U) < 2^{-i} \},$$

and the action is Borel with respect to σ, there are some A_i, U_i satisfying (iii) and (iv) with $x \in A_i^{\triangle U_i}$. Now that τ is regular by Lemma 4.4.4, and $A_i^{\triangle U_i}$ and M_i are τ-open, there is τ-open N_i such that $x_i \in N_i \subseteq \overline{N_i}^\tau \subseteq M_i \cap A_i^{\triangle U_i}$. Thus (i) and (ii) are also satisfied.

The strategy for Player II is therefore to keep the sets A_i and U_i on the side and play an N_i satisfying (i) through (iv). To see that this strategy is winning, let $g_i \in U_i$ with $g_i \cdot x_i \in A_i$, g be the unique element such that $\{g\} = \bigcap_i U_i$ and y be the unique element such that $\{y\} = \bigcap_i A_i$. Then $\lim_i g_i = g$ and $\lim_i^\sigma g_i \cdot x_i = y$. Since $\tau \subseteq \sigma$, we have $\lim_i^\tau g_i \cdot x_i = y$. By Lemma 4.4.2 the action is continuous with respect to τ, and thus $\lim_i^\tau x_i = \lim_i^\tau g_i^{-1} \cdot (g_i \cdot x_i) = g^{-1} \cdot y$. Let $x = g^{-1} \cdot y$. We claim that $x \in N_i$ for all $i \in \omega$. For this fix $i \in \omega$ and note that for all $j > i$, $x_j \in N_j \subseteq \overline{N_j}^\tau \subseteq M_j \subseteq N_i$. Thus $x = \lim_{j>i}^\tau x_j \in \overline{N_{i+1}}^\tau \subseteq N_i$. Thus the strategy is winning for Player II. $\quad\square$

Thus we have completed the proof of the following theorem.

Theorem 4.4.6 (Becker–Kechris)
Let G be a Polish group and X a Borel G-space. Then there is a Polish topology τ on X such that τ generates the Borel structure of X and (X, τ) is a Polish G-space.

Another formulation of the theorem states that any Borel G-space is Borel isomorphic to a Polish G-space.

The method given in this section can be easily adapted to give a proof of a theorem in the same spirit of Theorem 4.3.3. However, by the discussion preceding Lemma 4.4.5, the finer topology makes use of Borel sets of rank slightly higher than the given sets.

Combined with the technique of Theorem 4.2.5 the method of this section can also be adapted to show an analog of Theorem 4.2.5 in the G-space setting.

Exercise 4.4.1 Let G be a Polish group, X a Polish G-space, $\mathcal{U}$ a base for G, $\mathcal{A} \subseteq \boldsymbol{\Sigma}_1^1(X)$, and $\mathcal{D}$ a base for X. Show that the action is continuous with respect to the topology on X generated by $\mathcal{A}^{\triangle \mathcal{U}} \cup \mathcal{D}$.

Exercise 4.4.2 Let G be Polish group, X a Polish G-space, $\mathcal{U}$ a base for G, $\mathcal{A} \subseteq \boldsymbol{\Sigma}_1^1(X)$, and $\mathcal{D}$ a base for X. Show that the topology generated by $(\mathcal{A} \cup \mathcal{D})^{\triangle \mathcal{U}}$ is finer than the original topology on X.

Exercise 4.4.3 Let G be a Polish group with a countable base $\mathcal{U}$, X a Borel G-space, and $\mathcal{B}$ a countable base for a Polish topology τ on X generating the given Borel structure. Show that there is a countable Boolean algebra $\mathcal{A} \supseteq \mathcal{B}$ such that $\mathcal{A}^{\triangle \mathcal{U}} \subseteq \mathcal{A}$ and the topology generated by $\mathcal{A}$ is Polish.

Exercise 4.4.4 Prove the following uniform version of Theorem 4.4.6: Let G be a Polish group, X a Borel G-space, Y a standard Borel space, and $A \subseteq X \times Y$ a Borel set such that A_y is G-invariant for all $y \in Y$. Then there is a Borel set $B \subseteq X \times Y \times \omega$ such that $(x, y) \in A \iff (x, y, 0) \in B$, and for all $y \in Y$, the topology σ_y generated by the countable collection $\{B_{y,n} : n \in \omega\}$, where $B_{y,n} = \{x \in X : (x, y, n) \in B\}$, is a Polish topology giving the same Borel structure on X, and (X, σ_y) is a Polish G-space.

Part II

Theory of Equivalence Relations

Chapter 5

Borel Reducibility

In this part of the book we give an introduction to the theory of definable equivalence relations in the Borel reducibility hierarchy.

The notion of Borel reducibility is the most penetrating concept studied in the invariant descriptive set theory. It was originally borrowed from computability theory and structural complexity theory to help determine the relative complexity of equivalence relations. In this sense invariant descriptive set theory can be regarded as a complexity theory for equivalence relations. There are, however, two prominent features that are unique to invariant descriptive set theory. The first one is that almost no equivalence relations studied here are artificial. In fact almost all of them arose from other fields of mathematics and some of them have even been studied in those fields intensively before their study became the main focus of invariant descriptive set theory. Thus results in invariant descriptive set theory are relevant to a broad range of mathematical fields. The second feature, which might be more interesting, is that we are comparing objects or problems of different mathematical nature in the same framework, thus making new connections between mathematical fields. For instance, it is very common that a classification problem in algebra turns out to have the same complexity as a problem in topology or analysis.

5.1 Borel reductions

Definition 5.1.1
Let E be an equivalence relation on set X and F an equivalence relation on set Y. A function $f : X \to Y$ is called a **reduction** from E to F if

$$x_1 E x_2 \iff f(x_1) F f(x_2)$$

for all $x_1, x_2 \in X$. We say that E is **reducible** to F and denote $E \leq F$ if there is a reduction from E to F.

With the Axiom of Choice $E \leq F$ is equivalent to there being an injection of X/E into Y/F (Exercise 5.1.1). Thus the order $\leq$ on equivalence relations is exactly the order on cardinalities of the quotient spaces. The notion becomes much more interesting when we impose definability requirements on the spaces and functions.

Definition 5.1.2
Let X, Y be Polish spaces and E, F equivalence relations on X, Y respectively. E is said to be **continuously reducible** to F, denoted $E \leq_c F$, if there is a continuous reduction from E to F. E is **Borel reducible** to F, denoted $E \leq_B F$, if there is a Borel reduction from E to F.

The notion of Borel reducibility is the most useful in measuring the relative complexity of equivalence relations. Intuitively, if $E \leq_B F$ then E is considered to be at most as complex as F, since every inquiry about E can be transferred and answered through an inquiry about F. Of course there is a different reducibility notion associated with every class of definable functions. But the continuous and Borel reductions will be our main focus in this book. The notion of Borel reducibility also makes sense if X, Y are only standard Borel spaces.

One can also strengthen the notion of reduction as follows.

Definition 5.1.3
For equivalence relations E, F on sets X, Y respectively, a reduction $f : X \to Y$ is called an **embedding** if f is injective. We say that E **embeds** into F if there is an embedding of E into F, and denote $E \sqsubseteq F$. If X, Y are Polish spaces then E **continuously embeds** into F, denoted $E \sqsubseteq_c F$, if there is a continuous embedding from E into F, and E **Borel embeds** into F, denoted $E \sqsubseteq_B F$, if there is a Borel embedding from E into F.

Obviously the notion $E \sqsubseteq_B F$ also makes sense if X, Y are only standard Borel spaces. Finally we also have the notion of a strict order and various notions of equivalence.

Definition 5.1.4
Let E, F be equivalence relations on X, Y respectively.

(1) We write $E < F$ if $E \leq F$ and $F \nleq E$. The notation $<_c$ and $<_B$ is similarly defined.

(2) E and F are said to be **bireducible** (or **equivalent**) to each other, and denoted $E \sim F$, if both $E \leq F$ and $F \leq E$. The notation $\sim_c$ and $\sim_B$ is similarly defined.

(3) E and F are said to be **biembeddable** to each other if $E \sqsubseteq F$ and $F \sqsubseteq E$. The notions of continuous biembeddability and Borel biembeddability are similarly defined.

(4) E and F are said to be **isomorphic**, denoted $E \cong F$, if there is a bijective reduction $f : X \to Y$. E and F are **topologically isomorphic** if there is a reduction $f : X \to Y$ that is a homeomorphism between X

and Y. They are **Borel isomorphic**, denoted $E \cong_B F$, if there is a Borel bijective reduction $f : X \to Y$.

In the special case that the equivalence relations are induced by group actions, we have the following simple facts.

Proposition 5.1.5
Let G be a group and X, Y be G-spaces. Let E_G^X and E_G^Y be the induced orbit equivalence relations.

(a) *If $f : X \to Y$ is a G-embedding, then f is an embedding from E_G^X into E_G^Y.*

(b) *If there are G-embeddings $f : X \to Y$ and $g : Y \to X$, then E_G^X and E_G^Y are isomorphic.*

The proof is left as an exercise (Exercise 5.1.6). Note that in this proposition part (a) applies to any class of definable functions, but it is not the case with part (b). Nevertheless part (b) can be stated with Borel functions, from which the following proposition is a corollary.

Proposition 5.1.6
Let G be a Polish group and X, Y be Borel G-spaces. If there are Borel G-embeddings $f : X \to Y$ and $g : Y \to X$, then $E_G^X \cong_B E_G^Y$.

The existence of universal Borel G-spaces immediately gives rise to the following concept and result.

Definition 5.1.7
For any class $\mathcal{C}$ of equivalence relations, a **universal equivalence relation** for $\mathcal{C}$ is some $F \in \mathcal{C}$ such that for any $E \in \mathcal{C}$, $E \leq_B F$.

Theorem 5.1.8
For any Polish group G there is a universal G-orbit equivalence relation, that is, a universal equivalence relation in the class of all orbit equivalence relations induced by Borel actions of G.

Proof. By Theorem 3.3.4 and Proposition 5.1.5(b). $\quad\square$

If H is a Polish group and G is a closed subgroup of H, then Theorem 3.5.3 established that any G-orbit equivalence relation is Borel reducible to some H-orbit equivalence relation (see Exercise 5.1.7). Thus the existence of universal Polish groups immediately gives the following theorem.

Theorem 5.1.9
There is a universal equivalence relation for all orbit equivalence relations induced by Borel actions of Polish groups.

Proof. By Theorem 2.5.2 there is a universal Polish group H. Let G be a Polish group and X a Borel G-space. Then G is isomorphic to a closed subgroup of H. By Theorem 3.5.3 there is a Borel H-space Y so that X is a Borel subset of Y and that every H-orbit contains exactly one G-orbit. It follows that the identity map from X into Y is a Borel reduction of E_G^X to E_H^Y. Let E be a universal H-orbit equivalence relation given by Theorem 5.1.8 above. Then $E_H^Y \leq_B E$. It follows that $E_G^X \leq_B E$ and thus E is a universal equivalence relation for all E_G^X, where G is Polish and X a Borel G-space. ▯

Looking at the other end, the simplest equivalence relations are the identity relations.

Definition 5.1.10
Let X be a set. The **identity relation** (or the **equality relation**) on X, denoted by $\mathrm{id}(X)$, is the relation $\{(x, y) \in X \times X : x = y\}$.

Note that identity relations are also orbit equivalence relations. They are induced by trivial actions of the trivial group. Any reduction from an identity relation is necessarily an embedding. In addition, reductions involving the identity relation are closely related to the following concept.

Definition 5.1.11
Let X be a Polish space and E an equivalence relation on X. We say that there are **perfectly many** E-equivalence classes if there is a perfect set $A \subseteq X$ such that for any $x, y \in A$, xEy implies $x = y$.

The following proposition collects some easy facts about reductions involving the identity relation.

Proposition 5.1.12
Let E be an equivalence relation on a Polish space X. Then the following are equivalent:

(i) *There are perfectly many E-equivalence classes;*

(ii) *For some uncountable Polish space X, $\mathrm{id}(X) \leq_B E$;*

(iii) *For any Polish space X, $\mathrm{id}(X) \leq_B E$;*

(iv) *$\mathrm{id}(\omega^\omega) \leq_B E$;*

(v) *$\mathrm{id}(2^\omega) \leq_B E$;*

(vi) *$\mathrm{id}(2^\omega) \sqsubseteq_B E$;*

(vii) *$\mathrm{id}(2^\omega) \sqsubseteq_c E$.*

Proof. The implication (i)$\Rightarrow$(vii) follows from Exercise 1.3.4. It is obvious that (vii)$\Rightarrow$(vi)$\Rightarrow$(v). The equivalence of (ii) through (v) is by the Borel isomorphism theorem (Corollary 1.3.8). Finally (ii)$\Rightarrow$(i) by the perfect set theorem for analytic sets (Theorem 1.6.6). □

Note that conditions (ii) through (vi) make sense and are equivalent even if X is only a standard Borel space. In this case we say that there are perfectly many E-equivalence classes if any of the conditions (ii) through (vi) holds.

Exercise 5.1.1 Show that $E \leq F$ iff there is an injection from X/E into Y/F.

Exercise 5.1.2 Classify all equivalence relations on Polish spaces up to isomorphism, bireducibility, and biembeddability, respectively.

Exercise 5.1.3 Classify all Borel equivalence relations on Polish spaces with at most countably many equivalence classes up to Borel isomorphism, Borel bireducibility, and Borel biembeddability, respectively.

Exercise 5.1.4 Show that if $E \leq_c F$ and F is $\mathbf{\Pi}^0_\alpha$ (or $\mathbf{\Sigma}^0_\alpha$) for any $\alpha < \omega_1$ then so is E.

Exercise 5.1.5 Show that if $E \leq_B F$ and F is Borel (or analytic) then so is E.

Exercise 5.1.6 Prove Propositions 5.1.5 and 5.1.6.

Exercise 5.1.7 Let E^X_G and E^Y_H be given by Theorem 3.5.3. Show that $E^X_G \sim_B E^Y_H$.

Exercise 5.1.8 Let E be a Borel equivalence relation on a standard Borel space X. Suppose E has at most countably many equivalence classes. Show that there is a standard Borel space Y such that $E \sim_B \mathrm{id}(Y)$.

5.2 Faithful Borel reductions

Definition 5.2.1
Let X, Y be standard Borel spaces and E, F be equivalence relations on X, Y, respectively. We say that $f : X \to Y$ is a **faithful Borel reduction** from E to F, if f is a Borel reduction from E to F such that for any invariant Borel set $A \subseteq X$, the F-saturation of $f(A)$, $[f(A)]_F$ is Borel. We say that E is **faithfully Borel reducible** to F, denoted $E \leq_{fB} F$, if there is a faithful Borel reduction from E to F.

The notion of a faithful Borel reduction is a bit technical. However, it is related to several important problems in invariant descriptive set theory, and many natural reductions between equivalence relations turn out to be faithful.

Proposition 5.2.2
Let G be a Polish group and X, Y Borel G-spaces. If $f : X \to Y$ is a Borel G-embedding from X to Y then f is a faithful Borel reduction from E_G^X to E_G^Y.

Proof. Let $f : X \to Y$ be a Borel G-embedding and $A \subseteq X$ an invariant Borel set. Then $f(A)$ is invariant Borel. $\qquad\qquad$ ⬛

Theorem 5.2.3
Let G be a Polish group, X a Borel G-space, and Y a standard Borel space. Let $E = E_G^X$ and F a Borel equivalence relation on Y. Suppose f is a Borel reduction from E to F. Then f is a faithful Borel reduction from E to F.

Proof. Without loss of generality we may assume that $X = Y = \omega^\omega$. Let $A \subseteq X$ be invariant Borel and $B = [f(A)]_F$. By definition B is $\mathbf{\Sigma}_1^1$. We show that B is also $\mathbf{\Pi}_1^1$, hence it is Borel by Suslin's Theorem 1.6.3.

For this let α denote the action of G on X. Let z be a real such that $\alpha, f, A, F \in \Delta_1^1(z)$. We claim that

$$\forall y \in B \; \exists x \in \Delta_1^1(y, z) \; (x \in A \; \wedge \; f(x)Fy).$$

It follows from the claim that

$$B = \{ y \in Y : \exists x \in \Delta_1^1(y, z) \; (x \in A \; \wedge \; f(x)Fy) \},$$

and thus B is $\mathbf{\Pi}_1^1$ by Kleene's Theorem 1.7.5.

The rest of the proof is thus devoted to a proof of the claim. Fix $y \in B$ and $x_0 \in A$ with $f(x_0)Fy$. Since $[x_0]_G = f^{-1}([y]_F)$, there is a Borel code w for $[x_0]_G$ such that $w \in \Delta_1^1(y, z)$. We will show that $\exists x \in \Delta_1^1(w, z) \cap [x_0]_G$, which is enough for establishing the claim.

Recall that w is a labeled well-founded tree $\langle T, \lambda \rangle$ on ω. For the convenience of this proof we omit the definition of λ on nonterminal nodes of T and instead assume that nonterminal nodes of T on even levels correspond to countable intersections, those on odd levels correspond to countable unions, and for terminal nodes $t \in T$, $\lambda(t) \in \omega^\omega$ correspond to basic open sets $N_{\lambda(t)}$. With this convention we have that

$$x \in [x_0]_G \iff \forall i \exists j \ldots \exists k \; x \in N_{\lambda(\langle i, j, \ldots, k \rangle)}.$$

We consider a game $G(s, c)$, with $s \in \omega^{<\omega}$ and c a Borel code for a closed set, played as follows:

$$
\begin{array}{cccccc}
\text{I} & i_0 & & i_1 & & i_2 \\
& & & & & \cdots \\
\text{II} & & u, j_0 & & j_1 & & j_2
\end{array}
$$

where $i_0, j_0, i_1, j_1, \cdots \in \omega$ and $t \in \omega^{<\omega}$. The rules of the game are that $s \subsetneq u$ and that $\langle i_0, j_0, \dots \rangle \in T$. The play is finished when a terminal node t of T has been played. Then Player II wins if $\exists x \in [x_0]_G \exists^* g \in G$ $(g \cdot x \in N_u \cap B_c \cap N_{\lambda(t)})$ provided that $\exists x \in [x_0]_G \exists^* g \in G$ $(g \cdot x \in N_s \cap B_c)$.

It is easy to see that Player II has a winning strategy in the game $G(s, c)$. In fact, suppose that $\exists x \in [x_0]_G \exists^* g \in G$ $(g \cdot x \in N_s \cap B_c)$. Suppose also $[x_0]_G = \bigcap_{i_0} \bigcup_{j_0} C_{i_0, j_0}$ where each C_{i_0, j_0} is the Borel set coded by the tree $T_{\langle i_0, j_0 \rangle}$. Then for any i_0, we have that

$$[x_0]_G \cap N_s \cap B_c \subseteq \bigcup \{[x_0]_G \cap N_u \cap B_c \cap C_{i_0, j_0} : s \subsetneq u, j_0 \in \omega\}.$$

Since the right-hand side is a countable union it follows that for some $u \supsetneq s$ and $j_0 \in \omega$, $\exists x \in [x_0]_G \exists^* g \in G$ $(g \cdots x \in N_u \cap B_c \cap C_{i_0, j_0})$. This gives Player II's first move. Similar arguments apply to subsequent moves. And the winning condition for Player II is maintained throughout the play.

We also need to note that Player II has a winning strategy $\Gamma(s, c)$ which is in $\Delta_1^1(w, z, s, c)$. For this note that the winning condition is in fact Borel since $\exists x \in [x_0]_G \exists^* g \in G$ $(g \cdot x \in \dots)$ is equivalent to $\forall x \in [x_0]_G \exists^* g \in G$ $(g \cdot x \in \dots)$. Thus $\Gamma(s, c)$ can be defined by choosing the lexicographically least moves that maintain the winning condition.

Now let $P = \{u \in \omega^{<\omega} : \exists v \in \omega^{<\omega}$ (u, v) is a terminal node of $T\}$, where (u, v) denote the shuffle of u and v. Thus P consists of all legitimate moves for Player I in the above game. Let $<_P$ be a $\Delta_1^1(w, z)$ nonrepetitive enumeration of elements of P. For $i \in \omega$ let u_i be the i-th element of P under $<_P$.

Finally to find $x \in [x_0]_G \cap \Delta_1^1(w, z)$, we define in $\Delta_1^1(w, z)$ a sequence

$$s(0) \subseteq s(1) \subseteq s(2) \subseteq \cdots$$

of elements of $\omega^{<\omega}$ such that

$$\forall u \in P \exists v \in \omega^{<\omega} \exists i \in \omega \forall i' > i \exists x \in [x_0]_G \exists^* g \in G \ (g \cdot x \in N_{\lambda(u,v)} \cap N_{s(i')}).$$

Granting such a sequence, let $x = \bigcup_i s(i)$. Then $\forall u \in P \exists v \exists i \forall i' > i$ $(N_{s(i')} \cap N_{\lambda(u,v)} \neq \emptyset)$. It follows that $\forall u \in P \exists v$ (x is a limit point of $N_{\lambda(u,v)}$). Since $N_{\lambda(u,v)}$ is closed, we have that $\forall u \in P \exists v$ ($x \in N_{\lambda(u,v)}$), and hence $x \in [x_0]_G$.

To define the sequence we use induction and simultaneously define a sequence $c(i)$ of Borel codes for closed sets. To begin with let $s(0) = \emptyset$ and $B_{c(0)} = X$. In general suppose $s(i)$ and $c(i)$ have been defined and suppose $\Gamma(s(i), c(i))$ is the winning strategy for Player II in the game $G(s(i), c(i))$ defined in the preceding paragraph. Let Player I play u_i and apply $\Gamma(s(i), c(i))$ to obtain $s(i + 1) \supseteq s(i)$ as a part of Player II's response. Let v be the other part of Player II's moves. Then let $B_{c(i+1)} = B_c \cap N_{\lambda(u_i, v)}$. To verify that this is as required, let $u \in P$. Let $u_i = u$ and v be played by Player II according to the construction. Then for all $i' > i$, $B_{c(i')} \subseteq N_{\lambda(u,v)}$. Since $\Gamma(s(i), c(i))$ is winning for Player II, we have that $\exists x \in [x_0]_G \exists^* g \in G$ $(g \cdot x \in N_{\lambda(u,v)} \cap N_{s(i')})$.

This finishes the proof of the theorem. $\square$

Corollary 5.2.4
Let G be a Polish group, X a Borel G-space, Y a standard Borel space, and F a Borel equivalence relation on Y. Suppose $f : X \to Y$ is a Borel reduction from E_G^X to F such that $[f(X)]_F = Y$. Then $E_G^X \sim_B F$.

Proof. The above proof gives rise to a Borel function $g : Y \to X$ such that for any $y \in [f(X)]_F$, letting $x = g(y)$, then $f(x)Fy$. Note that the definition of x does not depend on the choice of x_0. It is easy to see that g is a reduction from F to E_G^X. □

Exercise 5.2.1 Show that the Borel reduction constructed in Corollary 5.2.4 is also faithful.

Exercise 5.2.2 Let E, F be Borel orbit equivalence relations on Polish spaces X, Y, respectively. Show that if $E \leq_B F$ then there is an invariant Borel set $Z \subseteq Y$ such that $E \sim_B F \upharpoonright Z$.

5.3   Perfect set theorems for equivalence relations

In this and the next sections we study Borel reductions involving the identity relations. Recall from Proposition 5.1.12 that reduction from the identity relation on 2^ω is equivalent to the existence of perfectly many equivalence classes. In this section we give various sufficient conditions for there being perfectly many equivalence classes for an equivalence relation on a Polish space. Again category arguments are playing the vital role in this study.

Theorem 5.3.1 (Mycielski)
Let X be a Polish space and E a meager equivalence relation on X. Then $id(2^\omega) \sqsubseteq_c E$.

Proof. First note that X is perfect. In fact, otherwise X must contain at least one isolated point x, and any equivalence relation on X must contain the open set $\{(x, x)\}$, and therefore is nonmeager.

Let $d < 1$ be a compatible complete metric on X. Let $E \subseteq \bigcup_n F_n$ with $F_n \subseteq X^2$ closed and nowhere dense for all $n \in \omega$. We may assume that $id(X) \subseteq F_0$ and $F_n \subseteq F_{n+1}$ for all $n \in \omega$.

We define a sequence $(U_s)_{s \in 2^{<\omega}}$ of open sets in X by induction on $\mathrm{lh}(s)$ such that, for all $s \in 2^{<\omega}$,

   (i) $\mathrm{diam}(U_s) \leq 2^{-\mathrm{lh}(s)}$;

   (ii) $\overline{U_{s^\frown 0}}, \overline{U_{s^\frown 1}} \subseteq U_s$; $\overline{U_{s^\frown 0}} \cap \overline{U_{s^\frown 1}} = \emptyset$;

(iii) for all $s, t \in 2^{<\omega}$ with $\mathrm{lh}(s) = \mathrm{lh}(t) = n+1$ and $s \neq t$, $(U_s \times U_t) \cap F_n = \emptyset$.

To begin let $U_\emptyset = X$. Assume now that for all s with $\mathrm{lh}(s) = n$ the set U_s has been defined. We define all $U_{s^\frown 0}$ and $U_{s^\frown 1}$. For notational convenience we enumerate all U_s for $\mathrm{lh}(s) = n$ as $U_1, \ldots, U_i, \ldots, U_k$ with $1 \leq i \leq k = 2^n$. Since X is perfect there are disjoint open subsets $V_i, V_{k+i} \subseteq U_i$ for all $1 \leq i \leq k$ with $\mathrm{diam}(V_i), \mathrm{diam}(V_{k+i}) \leq 2^{-n-1}$, and by regularity we may assume that $\overline{V_i}, \overline{V_{k+i}} \subseteq U_i$ and $\overline{V_i} \cap \overline{V_{k+i}} = \emptyset$. Thus we have obtained a collection $\{V_i\}_{1 \leq i \leq 2k}$ of disjoint open sets satisfying (i) and (ii). To guarantee (iii) we make the following observation. Since F_n is nowhere dense, for any disjoint nonempty open sets W_1 and W_2, $W_1 \times W_2 \not\subseteq F_n$. It follows that there are nonempty $W_1' \subseteq W_1$ and $W_2' \subseteq W_2$ with $(W_1' \times W_2') \cap F_n = \emptyset$. Applying this observation successively to all pairs (i, j) with $1 \leq i, j \leq 2k$ and $i \neq j$, we obtain open sets $V_i' \subseteq V_i$ such that, for all such pairs (i, j), $(V_i' \times V_j') \cap F_n = \emptyset$. Now condition (iii) is satisfied if we let $U_{s^\frown 0} = V_i'$ and $U_{s^\frown 1} = V_{k+i}'$ where $U_i = U_s$.

For $u \in 2^\omega$ we now let $f(u)$ to be the unique element of the set $\bigcap_n U_{u \restriction n}$. Then f is a continuous function from 2^ω into X. To finish the proof of the theorem we only need to verify that $(f(u), f(v)) \notin E$ for $u \neq v$. For this let $u, v \in 2^\omega$ with $u \neq v$. Let n be the least integer with $u(n) \neq v(n)$. Then for all $m > n$, $u \restriction m \neq v \restriction m$. By (iii), $(U_{u \restriction m} \times U_{v \restriction m}) \cap F_m = \emptyset$. Thus $(f(u), f(v)) \notin F_m$ for all $m > n$. Therefore $(f(u), f(v)) \notin E$ as required. $\quad\square$

Corollary 5.3.2
Let X be a Polish space and E an equivalence relation on X. If there is a nonempty open set $U \subseteq X$ such that E is meager on $U \times U$, then $\mathrm{id}(2^\omega) \sqsubseteq_c E$.

Corollary 5.3.3
Let X be a Polish space and E an equivalence relation on X. Suppose E has the Baire property. If every E-equivalence class is meager, then there are perfectly many E-equivalence classes.

Proof. The condition can be stated as $\forall x \; \forall^* y \; (x, y) \notin E$. By Kuratowski–Ulam Theorem 3.2.1 it follows that E is meager. $\quad\square$

Another corollary of Theorem 5.3.1 concerning the Gandy–Harrington topology is going to play an important role in the proof of Silver's theorem later in this section.

Corollary 5.3.4
Let τ be the Gandy–Harrington topology on the Baire space ω^ω and E an equivalence relation on ω^ω. If there is a Σ_1^1 set $V \subseteq \omega^\omega$ such that E is $\tau \times \tau$-meager in $V \times V$, then there are perfectly many E-equivalence classes.

Proof. Let $Y = X_{\mathrm{low}}$ be the Polish space of low elements with the Gandy–Harrington topology (Theorem 1.8.5). By the Gandy Basis Theorem 1.8.4

$U = Y \cap V$ is a nonempty open subset of Y. Now E is meager on $U \times U$, and thus by Corollary 5.3.2 there is a continuous reduction $f : 2^\omega \to Y$ witnessing $\mathrm{id}(2^\omega) \sqsubseteq E$. Since τ is finer than the usual topology f as a function from 2^ω into ω^ω is still continuous. Thus there are perfectly many E-equivalence classes. $\square$

The following theorem, often referred to as the **Silver dichotomy theorem**, is the first of a series of dichotomy theorems undoubtedly occupying the center of the invariant descriptive set theory of equivalence relations. It is also important because of its proof presented here, due to Harrington, which is an elegant example of the use of effective descriptive set theoretic methods in establishing an important result of classical character.

Theorem 5.3.5 (Silver)
Let X be a standard Borel space and E a coanalytic equivalence relation on X. Then either there are countably many E-equivalence classes or there are perfectly many E-equivalence classes.

Proof. Without loss of generality we may assume $X = \omega^\omega$. By relativization we also assume that E is Π_1^1. Let τ be the Gandy–Harrington topology on ω^ω. First we define

$$V = \{x \in X : \text{ there is no } \Delta_1^1 \text{ set } U \text{ such that } x \in U \subseteq [x]_E\}.$$

We claim that if $V = \emptyset$ then there are only countably many E-equivalence classes. This is because, if $V = \emptyset$, then every E-equivalence class contains some nonempty Δ_1^1 set. Since there are only countably many Δ_1^1 sets, it follows that there are only countably many E-equivalence classes.

For the rest of the proof we assume $V \neq \emptyset$. We claim that $V \in \Sigma_1^1$. Indeed,

$$x \in V \iff \forall U \in \Delta_1^1 \ (x \in U \to \exists y \in U \ (x, y) \notin E).$$

With the coding of Δ_1^1 sets given by Theorem 1.7.4 we get that

$$x \in V \iff \forall n \ [(n \in D \wedge x \in P_n^+) \to \exists y \ (y \notin P_n^- \wedge (x, y) \notin E)].$$

This shows that $V \in \Sigma_1^1$.

To finish the proof we claim that E is $\tau \times \tau$-meager in $V \times V$. Then by the above corollary there are perfectly many E-equivalence classes. We establish the claim in several steps. As the first step, we check that for every $x \in V$ there is no Σ_1^1 set U such that $x \in U \subseteq [x]_E$. To see this, assume toward a contradiction that $x \in V$ and $U \in \Sigma_1^1$ and $x \in U \subseteq [x]_E$. Then note that $[x]_E$ is Π_1^1 since

$$y \in [x]_E \iff \forall z (z \in U \to (y, z) \in E).$$

Now by the separation property for Σ_1^1 sets (see Theorem 1.7.1 and the remarks following it) there is $W \in \Delta_1^1$ such that $U \subseteq W \subseteq [x]_E$, thus $x \notin V$.

From this it follows immediately that every nonempty Σ_1^1 subset U of V meets more than one E-equivalence class.

Next we note that every set involved has the Baire property in the Gandy–Harrington topology or its products. Specifically, the equivalence relation E as a subset of $\omega^\omega \times \omega^\omega$ has the Baire property in $\tau \times \tau$, and each equivalence class $[x]_E$ has the Baire property in τ. This is because, by a theorem of Nikodym (see the remarks preceeding Proposition 2.3.1), the collection of all sets with the Baire property in any topological space is closed under the Suslin operation. Now both E and $[x]_E$ are Π_1^1, and therefore coanalytic, and therefore their complements are results of the Suslin operation applied to sequences of closed sets in the usual topology of ω^ω. Since the Gandy–Harrington topology is coarser, they continue to be the results of Suslin operation, and thus have the Baire property in the sense of τ. This allows us to use the Kuratowski–Ulam Theorem 3.2.1.

Now to show that E is $\tau \times \tau$-meager in $V \times V$, by the Kuratowski–Ulam theorem it suffices to show that for all $x \in V$, $[x]_E$ is τ-meager in V. Thus it suffices to show that for all $x \in V$, $[x]_E$ is not τ-comeager in any $U \subseteq V$ where $U \in \Sigma_1^1$. Again toward a final contradiction assume that for some $x \in V$ and $U \subseteq V$, $U \in \Sigma_1^1$, $[x]_E$ is τ-comeager in U. Now by Louveau's lemma Exercise 1.8.3, if $[x]_E$ is τ-comeager in $U \in \Sigma_1^1$, then $[x]_E \times [x]_E$ is comeager in $U \times U$ in the Gandy–Harrington topology of $\omega^\omega \times \omega^\omega$. Since $(U \times U) \cap (\omega^\omega \times \omega^\omega - E)$ is nonempty and Σ_1^1, we have that

$$([x]_E \times [x]_E) \cap (U \times U) \cap (\omega^\omega \times \omega^\omega - E) \neq \emptyset,$$

which is a contradiction. □

Exercise 5.3.1 Let X be a Polish space and E an equivalence relation on X with the Baire property. Suppose every E-equivalence class is countable. Show that there are perfectly many E-equivalence classes.

Exercise 5.3.2 Let X be a Polish space, $U \subseteq X$ open, and E an equivalence relation on X with the Baire property. Suppose τ is a strong Choquet topology on X finer than the Polish topology. Show that if E is $\tau \times \tau$-meager on $U \times U$, then there are perfectly many E-equivalence classes.

Exercise 5.3.3 Let E be a Π_1^1 equivalence relation on ω^ω. Show that the following are equivalent:

(i) There are only countably many E-equivalence classes.

(ii) For every $x \in \omega^\omega$ there exists a Δ_1^1 set U with $x \in U \subseteq [x]_E$.

(iii) For every $x \in \omega^\omega$ there exists a Σ_1^1 set U with $x \in U \subseteq [x]_E$.

(iv) Every E-equivalence class is Π_1^1.

Exercise 5.3.4 Let E be a Π_1^1 equivalence relation on ω^ω with only countably many classes. Show that E is a union of countably many Δ_1^1 relations. In particular E is Borel.

5.4 Smooth equivalence relations

Definition 5.4.1
Let E be an equivalence relation on a standard Borel space X. We call E **smooth** or **concretely classifiable** if $E \leq_B \mathrm{id}(2^\omega)$.

It is clear from the definition that smooth equivalence relations are Borel. Since any two uncountable standard Borel spaces are Borel isomorphic, E is smooth iff $E \leq_B \mathrm{id}(X)$ for any uncountable Polish space X. Among the countable Polish spaces we consider ω with the discrete topology and let $\mathrm{id}(\omega)$ denote its identity relation. The following is an immediate corollary of the Silver dichotomy theorem.

Theorem 5.4.2
Let E be a smooth equivalence relation. Then exactly one of the following holds:

(I) $E \sim_B \mathrm{id}(2^\omega)$;

(II) $E \leq_B \mathrm{id}(\omega)$.

Proof. By the Silver dichotomy theorem either there are perfectly many E-equivalence classes or else there are only countably many E-equivalence classes. In the first case we have both $\mathrm{id}(2^\omega) \leq_B E$ and, by smoothness, $E \leq_B \mathrm{id}(2^\omega)$. Hence $E \sim_B \mathrm{id}(2^\omega)$. For the second case let $C_0, C_1, \ldots$ enumerate all the E-equivalence classes. Then the function θ defined by $\theta(x) = n$ iff $x \in C_n$ gives a Borel reduction from E to $\mathrm{id}(\omega)$. $\Box$

The converse of this theorem is trivial. Easily we obtain a complete classification of smooth equivalence relations in terms of Borel bireducibility (see Exercise 5.4.2).

Definition 5.4.3
Let E be an equivalence relation on a space X and $\mathcal{S}$ a family of subsets of X. We say that E is **generated** by $\mathcal{S}$ if for any $x, y \in X$,

$$x E y \iff \text{for any } S \in \mathcal{S}, \, x \in S \text{ iff } y \in S.$$

In this situation $\mathcal{S}$ is also called a **generating family** or a **separating family** for E. If X is a Borel space then a generating family $\mathcal{S}$ is called **Borel** if every element of $\mathcal{S}$ is Borel.

Proposition 5.4.4
Let E be an equivalence relation on a standard Borel space X. Then E is smooth iff there is a countable Borel generating family for E.

Proof. Let $\mathcal{S} = \{S_0, S_1, \ldots, S_n, \ldots\}$ be a countable Borel generating family for E. Define $\theta : X \to 2^\omega$ by

$$\theta(x)(n) = 1 \iff x \in S_n.$$

Then θ is a Borel reduction from E to $\mathrm{id}(2^\omega)$. Conversely, suppose $\theta : X \to 2^\omega$ is a Borel reduction. Then define $S_n = \{z \in 2^\omega : \theta(z)(n) = 1\}$. We have that $\mathcal{S} = \{S_0, S_1, \ldots, S_n, \ldots\}$ is a countable Borel generating family for E. □

Corollary 5.4.5
Let E be a smooth equivalence relation on a Polish space X. Then there is a finer Polish topology σ on X such that E is closed as a subset of $(X^2, \sigma \times \sigma)$.

Proof. Let $S_0, S_1, \ldots, S_n, \ldots$ enumerate a countable Borel generating family for E. By Theorem 4.2.3 there is a finer Polish topology σ on X such that each of S_n is σ-clopen. Now

$$x E y \iff \forall n \, (x \in S_n \iff y \in S_n),$$

and thus E is closed in $(X^2, \sigma \times \sigma)$. □

Next we give some sufficient conditions for smoothness. First we show that closed equivalence relations are smooth. To do this we establish the following useful general lemma which is reminiscent of the Luzin separation theorem for Σ_1^1 sets.

Lemma 5.4.6
Let E be a Σ_1^1 equivalence relation on a standard Borel space X. Let $A, B \subseteq X$ be disjoint Σ_1^1 E-invariant sets. Then there is an E-invariant Borel set C with $A \subseteq C$ and $B \cap C = \emptyset$.

Proof. Let $A_0 = A$ and C_0 be a Borel set separating A_0 and B by the Luzin separation theorem. That is, $A_0 \subseteq C_0$ and $B \cap C_0 = \emptyset$. Let $A_1 = [C_0]_E$. Then A_1 is Σ_1^1 and $A_1 \cap B = \emptyset$. By the separation theorem again let C_1 be a Borel set separating A_1 and B. Continue this way, and we obtain

$$A = A_0 \subseteq C_0 \subseteq A_1 \subseteq C_1 \subseteq A_2 \subseteq \ldots C_n \subseteq A_n \subseteq \ldots$$

where C_n is Borel and A_n is E-invariant, Σ_1^1 and disjoint from B. Let now $C = \bigcup_n C_n = \bigcup_n A_n$. Then C is invariant Borel, $A \subseteq C$, and $B \cap C = \emptyset$ as required. □

Proposition 5.4.7
Let E be a closed equivalence relation on a Polish space X. Then E is smooth.

Proof. Since E is closed we can write $X^2 - E = \bigcup_n U_n \times V_n$ where U_n, V_n are basic open sets. Let $A_n = [U_n]_E$ and $B_n = [V_n]_E$. Then we have that A_n and B_n are disjoint Σ^1_1 E-invariant sets. By Lemma 5.4.6 there is E-invariant Borel C_n such that $A_n \subseteq C_n$ and $B_n \cap C_n = \emptyset$. We claim that $\{C_n : n \in \omega\}$ is a generating family for E, and thus E is smooth. To see this, let $x, y \in X$ be such that $x \in C_n \iff y \in C_n$ for all $n \in \omega$. Assume $(x, y) \notin E$. Then for some $n \in \omega$ we have that $x \in U_n$ and $y \in V_n$. However, this implies that $x \in A_n \subseteq C_n$ and $y \in B_n \subseteq X - C_n$, a contradiction. ⬚

In the next chapter we will show that any Borel equivalence relation whose equivalence classes are G_δ is smooth. For now we turn to the following easy fact.

Proposition 5.4.8
Let E be an equivalence relation on a standard Borel space. If E has a Borel selector then E is smooth.

Proof. The selector function is a reduction. ⬚

Note that not every closed equivalence relation has a Borel selector (Exercise 5.4.6), hence the converse of the above proposition is not true. Next we give a characterization for the existence of Borel selectors for smooth equivalence relations.

Definition 5.4.9
Let X be a standard Borel space and E an equivalence relation on X. We call E **idealistic** if there is an assignment $C \mapsto I_C$ that associates with each E-equivalence class C a σ-ideal I_C of subsets of C such that

(i) $C \notin I_C$;

(ii) for each Borel set $A \subseteq X^2$ the set A_I defined by

$$x \in A_I \iff \{y \in [x] : (x, y) \in A\} \in I_{[x]}$$

is Borel.

Clause (ii) in the above definition is often referred to as Borelness of the map $C \mapsto I_C$. The following proposition helps motivate the definition.

Proposition 5.4.10
Let G be a Polish group and X a Borel G-space. Then E^X_G is idealistic.

Proof. Let $x \in X$ and $C = [x]_G$. Then define I_C on C by

$$S \in I_C \iff \{g \in G : g \cdot x \in S\} \text{ is meager in } G.$$

It is easy to check that I_C is a σ-ideal and $C \notin I_C$. To see that $C \mapsto I_C$ is Borel, let $A \subseteq X^2$ be a Borel set. Then

$$
\begin{aligned}
x \in A_I &\iff \{y \in [x] : (x,y) \in A\} \in I_{[x]} \\
&\iff \{g \in G : (x, g \cdot x) \in A\} \text{ is meager in } G \\
&\iff \forall^* g \in G \ (x, g \cdot x) \notin A,
\end{aligned}
$$

hence A_I is Borel. □

Theorem 5.4.11 (Kechris)
Let E be an equivalence relation on a standard Borel space X. Then E has a Borel selector iff E is smooth and idealistic.

Proof. ($\Rightarrow$) Let s be a Borel selector for E. We show that E is idealistic. For any $x \in X$ and $C = [x]_E$, define I_C on C by

$$S \in I_C \iff s(x) \notin S.$$

Then I_C is a σ-ideal and $C \notin I_C$. To see that $C \mapsto I_C$ is Borel, let $A \subseteq X^2$ be a Borel set. Then

$$x \in A_I \iff \{y \in [x] : (x,y) \in A\} \in I_{[x]} \iff (x, s(x)) \notin A$$

is Borel.

($\Leftarrow$) Suppose E is smooth and idealistic. Let $f : X \to 2^\omega$ witness that $E \leq_B 2^\omega$. By Theorem 4.2.3 we may assume X is Polish and f is continuous. Let F be the graph of f, that is, $(x,y) \in F \iff f(x) = y$. Then $F \subseteq X \times 2^\omega$ is closed. By the usual argument we can obtain a sequence $(F_s)_{s \in \omega^{<\omega}}$ of nonempty closed subsets of $X \times 2^\omega$ so that $F_\emptyset = F$, $F_s = \bigcup_{n \in \omega} F_{s^\frown n}$, and $\operatorname{diam}(F_s) < 2^{-\operatorname{lh}(s)}$.

For each $y \in 2^\omega$ and $s \in \omega^{<\omega}$, let $F_s^y = \{x \in X : (x,y) \in F_s\}$. Then for any E-equivalence class $C = [x]_E$, if $f(x) = y$ then $F_\emptyset^y = C$ and for any $s \in \omega^{<\omega}$, $F_s^y = \bigcup_{n \in \omega} F_{s^\frown n}^y$. Now let $C \mapsto I_C$ witness that E is idealistic. Then $F_\emptyset^y \notin I_C$, and moreover if $F_s^y \notin I_C$ then for some $n \in \omega$, $F_{s^\frown n}^y \notin I_C$.

Thus if we define $T^y = \{s \in \omega^{<\omega} : F_s^y \notin I_C\}$, then T^y is a tree on ω. Let $g(y)$ be the leftmost branch of T^y. It is straightforward to check that g is a Borel function. Furthermore, $\bigcap_{n \in \omega} F_{g(y)\restriction n}^y$ is a singleton, and we name its unique element $h(y)$. It is clear that $(h(y), y) \in F$ and therefore $h(y) \in C$. Let H be the set of all $h(y)$ thus obtained. It is clear that H is a transversal for E on X. To finish the proof it suffices to show that H is Borel. For this, simply note that

$$x \in H \iff \forall n \in \omega \ (x \in F_{g(y)\restriction n}^y).$$

□

Corollary 5.4.12 (Burgess)
Let G be a Polish group and X a Borel G-space. If E_G^X is smooth then it has a Borel selector.

Exercise 5.4.1 Show that for any equivalence relation E on a Polish space, $\mathrm{id}(\omega) \leq_B E$ iff there are infinitely many E-equivalence classes.

Exercise 5.4.2 Show that if an equivalence relation E is smooth then either $E \sim_B \mathrm{id}(2^\omega)$, or $E \sim_B \mathrm{id}(\omega)$, or for some finite $n \in \omega$, $E \sim_B \mathrm{id}(n)$, where $\mathrm{id}(n)$ denotes the identity relation on any finite set of size n.

Exercise 5.4.3 Let E be an equivalence relation on a Polish space X. Show that E is smooth iff there is a finer Polish topology σ on X such that E is closed in $(X^2, \sigma \times \sigma)$.

Exercise 5.4.4 Show that an intersection of countably many smooth equivalence relations on a Polish space is smooth.

Exercise 5.4.5 Let G be a compact Polish group and X a Borel G-space. Show that E_G^X is smooth.

Exercise 5.4.6 Let $F \subseteq \omega^\omega \times \omega^\omega$ be closed so that $\{x : \exists y\, (x, y) \in F\}$ is Σ_1^1 but not Borel. Let $X = F$ and define the equivalence relation E on X by $(x_1, y_1) E(x_2, y_2) \iff x_1 = x_2$. Show that E is closed but has no Borel selector.

Exercise 5.4.7 Show that the orbit equivalence relation induced by the conjugacy action of S_∞ on itself is smooth.

Exercise 5.4.8 Let G be a Polish group and X be a Polish G-space. Suppose every orbit of E_G^X is G_δ. Show that E_G^X is smooth. (*Hint*: The map $x \mapsto \overline{[x]_G}$ is a Borel reduction.)

Chapter 6

The Glimm–Effros Dichotomy

The Glimm–Effros dichotomy is undoubtedly the most important theoretical result that helped form the subject of invariant descriptive set theory. In the context of operator algebra Glimm and Effros have obtained the dichotomy for locally compact group actions and more generally F_σ orbit equivalence relations. Nowadays the terminology usually refers to the remarkable theorem for all Borel equivalence relations proved by Harrington, Kechris, and Louveau. The theorem and its proof were so influential that for a while the main activity of the field was to prove new dichotomy theorems. For orbit equivalence relations the Glimm–Effros dichotomy have been obtained by Solecki, Hjorth, Becker, and others. Another remarkable connection with other fields of mathematics is the study of hyperfinite equivalence relations. To this date there are still intriguing open problems around which exciting research is actively going on.

6.1 The equivalence relation E_0

Definition 6.1.1
The equivalence relation E_0 is the relation of eventual agreement on 2^ω, that is, it is defined by

$$x E_0 y \iff \exists m \, \forall n \geq m \, x(n) = y(n).$$

We also consider the eventual agreement relation on ω^ω, denoted by $E_0(\omega)$.

Proposition 6.1.2
$E_0 \sim_B E_0(\omega)$.

Proof. The identity embedding from 2^ω into ω^ω is a reduction from E_0 to $E_0(\omega)$. To define a reduction f to witness $E_0(\omega) \leq_B E_0$, fix a computable bijection $\langle \cdot, \cdot \rangle : \omega \times \omega \to \omega$. Given $x \in \omega^\omega$, let $f(x) \in 2^\omega$ be defined by

$$f(x)(\langle n, k \rangle) = \text{ the } (k+1)\text{-th least significant digit of} \\ \text{the binary expansion of } x(n).$$

f is actually continuous, even computable.

To see that it is a desired reduction, assume that $xE_0(\omega)y$. Suppose $x(n) = y(n)$ for $n \geq m$. Then for all $n \geq m$ and $k \in \omega$, $f(x)(\langle n, k \rangle) = f(y)(\langle n, k \rangle)$. For $n < m$, since the binary expansions of $x(n)$ and $y(n)$ are of finite lengths, there is k_n such that for all $k \geq k_n$, $f(x)(\langle n, k \rangle) = 0 = f(y)(\langle n, k \rangle)$. Thus for any $n, k \in \omega$, if $f(x)(\langle n, k \rangle) \neq f(y)(\langle n, k \rangle)$ we must have that $n < m$ and $k < k_n$, and there are only finitely many such pairs n, k. This shows that $f(x)E_0 f(y)$.

Conversely, if $x \not\mathrel{E_0} y$, then the set $A = \{n \in \omega : x(n) \neq y(n)\}$ is infinite. For each $n \in A$, let k_n be such that the $(k_n + 1)$-th least significant digits of binary expansions of $x(n)$ and $y(n)$ differ. Then we have that $f(x)(\langle n, k_n \rangle) \neq f(y)(\langle n, k_n \rangle)$ for all $n \in A$. This means that $f(x)$ and $f(y)$ are not $E_0(\omega)$-equivalent. ☐

Definition 6.1.3
The **Vitali equivalence relation** E_v is the equivalence relation on $\mathbb{R}$ defined by

$$xE_v y \iff x - y \in \mathbb{Q}.$$

The terminology is motivated by the theorem of Vitali in real analysis that any transversal of E_v is not Lebesgue measurable.

Proposition 6.1.4
$E_0 \sim_B E_v$.

Proof.   We first show that $E_v \leq_B E_0$. Given $x \in \mathbb{R}$, express x as

$$x = a_0 + \frac{a_1}{2!} + \frac{a_2}{3!} + \cdots + \frac{a_n}{(n+1)!} + \cdots$$

where $a_0 = \lfloor x \rfloor$ and for $n > 0$, $a_n \in \{0, 1, \ldots, n\}$. Note that such an expression is uniquely determined. In fact, one can inductively define that

$$a_n = \lfloor (x - a_0 - \frac{a_1}{2!} - \cdots - \frac{a_{n-1}}{n!})(n+1)! \rfloor.$$

This allows us to define a function $f : \mathbb{R} \to \omega^\omega$ by $f(x)(n) = a_n$. It is clear that f is Borel (but not continuous). To see that f is a reduction, let $x, y \in \mathbb{R}$. First assume that $xE_v y$. Without loss of generality assume $x > y$ and let $r = x - y$. Let $m \in \omega$ be the least such that $r = \dfrac{k}{(m+1)!}$ for some $k \in \omega$. Then for all $n > m$, $f(x)(n) = f(y)(n)$ from the definition of f. Conversely, assume $f(x)E_0 f(y)$. Then the expressions for x and y differ by only finitely many terms, and it is clear that $x - y$ is rational.

Thus we have shown that $E_v \leq_B E_0(\omega)$. But since $E_0(\omega) \sim_B E_0$, we have that $E_v \leq_B E_0$.

To show the converse, that $E_0 \leq_B E_v$, we follow the same idea. Let $x \in 2^\omega$ be given. We define

$$g(x) = x(0) + \frac{x(1)}{2!} + \frac{x(2)}{3!} + \cdots + \frac{x(n)}{(n+1)!} + \cdots .$$

The series is absolutely convergent, and thus $g(x)$ is well-defined. It follows from the above argument that $x E_0 y$ iff $g(x) - g(y) \in \mathbb{Q}$. Therefore g is a Borel reduction from E_0 to E_v. □

Note that every E_0-equivalence class is countable. In fact, E_0 can be viewed as an orbit equivalence relation of a countable group action, where the acting group is $\mathbb{Z}_2^{<\omega}$, the direct sum of infinitely many copies of the two-element group $\mathbb{Z}_2$. For this reason we also refer to the E_0-equivalence classes as E_0-orbits.

Also note that every E_0-orbit is dense. It follows that every E_0-orbit is not G_δ. In fact, a dense G_δ orbit would be comeager, whereas every E_0-orbit is countable, and therefore meager. These obvious facts distinguish E_0 from all smooth equivalence relations. In the following we give two proofs of the nonsmoothness of E_0.

Definition 6.1.5
Let X be a standard Borel space and E an equivalence relation on X. A σ-finite nonzero Borel measure μ on X is $(E-)$**ergodic** if for any E-invariant Borel set $A \subseteq X$, either $\mu(A) = 0$ or $\mu(X - A) = 0$. The measure μ is $(E-)$**nonatomic** if $\mu(C) = 0$ for any E-equivalence class C.

Proposition 6.1.6
Let E be an equivalence relation on a standard Borel space. If there is an E-nonatomic, E-ergodic measure on X, then E is not smooth.

Proof. Let μ be E-nonatomic and E-ergodic. Assume E is smooth, and $\{S_n\}_{n \in \omega}$ a countable Borel generating family for E. Since each S_n is E-invariant, either $\mu(S_n) = 0$ or $\mu(X - S_n) = 0$. Consider the set

$$C = \bigcap \{X - S_n : \mu(S_n) = 0\} \cap \bigcap \{S_n : \mu(X - S_n) = 0\}.$$

Then $\mu(X - C) = 0$. But C is an equivalence class, hence $\mu(C) = 0$. It follows that $\mu(X) = 0$, a contradiction. □

Proposition 6.1.7
The usual product measure on 2^ω is E_0-nonatomic and E_0-ergodic. Hence E_0 is not smooth.

Proof. Let μ be the product measure on 2^ω. Since μ is atomless, that is, $\mu(\{x\}) = 0$ for any $x \in 2^\omega$, and every E_0-orbit is countable, μ is E_0-nonatomic.

To see that μ is also E_0-ergodic, let $A \subseteq 2^\omega$ be an E_0-invariant Borel set. Suppose $\mu(A) > 0$. Let $\epsilon > 0$. Then there exists a basic open set N_s such that $\mu(A \cap N_s)/\mu(N_s) > 1 - \epsilon$. We claim then that for any $t \in 2^{<\omega}$ with $\mathrm{lh}(t) = \mathrm{lh}(s)$, $\mu(A \cap N_t)/\mu(N_t) > 1 - \epsilon$. In fact, let $\varphi_{s,t} : 2^\omega \to 2^\omega$ be given by

$$\varphi_{s,t}(x) = \begin{cases} x, & \text{if } s \not\subseteq x \text{ and } t \not\subseteq x, \\ t^\frown y, & \text{if } x = s^\frown y, \\ s^\frown y, & \text{if } x = t^\frown y. \end{cases}$$

Then $\varphi_{s,t}(N_s) = N_t$ and $\varphi_{s,t}(A \cap N_s) = A \cap N_t$, and moreover $\varphi_{s,t}$ preserves the measure μ. It follows that $\mu(A) > 1 - \epsilon$ for any $\epsilon > 0$, and therefore indeed $\mu(A) = 1$. □

Definition 6.1.8
Let X be a Polish space and E an equivalence relation on X. We say that E is **generically ergodic** if every E-invariant Borel set is either meager or comeager.

Proposition 6.1.9
Let G be a group of homeomorphisms on a Polish space X. Then the following are equivalent:

(i) *E_G is generically ergodic;*

(ii) *Every nonempty E_G-invariant open set is dense;*

(iii) *There is an invariant dense G_δ set $Y \subseteq X$ all of whose orbits are dense in X;*

(iv) *There exists a dense orbit;*

(v) *Every invariant set with the Baire property is meager or comeager.*

Proof. The implications (i)$\Rightarrow$(ii), (iii)$\Rightarrow$(iv), and (v)$\Rightarrow$(i) are obvious. The implication (iv)$\Rightarrow$(v) follows from Exercise 3.3.2. It remains to show (ii)$\Rightarrow$(iii). Let $\{U_n\}_{n\in\omega}$ be a countable base for the topology of X. Every saturation $[U_n]_G$ is E_G-invariant open, and therefore dense. Let $Y = \bigcap_n [U_n]_G$. Then Y is an E_G-invariant dense G_δ set. If $x \in Y$, then $[x]_G \subseteq [U_n]_G$ and therefore $[x]_G \cap U_n \neq \emptyset$. Thus every orbit in Y is dense. □

Now we have an analog of Proposition 6.1.6 for generically ergodic equivalence relations, whose proof we leave as an exercise (Exercise 6.1.4).

Proposition 6.1.10
Let E be an equivalence relation on a Polish space X. If E is generically ergodic and has no comeager orbits, then E is not smooth.

Exercise 6.1.1 Define the **tail equivalence** relation E_t on 2^ω by

$$x E_t y \iff \exists n \exists m \forall k\, (\, x(n+k) = y(m+k)\,).$$

Show that E_t is not smooth.

Exercise 6.1.2 Show that the Lebesgue measure on $\mathbb{R}$ is E_v-nonatomic and E_v-ergodic.

Exercise 6.1.3 Show that if $E_0 \leq_B E$ then there is an E-nonatomic, E-ergodic measure.

Exercise 6.1.4 Prove Proposition 6.1.10.

Exercise 6.1.5 Show that both E_0 and E_v are generically ergodic.

Exercise 6.1.6 Let X, Y be Polish spaces and G, H be groups of homeomorphisms on X, Y respectively. Show that if E_G^X is generically ergodic and $E_G^X \leq_c E_H^Y$ then there is a closed invariant subspace Z of Y so that E_H^Z is generically ergodic.

Exercise 6.1.7 Let G be a countably infinite group acting by shift on 2^G. Let E_G be the induced orbit equivalence relation. Show that E_G is generically ergodic.

6.2 Orbit equivalence relations embedding E_0

In this section we give a construction of an embedding of E_0 into a given equivalence relation. In particular we will derive the original Glimm–Effros dichotomy theorems.

Most theorems about E_0-embeddability require that the equivalence relation is dense and meager. However, we first note that more needs to be assumed. Take the example of the following equivalence relation E on $\mathbb{R}$:

$$x E y \iff x = y \lor (x, y \in \mathbb{Q}).$$

Then E is dense and meager as a subset of $\mathbb{R}^2$, but E is smooth. The following theorem assumes a condition weaker than a continuous action of a Polish group.

Theorem 6.2.1 (Becker–Kechris)
Let X be a Polish space and G a group of homeomorphisms of X. Suppose E_G is meager and that there is a dense orbit. Then $E_0 \sqsubseteq_c E_G$.

Proof. Let D be a dense orbit and $\{W_n\}_{n\in\omega}$ be a decreasing sequence of dense open subsets of X^2 with $E_G \cap \bigcap_n W_n = \emptyset$. Without loss of generality assume that $W_0 \cap \{(x,x) : x \in X\} = \emptyset$. We construct a sequence $\{U_s\}_{s\in 2^{<\omega}}$ of open sets and elements $g_{s,t} \in G$ for $s,t \in 2^{<\omega}$ and $\mathrm{lh}(s) = \mathrm{lh}(t)$, such that, for all $s,t,u \in 2^{<\omega}$ with $\mathrm{lh}(s) = \mathrm{lh}(t) = \mathrm{lh}(u)$:

(a) $U_\emptyset = X$; $\mathrm{diam}(U_s) < 2^{-\mathrm{lh}(s)}$; $\overline{U_{s^\frown 0}}, \overline{U_{s^\frown 1}} \subseteq U_s$; $U_{s^\frown 0} \cap U_{s^\frown 1} = \emptyset$;

(b) If $\mathrm{lh}(s) = \mathrm{lh}(t) = n$ and $s(n-1) \neq t(n-1)$, then $U_s \times U_t \subseteq W_n$;

(c) $g_{s,t} \cdot U_s = U_t$; $g_{s,s} = 1_G$; $g_{t,s} = g_{s,t}^{-1}$; $g_{s,u} = g_{t,u} g_{s,t}$;

(d) If $n \leq \mathrm{lh}(s) = \mathrm{lh}(t)$ is the largest with $s(n-1) \neq t(n-1)$, then $g_{s,t} = g_{s\restriction n, t\restriction n}$.

Granting the construction, we may continuously embed E_0 into E_G as follows. For $x \in 2^\omega$, we let $f(x)$ be the unique element of the singleton $\bigcap_n U_{x\restriction n}$. Then f is a continuous embedding of 2^ω into X. We claim that f is also a reduction from E_0 into E_G. Suppose first $x \not\mathrel{E_0} y$. Then for infinitely many n, $x(n-1) \neq y(n-1)$. By (c), $U_{x\restriction n} \times U_{y\restriction n} \subseteq W_n$ for infinitely many n, and therefore $(f(x), f(y)) \in \bigcap_n W_n \subseteq X^2 - E_G$. On the other hand, suppose $x E_0 y$ and n is the largest such that $x(n-1) \neq y(n-1)$. Let $g = g_{x\restriction n, y\restriction n}$. Then by (d) $g_{x\restriction m, y\restriction m} = g$ for any $m \geq n$. Since $g \cdot U_{x\restriction m} = U_{y\restriction m}$ for all $m \geq n$, we have that $g \cdot f(x) = f(y)$, and thus $f(x) E_G f(y)$.

The construction is by induction on $\mathrm{lh}(s)$. We start with $\mathrm{lh}(s) = 1$. At this stage we need to find disjoint open sets $U_{\langle 0 \rangle}, U_{\langle 1 \rangle}$ and $g_{\langle 0 \rangle, \langle 1 \rangle} \in G$ such that $\mathrm{diam}(U_{\langle 0 \rangle}), \mathrm{diam}(U_{\langle 1 \rangle}) < 1/2$, $g_{\langle 0 \rangle, \langle 1 \rangle} \cdot U_{\langle 0 \rangle} = U_{\langle 1 \rangle}$, and $U_{\langle 0 \rangle} \times U_{\langle 1 \rangle} \subseteq W_1$. To do this we first let $V_0, V_1 \subseteq X$ be open sets such that $V_0 \times V_1 \subseteq W_1$. Without loss of generality we may assume $\mathrm{diam}(V_0), \mathrm{diam}(V_1) < 1/2$. By the density of D we can find $x_0 \in D \cap V_0$ and $x_1 \in D \cap V_1$. Let $g_{\langle 0 \rangle, \langle 1 \rangle}$ be an arbitrary $g \in G$ such that $g \cdot x_0 = x_1$. Let now $U_{\langle 0 \rangle} = g^{-1}(g \cdot V_0 \cap V_1)$ and $U_{\langle 1 \rangle} = g \cdot U_{\langle 0 \rangle}$. Since $U_{\langle 0 \rangle} \subseteq V_0$ and $U_{\langle 1 \rangle} \subseteq V_1$, we have the required properties.

In the inductive step assume U_s and $g_{s,t}$ have been defined for all s,t with $\mathrm{lh}(s) = \mathrm{lh}(t) \leq n$. We need to define $U_{s^\frown i}$ and $g_{s^\frown i, t^\frown j}$ for all s,t with $\mathrm{lh}(s) = \mathrm{lh}(t) = n$ and $i,j \in \{0,1\}$. First note that it is required by the condition (d) that $g_{s^\frown i, t^\frown i} = g_{s,t}$ for all s,t with $\mathrm{lh}(s) = \mathrm{lh}(t) = n$ and $i \in \{0,1\}$. Therefore in view of (c) all $g_{s^\frown i, t^\frown j}$ will be determined once we have defined $g_{\vec{0}^\frown 0, \vec{0}^\frown 1}$, where $\vec{0}$ here stands for the element s with $\mathrm{lh}(s) = n$ and $s(k) = 0$ for all $k < n$. Similarly, all $U_{s^\frown i}$ will be determined once we have defined $U_{\vec{0}^\frown 0}$.

As in the base step we start with choosing disjoint open sets $V_{\vec{0}^\frown 0}$ and $V_{\vec{0}^\frown 1}$, both of diameter $< 2^{-n-1}$, whose closures are contained in $U_{\vec{0}}$, and so that $V_{\vec{0}^\frown 0} \times V_{\vec{0}^\frown 1} \subseteq W_{n+1}$. The last requirement can be fulfilled since W_{n+1} is dense open. For any other s with $\mathrm{lh}(s) = n$, we can then let $V_{s^\frown 0} = g_{\vec{0},s} \cdot V_{\vec{0}^\frown 0}$ and $V_{s^\frown 1} = g_{\vec{0},s} \cdot V_{\vec{0}^\frown 1}$. Note that $V_{s^\frown 0}$ and $V_{s^\frown 1}$ are disjoint, open, and their closures are contained in U_s. By shrinking we may assume also that each of them has diameter $< 2^{-n-1}$. Note that in the shrinking process we always maintain that $g_{\vec{0},s} \cdot V_{\vec{0}^\frown i} = V_{s^\frown i}$.

We next go through a finite number of steps to shrink the sets $V_{s^\frown i}$ so as to guarantee that $V_{s^\frown 0} \times V_{t^\frown 1} \subseteq W_{n+1}$. Again in this shrinking process we always maintain that $g_{\vec{0},s} \cdot V_{\vec{0}^\frown i} = V_{s^\frown i}$. To illustrate the argument we consider $V_{\vec{0}^\frown 0}$ and $V_{\vec{0}^\frown 1}$. Since W_{n+1} is dense open, it meets the set $V_{\vec{0}^\frown 0} \times V_{\vec{0}^\frown 1}$. Thus we could shrink both of them so as to guarantee that $V_{\vec{0}^\frown 0} \times V_{\vec{0}^\frown 1} \subseteq W_{n+1}$. Remember this could result in other sets being shrunken too. In general, as we go through all possible pairs $V_{s^\frown 0}$ and $V_{t^\frown 1}$, each of the sets in the collection could be shrunken, but the requirements fulfilled in the earlier process are maintained. Thus at the end of this shrinking process, we obtained open sets satisfying the properties prescribed by (a) and (b). Now by the density of D we may find $x_{\vec{0}^\frown 0} \in V_{\vec{0}^\frown 0}$ and $x_{\vec{0}^\frown 1} \in V_{\vec{0}^\frown 1}$. Let $g_{\vec{0}^\frown 0,\vec{0}^\frown 1}$ be an arbitrary $g \in G$ such that $g \cdot x_{\vec{0}^\frown 0} = x_{\vec{0}^\frown 1}$. Finally let

$$U_{\vec{0}^\frown 0} = g^{-1}(g \cdot V_{\vec{0}^\frown 0} \cap V_{\vec{0}^\frown 1}).$$

As remarked above, all other sets $U_{s^\frown i}$ and elements $g_{s^\frown i,t^\frown j}$ are determined.

This finishes the proof of the theorem. ∎

Corollary 6.2.2
Let G be a Polish group and X a Polish G-space. Suppose there is no G_δ orbit. Then $E_0 \sqsubseteq_c E_G^X$.

Proof. Let $x \in X$ be arbitrary and consider $Y = \overline{[x]_G}$. Then Y is a Polish G-space and $E_G^Y \sqsubseteq_c E_G^X$ via the identity embedding. In Y there is obviously a dense orbit. Since there is no G_δ orbit in X, the same is true for Y. It follows from Effros' theorem that every orbit is meager in itself, and in particular meager in Y. Since E_G^Y is analytic, it has the Baire property. By the Kuratowski–Ulam theorem E_G^Y is meager. Thus $E_0 \sqsubseteq_c E_G^Y$ and so $E_0 \sqsubseteq_c E_G^X$. ∎

The corollary is very useful since it explicitly relates the descriptive complexity of orbits to the complexity of the entire equivalence relation. It is usually easier to investigate the complexity of a particular orbit. For instance, in a Baire space a dense F_σ set with a dense complement is not G_δ, since otherwise both the set and its complement would be dense G_δ, hence having a nonempty intersection. Using this it is easy to show that E_0 and E_v have no G_δ orbits.

The following theorem establishes a strong dichotomy if the Polish action meets some strong requirement. From it we can deduce the original Glimm–Effros dichotomy theorems.

Theorem 6.2.3
Let G be a Polish group and X a Polish G-space. Suppose every G_δ orbit is also F_σ. Then either every orbit is G_δ or else $E_0 \sqsubseteq_c E_G^X$. In particular, either E_G^X is smooth or else $E_0 \sqsubseteq_c E_G^X$.

Proof. Suppose $E_0 \not\sqsubseteq_c E_G^X$. We show that every orbit is G_δ. Let $x \in X$. Consider
$$Y = \{y \in X : \overline{[y]_G} = \overline{[x]_G}\}.$$

It is easy to see that Y is invariant. We also note that Y is G_δ. To see this, let $\mathcal{U}$ be a countable base for the topology of X. Then

$$y \in Y \iff \forall U \in \mathcal{U}\,(U \cap [x]_G \neq \emptyset \Rightarrow U \cap [y]_G \neq \emptyset) \land$$
$$\forall U \in \mathcal{U}\,(U \cap [y]_G \neq \emptyset \Rightarrow U \cap [x]_G \neq \emptyset)$$

$$\iff \forall U \in \mathcal{U}\,(U \cap [x]_G \neq \emptyset \Rightarrow y \in [U]_G) \land$$
$$\forall U \in \mathcal{U}\,(U \cap [x]_G = \emptyset \Rightarrow y \notin [U]_G).$$

Since $[U]_G$ is an open set, the first condition is G_δ and the second is closed. It follows that Y is Polish, and in fact a Polish G-space with the inherited action of G. Since $E_G^Y \sqsubseteq_c E_G^X$ via the identity embedding, we have that $E_0 \not\sqsubseteq_c E_G^Y$. Now by Corollary 6.2.2 there is a G_δ orbit in Y. Let $[y]_G \subseteq Y$ be G_δ.

It is also true that every orbit of Y is dense. Hence Y cannot contain more than one orbit. In fact, by our assumption $Y - [y]_G$ is also G_δ, hence if it were nonempty then it would be dense G_δ and meeting $[y]_G$, a contradiction. It follows that $Y = [x]_G$, and hence $[x]_G$ is G_δ.

Finally by Exercise 5.4.8 if every orbit of E_G^X is G_δ then E_G^X is smooth. ∎

Corollary 6.2.4 (Effros)
Let G be a Polish group and X a Polish G-space. Suppose E_G^X is F_σ. Then either E_G^X is smooth or else $E_0 \sqsubseteq_c E_G^X$.

Proof. Since the equivalence relation is F_σ every orbit is F_σ too; in particular, every G_δ orbit is also F_σ. ∎

Corollary 6.2.4 is the original **Glimm–Effros dichotomy theorem**, proved by Effros, strengthening earlier work of Glimm for locally compact Polish group actions (see Exercise 6.2.2).

Exercise 6.2.1 Consider the following equivalence relation E on $\mathbb{R}^\omega$:

$$xEy \iff x - y \in \ell_1 \iff \sum_{n=0}^{\infty} |x(n) - y(n)| < \infty.$$

Show that $E_0 \sqsubseteq_c E$.

Exercise 6.2.2 (Glimm) Let G be a locally compact Polish group and X a Polish G-space. Then either E_G^X is smooth or else $E_0 \sqsubseteq_c E_G^X$.

Exercise 6.2.3 Let G be a Polish group and X a Polish G-space. Define an equivalence relation F on X by

$$xFy \iff \overline{[x]_G} = \overline{[y]_G}.$$

(a) Verify that F is an equivalence relation and show that $E_G^X \subseteq F$.

(b) Show that F is G_δ.

(c) Show that $xFy \iff$ for all E_G^X-invariant open (or closed) sets C, $x \in C$ iff $y \in C$.

6.3 The Harrington–Kechris–Louveau theorem

In this section we prove the Glimm–Effros dichotomy theorem for Borel equivalence relations [64]. This is again a theorem of classical character for which the only known proof uses effectively descriptive set theory.

Theorem 6.3.1 (Harrington–Kechris–Louveau)
Let E be a Borel equivalence relation on a Polish space X. Then either E is smooth or else $E_0 \sqsubseteq_c E$.

The real theorem being proved is the following effective version of the Glimm–Effros dichotomy. Theorem 6.3.1 follows immediately by relativization.

Theorem 6.3.2
Let E be a Δ_1^1 equivalence relation on ω^ω. Then either there is a Δ_1^1 generating family for E or else $E_0 \sqsubseteq_c E$.

The rest of this section is devoted to a proof of Theorem 6.3.2. We leave further remarks and corollaries to the next section. First we state an effective version of Lemma 5.4.6. The proof is left as an exercise (Exercise 6.3.1).

Lemma 6.3.3
Let E be a Σ_1^1 equivalence relation on ω^ω. Let A, B be disjoint Σ_1^1 E-invariant sets. Then there is a Δ_1^1 E-invariant set C with $A \subseteq C$ and $B \cap C = \emptyset$.

Next for any Σ_1^1 equivalence relation E on ω^ω we define a derived equivalence relation E^* by

$$xE^*y \iff \text{for all } \Delta_1^1 \text{ } E\text{-invariant sets } C, x \in C \text{ iff } y \in C.$$

Then the following lemma collects some basic properties of E^*.

Lemma 6.3.4
$E^* \supseteq E$ is a Σ_1^1 equivalence relation with a Δ_1^1 generating family.

Proof. It follows straightforward from the definition of E^* that $E \subseteq E^*$, E^* is an equivalence relation, and that E^* has a Δ_1^1 generating family, that is, the family of all Δ_1^1 E-invariant sets. We show that E^* is Σ_1^1. Using the coding in Theorem 1.7.4 we have

$$xE^*y \iff \forall n \{\exists w \exists z \, (w \notin P_n^- \wedge z \notin P_n^+ \wedge wEz) \vee \\ [(x \in P_n^+ \Rightarrow y \notin P_n^-) \wedge (y \in P_n^+ \Rightarrow x \notin P_n^-)]\}.$$

Thus E^* is Σ_1^1. □

To consider further properties of E^* we take into account the Gandy–Harrington topologies. Let τ be the Gandy–Harrington topology on ω^ω. Let $\tau^2 = \tau \times \tau$ be the product topology on $(\omega^\omega)^2$ and τ_2 be the Gandy–Harrington topology on $(\omega^\omega)^2$. Then $\tau^2 \subseteq \tau_2$.

Lemma 6.3.5
E^* is G_δ in τ^2.

Proof. Let $\mathcal{C}$ be the (countable) family of all Δ_1^1 E-invariant sets. Then

$$E^* = \bigcap_{C \in \mathcal{C}} \{(C \times C) \cup [(\omega^\omega - C) \times (\omega^\omega - C)]\}.$$

□

Since (ω^ω, τ) is a second countable Baire space, it follows that E^* is Baire.

Lemma 6.3.6
E^* is the closure of E in τ^2.

Proof. We first show that E is dense in E^*. For this we show that, for any Σ_1^1 $A, B \subseteq \omega^\omega$, if $(A \times B) \cap E = \emptyset$, then $(A \times B) \cap E^* = \emptyset$. Assume A, B are Σ_1^1 sets with $(A \times B) \cap E = \emptyset$. Then $[A]_E \cap [B]_E = \emptyset$. Since both $[A]_E$ and $[B]_E$ are Σ_1^1, by Lemma 6.3.3 there is a Δ_1^1 E-invariant set C with $A \subseteq C$ and $B \cap C = \emptyset$. Now for any $x \in A$ and $y \in B$, $x \in C$ but $y \notin C$, hence $x \not\!\!E^* y$. This means $(A \times B) \cap E^* = \emptyset$ as required.

Next we show that E^* is closed in τ^2. For this let $(x, y) \notin E^*$, then there is a Δ_1^1 E-invariant set C with $x \in C$ and $y \notin C$. Let $A = C$ and $B = \omega^\omega - C$. Then both A and B are Σ_1^1 sets, and $(A \times B) \cap E^* = \emptyset$. This shows that $(\omega^\omega)^2 - E^*$ is open in τ^2, and therefore E^* is closed. □

If $E^* = E$ then E has a Δ_1^1 generating family and the first alternative of the conclusion of Theorem 6.3.2 is obtained. From now on we assume $E^* \neq E$. Consider the set

$$X = \{x \in \omega^\omega : [x]_{E^*} \neq [x]_E\}.$$

Then X is nonempty. If in addition E is Δ_1^1 then X is Σ_1^1, since

$$x \in X \iff \exists y\, (yE^*x \wedge y \not\!\!E\, x).$$

From now on we also assume that E is Δ_1^1.

We are now ready to take category arguments into account.

Lemma 6.3.7
If $A, B \subseteq X$ are nonempty Σ_1^1 sets such that E is comeager in $(A \times B) \cap E^*$, then $(A \times A) \cap E^* \subseteq E$.

Proof. Consider $Y = \{(x, y, z) \in A \times A \times B : xE^*yE^*z\}$ with the $\tau_2 \times \tau$ topology. Similar to Lemma 6.3.5 we have that

$$Y = (A \times A \times B) \cap \bigcap_{C \in \mathcal{C}} \{(C \times C \times C) \cup [(\omega^\omega - C) \times (\omega^\omega - C) \times (\omega^\omega - C)]\},$$

where $\mathcal{C}$ is the countable family of all Δ_1^1 E-invariant sets. Thus Y is G_δ in $\tau^2 \times \tau$ as well as in $\tau_2 \times \tau$. Let $Y_1 = \{(x, y, z) \in Y : xEz\}$ and $Y_2 = \{(x, y, z) \in Y : yEz\}$. Since E is comeager in $(A \times B) \cap E^*$, both Y_1 and Y_2 are comeager in Y.

Now suppose $(A \times A) \cap E^* \not\subseteq E$. Consider $Y_3 = \{(x, y, z) \in Y : x \not\!\!E\, y\}$. Then Y_3 is nonempty. Since E is Δ_1^1, Y_3 is Σ_1^1 and open in $\tau_2 \times \tau$. It follows that $Y_1 \cap Y_2 \cap Y_3 \neq \emptyset$. But if $(x, y, z) \in Y_1 \cap Y_2 \cap Y_3$, then xEz, yEz, and $x \not\!\!E\, y$, a contradiction. $\quad\square$

Lemma 6.3.8
E is dense and meager in $X^2 \cap E^*$ in τ^2.

Proof. Since X^2 is Σ_1^1, and thus open in τ^2, the density is immediate from Lemma 6.3.6. Assume toward a contradiction that E is nonmeager in $X^2 \cap E^*$. Then there exist nonempty Σ_1^1 sets $A, B \subseteq X$ such that E is comeager in $(A \times B) \cap E^*$. By the preceding lemma we have that $(A \times A) \cap E^* \subseteq E$. Therefore in fact $([A]_E \times [A]_E) \cap E^* \subseteq E$. We claim that $[A]_E = [A]_{E^*}$. To see this, assume that $[A]_E \neq [A]_{E^*}$. Then the set $Z = \{x \in [A]_{E^*} : \exists y \in [A]_E\,(xE^*y \wedge x \not\!\!E\, y)\}$ is nonempty, Σ_1^1, and $(Z \times [A]_E) \cap E^* \neq \emptyset$. By the density of E we have that $(Z \times [A]_E) \cap E \neq \emptyset$. But if $z \in Z$, $x \in [A]_E$, and xEz then there is $y \in [A]_E$ with zE^*y and $z \not\!\!E\, y$; since $(x, y) \in ([A]_E \times [A]_E) \cap E^* \subseteq E$, we have xEy, a contradiction.

Finally we claim that for any $x \in A$, $[x]_E = [x]_{E^*}$, a contradiction to the assumption that $x \in X$. So let $x \in A$ and yE^*x. Then $y \in [A]_{E^*} = [A]_E$, and hence $(x, y) \in ([A]_E \times [A]_E) \cap E^* \subseteq E$. This means that xEy and $y \in [x]_E$. $\quad\square$

We are finally ready to embed E_0 into E. Let us comment on the main ideas before giving the construction itself. As usual we will construct a sequence $\{U_s\}_{s \in 2^{<\omega}}$ of open sets and define the embedding from 2^ω by letting $f(z)$ to be the unique element of the singleton $\bigcap_n U_{z \restriction n}$. However, since we are working with the Gandy–Harrington topology, the strong Choquet property is used to guarantee the nonemptiness of the intersection $\bigcap_n U_{z \restriction n}$ in place of complete metrizability. For this we need to play the strong Choquet game and let U_s be coming from a winning strategy for Player II. Thus for each $s \in 2^{<\omega}$ we need to find x_s and V_s as Player I's move.

To make sure we are creating a reduction of E_0 into E a scheme similar to the proof of Theorem 6.2.1 is used. To ensure that $z \not\mathrel{E}_0 y \Rightarrow f(z) \not\mathrel{E} f(y)$ we ask that $U_s \times U_t$ be contained in some dense open set avoiding the meager relation E if s, t have the same length but their last entries differ. For the other implication, namely $z E_0 y \Rightarrow f(z) E f(y)$, the difficulty now is that no group action is present to provide the uniform linkage we had in the proof of Theorem 6.2.1. The alternative is also making use of the strong Choquet property, this time on the space of pairs, to guarantee that appropriate linkage exists among the elements x_s. When $z E_0 y$, the pairs $(x_{z \restriction n}, x_{y \restriction n})$ will converge to a unique linked pair, forcing that $f(z) E f(y)$.

To deal with the combinatorics associated with the linkage of elements x_s, we introduce the following notation. For $s, t \in 2^{<\omega}$ with $\mathrm{lh}(s) = \mathrm{lh}(t)$ and $k \in \omega$, define a relation R_k by

$$s R_k t \iff \forall i < k \; s(i) = t(i) = 0 \; \wedge \; s(k) \neq t(k) \; \wedge \; \forall k < i < \mathrm{lh}(s) \; s(i) = t(i).$$

Note that if $k \neq k'$ then $R_k \cap R_{k'} = \emptyset$. If $s R_k t$ then $(s^\frown u) R_k (t^\frown u)$ for any $u \in 2^{<\omega}$. Also if $s R_k t$ and $k < n < \mathrm{lh}(s)$ then $(s \restriction n) R_k (t \restriction n)$. Let also $R = \bigcup_k R_k$. The following lemma collects some other basic properties of the relation R. The proof is an easy induction on $n \in \omega$, which we leave as an exercise (Exercise 6.3.3).

Lemma 6.3.9
For each $n \in \omega$ the following are true:

(i) $R \cap (2^n \times 2^n) \subseteq \bigcup_{k<n} R_k$;

(ii) $R_{n-1} \cap (2^n \times 2^n)$ *is a singleton;*

(iii) *The transitive closure of $R \cap (2^n \times 2^n)$ is $2^n \times 2^n$.*

It will be the pairs in R that we keep track of their linkage. In particular, we will require that if $s R t$ then $x_s E x_t$. By property (iii) above we are in effect requiring that $x_s E x_t$ for all s, t with the same length. The unique element in $R_{n-1} \cap (2^n \times 2^n)$ is in fact $(\vec{0}^\frown 0, \vec{0}^\frown 1)$.

Lemma 6.3.10
If $E \neq E^*$ then $E_0 \sqsubseteq_c E$.

Proof. Fix winning strategies for Player II in the strong Choquet games played on (X, τ) and $((X \times X) \cap E, \tau_2)$ respectively. Since E is meager in $(X \times X) \cap E^*$ in τ^2 we can find a decreasing sequence of dense open subsets of $(X \times X) \cap E^*$ in τ^2 such that $E \cap \bigcap_n W_n = \emptyset$. Without loss of generality we may assume that $W_0 \cap \{(x, x) : x \in X\} = \emptyset$.

We define sequences $\{x_s\}_{s \in 2^{<\omega}}$ of elements of X, $\{V_s\}_{s \in 2^{<\omega}}$, $\{U_s\}_{s \in 2^{<\omega}}$ of τ-open subsets of X, and $\{F_{s,t}\}_{sRt}$, $\{E_{s,t}\}_{sRt}$ of τ_2-open subsets of $(X \times X) \cap E$ such that, for any $s, t \in 2^{<\omega}$ with $\mathrm{lh}(s) = \mathrm{lh}(t)$, the following requirements are fulfilled:

(i) $V_\emptyset = U_\emptyset = X$;

(ii) The following is a play in the strong Choquet game on X according to Player II's winning strategy:

$$
\begin{array}{lllll}
\text{I} & x_{s\restriction 1}, V_{s\restriction 1} & x_{s\restriction 2}, V_{s\restriction 2} & \cdots & x_s, V_s \\[2mm]
\text{II} & \quad\ U_{s\restriction 1} & \quad\ U_{s\restriction 2} & \cdots & \ U_s
\end{array}
$$

In particular, $x_s \in U_s \subseteq V_s$;

(iii) $\mathrm{diam}(V_s) \le 2^{-\mathrm{lh}(s)}$ (in the usual topology of ω^ω);

(iv) If $\mathrm{lh}(s) = \mathrm{lh}(t) = n$ and $s(n-1) \ne t(n-1)$ then $V_s \times V_t \subseteq W_n$;

(v) If $\mathrm{lh}(s) = \mathrm{lh}(t) = n$ and $sR_k t$ then the following is a play in the strong Choquet game on $(X \times X) \cap E$ according to Player II's winning strategy:

$$
\begin{array}{llll}
\text{I} & (x_{s\restriction k}, x_{t\restriction k}), F_{s\restriction k, t\restriction k} & \cdots & (x_s, x_t), F_{s,t} \\[2mm]
\text{II} & \quad\quad E_{s\restriction k, t\restriction k} & \cdots & \ E_{s,t}
\end{array}
$$

(vi) $\mathrm{diam}(F_{s,t}) \le 2^{-\mathrm{lh}(s)}$ for sRt (in the usual topology of $(\omega^\omega)^2$).

The construction is by induction on $\mathrm{lh}(s)$. We start with $\mathrm{lh}(s) = 1$. At this stage we need Σ_1^1 sets $V_{\langle 0 \rangle}, V_{\langle 1 \rangle} \subseteq X$ of diameter $\le 1/2$ such that $V_{\langle 0 \rangle} \times V_{\langle 1 \rangle} \subseteq W_1$ and elements $x_{\langle 0 \rangle} \in V_{\langle 0 \rangle}$, $x_{\langle 1 \rangle} \in V_{\langle 1 \rangle}$ so that $(x_{\langle 0 \rangle}, x_{\langle 1 \rangle}) \in E$. Since W_1 is dense open in $(X \times X) \cap E^*$ in τ^2 such $V_{\langle 0 \rangle}, V_{\langle 1 \rangle}$ exist; and since E is dense in $(X \times X) \cap E^*$, $E \cap (V_{\langle 0 \rangle} \times V_{\langle 1 \rangle}) \ne \emptyset$. Let $F_{\langle 0 \rangle, \langle 1 \rangle}$ be any Σ_1^1 subset of $(X \times X) \cap E$ containing $(x_{\langle 0 \rangle}, x_{\langle 1 \rangle})$ and of diameter $< 1/2$. The sets $U_{\langle 0 \rangle}, U_{\langle 1 \rangle}$ and $E_{\langle 0 \rangle, \langle 1 \rangle}$ are obtained by playing the strong Choquet games. This finishes the construction for $\mathrm{lh}(s) = 1$.

In the inductive step assume all x_s, V_s, U_s, $F_{s,t}$, and $E_{s,t}$ have been defined for $\mathrm{lh}(s) \le n$ and sRt. We need to define $x_{s^\frown i}$, $V_{s^\frown i}$, $F_{s^\frown i, t^\frown i}$ for $\mathrm{lh}(s) = n$, sRt, and $i \in \{0, 1\}$, as well as $F_{\bar{0}^\frown 0, \bar{0}^\frown 1}$. Once these are defined the sets $U_{s^\frown i}$, $E_{s^\frown i, t^\frown i}$, and $E_{\bar{0}^\frown 0, \bar{0}^\frown 1}$ are obtained by playing the strong Choquet games.

To satisfy the requirements we need at least to maintain, for all s, t with $\mathrm{lh}(s) = \mathrm{lh}(t) = n$ and sRt, that $x_{s^\frown 0}, x_{s^\frown 1} \in U_s$, $x_{t^\frown 0}, x_{t^\frown 1} \in U_t$, and $(x_{s^\frown 0}, x_{t^\frown 0}), (x_{s^\frown 1}, x_{t^\frown 1}) \in E_{s,t}$. We consider the set of all such tuples

$$Y = \{((y_{s^\frown 0})_{s \in 2^n}, (y_{s^\frown 1})_{s \in 2^n}) : \forall s, t \in 2^n \, [y_{s^\frown i} \in U_s \wedge$$

$$(y_{s^\frown i}, y_{t^\frown i}) \in E^* \wedge (sRt \Rightarrow (y_{s^\frown i}, y_{t^\frown i}) \in E_{s,t})], i = 0, 1\}.$$

As a subset of $X^{2^n} \times X^{2^n}$ the set Y is obviously of the form $Z \times Z$ for some $Z \subseteq X^{2^n}$ open in the Gandy–Harrington topology on X^{2^n}. And it is nonempty since $(x_s)_{s \in 2^n} \in Z$ and therefore $((x_s)_{s \in 2^n}, (x_s)_{s \in 2^n}) \in Y$.

To further satisfy requirement (iv) we fix any $s_0, t_0 \in 2^n$. Since W_{n+1} is dense open in $(X \times X) \cap E^*$, the set

$$Y_{s_0, t_0} = \{((y_{s^\frown 0})_{s \in 2^n}, (y_{s^\frown 1})_{s \in 2^n}) \in Y : (y_{s_0^\frown 0}, y_{t_0^\frown 1}) \in W_{n+1}\}$$

is dense open in Y. It follows that the finite intersection $\bigcap_{s,t \in 2^n} Y_{s,t}$ is still dense open, and thus there are Σ_1^1 sets $V_{s^\frown 0}, V_{t^\frown 1}$ such that

$$\prod_s V_{s^\frown 0} \times \prod_t V_{t^\frown 1} \subseteq \bigcap_{s,t} Y_{s,t}.$$

With possible shrinking we may require that $\mathrm{diam}(V_{s^\frown 0}), \mathrm{diam}(V_{t^\frown 1}) \leq 2^{-\mathrm{lh}(s)-1}$. Thus (iii) and (iv) are satisfied.

Furthermore since E is dense in $(X \times X) \cap E^*$, the set

$$Y' = \{((y_{s^\frown 0})_{s \in 2^n}, (y_{s^\frown 1})_{s \in 2^n}) \in Y : (y_{\vec{0}^\frown 0}, y_{\vec{0}^\frown 1}) \in E\}$$

is dense in Y. Let $((x_{s^\frown 0})_{s \in 2^n}, (x_{s^\frown 1})_{s \in 2^n})$ be any element of the nonempty intersection

$$Y' \cap \prod_s V_{s^\frown 0} \times \prod_t V_{t^\frown 1}.$$

We have defined the required elements $(x_u)_{u \in 2^{n+1}}$ satisfying (ii).

Finally for each pair (s, t) with sRt and $i \in \{0, 1\}$ let $F_{s^\frown i, t^\frown i}$ be any Σ_1^1 subset of $(X \times X) \cap E$ containing $(x_{s^\frown 0}, x_{t^\frown 1})$ and with diameter $\leq 2^{-\mathrm{lh}(s)-1}$. Let also $F_{\vec{0}^\frown 0, \vec{0}^\frown 1}$ be any Σ_1^1 subset of $(X \times X) \cap E$ containing $(x_{\vec{0}^\frown 0}, x_{\vec{0}^\frown 1})$ and with diameter $\leq 2^{-\mathrm{lh}(s)-1}$. We have thus finished the definition of all elements and sets involved, and conditions (i) through (vi) are all satisfied.

It follows from (ii) that for any $z \in 2^\omega$ the objects $x_{z \restriction n}, V_{z \restriction n}, U_{z \restriction n}$ form a play in the strong Choquet game on X according to Player II's winning strategy. Thus $\bigcap_n U_{z \restriction n}$ is nonempty, and by (iii), is a singleton. We define $f(z)$ to be this unique element. This defines a continuous embedding of 2^ω into X. It remains to verify that f is a reduction from E_0 to E.

First suppose $z \not\mathrel{E_0} y$. Then for infinitely many n, $z(n - 1) \neq y(n - 1)$, and by (iv), for infinitely many n, $U_{z \restriction n} \times U_{y \restriction n} \subseteq W_n$. It follows that $(f(z), f(y)) \in \bigcap_n W_n$, and hence $(f(z), f(y)) \notin E$.

For the other direction we show that for any s, t with $\mathrm{lh}(s) = \mathrm{lh}(t) = n$ and $z \in 2^\omega$, $f(s^\frown 0^\frown z) E f(t^\frown 1^\frown z)$. This is by induction on n. For $n = 0$ we note that $(0^\frown z \restriction k) R_1 (1^\frown z \restriction k)$ for any $k \in \omega$. Thus the objects $(x_{0^\frown z \restriction k}, x_{1^\frown z \restriction k}), F_{0^\frown z \restriction k, 1^\frown z \restriction k}, E_{0^\frown z \restriction k, 1^\frown z \restriction k}$ form a play in the strong Choquet game on $(X \times X) \cap E$ according to Player II's winning strategy. It follows from (v) and (vi) that $\bigcap_k E_{0^\frown z \restriction k, 1^\frown z \restriction k}$ is a singleton. By our construction this unique pair must be $(f(0^\frown z), f(1^\frown z))$. Therefore $f(0^\frown z) E f(1^\frown z)$. For the inductive step when $n > 0$, consider the elements $f(\vec{0}^\frown 0^\frown z)$ and $f(\vec{0}^\frown 1^\frown z)$. By the inductive hypothesis $f(\vec{0}^\frown 0^\frown z) E f(s^\frown 0^\frown z)$ and $f(\vec{0}^\frown 1^\frown z) E f(t^\frown 1^\frown z)$. Thus it suffices to show that $f(\vec{0}^\frown 0^\frown z) E f(\vec{0}^\frown 1^\frown z)$. However this is similar to the argument for the case $n = 0$. For any $k \in \omega$ let $u_k = \vec{0}^\frown 0^\frown z \restriction k$ and $v_k = \vec{0}^\frown 1^\frown z \restriction k$. By our construction the objects $(x_{u_k}, x_{v_k}), F_{u_k, v_k}, E_{u_k, v_k}$ form a play in the strong Choquet game on $(X \times X) \cap E$ according to Player II's winning strategy. It follows again from (v) and (vi) that $\bigcap_k E_{u_k, v_k}$ is a singleton and its unique element must be $(f(\vec{0}^\frown 0^\frown z), f(\vec{0}^\frown 1^\frown z))$. Therefore $f(\vec{0}^\frown 0^\frown z) E f(\vec{0}^\frown 1^\frown z)$. □

Exercise 6.3.1 Prove Lemma 6.3.3. (*Hint:* Δ^1_1 is closed under effective countable unions but not arbitrary countable unions.)

Exercise 6.3.2 Show that if E is a Σ^1_1 equivalence relation on ω^ω then E^* is clopen in the Gandy–Harrington topology τ_2.

Exercise 6.3.3 Prove Lemma 6.3.9.

6.4 Consequences of the Glimm–Effros dichotomy

In this section we derive some consequences of the Glimm–Effros dichotomy for Borel equivalence relations. The main results of this section are due to Harrington, Kechris, and Louveau [64]. We first recall the following definition.

Definition 6.4.1
Let X, Y be standard Borel spaces. A subset $A \subseteq X$ is called **universally measurable** if it is μ-measurable for any σ-finite Borel measure μ on X. A function $f : X \to Y$ is **universally measurable** if it is μ-measurable for any σ-finite Borel measure μ on X.

A theorem of Luzin states that every analytic set is universally measurable (see Reference [97] Theorem 21.10). Let $\sigma(\Sigma^1_1)$ be the σ-algebra generated by the analytic sets. Then it follows that every set in $\sigma(\Sigma^1_1)$ is universally measurable. If X and Y are standard Borel spaces, then a function $f : X \to Y$

is $\sigma(\mathbf{\Sigma}_1^1)$-**measurable** if for any Borel set $B \subseteq Y$, $f^{-1}(B)$ is a $\sigma(\mathbf{\Sigma}_1^1)$ subset of X.

We also recall the following definition.

Definition 6.4.2

Let X, Y be sets and $P \subseteq X \times Y$. Let $\mathrm{proj}_X(P) = \{x \in X : \exists y \in Y \ (x, y) \in P\}$. A **uniformization** of P is a function $f : \mathrm{proj}_X(P) \to Y$ such that for all $x \in \mathrm{proj}_X(P)$, $(x, f(x)) \in P$.

We will use the following theorem known as the **Jankov–von Neumann uniformization theorem**.

Theorem 6.4.3 (Jankov–von Neumann)

Let X, Y be standard Borel spaces and $P \subseteq X \times Y$ an analytic set. Then there is a uniformization of P that is $\sigma(\mathbf{\Sigma}_1^1)$-measurable.

For a proof of this theorem see Reference [97] Theorem 18.1.

We are now ready to derive some consequences of the Glimm–Effros dichotomy.

Theorem 6.4.4

Let X be a standard Borel space and E a Borel equivalence relation on X. Then the following are equivalent:

(i) E is smooth;

(ii) There is a compatible Polish topology τ on X such that E is closed in (X^2, τ^2);

(iii) There is a compatible Polish topology τ on X such that E is G_δ in (X^2, τ^2);

(iv) There is a compatible Polish topology τ on X such that every E-equivalence class is G_δ;

(v) There is a countable generating family for E consisting of analytic sets;

(vi) There is a countable generating family for E consisting of universally measurable sets;

(vii) There is a universally measurable reduction from E to $\mathrm{id}(2^\omega)$;

(viii) E has a $\sigma(\mathbf{\Sigma}_1^1)$-measurable selector;

(ix) E has a universally measurable selector.

Proof. The implication (i)⇒(ii) was proved in Corollary 5.4.5. The implications (ii)⇒(iii)⇒(iv), (v)⇒(vi), and (viii)⇒(ix) are obvious. The equivalence between (vi) and (vii) follows from the proof of Proposition 5.4.4. Also (ix)⇒(vi) by a similar argument. We prove the other implications.

For (iv)⇒(v), let $\{U_n\}_{n\in\omega}$ enumerate a countable base for (X,τ). We claim that $\{[U_n]_E\}_{n\in\omega}$ is a countable generating family for E consisting of analytic sets. It is clear that each $[U_n]_E$ is analytic. To see that they form a generating family for E, let $x,y \in X$ be such that $x \in [U_n]_E \iff y \in [U_n]_E$ for all $n \in \omega$. Since for any $n \in \omega$,

$$\overline{[x]_E} \cap U_n \neq \emptyset \iff [x]_E \cap U_n \neq \emptyset \iff x \in [U_n]_E,$$

we have that for all $n \in \omega$,

$$\overline{[x]_E} \cap U_n \neq \emptyset \iff \overline{[y]_E} \cap U_n \neq \emptyset,$$

which in turn implies that $\overline{[x]_E} = \overline{[y]_E}$. Now $[x]_E, [y]_E$ are both dense G_δ subsets of $\overline{[x]_E}$, thus $[x]_E \cap [y]_E \neq \emptyset$, and therefore in fact xEy.

Next we show that (vi)⇒(i) from the Glimm–Effros dichotomy. Let $\{A_n\}_{n\in\omega}$ be a countable generating family for E where each A_n is universally measurable. Assume that E is not smooth. Then there is a Borel embedding $\theta : 2^\omega \to X$ of E_0 into E. It is easy to verify that the family $\{\theta^{-1}(A_n)\}_{n\in\omega}$ is a generating family for E_0. Note also that each $\theta^{-1}(A_n)$ is universally measurable. To see this let μ be a σ-finite Borel measure on 2^ω. Then θ_μ defined by $\theta_\mu(A) = \mu(\theta^{-1}(A))$ is a σ-finite Borel measure on X. Since each A_n is universally measurable, A_n is θ_μ-measurable, and hence $\theta^{-1}(A_n)$ is μ-measurable. To summarize, we now have a countable separating family for E_0 consisting of universally measurable sets. By the proofs of Propositions 6.1.6 and 6.1.7 show that E_0 does not admit such families.

Finally we show that (i)⇒(viii). For this let $\varphi : X \to 2^\omega$ be a Borel reduction of E to $\mathrm{id}(2^\omega)$. Consider $A = \{(z,x) \in 2^\omega \times X : z = \varphi(x)\}$. Then A is Borel, in particular Σ^1_1, and by the Jankov–von Neumann uniformization theorem, allows a $\sigma(\Sigma^1_1)$-measurable uniformization. Let $\psi : \varphi(X) \to X$ be a $\sigma(\Sigma^1_1)$-measurable uniformization of A. Then $\psi \circ \varphi$ is $\sigma(\Sigma^1_1)$-measurable. But $\psi \circ \varphi$ is clearly a selector for E. □

Recall that smooth equivalence relations need not have Borel selectors. In contrast, clauses (viii) and (ix) provide characterizations for smoothness in terms of $\sigma(\Sigma^1_1)$-measurable and universally measurable selectors. Also for Borel equivalence relations the existence of Borel selectors and of Borel transversals are equivalent (see the remarks preceding Proposition 3.4.6), therefore smooth equivalence relations need not have Borel transversals either.

On the flip side we also get characterizations for E_0-embeddability.

Theorem 6.4.5
Let X be a Polish space and E a Borel equivalence relation on X. Then the following are equivalent:

(a) $E_0 \sqsubseteq_c E$;

(b) $E_0 \leq_B E$;

(c) There is a universally measurable reduction from E_0 to E;

(d) There is an E-nonatomic, E-ergodic probability Borel measure on X.

Proof. The implications (a)$\Rightarrow$(b)$\Rightarrow$(c) are obvious. To see that (c)$\Rightarrow$(a), suppose there is a universally measurable reduction θ from E_0 to E. By the proofs of Propositions 6.1.6 and 6.1.7 we know that there are no countable generating families of E_0 consisting of universally measurable sets. However, if (a) fails then by the Glimm–Effros dichotomy we have that E is smooth and thus admitting a countable Borel generating family $\{A_n\}_{n\in\omega}$. It follows that $\{\theta^{-1}(A_n)\}_{n\in\omega}$ is a generating family for E_0 consisting of universally measurable sets, a contradiction. A similar argument shows that (d)$\Rightarrow$(a). We finally show that (a)$\Rightarrow$(d). Let μ be the product measure on 2^ω and $\theta : 2^\omega \to X$ be a continuous embedding of E_0 into E. Then the measure θ_μ defined by $\theta_\mu(A) = \mu(\theta^{-1}(A))$ is a probability Borel measure on X. It is E-nonatomic because for any $x \in X$, $\theta_\mu([x]_E) = \mu(\theta^{-1}([x]_E)) = 0$ since $\theta^{-1}([x]_E)$ is contained in at most one E_0-equivalence class. It is also E-ergodic since for any invariant Borel set $A \subseteq X$, $\theta^{-1}(A)$ is also invariant Borel. $\qquad\square$

Exercise 6.4.1 Give a direct proof that an equivalence relation E on a Polish space X with every E-equivalence class G_δ admits a $\sigma(\Sigma_1^1)$-measurable selector.

Exercise 6.4.2 Show that a Borel equivalence relation with a Σ_1^1 transversal is smooth.

Exercise 6.4.3 Let G be a Polish group and X a Polish G-space. Suppose E_G^X is F_σ. Show that the following are equivalent:

(1) E_G^X is smooth;

(2) Every orbit is G_δ;

(3) E_G^X is G_δ;

(4) There is a Borel selector for E_G^X.

Exercise 6.4.4 Let G be a Polish group and H a Borel subgroup of G. Let E be the coset equivalence relation on G defined by $xEy \iff xH = yH$. Show that E is smooth iff H is closed.

6.5 Actions of cli Polish groups

In this section we prove a theorem of Becker [5] that establishes the Glimm–Effros dichotomy for orbit equivalence relations induced by Borel actions of cli Polish groups. The main theorem of this section is the following.

Theorem 6.5.1 (Becker)
Let G be a cli Polish group and X a Borel G-space. Then there is a Polish topology on X such that the action is continuous and every G_δ orbit is closed.

By Theorem 6.2.3, the Glimm–Effros dichotomy follows immediately.

Theorem 6.5.2 (Becker)
Let G be a cli Polish group and X a Borel G-space. Then either E_G^X is smooth or else $E_0 \sqsubseteq_B E_G^X$.

The rest of the section is devoted to a proof of Theorem 6.5.1. We fix a cli Polish group G and a compatible complete left-invariant metric d_G on G. Also fix a countable dense subgroup G_0 of G and a countable base $\mathcal{U}$ for the topology of G such that $\mathcal{U}$ is closed under left-translates by elements of G_0, that is, for $g \in G_0$ and $U \in \mathcal{U}$, $gU \in \mathcal{U}$. Without loss of generality assume that $G \in \mathcal{U}$ and $\emptyset \notin \mathcal{U}$. For $U, V \in \mathcal{U}$, define

$$G_0(U, V) = \{g \in G_0 : gU \cap V \neq \emptyset\}.$$

Then note that $G_0(U, V)$ is nonempty.

Lemma 6.5.3
For any open $U' \subseteq U$, $\bigcup\{gU' : g \in G, \ gU' \cap V \neq \emptyset\} \subseteq G_0(U, V)U$. In particular, $V \subseteq G_0(U, V)U$.

Proof. Let $g \in G$ and $h, k \in U'$ with $gh \in V$. We need to show that $gk \in G_0(U, V)U$. By the continuity of the group operations there are open sets U_0, U_1, U_2, N such that $g \in U_0$, $1_G \in N$, $h \in U_1$, $k \in U_2$, $NU_1 \subseteq U'$, $NU_2 \subseteq U$, and $U_0NU_1 \subseteq V$. Note that $gN^{-1} \cap U_0$ is nonempty open. We let $g_0 \in G_0 \cap gN^{-1} \cap U_0$. Then $g_0h \in U_0NU_1 \subseteq V$, and therefore $g_0 \in G_0(U, V)$. Also $gk = g_0(g_0^{-1}g)k$, where $(g_0^{-1}g)k \in NU_2 \subseteq U$. Thus $gk \in G_0(U, V)U$. This proves the first half of the lemma. The second half follows since $V \subseteq \{gU : g \in G, \ gU \cap V \neq \emptyset\}$. □

Next we consider the action of G on X and find a Polish topology on X by the method of Section 4.4.

Lemma 6.5.4
There is a countable collection $\mathcal{B}$ of Borel sets on X such that

(i) the topology τ generated by $\mathcal{B}$ is Polish,

(ii) the action of G on (X, τ) is continuous, and

(iii) for all $A, B \in \mathcal{B}$ and $U, V \in \mathcal{U}$, the set $[A - G_0(U, V) \cdot B]^{\triangle G}$ is τ-open.

Proof. Let $\mathcal{B}_0$ be a countable base for a Polish topology on X generating the given Borel structure. Let $\mathcal{A}$ be a countable Boolean algebra of Borel subsets of X with the following closure properties:

(a) $\mathcal{B}_0 \subseteq \mathcal{A}$;

(b) For all $A \in \mathcal{A}$ and $U, V \in \mathcal{U}$, $G_0(U, V) \cdot A \in \mathcal{A}$;

(c) For all $A \in \mathcal{A}$ and $U \in \mathcal{U}$, $A^{\triangle U} \in \mathcal{A}$;

(d) The topology generated by $\mathcal{A}$ is Polish.

This can be easily done by the remarks preceding Lemma 4.4.5. Now let τ be the topology on X generated by $\mathcal{B} = \mathcal{A}^{\triangle \mathcal{U}}$. By the lemmas in Section 4.4 (X, τ) becomes a Polish G-space. To check (iii) note that $\mathcal{B} \subseteq \mathcal{A}$, and for any $A, B \in \mathcal{A}$ and $U, V \in \mathcal{U}$, $[A - G_0(U, V) \cdot B]^{\triangle G} \in \mathcal{A}^{\triangle \mathcal{U}}$ by condition (b) and the fact that $\mathcal{A}$ is a Boolean algebra. $\square$

We fix a countable base $\mathcal{B}$ and the Polish topology τ on X with the properties described above, and investigate the G_δ orbits. For this we also fix $x \in X$ so that $[x]_G$ is G_δ. By Effros' Theorem 3.2.4 the canonical map $g \mapsto g \cdot x$ is open from G onto $[x]_G$. It follows that for any open $U \subseteq G$, $U \cdot x$ is a relatively open subset of $[x]_G$. We define for open $U \subseteq G$,

$$\Omega(U) = \bigcup \{A \subseteq X \ : \ A \in \tau, \ A \cap [x]_G \subseteq U \cdot x\}.$$

It is clear that $\Omega(U)$ is the largest open subset O of X such that $O \cap [x]_G = U \cdot x$. It is also easy to see that $U \subseteq V$ implies $\Omega(U) \subseteq \Omega(V)$, and for any $g \in G$, $g \cdot \Omega(U) = \Omega(gU)$. Note as well that for any sequence (U_n) of open sets in G, $\bigcup_n \Omega(U_n) \subseteq \Omega(\bigcup_n U_n)$. We also define, for $U, V \in \mathcal{U}$,

$$\Lambda(U, V) = [\Omega(V) - \bigcup \{\Omega(hU) \ : \ h \in G_0(U, V)\}]^{\triangle G}.$$

Then $\Lambda(U, V)$ is an invariant Borel subset of X. The main issue is whether it is empty. We first show that it is empty on the orbit $[x]_G$.

Lemma 6.5.5
For any $U, V \in \mathcal{U}$, $\Lambda(U, V) \cap [x]_G = \emptyset$.

Proof. Assume $\Lambda(U, V) \cap [x]_G \neq \emptyset$. Since $\Lambda(U, V)$ is invariant, we have $x \in \Lambda(U, V)$. By definition there is $g \in G$ such that $g \cdot x \in \Omega(V)$ but for all

$h \in G_0(U, V)$, $g \cdot x \notin \Omega(hU)$. From $g \cdot x \in \Omega(V)$ we get that $g \cdot x \subseteq V \cdot x$. Thus we may assume without loss of generality that $g \in V$. But by Lemma 6.5.3 $g \in G_0(U, V)U$. Thus for some $h \in G_0(U, V)$, $g \in hU$, and $g \cdot x \in hU \cdot x \subseteq \Omega(hU)$, a contradiction. □

The following lemma is one of two key lemmas for the proof of Theorem 6.5.1.

Lemma 6.5.6
Suppose $y \in \Lambda(U, V)$. *Then there are* $A, B \in \mathcal{B}$ *such that* $[y]_G \subseteq [A - G_0(U, V) \cdot B]^{\triangle G}$ *and* $[x]_G \cap [A - G_0(U, V) \cdot B]^{\triangle G} = \emptyset$.

Proof. Since $y \in \Lambda(U, V)$ there is a nonmeager set $N_0 \subseteq G$ such that for all $g \in N_0$,

$$g \cdot y \in \Omega(V) - \bigcup \{\Omega(hU) : h \in G_0(U, V)\}.$$

Thus for all $g \in N_0$, $g \cdot y \in \Omega(V)$ but for all $h \in G_0(U, V)$, $g \cdot y \notin \Omega(hU)$. Since $\Omega(V)$ is open in X and $\mathcal{B}$ is a countable base, $\Omega(V) = \bigcup \{A : A \in \mathcal{B} \wedge A \subseteq \Omega(V)\}$. It follows that there is $A \in \mathcal{B}$ and a nonmeager set $N_1 \subseteq N_0$ such that for all $g \in N_1$, $g \cdot y \in A \subseteq \Omega(V)$. Let $B \in \mathcal{B}$ be such that $B \subseteq \Omega(U)$ and $B \cap U \cdot x \neq \emptyset$. Then $B \cap [x]_G \subseteq U \cdot x$. It follows that for $h \in G_0(U, V)$, $h \cdot B \cap [x]_G \subseteq h \cdot (U \cdot x) = (hU) \cdot x$, and thus $h \cdot B \subseteq \Omega(hU)$. This implies that for all $g \in N_1$, $g \cdot y \notin h \cdot B$. We have shown that $y \in [A - G_0(U, V) \cdot B]^{\triangle G}$.
    It remains to show $[x]_G \cap [A - G_0(U, V) \cdot B]^{\triangle G} = \emptyset$. Assume this is not the case. Then there is $k \in G$ such that $k \cdot x \in A - G_0(U, V) \cdot B$. Since $A \subseteq \Omega(V)$ we may assume without loss of generality that $k \in V$. Since $B \subseteq \Omega(U)$ is open there is an open $U' \subseteq U$ such that $B \cap [x]_G = U' \cdot x$. $U' \neq \emptyset$ since $B \cap U \cdot x \neq \emptyset$. Note that $G_0(U', V) \subseteq G_0(U, V)$, and hence $k \cdot x \notin G_0(U', V) \cdot B$. However, by Lemma 6.5.3 $V \subseteq G_0(U', V)U'$. Thus $k \cdot x \in G_0(U', V) \cdot (U' \cdot x) \subseteq G_0(U', V) \cdot B$, a contradiction. □

Note that so far we have not used the assumption that G is a cli Polish group. The next key lemma is the only place where we essentially use this assumption.

Lemma 6.5.7
If $y \in \overline{[x]_G} - [x]_G$, *then there are* $U, V \in \mathcal{U}$ *such that* $y \in \Lambda(U, V)$.

Proof. Suppose $y \in \overline{[x]_G}$ but assume there are no $U, V \in \mathcal{U}$ such that $y \in \Lambda(U, V)$. We construct sequences $(U_n)_{n \in \omega}$ of elements of $\mathcal{U}$ and $(g_n)_{n \in \omega}$ of elements of G such that, for all $n \in \omega$,

(i) $\operatorname{diam}(U_n) < 2^{-n}$,

(ii) $U_n \cap U_{n+1} \neq \emptyset$,

(iii) $d_G(g_n, g_{n+1}) < 2^{-n}$,

(iv) $g_n \cdot y \in \Omega(U_n)$.

The construction is by induction on n. To begin with, let $U_0 = G$ and $g_0 = 1_G$. Since $\Omega(G) = X$ we have that $g_0 \cdot y \in \Omega(U_0)$. In general, suppose U_n and g_n have been defined to satisfy the requirements. We now define $U_{n+1} \in \mathcal{U}$ and $g_{n+1} \in G$. Let $V = U_n$ and $U \in \mathcal{U}$ such that $\mathrm{diam}(U) < 2^{-(n+1)}$. Since $y \notin \Lambda(U, V)$ and $\Lambda(U, V)$ is invariant, we have that $g_n \cdot y \notin \Lambda(U, V)$. Thus there is a comeager set $C \subseteq G$ such that for all $g \in C$,

$$g \cdot (g_n \cdot y) \notin \Omega(U_n) - \bigcup \{\Omega(hU) : h \in G_0(U, U_n)\}.$$

Let

$$W = \{g \in G : g \cdot (g_n \cdot y) \in \Omega(U_n) \wedge d_G(gg_n, g_n) < 2^{-n}\}.$$

By (iv) $g_n \cdot y \in \Omega(U_n)$, and since $\Omega(U_n)$ is open in X, W is nonempty open. Let $g \in C \cap W$ and $g_{n+1} = gg_n$. Then $d_G(g_n, g_{n+1}) = d_G(g_n, gg_n) < 2^{-n}$. Now $g_{n+1} \cdot y = g \cdot (g_n \cdot y) \in \Omega(U_n)$ since $g \in W$, so by $g \in C$ we obtain an $h \in G_0(U, U_n)$ such that $g_{n+1} \cdot y = g \cdot (g_n \cdot y) \in \Omega(hU)$. Let $U_{n+1} = hU$. Since $\mathrm{diam}(U) < 2^{-(n+1)}$, the left invariance gives that $\mathrm{diam}(U_{n+1}) = \mathrm{diam}(hU) < 2^{-(n+1)}$. Also $U_n \cap U_{n+1} = U_n \cap hU \neq \emptyset$ since $h \in G_0(U, U_n)$. Lastly, $g_{n+1} \cdot y = g \cdot (g_n \cdot y) \in \Omega(hU) = \Omega(U_{n+1})$. Thus the construction is finished with all the requirements fulfilled.

By the construction we have that the sequence (g_n) is d_G-Cauchy, and therefore by the completeness of d_G there is $g_\infty \in G$ such that $g_n \to g_\infty$ as $n \to \infty$. Let $y_\infty = g_\infty \cdot y$. Let d_X be a complete metric on X compatible with τ. Let $(M_n)_{n \in \omega}$ be a countable open nbhd base of y_∞ with $g_n \cdot y \in M_n$. We claim that $U_n \cdot x \cap M_n \neq \emptyset$ for all $n \in \omega$. To see this fix an n and let $A = \Omega(U_n) \cap M_n$. Then A is open and $g_n \cdot y \in A$ by (iv). Since $y \in \overline{[x]_G}$, we get that $g_n \cdot y \in \overline{[x]_G}$ and thus $A \cap [x]_G \neq \emptyset$. It follows that $A \cap [x]_G \neq \emptyset$, and therefore $U_n \cdot x \cap M_n \neq \emptyset$.

For each $n \in \omega$ let $x_n \in U_n \cdot x \cap M_n$ and $h_n \in U_n$ be such that $h_n \cdot x = x_n$. By the above construction (h_n) is a d_G-Cauchy sequence, and therefore there is $h_\infty \in G$ such that $h_n \to h_\infty$ as $n \to \infty$. By the continuity of the action we thus have that $x_n = h_n \cdot x \to h_\infty \cdot x$ as $n \to \infty$. On the other hand, $x_n \to y_\infty$ as (M_n) is a nbhd base for y_∞. Thus $y_\infty = h_\infty \cdot x$, and $y \in [x]_G$. ∎

We are now ready to give the proof of Theorem 6.5.1.

Proof. Following the notation of this section let $x \in X$ be such that $[x]_G$ is G_δ. We show that $[x]_G$ is closed. Assume that $y \in \overline{[x]_G} - [x]_G$. By Lemma 6.5.7 there are $U, V \in \mathcal{U}$ such that $y \in \Lambda(U, V)$. And so by Lemma 6.5.6 there are $A, B \in \mathcal{B}$ such that, letting $D = [A - G_0(U, V) \cdot B]^{\triangle G}$, we have $[y]_G \subseteq D$ and $[x]_G \cap D = \emptyset$. By Lemma 6.5.4 D is open, and so $y \notin \overline{[x]_G}$, contradicting our assumption. ∎

Becker's theorem implies earlier results of Sami (for abelian Polish groups [134]; see Exercises 9.5.3 through 9.5.5), Hjorth–Solecki (for nilpotent Polish groups [83]), and Solecki (for Polish groups with two-sided invariant metrics [83]).

The following exercise problems follow the notation of this section, where G is an arbitrary Polish group.

Exercise 6.5.1 Show that for any $g \in G$ and open $U \subseteq G$, $g \cdot \Omega(U) = \Omega(gU)$.

Exercise 6.5.2 Show that for any $A \in \mathcal{B}$, $[A]_G$ is clopen.

Exercise 6.5.3 Define an equivalence relation $\sim$ on X by $x \sim y$ iff $\overline{[x]_G} = \overline{[y]_G}$. Show that every $\sim$-equivalence class is closed.

Chapter 7

Countable Borel Equivalence Relations

The topic of Borel equivalence relations is so vast that it certainly cannot fit into two chapters of any book. In this and the next chapters we just give an introduction of the topic and leave further explorations of the literature to the interested reader. There are many theoretical results about Borel equivalence relations, and at the same time we have also identified many classification problems and equivalence relations from other fields of mathematics that turn out to be Borel equivalence relations on standard Borel spaces. In this chapter we will discuss countable Borel equivalence relations. For reasons that will become clear the study of countable Borel equivalence relations have become intertwined with the study of countable group theory and ergodic theory. Thus most of the important results about countable Borel equivalence relations cannot be proved using set theoretic or general topological tools alone. Our objective in this chapter is to provide a self-contained inroad into the subject.

7.1 Generalities of countable Borel equivalence relations

Definition 7.1.1
Let X be a standard Borel space. An equivalence relation E on X is called **finite** if every E-equivalence class is finite. E is called **countable** if every E-equivalence class is countable.

Any finite Borel equivalence relation is smooth (see Exercise 7.1.1). The following theorem of Luzin–Novikov on Borel uniformizations will be useful in proving theorems about countable Borel equivalence relations. The proof can be found in Reference [97] Theorem 18.10.

Theorem 7.1.2 (Luzin–Novikov)
Let X, Y be standard Borel spaces and let $P \subseteq X \times Y$ be Borel. Suppose every section $P_x = \{y \in Y : (x, y) \in P\}$ is countable. Then $\text{proj}_X(P) = \{x \in X : \exists y \ (x, y) \in P\}$ is Borel, and P has a Borel uniformization, that is, there is a Borel function $f : \text{proj}_X(P) \to Y$ such that $(x, f(x)) \in P$ for any $x \in \text{proj}_X(P)$.

Moreover, P can be written as $\bigcup_n P_n$ where each P_n is the graph of some Borel function $f_n : A_n \to Y$ for A_n Borel, that is, $P_n = \{(x, f_n(x)) : x \in A_n\}$.

This uniformization theorem greatly simplifies the study of reductions among countable Borel equivalence relations, as the following lemma shows.

Lemma 7.1.3
Let E be a countable Borel equivalence relation on a standard Borel space X, Y a standard Borel space, and $f : X \to Y$ a Borel function such that $f(x) = f(y)$ implies xEy. Then $f(X)$ is Borel in Y and there is a Borel function $g : f(X) \to X$ such that $f \circ g = \mathrm{id}$.

Proof. Consider $P = \{(z, x) \in Y \times X : f(x) = z\}$. Then P is a Borel subset of $Y \times X$ with each P_z countable. By the Luzin–Novikov uniformization theorem $f(X) = \mathrm{proj}_Y(P)$ is Borel. Let $g : f(X) \to X$ be a Borel uniformization function. Then $f \circ g = \mathrm{id}$. ∎

The following is another application of the Luzin–Novikov uniformization theorem. The theorem states that all countable Borel equivalence relations are orbit equivalence relations of countable group actions.

Theorem 7.1.4 (Feldman–Moore)
Let E be a countable Borel equivalence relation on a standard Borel space X. Then there is a countable group G and a Borel action of G on X such that $E = E_G^X$.

Proof. Consider E as a Borel subset of $X \times X$. Since E is countable, each section E_x is countable. By the Luzin–Novikov uniformization theorem we can write E as $\bigcup_n P_n$ where each P_n is the graph of a Borel function $f_n : A_n \to X$ for A_n Borel. Without loss of generality we may assume that $P_n \cap P_m = \emptyset$ for $n \neq m$.

Let $\Delta = \{(x, x) : x \in X\}$. First note that $X - \Delta$ can be partitioned into countably many sets of the form $A \times B$, where A, B are disjoint Borel subsets of X (see Exercise 7.1.3). We write $X - \Delta = \bigcup_k R_k$, where $\{R_k\}_{k \in \omega}$ is such a partition. Next for $n, m, k \in \omega$ we let

$$E_{n,m,k} = \{(x, y) \in E : (x, y) \in P_n, \; (y, x) \in P_m, \; x \neq y, \; (x, y) \in R_k\}.$$

Then $\{E_{n,m,k}\}$ is a partition of $E - \Delta$. And since each P_n is the graph of a Borel function, so is each $E_{n,m,k}$. We let $h_{n,m,k} : A_{n,m,k} \to B_{n,m,k}$ be a Borel isomorphism between Borel subsets $A_{n,m,k}$ and $B_{n,m,k}$ such that $E_{n,m,k} = \{(x, h_{n,m,k}(x)) : x \in A_{n,m,k}\}$. Define a Borel automorphism $g_{n,m,k}$ of X by

$$g_{n,m,k}(x) = \begin{cases} h_{n,m,k}(x), & \text{if } x \in A_{n,m,k}, \\ h_{n,m,k}^{-1}(x), & \text{if } x \in B_{n,m,k}, \\ x, & \text{otherwise.} \end{cases}$$

Let G be the group of Borel automorphisms generated by $g_{n,m,k}$ for $n, m, k \in \omega$. Then

$$x E y \iff \exists n, m, k \ (g_{n,m,k}(x) = y) \lor (x = y) \iff \exists g \in G(g \cdot x = y).$$

Thus $E = E_G^X$. $\qquad\qquad\qquad\qquad\qquad\qquad\qquad\qquad\qquad\qquad\qquad\qquad$ ◻

The following lemma is another basic and useful result about countable Borel equivalence relations. It is often referred to as the **marker lemma**.

Lemma 7.1.5 (Slaman–Steel)
Let E be a countable Borel equivalence relation on a standard Borel space X. Let $A = \{x \in X : [x]_E \text{ is finite}\}$ and $B = X - A$. Then there are Borel sets $B \supseteq S_0 \supseteq S_1 \supseteq S_2 \supseteq \dots$ such that $[S_n]_E = B$ for all n, and $\bigcap_n S_n = \emptyset$.

Proof. Without loss of generality we may assume that $X = 2^\omega$. Let G be a countable group acting on X in a Borel manner with $E = E_G^X$. Enumerate G as $\{g_n\}_{n \in \omega}$. For any E-equivalence class C, let $\overline{C}$ be the closure of C and x_C be the lexicographically least element of $\overline{C}$. Then map $y \mapsto x_{[y]_E}$ is Borel, since for any $s \in 2^{<\omega}$,

$$x_{[y]_E} \in N_s \iff \exists n \ (g_n \cdot y \in N_s) \land \forall s' <_{\text{lex}} s \forall n(g_n \cdot y \notin N_{s'}).$$

Let $A' = \{y : x_{[y]_E} \in [y]_E\}$. Then A' is invariant Borel and $y \mapsto x_{[y]_E}$ is a Borel selector of $E \upharpoonright A'$. Moreover, $A \subseteq A'$. Thus let T be a Borel transversal for $E \upharpoonright A'$. Define

$$y \in S_n \iff (y \in A' - A \land \forall m < n \ g_m \cdot y \notin T) \lor (y \notin A' \land y \upharpoonright n = x_{[y]_E} \upharpoonright n).$$

Then the sets S_n are as required. $\qquad\qquad\qquad\qquad\qquad\qquad\qquad\qquad$ ◻

Exercise 7.1.1 Show that any finite Borel equivalence relation has a Borel selector.

Exercise 7.1.2 Give an example of an analytic, non-Borel, finite equivalence relation. (*Hint:* Let $A \subseteq \omega^\omega$ be a Σ_1^1 non-Borel set. Consider the equivalence relation E on $\omega^\omega \times \{0, 1\}$ defined by $(x, i)E(y, j)$ iff $(x, i) = (y, j)$ or $x = y \in A$.)

Exercise 7.1.3 For $s, t \in \omega^{<\omega}$, define sRt if $\text{lh}(s) = \text{lh}(t) = n + 1$, $s \upharpoonright n = t \upharpoonright n$, but $s(n) \neq t(n)$. Let $\Delta = \{(x, x) : x \in \omega^\omega\}$. Show that

(i) $\omega^\omega - \Delta = \bigcup \{N_s \times N_t : sRt\}$.

(ii) If sRt and $s'Rt'$ but $(s, t) \neq (s', t')$, then $(N_s \times N_t) \cap (N_{s'} \times N_{t'}) = \emptyset$.

Exercise 7.1.4 Show that the sets S_n defined in the proof of Lemma 7.1.5 are as required.

Exercise 7.1.5 Show that every countable Borel equivalence relation is idealistic.

Exercise 7.1.6 Show that the following are equivalent for a countable Borel equivalence relation E:

(i) E is smooth;

(ii) E has a Borel transversal;

(iii) E has a Borel selector.

Exercise 7.1.7 Show that for every countable Borel equivalence relation E there exist finite equivalence relations E_n such that $E = \bigcup_n E_n$.

Exercise 7.1.8 Let X, Y be standard Borel spaces and E, F countable Borel equivalence relations on X, Y, respectively. Show that if $E \leq_B F$ then there is an F-invariant Borel subset B of Y such that $E \sim_B F \restriction B$.

7.2 Hyperfinite equivalence relations

Definition 7.2.1
Let X be a standard Borel space. A Borel equivalence relation E on X is **hyperfinite** if there are finite Borel equivalence relations E_n, $n \in \omega$, with $E_n \subseteq E_{n+1}$ for all n, such that $E = \bigcup_n E_n$.

By definition every hyperfinite equivalence relation is a countable Borel equivalence relation. Note that the equivalence relation E_0 is hyperfinite. In fact, we can define, for $n \geq 1$, equivalence relations E_n on 2^ω by

$$x E_n y \iff \forall m \geq n \; x(m) = y(m).$$

Then each E_n is apparently Borel and finite, $E_n \subseteq E_{n+1}$, and $E_0 = \bigcup_{n \geq 1} E_n$. The theorems proved in this section will show that E_0 is a typical hyperfinite equivalence relation in a strong sense.

The following proposition collects some basic closure properties of hyperfinite equivalence relations.

Proposition 7.2.2
Let E and F be countable Borel equivalence relations on standard Borel spaces X and Y, respectively. Then the following hold:

(1) If $X = Y$, F is hyperfinite and $E \subseteq F$, then E is hyperfinite.

(2) If E is hyperfinite and $A \subseteq X$ is Borel, then $E \restriction A$ is hyperfinite.

(3) If $A \subseteq X$ is Borel, $[A]_E = X$, and $E \restriction A$ is hyperfinite, then E is hyperfinite.

(4) If $E \leq_B F$ and F is hyperfinite, then E is hyperfinite.

(5) If both E and F are hyperfinite, then $E \times F$ is hyperfinite.

Proof. Properties (1), (2) and (5) follow easily from the definition. We only show (3) and (4). For (3) suppose $E \restriction A = \bigcup_n F_n$, where F_n are finite Borel equivalence relations and $F_n \subseteq F_{n+1}$ for all n. By Theorem 7.1.4 $E = E_G^X$ for some countable group G acting in a Borel manner on X. Enumerate G by $\{g_m\}_{m \in \omega}$. For each $x \in X$, define $m(x)$ to be the least m such that $g_m \cdot x \in A$. By the assumption $[A]_E = X$, $m(x)$ is well defined for all $x \in X$. We then define

$$x E_n y \iff (m(x), m(y) < n \wedge g_{m(x)} \cdot x \, F_n \, g_{m(y)} \cdot y) \vee x = y.$$

Then E_n are finite Borel equivalence relations, $E_n \subseteq E_{n+1}$, and $E = \bigcup_n E_n$. Thus E is hyperfinite.

For (4) we let $f : X \to Y$ be a Borel reduction of E to F. Then $f(x) = f(y)$ implies $x E y$. By Lemma 7.1.3 there is a Borel function $g : f(X) \to X$ so that $f \circ g = \mathrm{id}$. Now g is a Borel injection, thus $g(f(X))$ is Borel. Let $A = g(f(X))$. Then $[A]_E = X$. Moreover $E \restriction A$ is Borel isomorphic (via g^{-1}) to $F \restriction f(X)$. Therefore by (2) $F \restriction f(X)$ is hyperfinite. It follows that $E \restriction A$ is hyperfinite and by (3) E is hyperfinite. □

The following theorem gives various useful characterizations of hyperfiniteness. Especially of interest is the characterization in terms of Borel reducibility to E_0. The theorem was based on earlier work by Weiss and by Slaman–Steel, and appeared in this form in Reference [30].

Theorem 7.2.3 (Dougherty–Jackson–Kechris)
Let X be a standard Borel space and E a countable Borel equivalence relation. Then the following are equivalent:

(i) $E \leq_B E_0$.

(ii) E is hyperfinite.

(iii) $E = \bigcup_n E_n$, where E_n are finite Borel equivalence relations, $E_n \subseteq E_{n+1}$, and every E_n-equivalence class has at most n elements.

(iv) There is a Borel assignment $C \mapsto <_C$ associating with each E-equivalence class C a linear order $<_C$ on C so that there is an order-preserving map from $(C, <_C)$ into $(\mathbb{Z}, <)$. Here $C \mapsto <_C$ is Borel when the relation $R(x, y, z) \iff y <_{[x]_E} z$ is Borel.

(v) *There is a Borel automorphism T of X such that $xEy \iff \exists n \in \mathbb{Z}\ T^n(x) = y$.*

(vi) *There is a Borel action of the additive group $\mathbb{Z}$ on X such that $E = E_{\mathbb{Z}}^X$.*

Proof. The implication (i)$\Rightarrow$(ii) follows immediately from Proposition 7.2.2 (4). The implication (iii)$\Rightarrow$(ii) and the equivalence of (iv), (v), and (vi) are obvious. We show the other implications.

(ii)$\Rightarrow$(iii): Let $E = \bigcup_n F_n$, where F_n are finite Borel equivalence relations and $F_n \subseteq F_{n+1}$. Without loss of generality we may assume $F_1 = \mathrm{id}(X)$. We define a sequence of finite Borel equivalence relations E_n, $n \geq 1$, such that $E_n \subseteq E_{n+1}$ and each E_n-equivalence class has at most n elements. Define $E_1 = \mathrm{id}(X) = F_1$. For $n > 1$, inductively define the following sequence of sets:

$$X_n = \{x \in X : [x]_{F_n} \text{ has at most } n \text{ elements}\},$$

$$X_{n-1} = \{x \notin X_n : [x]_{F_{n-1}} \text{ has at most } n \text{ elements}\},$$

$$\ldots\ldots$$

$$X_i = \{x \notin \textstyle\bigcup_{i<j\leq n} X_j : [x]_{F_i} \text{ has at most } n \text{ elements}\},$$

$$\ldots\ldots$$

$$X_2 = \{x \notin X_n \cup \cdots \cup X_3 : [x]_{F_2} \text{ has at most } n \text{ elements}\},$$

$$X_1 = X - \textstyle\bigcup_{1<j\leq n} X_j.$$

Note that the definition of X_1 can be formulated in the same manner as those of the other sets. Since $F_1 \subseteq F_2 \subseteq \cdots \subseteq F_n$, each X_i is F_j-invariant if $j \leq i$. Let $E_n = (F_n \restriction X_n) \cup (F_{n-1} \restriction X_{n-1}) \cup \cdots \cup (F_2 \restriction X_2) \cup (F_1 \restriction X_1)$. Then E_n is a finite Borel equivalence relation with each E_n-equivalence class containing at most n elements. To see that $E_n \subseteq E_{n+1}$ let $Y_{n+1}, Y_n, \ldots, Y_2, Y_1$ be the sequence of sets defined in the definition of E_{n+1}. Suppose $xE_n y$. Then for some $i \leq n$, $x, y \in X_i$ and $xF_i y$. Since $X_i \subseteq Y_{n+1} \cup Y_n \cup \cdots \cup Y_i$ and each of $Y_{n+1}, Y_n, \ldots, Y_i$ is F_i-invariant, there is $j \geq i$ such that $x, y \in Y_j$ and $xF_j y$, and thus $xE_{n+1}y$. This shows that $E_n \subseteq E_{n+1}$. It is easy to see that $E = \bigcup_n E_n$.

(ii)$\Rightarrow$(iv): Let $E = \bigcup_n E_n$ where E_n are finite Borel equivalence relations and $E_n \subseteq E_{n+1}$. Fix a Borel linear order $<$ on X. We define $<^n_C$ inductively on each E_n-equivalence class C as follows. Without loss of generality assume $E_1 = \mathrm{id}(X)$ and so let $<^1_C = \emptyset$ for all E_1-equivalence classes C. In general suppose $<^n_C$ has been defined on all E_n-equivalence classes C. Consider an E_{n+1}-equivalence class D. Then there are finitely many E_n-equivalence classes $C_1, \ldots, C_k$ with $D = C_1 \cup \cdots \cup C_k$ and moreover the linear orders $<^n_{C_1}, \ldots, <^n_{C_k}$ are defined. For each $1 \leq i \leq k$ let x_i be the $<^n_{C_i}$-least element of C_i. Then define $<^{n+1}_D$ by

$$x <^{n+1}_D y \iff \exists i\, (x, y \in C_i \wedge x <^n_{C_i} y) \vee \exists i \neq i'\, (x \in C_i \wedge y \in C_{i'} \wedge x_i < x_{i'}).$$

It follows from this construction that if C is an E_n-equivalence class and D is an E_{n+1}-equivalence class with $C \subseteq D$, then $<^n_C \subseteq <^{n+1}_D$. We then define, for any E-equivalence class C,

$$<_C = \bigcup \{<^n_D : \ D \text{ is an } E_n\text{-equivalence class and } D \subseteq C\}.$$

Then $C \mapsto <_C$ is a Borel assignment and $<_C$ is a linear order from which there is an order-preserving map into $(\mathbb{Z}, <)$.

(iv)$\Rightarrow$(ii): It suffices to find a Borel partition $\{X_n\}_{n\in\omega}$ of X so that $E \restriction X_n$ is hyperfinite for all $n \in \omega$. Note that by Proposition 7.2.2 (4) smooth countable equivalence relations are hyperfinite. Let $X_0 = \{x \in X : \ [x]_E \text{ is finite}\}$. Then $E \restriction X_0$ is finite and therefore smooth. Next let $B = X - X_0$. By Lemma 7.1.5 there are Borel sets $B \supseteq S_0 \supseteq S_1 \supseteq S_2 \supseteq \ldots$ such that $[S_n]_E = B$ for all n and $\bigcap_n S_n = \emptyset$. Let $C \mapsto <_C$ be a Borel assignment associating with each E-equivalence class C a Borel linear order of C of order type $\mathbb{Z}$. Let

$$X_1 = \{x \in X : \ \text{there is a } <_{[x]_E}\text{-least element or a } <_{[x]_E}\text{-greatest element}$$
$$\text{of } S_n \cap [x]_E \text{ for some } n \in \omega\}.$$

Then X_1 is invariant Borel. It is also obvious that $E \restriction X_1$ has a Borel selector, namely the function choosing the $<_{[x]_E}$-least or $<_{[x]_E}$-greatest element of $[x]_E \cap S_n$ for least such n. So $E \restriction X_1$ is smooth.

Finally let $X_2 = X - X_0 - X_1$. Then for each $n \in \omega$ and $x \in X_2$, $<_{[x]_E} \restriction S_n$ has order type $\mathbb{Z}$, and we let $r_n(x)$ be the $<^n_{[x]_E}$-least element y of $S_n \cap [x]_E$ with $x = y$ or $x <_{[x]_E} y$. The function r_n is well defined and Borel. We then define a sequence of finite Borel equivalence relations E_n by

$$x E_n y \iff x E y \wedge r_n(x) = r_n(y).$$

It is clear that E_n is a finite Borel equivalence relation. $E_n \subseteq E_{n+1}$ follows from the property that $S_n \supseteq S_{n+1}$, and $E \restriction X_2 = \bigcup_n E_n$ since $\bigcap_n S_n = \emptyset$.

(vi)$\Rightarrow$(i): This would make the above proof of (iv)$\Rightarrow$(ii) appear redundant, but in fact we are going to give a proof based on the above one. First by Theorem 3.3.4 the orbit equivalence relation induced by the $\mathbb{Z}$-action on $2^{\omega \times \mathbb{Z}}$ is a universal equivalence relation for all $\mathbb{Z}$-orbit equivalence relations. Denote this equivalence relation by $E_\mathbb{Z}$. It is thus sufficient to show that $E_\mathbb{Z} \leq_B E_0$. For convenience we use the variation of E_0 on the Baire space ω^ω and continue to denote $E_\mathbb{Z}$ by E.

By Exercise 7.2.1 it again suffices to find a Borel partition $\{X_n\}_{n\in\omega}$ so that $E \restriction X_n \leq_B E_0$ for all $n \in \omega$. We repeat the above proof of (iv)$\Rightarrow$(ii) to define the invariant Borel sets X_0, X_1, and X_2. It suffices to show that $E \restriction X_2 \leq_B E_0$. And we also have the finite equivalence relations E_n defined. Now for each $x \in X_2$ and $n \in \omega$ we let $\pi_n(x) = \{g \in \mathbb{Z} : g \cdot x E_n x\}$. We note that $\pi_n(x)$ is a finite set of consecutive integers and for $x E_n y$, $\pi_n(x) \cdot x = \pi_n(y) \cdot y$. Let $m = |\pi_n(x)|$ and $g_n(x) = \inf \pi_n(x)$. Then define an $n \times m$ matrix $M_n(x) = (a_{i,j})_{i<n, j<m}$ by $a_{i,j} = [g_n(x) \cdot x](i, j)$. Then for $x E_n y$, we

have that $M_n(x) = M_n(y)$. Moreover, $M_{n-1}(x)$ is a submatrix of $M_n(x)$, and its relative position in $M_n(x)$ can be determined by $g_{n-1}(x) - g_n(x)$. It is also true that if $xE_{n-1}y$ then $M_{n-1}(x) = M_{n-1}(y)$, $M_n(x) = M_n(y)$, and $g_{n-1}(x) - g_n(x) = g_{n-1}(y) - g_n(y)$. Let $c_n(x)$ be an integer coding the matrix $M_{n,m}(x)$ and the integer $g_{n-1}(x) - g_n(x)$. Then we have that if $xE_{n-1}y$, then $r_n(x) = r_n(y)$. We claim that xEy iff $(c_n(x))E_0(c_n(y))$. If xEy then for some n we have that $xE_{n'}y$ for all $n' > n$. By the construction we have that $\pi_{n'}(x) \cdot x = \pi_{n'}(y) \cdot y$ and eventually $c_{n'}(x) = c_{n'}(y)$. Thus $(c_n(x))E_0(c_n(y))$. Conversely, from the sequence $(c_n(x))$ we can determine $x \in 2^{\omega \times \mathbb{Z}}$ up to a shift by an element of $\mathbb{Z}$. This is because, from the information of the sequence $M_n(x)$ and $g_{n-1}(x) - g_n(x)$ we may recover x if $g_0(x)$ is given. Thus if $(c_n(x)) = (c_n(y))$ then x and y are in the same orbit of the $\mathbb{Z}$-action, and so xEy. $\qquad\square$

Example 7.2.4
Let X be a standard Borel space and let $\mathbb{Z}$ act on $X^{\mathbb{Z}}$ by

$$(n \cdot x)(m) = x(m - n).$$

Then the orbit equivalence relation is hyperfinite.

In view of the Glimm–Effros dichotomy, we have the following immediate corollary of the theorem.

Corollary 7.2.5
Let X be a standard Borel space and E a hyperfinite equivalence relation on X. Then either E is smooth or else $E \sim_B E_0$.

Thus E_0 is a typical nonsmooth hyperfinite equivalence relation and in fact the unique one up to Borel bireducibility. The theorem also motivates the following definitions.

Definition 7.2.6
Let X be a standard Borel space. An equivalence relation E is **essentially hyperfinite** if $E \leq_B E_0$. E is **essentially countable** if there is a countable Borel equivalence relation F such that $E \leq_B F$.

However, no good characterizations of essential hyperfiniteness are known.

Exercise 7.2.1 Let E be a countable Borel equivalence relation on a standard Borel space X. Suppose $\{X_n\}_{n\in\omega}$ is a partition of X into disjoint Borel sets and $E \restriction X_n \leq_B E_0$ for all $n \in \omega$. Show that $E \leq_B E_0$.

Exercise 7.2.2 (Jackson) Let E, F be countable Borel equivalence relations on a standard Borel space X. Suppose $E \subseteq F$, E is hyperfinite, and every F-equivalence class contains only finitely many E-equivalence classes. Show that F is hyperfinite.

7.3 Universal countable Borel equivalence relations

In this section we investigate countable Borel equivalence relations that are most complicated in the Borel reducibility hierarchy. Such equivalence relations are called **universal countable Borel equivalence relations**. All results in this section are due to Dougherty, Jackson, and Kechris [30].

By the Feldman–Moore theorem (Theorem 7.1.4) every countable Borel equivalence relation is the orbit equivalence relation of a Borel action of a countable group. If G is a fixed countable group, then among all Borel actions of G one is identified as universal by Theorem 3.3.4. The action is the shift action of G on the power of the Effros Borel space $F(G)^\omega$. Here, however, G has the discrete topology, and hence every subset of G is closed. It is thus equivalent to consider the shift action of G on the space $(2^G)^\omega$ or $2^{G \times \omega}$ defined by

$$(g \cdot f)(h, n) = f(g^{-1}h, n)$$

for $g \in G$, $f \in 2^{G \times \omega}$, $h \in G$, and $n \in \omega$. Note that this action is continuous. We have the following immediate corollary.

Proposition 7.3.1
Let G be a countable group and Let X be the Polish G-space $2^{G \times \omega}$. Then X is a universal Borel G-space and E_G^X is a universal G-orbit equivalence relation.

Also by Theorem 3.5.2 and a trivial analog of it for quotient groups we have the following extension result.

Proposition 7.3.2
Let G_1, G_2 be countable groups, $X_i = 2^{G_i \times \omega}$, and $E_i = E_{G_i}^{X_i}$ for $i = 1, 2$. If $G_1 \leq_w G_2$, then $E_1 \sqsubseteq_B E_2$.

Thus the Borel complexity of the orbit equivalence relations of the shift actions reflect the complexity of the structures of the acting groups. It turns out that this connection between the group structures and the dynamics they produce is much stronger than we would suspect. This strong bondage is now refered to as the rigidity or superrigidity of the groups. It is therefore not surprising that ergodic theory, especially the superrigidity theory, is playing a central role in the study of countable Borel equivalence relations nowadays. Here we only provide an inroad to the study by focusing on the universal countable Borel equivalence relations.

For $1 < n < \omega$ let $\mathbb{F}_n$ be the free group with n generators. Let $\mathbb{F}_\omega$ be the free group with countably infinitely many generators. It is an easy fact of group theory that $\mathbb{F}_\omega$ is a surjectively universal countable group, that is, every countable group is a quotient group of $\mathbb{F}_\omega$. Also $\mathbb{F}_\omega$ can be embedded as a subgroup of $\mathbb{F}_2$. This implies that each of $\mathbb{F}_n$, for $1 < n \leq \omega$, is a weakly universal countable group. Therefore we have the following examples of universal countable Borel equivalence relations.

Proposition 7.3.3
Let $1 < n \leq \omega$, $X = 2^{\mathbb{F}_n \times \omega}$, and $E = E_{\mathbb{F}_n}^X$. Then E is a universal countable Borel equivalence relation.

In search for a simpler representation of the universal countable Borel equivalence relation, we prove the following sharper results.

Proposition 7.3.4
Let G_1, G_2 be countable groups and Y a Polish space. For $i = 1, 2$ let $X_i = Y^{G_i}$ with the shift action of G_i and $E_i = E_{G_i}^{X_i}$. If $G_1 \leq_w G_2$ then $E_1 \sqsubseteq_B E_2$.

Proof. First suppose $G_1 \leq G_2$. Let $y_0 \in Y$. Define an embedding $\varphi : X_1 \to X_2$ by
$$\varphi(f)(g) = \begin{cases} f(g), & \text{if } g \in G_1, \\ y_0, & \text{if } g \notin G_1. \end{cases}$$
Then for $g_1 \in G_1$, $\varphi(g_1 \cdot f) = g_1 \cdot \varphi(f)$. Thus $f E_1 f'$ implies $\varphi(f) E_2 \varphi(f')$. Conversely, suppose $g \cdot \varphi(f) = \varphi(f')$ for some $g \in G_2$. If $g \in G_1$ then $g \cdot f = f'$. Otherwise it must be the case that both f and f' take constant value of y_0, and thus $f = f'$.

Next suppose G_1 is a quotient of G_2, and let $\pi : G_2 \to G_1$ be an onto homomorphism. Then define $\psi : X_1 \to X_2$ by
$$\psi(f)(g) = f(\pi(g)).$$
Then for any $g_2 \in G_2$, $g_2 \cdot \psi(f) = \psi(\pi(g_2) \cdot f)$. It follows easily that $E_1 \sqsubseteq_B E_2$.

The general case $G_1 \leq_w G_2$ is a composition of the above cases considered, and we are done by the transitivity of $\sqsubseteq_B$. □

In several steps we will build up a reduction of the shift action of $\mathbb{F}_2$ on $2^{\mathbb{F}_2 \times \omega}$ to simply its shift action on $2^{\mathbb{F}_2}$.

Proposition 7.3.5
Let G be a countable group, $X = 2^{G \times \omega}$, and $E = E_G^X$. Let $Y = 3^{G \times \mathbb{Z}}$ with the shift action of $G \times \mathbb{Z}$ and let F be the orbit equivalence relation. Then $E \sqsubseteq_B F$.

Proof. Fix a bijection of ω with $\mathbb{Z} - \{0\}$. We can then view X as $2^{G \times (\mathbb{Z} - \{0\})}$ since they are Borel isomorphic as Borel G-spaces. Define an embedding $\varphi : 2^{G \times (\mathbb{Z} - \{0\})} \to Y$ by
$$\varphi(f)(h, n) = \begin{cases} f(h, n), & \text{if } n \neq 0, \\ 2, & \text{if } n = 0. \end{cases}$$

Then for any $g \in G$, $\varphi(g \cdot f) = (g, 0) \cdot \varphi(f)$. Conversely, if $\varphi(f') = (g, m) \cdot \varphi(f)$, then $m = 0$ and $f' = g \cdot f$. This is because, suppose $m \neq 0$, then $\varphi(f)(h, 0) = 2$ for all $h \in G$ and therefore $\varphi(f')(h, m) = ((g, m) \cdot \varphi(f))(h, m) = \varphi(f)(g^{-1}h, 0) = 2$ for all $h \in G$, but this is a contradiction since $\varphi(f')(h, m) = f'(h, m) \in \{0, 1\}$ by definition. □

Proposition 7.3.6
Let G be a countable group, $X = 3^G$, and $E = E_G^X$. Let F be the orbit equivalence relation of the shift action of $G \times \mathbb{Z}_2$ on $2^{G \times \mathbb{Z}_2}$. Then $E \sqsubseteq_B F$.

Proof. View 0 as encoded by 00, 1 by 01, and 2 by 11. Define $\varphi : X \to 2^{G \times \mathbb{Z}_2}$ by

$$\varphi(f)(h, i) = \begin{cases} 0, & \text{if } f(h) = 0 \text{ or if } f(h) = 1 \text{ and } i = 0, \\ 1, & \text{if } f(h) = 1 \text{ and } i = 1 \text{ or if } f(h) = 2. \end{cases}$$

Then for any $g \in G$, $\varphi(g \cdot f) = (g, 0) \cdot \varphi(f)$. Conversely, if $\varphi(f') = (g, i)\varphi(f)$, then we claim that $\varphi(f') = (g, 0) \cdot \varphi(f)$ and $f' = g \cdot f$. Suppose $\varphi(f') = (g, 1) \cdot \varphi(f)$, then $\varphi(f') = (1_G, 1) \cdot \varphi(g \cdot f)$, and thus without loss of generality we may assume $\varphi(f') = (1_G, 1) \cdot \varphi(f)$. It is a property of the coding used that if $\varphi(f')(h, 0) = 1$ then $\varphi(f')(h, 1) = 1$ as well. Thus by our assumption if $\varphi(f)(h, 1) = \varphi(f')(h, 0) = 1$ then $\varphi(f)(h, 0) = \varphi(f')(h, 1) = 1$. Similarly we also have that if $\varphi(f)(h, 1) = 0$ then $\varphi(f)(h, 0) = 0$. It follows that for any $h \in G$, $\varphi(f)(h, 0) = \varphi(f)(h, 1) = \varphi(f')(h, 0) = \varphi(f')(h, 1)$. So $f = f'$. □

Notation 7.3.7
Let E_∞ denote the orbit equivalence relation induced by the shift action of $\mathbb{F}_2$ on $2^{\mathbb{F}_2}$.

Theorem 7.3.8
E_∞ is a universal countable Borel equivalence relation.

Proof. By Proposition 7.3.3 the orbit equivalence relation of $\mathbb{F}_2$ shift action on $2^{\mathbb{F}_2 \times \omega}$ is already a universal countable Borel equivalence relation. By Proposition 7.3.5 it is Borel embeddable into orbit equivalence relation of the shift action of $\mathbb{F}_2 \times \mathbb{Z}$ on $3^{\mathbb{F}_2 \times \mathbb{Z}}$, and by Proposition 7.3.6 the latter orbit equivalence relation is in turn Borel embeddable into that of the shift action of $\mathbb{F}_2 \times \mathbb{Z} \times \mathbb{Z}_2$ on $2^{\mathbb{F}_2 \times \mathbb{Z} \times \mathbb{Z}_2}$. The group $\mathbb{F}_2 \times \mathbb{Z} \times \mathbb{Z}_2$ is a quotient of $\mathbb{F}_\omega$, and thus by Proposition 7.3.4, the last equivalence relation is Borel embeddable into that of the shift action of $\mathbb{F}_\omega$ on $2^{\mathbb{F}_\omega}$. This shows that the shift action of $\mathbb{F}_\omega$ on $2^{\mathbb{F}_\omega}$ is a universal countable Borel equivalence relation. Now since $\mathbb{F}_\omega$ is a subgroup of $\mathbb{F}_2$, by Proposition 7.3.4 again E_∞ is also universal. □

In the next section we will show that $E_0 <_B E_\infty$. It is now known that the Borel reducibility among countable Borel equivalence relations is very complicated (see Reference [1] by Adams and Kechris).

Exercise 7.3.1 Give a direct proof of Proposition 7.3.1.

Exercise 7.3.2 Prove Proposition 7.3.2.

Exercise 7.3.3 Show that for any $n \geq 2$ the shift action of $\mathbb{F}_n$ on $2^{\mathbb{F}_n}$ gives rise to a universal countable Borel equivalence relation.

Consider the shift action of $\mathbb{F}_n$ on $2^{\mathbb{F}_n}$. Let F_n be the free part of this action, that is,

$$F_n = \{ x \in 2^{\mathbb{F}_n} : \forall g \in \mathbb{F}_n \, (g \neq 1_G \Rightarrow g \cdot x \neq x) \}.$$

Exercise 7.3.4 Show that F_n is an invariant dense G_δ subset of $2^{\mathbb{F}_n}$.

Exercise 7.3.5 Let μ_n be the usual product measure on $2^{\mathbb{F}_n}$. Show that $\mu_n(F_n) = 1$.

Exercise 7.3.6 Let E_n be the orbit equivalence relation on F_n of the induced $\mathbb{F}_n$ action. Show that $E_n \sqsubseteq_B E_2$ for all $n \geq 2$.

7.4 Amenable groups and amenable equivalence relations

In this section we show that the universal countable Borel equivalence relation E_∞ defined in the preceding section is not hyperfinite. To do this we discuss the concept of amenability for countable groups and their orbit equivalence relations. The study of amenable groups and amenable equivalence relations is interesting in its own right and is a well-established subject in ergodic theory. We first give the definitions.

Definition 7.4.1
Let X be a set. A **finitely additive probability measure** on X is a map

$$\mu : P(X) \rightarrow [0, 1],$$

where $P(X) = \{A : A \subseteq X\}$ is the power set of X, such that $\mu(\emptyset) = 0$, $\mu(X) = 1$, and $\mu(A \cup B) = \mu(A) + \mu(B)$ if $A \cap B = \emptyset$.

If G is a group, a finitely additive probability measure μ on G is **left-invariant** if for all $g \in G$ and $A \subseteq G$, $\mu(gA) = \mu(A)$.

A countable group G is **amenable** if there is a left-invariant finitely additive probability measure μ on G.

Definition 7.4.2
A countable group G satisfies the **Reiter condition** if for all $\epsilon > 0$ and $g_1, \ldots, g_n \in G$ there is $f : G \to \mathbb{R}_{\geq 0}$ such that $\sum_{g \in G} f(g) = 1$ and for any $1 \leq i \leq n$, $\sum_{g \in G} |f(g) - f(g_i^{-1}g)| < \epsilon$.

   G satisfies the **Følner condition** if for all $\epsilon > 0$ and $g_1, \ldots, g_n \in G$ there is a finite $F \subseteq G$ such that for any $1 \leq i \leq n$, $|g_i F \triangle F| < \epsilon |F|$.

It is known that the Reiter condition and the Følner condition are equivalent to amenability for countable groups. Here, to keep our exposition self-contained, we only prove the following weaker theorem.

Theorem 7.4.3
Let G be a countable group. If G satisfies the Reiter condition then it also satisfies the Følner condition. If G satisfies the Følner condition then it is amenable.

Proof. First we assume that G satisfies the Reiter condition. Let $\epsilon > 0$ and $g_1, \ldots, g_n \in G$. Let $f : G \to \mathbb{R}_{\geq 0}$ be such that $\sum_{g \in G} f(g) = 1$ and for any $1 \leq i \leq n$, $\sum_{g \in G} |f(g) - f(g_i^{-1}g)| < \epsilon/n$. For each $0 < r \leq 1$ we let $F_r = \{g \in G : f(g) \geq r\}$ and χ_r be the characteristic function of F_r. Since $\sum_{g \in G} f(g) = 1$ the set F_r is finite for any $r > 0$. Note that $f(g) = \int_0^1 \chi_r(g) dr$, $|F_r| = \sum_{g \in G} \chi_r(g)$, and $|F_r \triangle g_i F_r| = |F_r \triangle g_i^{-1} F_r| = \sum_{g \in G} |\chi_r(g) - \chi_r(g_i^{-1}g)|$ for any $1 \leq i \leq n$. Then for any $1 \leq i \leq n$,

$$\int_0^1 |F_r \triangle g_i F_r| dr = \int_0^1 \sum_{g \in G} |\chi_r(g) - \chi_r(g_i^{-1}g)| dr$$

$$= \sum_{g \in G} \int_0^1 |\chi_r(g) - \chi_r(g_i^{-1}g)| dr = \sum_{g \in G} |f(g) - f(g_i^{-1}g)|$$

$$< \frac{\epsilon}{n} = \frac{\epsilon}{n} \sum_{g \in G} f(g) = \frac{\epsilon}{n} \sum_{g \in G} \int_0^1 \chi_r(g) dr = \frac{\epsilon}{n} \int_0^1 \sum_{g \in G} \chi_r(g) dr$$

$$= \frac{\epsilon}{n} \int_0^1 |F_r| dr.$$

It follows that

$$\int_0^1 \sum_{i=1}^n |F_r \triangle g_i F_r| dr < \epsilon \int_0^1 |F_r| dr,$$

and thus there is $0 < r \leq 1$ such that $\sum_{i=1}^n |F_r \triangle g_i F_r| < \epsilon |F_r|$. This F_r witnesses the Følner condition for G.

   Now for the second part of the theorem assume that G satisfies the Følner condition. Enumerate G as $g_0, g_1, \ldots, g_n, \ldots$ and for each $n \in \omega$ let $F_n \subseteq G$ be a finite subset of G such that $|F_n \triangle g_i F_n| < 2^{-n}|F_n|$. Let $\mathcal{U}$ be a nonprincipal ultrafilter on ω. For $A \subseteq G$, define

$$\mu(A) = \lim_{n \in \mathcal{U}} \frac{|A \cap F_n|}{|F_n|}.$$

Then μ is a finitely additive probability measure on G. It remains to check that μ is left-invariant. For this let $g \in G$ and $A \subseteq G$. Then

$$|\mu(gA) - \mu(A)| = \lim_{n \in \mathcal{U}} \frac{||gA \cap F_n| - |A \cap F_n||}{|F_n|}$$

$$= \lim_{n \in \mathcal{U}} \frac{||A \cap g^{-1}F_n| - |A \cap F_n||}{|F_n|}$$

$$\leq \lim_{n \in \mathcal{U}} \frac{|g^{-1}F_n \triangle F_n|}{|F_n|} = 0.$$

Thus G is amenable. ☐

It is easy to see that $\mathbb{Z}$ satisfies the Reiter condition. In fact for any integer $N > 1$ let $f_N : \mathbb{Z} \to \mathbb{R}_{\geq 0}$ be defined as

$$f_N(n) = \begin{cases} 1/N, & \text{if } 0 < n \leq N, \\ 0, & \text{otherwise.} \end{cases}$$

Then for any finite collection of elements of $\mathbb{Z}$ one of the functions f_N witnesses the Reiter condition. It follows that $\mathbb{Z}$ is amenable.

Proposition 7.4.4
$\mathbb{F}_2$ *is not amenable.*

Proof. Assume that μ is a left-invariant finitely additive probability measure on $\mathbb{F}_2$. Let a and b be the generators of $\mathbb{F}_2$. For each $s \in \{a, b, a^{-1}, b^{-1}\}$ let $A(s)$ be the set of all elements of $\mathbb{F}_2$ whose reduced form starts with s. Then $A(a), A(b), A(a^{-1}), A(b^{-1})$ form a partition of $\mathbb{F}_2$. Also we have that $A(a) = a(A(b) \cup A(b^{-1}) \cup A(a))$. Since μ is left-invariant, we have that

$$\mu(A(a)) = \mu(A(b) \cup A(b^{-1}) \cup A(a)) = \mu(A(b)) + \mu(A(b^{-1})) + \mu(A(a)).$$

It follows that $\mu(A(b)) = \mu(A(b^{-1})) = 0$. Similarly we also obtain that $\mu(A(a)) = \mu(A(a^{-1})) = 0$. But then $\mu(\mathbb{F}_2) = 0$, a contradiction. ☐

We now turn to equivalence relations. The following definition of amenable equivalence relations is inspired by the Reiter condition.

Definition 7.4.5

Let X be a standard Borel space, μ a probability Borel measure on X, and E a countable equivalence relation. We say that (X, E, μ) is **amenable**, or E is μ-**amenable**, if there is a sequence of Borel functions

$$\varphi_n : E \to \mathbb{R}_{\geq 0}$$

such that

(a) for any $x \in X$, $\sum_{yEx} \varphi_n(x, y) = 1$, and

(b) there is a Borel E-invariant set $A \subseteq X$ with $\mu(A) = 1$ such that for all $(x, x') \in E \cap (A \times A)$,

$$\sum_{yEx} |\varphi_n(x, y) - \varphi_n(x', y)| \to 0 \text{ as } n \to \infty.$$

Proposition 7.4.6

Let G be a countable group, X a Borel G-space, and μ a Borel probability measure on X. If G satisfies the Reiter condition, then E_G^X is μ-amenable.

Proof.   Enumerate G as $g_0, g_1, \ldots, g_n, \ldots$ and for each $n \geq 1$ let $f_n : G \to \mathbb{R}_{\geq 0}$ be such that $\sum_{g \in G} f(g) = 1$ and $\sum_{g \in G} |f(g) - f(g_i^{-1}g)| < 1/n$ for all $i < n$. Then for $x, y \in X$ let

$$\varphi_n(x, y) = \sum_{g \cdot y = x} f_n(g).$$

Then for every $x \in X$, $\sum_{yEx} \varphi_n(x, y) = \sum_{g \in G} f_n(g) = 1$. If xEx' and $x = h \cdot x'$, then for any yEx,

$$\varphi_n(x, y) - \varphi_n(x', y) = \sum_{g \cdot y = x} f_n(g) - \sum_{g' \cdot y = x'} f_n(g') = \sum_{g \cdot y = x} f_n(g) - f_n(h^{-1}g),$$

and

$$\sum_{yEx} |\varphi_n(x, y) - \varphi_n(x', y)| \leq \sum_{g \in G} |f_n(g) - f_n(h^{-1}g)| \to 0$$

as $n \to \infty$. Thus E_G^X is μ-amenable.   ☐

In particular, every hyperfinite equivalence relation is μ-amenable for any Borel probability measure μ. We will show below that the amenability of an orbit equivalence relation can be pulled back to conclude the amenability of the group. For this we need the notion of G-invariance for measures, as follows.

Definition 7.4.7
Let G be a countable group, X a Borel G-space, and μ a Borel probability measure on X. We say that μ is **G-invariant** if for all $g \in G$ and Borel $S \subseteq X$, $\mu(S) = \mu(g \cdot S)$.

Theorem 7.4.8
Let G be a countable group, X a Borel G-space, and μ a Borel probability measure on X. Suppose the action is free and μ is G-invariant. If E_G^X is μ-amenable, then G is amenable.

Proof. It suffices to show that G satisfies the Reiter condition. Let φ_n be Borel functions witnessing the μ-amenability of E_G^X. Define

$$f_n(g) = \int \varphi_n(x, g \cdot x) d\mu(x).$$

Then $\sum_{g \in G} f_n(g) = \int \sum_{g \in G} \varphi_n(x, g \cdot x) d\mu(x) = 1$. In addition, for every $h \in G$, by the assumptions we have

$$\sum_{g \in G} |f_n(g) - f_n(h^{-1}g)|$$

$$= \sum_{g \in G} |\int \varphi_n(x, g \cdot x) d\mu(x) - \int \varphi_n(x, h^{-1}g \cdot x) d\mu(x)|$$

$$= \sum_{g \in G} |\int \varphi_n(x, g \cdot x) d\mu(x) - \int \varphi_n(h \cdot x, g \cdot x) d\mu(x)|$$

$$\leq \sum_{g \in G} \int |\varphi_n(x, g \cdot x) - \varphi(h \cdot x, g \cdot x)| d\mu(x)$$

$$= \int \sum_{g \in G} |\varphi_n(x, g \cdot x) - \varphi_n(h \cdot x, g \cdot x)| d\mu(x)$$

$$= \int \sum_{y E x} |\varphi_n(x, y) - \varphi_n(h \cdot x, y)| d\mu(x) \to 0$$

as $n \to \infty$. Now if $\epsilon > 0$ and $g_1, \ldots, g_m \in G$, then for large enough n we have $\sum_{g \in G} |f_n(g) - f_n(g_i^{-1}g)| < \epsilon$ for all $1 \leq i \leq m$. This shows that G satisfies the Reiter condition, and is therefore amenable. $\square$

The usual product measure on 2^G is a canonical example of a G-invariant measure (Exercise 7.4.5). To apply the above theorem we check below that its free part has full measure.

Lemma 7.4.9
Let G be a countably infinite group, $X = 2^G$ with the shift action of G, and μ the usual product measure on X. Let F be the free part of X. Then $\mu(F) = 1$.

Proof. We show that for any $g \in G$, the set $\{x \in 2^G : g \cdot x = x\}$ has measure 0. Consider two cases. Case 1: g is of infinite order, that is, for any $n \neq m \in \mathbb{Z}$, $g^n \neq g^m$. In this case note that if $g \cdot x = x$ then for all $n \in \mathbb{Z}$, we have $x(g^{2n}) = x(g^{2n+1})$. Thus

$$\mu\left(\{x : g \cdot x = x\}\right) \leq \mu\left(\{x : \forall n \in \mathbb{Z}\ x(g^{2n}) = x(g^{2n+1})\}\right)$$

$$= \prod_{n \in \mathbb{Z}} \mu\left(\{x : x(g^{2n}) = x(g^{2n+1})\}\right) = \prod_{n \in \mathbb{Z}} \frac{1}{2} = 0.$$

Case 2: g has finite order, that is, for some $n \in \mathbb{Z}$, $g^n = 1_G$. In this case let $H = \langle g \rangle \leq G$. Then H is a finite subgroup of G. Since G is infinite, there are infinitely many right-cosets of H in G. Let $g_0, g_1, \ldots$ be chosen from infinitely many distinct cosets of H in G. Note that if $g \cdot x = x$ then for all $n \in \omega$, we have $x(g_n) = x(gg_n)$. Moreover, for distinct $n, m \in \omega$, $\{g_n, gg_n\} \cap \{g_m, gg_m\} = \emptyset$. Thus similar to Case 1 we have

$$\mu\left(\{x : g \cdot x = x\}\right) \leq \mu\left(\{x : \forall n \in \omega\ x(g_n) = x(gg_n)\}\right)$$

$$= \prod_{n \in \omega} \mu\left(\{x : x(g_n) = x(gg_n)\}\right) = \prod_{n \in \mathbb{Z}} \frac{1}{2} = 0.$$

□

We are finally ready to prove the main theorem of this section.

Theorem 7.4.10
E_∞ is not hyperfinite. In particular, $E_0 <_B E_\infty$.

Proof. Let $G = \mathbb{F}_2$, $X =$ the free part of $2^{\mathbb{F}_2}$ with the shift action, and μ be the usual product measure restricted on X. If E_G^X is hyperfinite then by Theorem 7.2.3 (vi) there is a Borel action of $\mathbb{Z}$ on X such that $E_\mathbb{Z}^X = E_G^X$, and by Proposition 7.4.6 E_G^X is μ-amenable. This implies that $\mathbb{F}_2$ is amenable by Theorem 7.4.8, contradicting Proposition 7.4.4. □

Exercise 7.4.1 Show that every finite group is amenable.

Exercise 7.4.2 Let G be a countable group. A finitely additive probability measure μ on G is **right-invariant** if for all $g \in G$ and $A \subseteq G$, $\mu(Ag) = \mu(A)$. μ is **two-sided invariant** if it is both left-invariant and right-invariant. Show that G is amenable iff there is a two-sided invariant finitely additive probability measure on G.

Exercise 7.4.3 Prove that $\mathbb{Z}^n$ satisfies the Reiter condition.

Exercise 7.4.4 Prove that if G and H both satisfy the Reiter condition then so does $G \times H$.

Exercise 7.4.5 Let G be a countable group, $X = 2^G$ with the shift action of G, and μ the usual product measure on X. Show that μ is G-invariant.

Exercise 7.4.6 Let F_2 be the free part of $2^{\mathbb{F}_2}$ with the shift action of $\mathbb{F}_2$. Let $E_{\infty T}$ denote the equivalence relation E_∞ restricted to F_2. Show that $E_{\infty T}$ is not hyperfinite.

7.5 Actions of locally compact Polish groups

In this section we prove a theorem of Kechris [92] that any orbit equivalence relation of an action of a locally compact Polish group is essentially countable, that is, Borel reducible to a countable Borel equivalence relation. We will use the following theorem, which is a generalization of the Luzin–Novikov uniformization theorem (Theorem 7.1.2) to sets with K_σ sections. Recall that a subset of topological space is K_σ if it is a union of countably many compact sets.

Theorem 7.5.1 (Arsenin–Kunugui)
Let Y be a Polish space, X a standard Borel space, and $P \subseteq X \times Y$ Borel. Suppose each section P_x is K_σ. Then $\mathrm{proj}_X(P)$ is Borel and P has a Borel uniformization.

A proof can be found in Reference [97] Theorem 35.46. Also recall that a complete section of an equivalence relation E on X is a subset Y of X which meets every E-equivalence class. We say that a complete section Y is **countable** if $Y \cap [x]_E$ is countable for every $x \in X$.

Theorem 7.5.2 (Kechris)
Let G be a locally compact Polish group and X a Borel G-space. Then there is a Borel countable complete section of E_G^X.

Proof. In view of Theorem 4.4.6 we may assume X is a Polish G-space without loss of generality. Let d be a compatible metric on X. Let K_0 be a compact nbhd of 1_G and K a compact symmetric nbhd (that is, $K = K^{-1}$) of 1_G with $K^2 \subseteq K_0$. Let $N = \mathrm{Int}(K)$. Consider the following relation defined on X:

$$R(x, y) \iff \exists g \in K \ (g \cdot x = y).$$

R is symmetric and reflexive, but not necessarily an equivalence relation on X. However, we will construct a sequence of Borel sets $X_n \subseteq X$ with $X = \bigcup_n X_n$ such that $R \upharpoonright X_n$ is an equivalence relation.

For $\epsilon > 0$ put

$$A_\epsilon = \{x \in X : \forall g \in K_0 \, (d(x, g \cdot x) \le \epsilon \Rightarrow g \in NG_x) \}.$$

Recall that G_x is the stabilizer of x.

We check that each A_ϵ is Borel. Note that

$$x \notin A_\epsilon \iff \exists g \, P(x, g),$$

where

$$P(x, g) \iff g \in K_0 - NG_x \text{ and } d(x, g \cdot x) \le \epsilon.$$

By the Arsenin–Kunugui uniformization theorem it suffices to show that P is Borel and that each section P_x is K_σ. It is clear that every section P_x is K_σ since G as a locally compact Polish group is K_σ and each P_x is closed in G. To see that P is Borel, note that the conditions $g \in K_0$ and $d(x, g \cdot x) \le \epsilon$ are closed. Hence it is enough to see that $g \notin NG_x$ is a Borel condition. For this we note that

$$g \in NG_x \iff \exists h \in G_x \exists k \in N \, (g = hk) \iff \exists h \exists k \, Q(g, x, h, k),$$

where

$$Q(g, x, h, k) \iff h \cdot x = x \wedge k \in N \wedge g = hk.$$

Q is obviously Borel, with sections $Q_{g,x}$ F_σ in $G \times G$ and hence K_σ. Thus again by the Arsenin–Kunugui theorem we have that $g \in NG_x$ is Borel, which in turn implies that A_ϵ is Borel.

Next we claim that $X = \bigcup_{n \ge 1} A_{1/n}$. Toward a contradiction assume $x \notin A_{1/n}$ for any $n \ge 1$. We obtain a sequence $g_n \in K_0$ such that $d(x, g_n \cdot x) \le 1/n$ but $g_n \notin NG_x$. By compactness of K_0 there is an accumulation point $g_\infty \in K_0$ of g_n. Then $g_\infty \cdot x = x$ but $g_\infty \notin NG_x$ since NG_x is open. This is a contradiction since $1_G \in N$.

Next we note that if $B \subseteq A_\epsilon$ has diameter $\le \epsilon$ then $R \upharpoonright B$ is an equivalence relation. If suffices to check that $R \upharpoonright B$ is transitive. For this let $x, y, z \in B$ with $R(x, y)$ and $R(y, z)$. So there are $g, h \in K$ with $g \cdot x = y$ and $h \cdot y = z$. Since $K^2 \subseteq K_0$, $hg \in K_0$ and $d(x, hg \cdot x) = d(x, z) \le \text{diam}(B) \le \epsilon$. Thus $hg \in NG_x$. This means that there is $k \in N \subseteq K$ with $z = k \cdot x$, and hence $R(x, z)$.

For each $n \ge 1$ let $\{B_{n,m} : m \in \omega\}$ be a countable open cover of X with $\text{diam}(B_{n,m}) \le 1/n$. Then the collection

$$\{A_{1/n} \cap B_{n,m} : n \ge 1, m \in \omega\}$$

is a countable Borel cover of X so that the restriction of R on each set is an equivalence relation. Let X_n enumerate this collection of Borel sets.

We claim that $R \upharpoonright X_n$ is smooth. To see this note that R is closed in $X \times X$. Let τ be a Polish topology on X extending the topology induced by d so that X_n is clopen in (X, τ). Then R is still closed in $\tau \times \tau$. Now $R \upharpoonright X_n$ is a closed equivalence relation on the Polish space $(X_n, \tau \upharpoonright X_n)$, therefore it is smooth.

By Proposition 5.4.10 E_G^X is idealistic, that is, there is a Borel map $C \mapsto I_C$ assigning to each E_G^X-equivalence class C a σ-ideal I_C so that $C \notin I_C$. Now since $X = \bigcup_n X_n$, for each E_G^X equivalence class C we also have $C = \bigcup_n (C \cap X_n)$. It follows that there is an $n \in \omega$ such that $C \cap X_n \notin I_C$. Let $n(C)$ be the least such n for the class C. Put

$$Y_0 = \bigcup_{x \in X} [x]_G \cap X_{n([x]_G)}.$$

Then Y_0 is Borel since for $C = [x]_G$,

$$x \in Y_0 \iff \exists n \left[\forall m < n \left(C \cap X_m \in I_C \right) \wedge C \cap X_n \notin I_C \wedge x \in X_n \right].$$

It is clear that Y_0 is a complete section for E_G^X. Let $Y_{0,n} = Y_0 \cap X_n$. Then $\{Y_{0,n} : n \in \omega\}$ is a partition of Y_0. Since $R \upharpoonright X_n$ is a smooth equivalence relation, it follows that $R \upharpoonright Y_{0,n}$ is also a smooth equivalence relation, and therefore so is $R \upharpoonright Y_0$. It is clear that each $E_G^X \upharpoonright Y_0$ equivalence class in Y_0 contains only countably many $R \upharpoonright Y_0$-classes.

Define $Y \subseteq Y_0$ by

$$x \in Y \iff x \in Y_0 \wedge [x]_{R \upharpoonright Y_0} \notin I_{[x]_G}.$$

Then again Y is a Borel complete section of E_G^X. We still have that $R \upharpoonright Y$ is smooth. Now $R \upharpoonright Y$ is idealistic, witnessed by the assignment

$$[x]_{R \upharpoonright Y} \longmapsto I_{[x]_G} \upharpoonright [x]_{R \upharpoonright Y}.$$

By Theorem 5.4.11 $R \upharpoonright Y$ has a Borel selector as well as a Borel transversal. Let A be a Borel transversal of $R \upharpoonright Y$ on Y. Since $R \subseteq E_G^X$, A is a complete section of E_G^X. Each $E_G^X \upharpoonright Y$ class in Y contains only countably many $R \upharpoonright Y$ classes. Thus A is a Borel countable complete section for E_G^X, as required. ☐

Corollary 7.5.3
Let G be a locally compact Polish group and X a Borel G-space. Then E_G^X is essentially countable, that is, $E_G^X \leq_B E_\infty$.

Proof. Let $A \subseteq X$ be a Borel countable complete section of E_G^X. Let $F = E_G^X \upharpoonright A$. Then F is a countable Borel equivalence relation on the standard Borel space A. By Theorem 7.1.2 there is a Borel function $f : X \to A$ so that $f(x) E_G^X x$ for all $x \in X$. Then f witnesses that $E_G^X \leq_B F$. ☐

Exercise 7.5.1 Show that the following are equivalent for a Borel equivalence relation E:

(a) E is essentially countable;

(b) there is a Borel countable complete section of E;

(c) $E = \bigcup_n E_n$ where each E_n is smooth and idealistic;

(d) $E = \bigcup_n E_n$ where each E_n has a Borel selector.

Exercise 7.5.2 Let $n \geq 1$. Consider the space $F(\mathbb{R}^n)$ of all closed subsets of $\mathbb{R}^n$. If $A, B \in F(\mathbb{R}^n)$ we say that A is **congruent** to B, denoted $A \sim B$, if there is an isometry from A onto B. Show that the congruence relation on $F(\mathbb{R}^n)$ is essentially countable.

Chapter 8

Borel Equivalence Relations

In this chapter we continue our introduction of Borel equivalence relations. The primary classification here is orbit equivalence relations versus equivalence relations not reducible to orbit equivalence relations. Again we only give a self-contained account of the basic facts. The more interested reader should go beyond this book and find a vast literature about the subject. Along with many theorems there are also a large number of open problems. Many of the problems are fundamental, the answers of which might potentially redefine the subject.

8.1 Hypersmooth equivalence relations

The concept of a hypersmooth equivalence relation is a natural generalization of hyperfiniteness.

Definition 8.1.1
Let X be a standard Borel space. An equivalence relation E on X is **hypersmooth** if $E = \bigcup_n E_n$, where E_n are smooth equivalence relations and $E_n \subseteq E_{n+1}$ for all n.

Similar to the equivalence relation E_0 we define a canonical hypersmooth equivalence relation E_1.

Definition 8.1.2
The equivalence relation E_1 is defined on $2^{\omega \times \omega}$ as

$$x E_1 y \iff \exists m \; \forall n \geq m \; \forall k \; x(n,k) = y(n,k).$$

It is easy to check that E_1 is hypersmooth. We can also consider the eventual agreement for sequences of actual real numbers. Let $E_1(\mathbb{R})$ denote the equivalence relation on $\mathbb{R}^\omega$ defined by

$$(x_n) E_1(y_n) \iff \exists m \; \forall n \geq m \; x_n = y_n.$$

It is easy to see that $E_1(\mathbb{R})$ is also hypersmooth and that $E_1 \sim_B E_1(\mathbb{R})$. For notational simplicity we also write $x = (x_n)$ for $x \in 2^{\omega \times \omega}$ and $x_n \in 2^\omega$, where $x(n,k) = x_n(k)$ for all $n, k \in \omega$.

We have the following closure properties for hypersmooth equivalence relations. The proof is left to the reader (Exercise 8.1.1).

Proposition 8.1.3
Let X, Y be standard Borel spaces and E, F equivalence relations on X, Y, respectively. Then the following are true:

(1) *If F is hypersmooth and $E \leq_B F$, then E is hypersmooth.*

(2) *If E is hypersmooth and $A \subseteq X$ is Borel, then $E \upharpoonright A$ is hypersmooth.*

(3) *If E and F are hypersmooth, then so is $E \times F$.*

The following characterization of hypersmoothness is analogous to the one on hyperfiniteness.

Proposition 8.1.4
Let X be a standard Borel space and E an equivalence relation on X. Then the following are equivalent:

(i) *E is hypersmooth.*

(ii) *$E \leq_B E_1$.*

(iii) *$E \sqsubseteq_B E_1$.*

Proof. By Proposition 8.1.3 (1) we have that (iii)$\Rightarrow$(ii)$\Rightarrow$(i). It suffices to show (i)$\Rightarrow$(iii). For this let E be hypersmooth. Let $E = \bigcup_n F_n$ where F_n is smooth and $F_n \subseteq F_{n+1}$ for all $n \in \omega$. Without loss of generality assume that $F_0 = \mathrm{id}(X)$. Let $f_n : X \to 2^\omega$ witness that $F_n \leq_B \mathrm{id}(2^\omega)$. Then in particular f_0 is a Borel injection from X into 2^ω. Define $f : X \to 2^{\omega \times \omega}$ by

$$f(x)(n, k) = f_n(x)(k).$$

Then f is a Borel injection and $x E y \iff f(x) E_1 f(y)$, as required. ⬜

Hyperfinite equivalence relations are of course hypersmooth. Conversely, we have the following theorem.

Theorem 8.1.5 (Dougherty–Jackson–Kechris)
Let X be a standard Borel space and E a countable Borel equivalence relation on X. If E is hypersmooth, then it is hyperfinite.

Proof. Let $E = \bigcup_n F_n$, where each F_n is smooth, $F_n \subseteq F_{n+1}$ for all $n \in \omega$. Without loss of generality assume $F_0 = \mathrm{id}(X)$. Each F_n is a countable equivalence relation. Hence by the Feldman–Moore Theorem 7.1.4 there are countable groups G_n with Borel actions on X so that $F_n = E_{G_n}^X$. By Proposition 5.4.10 and Theorem 5.4.11 there are Borel selectors s_n for F_n, that is,

$s_n(x)F_n x$ and $s_n(x) = s_n(y)$ for all $x, y \in X$ with $x E_n y$. Fix an enumeration for G_n as $\{g_{n,k}\}_{k \in \omega}$.

We define a sequence of relations R_n as follows:

$$x R_n y \iff \exists m \le n \ (\ x F_m y \wedge$$
$$\exists k_0, \dots, k_m \le n \ [x = g_{0,k_0} s_0 g_{1,k_1} s_1 \cdots g_{m,k_m} s_m(x)] \wedge$$
$$\exists l_0, \dots, l_m \le n \ [y = g_{0,l_0} s_0 g_{1,l_1} s_1 \cdots g_{m,l_m} s_m(y)]).$$

Then $R_n \subseteq F_n$ and $R_n \subseteq R_{n+1}$.

We claim that $E = \bigcup_n R_n$. Suppose $x E y$. Then there is $m \in \omega$ such that $x F_m y$. For every $j \le m$, there is $g_{j,k_j} \in G$ with $x = g_{j,k_j} s_j(x)$. Similarly, for each $j \le m$ there is $g_{j,l_j} \in G$ with $y = g_{j,l_j} s_j(y)$. Then we have that $x = g_{0,j_0} s_0 g_{1,j_1} s_1 \cdots g_{m,k_m} s_m(x)$ and $y = g_{0,l_0} s_0 g_{1,l_1} s_1 \cdots g_{m,l_m} s_m(y)$. Let $n = \max\{m, k_0, k_1, \dots, k_m, l_0, \dots, l_m\}$. Then $x R_n y$.

Next we check that R_n is an equivalence relation. Clearly R_n is reflexive and symmetric. We only show its transitivity. For this let $x R_n y R_n z$. Let $m, k_0, \dots, k_m, l_0, \dots, l_m \le n$ witness that $x R_n y$ and, without loss of generality, let $p \le m$ and $h_0, \dots, h_p \le n$ be such that $y E_p z$ and

$$g_{0,h_0} s_0 g_{1,h_1} s_1 \cdots g_{p,h_p} s_p(z) = z.$$

If $m = p$ then there is nothing to prove. We assume $p < m$. Then we have that $x F_m y F_p z$ and thus $x F_m z$. Note that $s_m(y) = s_m(z)$ since $y F_p z$ and $p < m$. Let

$$v = g_{p+1,l_{p+1}} s_{p+1} \cdots g_{m,l_m} s_m(y) = g_{p+1,l_{p+1}} s_{p+1} \cdots g_{m,l_m} s_m(z).$$

Then

$$y = g_{0,l_0} s_0 g_{1,l_1} s_1 \cdots g_{p,l_p} s_p(v)$$

and hence $v F_p y$. Since $y F_p z$ it follows that $v F_p z$, and thus $s_p(v) = s_p(z)$. Therefore

$$z = g_{0,h_0} s_0 g_{1,h_1} s_1 \cdots g_{p,h_p} s_p(z)$$
$$= g_{0,h_0} s_0 g_{1,h_1} s_1 \cdots g_{p,h_p} s_p(v)$$
$$= g_{0,h_0} s_0 g_{1,h_1} s_1 \cdots g_{p,h_p} s_p g_{p+1,l_{p+1}} s_{p+1} \cdots g_{m,l_m} s_m(y)$$
$$= g_{0,h_0} s_0 g_{1,h_1} s_1 \cdots g_{p,h_p} s_p g_{p+1,l_{p+1}} s_{p+1} \cdots g_{m,l_m} s_m(z).$$

This shows that $x R_n z$, and thus R_n is an equivalence relation.

Finally note that R_n is finite, since $s_m(x) = s_m(y)$ for any $y F_m x$. We have thus obtained an increasing sequence of finite Borel equivalence relations R_n with $E = \bigcup_n R_n$, and therefore E is hyperfinite. ☐

In the rest of this section we analyze the equivalence relation E_1 and show that it is not essentially countable. This implies in particular that it is not essentially hyperfinite, and hence $E_0 <_B E_1$. It also follows that E_1 is not idealistic (see Exercise 8.1.3), and therefore in particular E_1 is not the orbit equivalence relation of a Polish group action.

Theorem 8.1.6 (Kechris–Louveau)

Let X be a standard Borel space, F a countable Borel equivalence relation on X, and $f : 2^{\omega \times \omega} \to X$ a Borel map such that $x E_1 y \Rightarrow f(x) F f(y)$. Then there are $x, y \in 2^{\omega \times \omega}$ such that

(a) for $n \neq m$, $x_n \neq x_m$ and $y_n \neq y_m$,

(b) for $n, m \in \omega$, $x_n \neq y_m$, and

(c) $f(x) = f(y)$.

In particular, $E_1 \not\leq_B F$.

Proof. Let $C = \bigcap_n U_n \subseteq 2^{\omega \times \omega}$ where $U_n \supseteq U_{n+1}$ are dense open such that $f \upharpoonright C$ is continuous. We construct sequences $(l_n), (k_n), (p_n), (q_n), (x_n), (y_n)$ such that the following conditions are met:

(i) $l_n, k_n \in \omega$, $l_n < l_{n+1}$, $k_n < k_{n+1}$ for all $n \in \omega$;

(ii) $p_n, q_n \in 2^{l_n \times k_n}$;

(iii) $N_{p_n}, N_{q_n} \subseteq U_n$, where $N_s = \{x \in 2^{\omega \times \omega} : s \subseteq x\}$ for any $s \in 2^{l \times k}$ with $l, k \in \omega$;

(iv) $x_n, y_n \in 2^{l_n \times \omega}$ with $p_n \subseteq x_n \subseteq x_{n+1}$, $q_n \subseteq y_n \subseteq y_{n+1}$; for $i \neq j < l_n$, $(x_n)_i \neq (x_n)_j$, $(y_n)_i \neq (y_n)_j$; for any $i, j < l_n$, $(x_n)_i \neq (x_n)_j$;

(v) $\forall^* z \in 2^{\omega \times \omega}$ $(x_n {}^\frown z \in C \wedge y_n {}^\frown z \in C)$;

(vi) $\exists^* z \in 2^{\omega \times \omega}$ $f(x_n {}^\frown z) = f(y_n {}^\frown z)$.

Granting this construction, we can let $x = \bigcup_n x_n$ and $y = \bigcup_n y_n$. Then x and y satisfy conditions (a) and (b) of the theorem, and by (iii) $x, y \in C$. Let $z_n \in 2^{\omega \times \omega}$ be such that $x_n {}^\frown z_n, y_n {}^\frown z_n \in C$ and $f(x_n {}^\frown z_n) = f(y_n {}^\frown z_n)$. Such z_n exists by (v) and (vi). Then $x_n {}^\frown z_n \to x$ and $y_n {}^\frown z_n \to y$ as $n \to \infty$, and by the continuity of f on C we get that $f(x) = f(y)$.

To show that the construction is possible, we need to use the following claim.

Claim. Let $p \in 2^{l \times k}$ and $\varphi : D \to X$, where $D \subseteq N_p$ is comeager in N_p, such that for $x, y \in D$, $x E_1 y \Rightarrow \varphi(x) F \varphi(y)$. Then there is $p' \geq p$ and r both in $2^{<\omega \times <\omega}$ such that

$$\forall^* u \in N_{p'} \ \forall^* v \in N_{p'} \ \forall^* z \in N_r \ \varphi(u {}^\frown z) = \varphi(v {}^\frown z).$$

To prove the claim, consider a family of functions defined as follows. For $x \in 2^{\omega \times \omega}$ define $\psi_x : 2^{l \times \omega} \to X$ by letting $\psi_x(u) = \varphi(u {}^\frown x)$ wherever φ is defined. For a comeager set of x in $2^{\omega \times \omega}$, ψ_x is well defined on a comeager set of u in N_p. Since for $u, v \in 2^{l \times \omega}$, $u^x E_1 v^x$ for any $x \in 2^{\omega \times \omega}$, we also get that

$$\psi_x(u) = \varphi(u {}^\frown x) F \varphi(v {}^\frown x) = \psi_x(v).$$

Therefore, whenever ψ_x is defined, its range is a countable set. Thus there are $p_x \supseteq p$ in $2^{l \times < \omega}$ such that ψ_x is constant on a comeager subset of N_{p_x} in $2^{l \times \omega}$. A similar argument shows that there is $r \in 2^{< \omega \times < \omega}$ such that for a comeager set of x in N_r the map $x \mapsto p_x$ is constant. Let p' be this constant value of p_x for comeager many $x \in N_r$. Thus

$$\forall^* x \in N_r \ \forall^* u \in N_{p'} \ \forall^* v \in N_{p'} \ \varphi(u^\frown x) = \varphi(v^\frown x).$$

The claim follows by the Kuratowski–Ulam theorem.

We now turn to the construction by induction on n. Assume that l_n, k_n, p_n, q_n, x_n, and y_n have been constructed to satisfy (i) through (vi). By (v) and (vi) there is $r \in 2^{< \omega \times < \omega}$ such that

$$\forall^* z \in N_r \ \left(x_n^\frown z, y_n^\frown z \in C \wedge f(x_n^\frown z) = f(y_n^\frown z) \right).$$

Without loss of generality assume $r \in 2^{l' \times k'}$ for some $l' > 0$ and $k' > k_n$. Also by extending we may assume that if we let $s = x_n \restriction (l_n \times k')^\frown r$ then $N_s \subseteq U_{n+1}$. Then it follows that $\forall^* u \in N_r \cap 2^{l' \times \omega} \ \forall^* z \in 2^{\omega \times \omega}$

$$f(x_n^\frown u^\frown z) = f(y_n^\frown u^\frown z).$$

Now define $\varphi(x) = f(x_n^\frown x) = f(y_n^\frown x)$ for $x = u^\frown z$ where $u \in N_r$ and $z \in 2^{\omega \times \omega}$. Then on a comeager set of N_r φ is well defined. It certainly satisfies that $x E_1 y \Rightarrow \varphi(x) F \varphi(y)$ since f satisfies it. Thus by the above claim we can find extensions $r' \supseteq r$ such that $\forall^* u \in N_{r'} \ \forall^* v \in N_{r'} \ \forall^* z \in 2^{\omega \times \omega}$:

$$f(x_n^\frown u^\frown z) = f(y_n^\frown v^\frown z).$$

Now it is easy to extend r' to p_{n+1} and q_{n+1} so that for any $u \supseteq p_{n+1}$ and $v \supseteq q_{n+1}$, condition (iv) holds for u, v. Let $x_{n+1} \supseteq p_{n+1}$ and $y_{n+1} \supseteq q_{n+1}$ be witnesses in the appropriate comeager set. Then (v) and (vi) are satisfied. This completes the construction as well as the proof of the theorem. □

Kechris and Louveau [102] have shown a dichotomy theorem that any hypersmooth equivalence relation is either essentially hyperfinite or else Borel bireducible with E_1. The proof uses the Gandy–Harrington topology as in the proof of the Glimm–Effros dichotomy theorem. This penetrating theorem completely determines the Borel reducibility structure of hypersmooth equivalence relations. A by-product of their proof is that E_1 is not Borel reducible to any idealistic Borel equivalence relation. A further result of theirs states that E_1 is not Borel reducible to any orbit equivalence relation E_G^X (even if it is non-Borel). These results suggest the possibility that idealistic equivalence relations coincide with the orbit equivalence relations.

We will give a proof of this last result, that E_1 is not Borel reducible to any orbit equivalence relation, later in Section 10.6 (Theorem 10.6.1) as an application of the techniques developed in Chapter 10.

Exercise 8.1.1 Prove Proposition 8.1.3.

Exercise 8.1.2 Recall that the tail equivalence relation E_t on 2^ω is defined by

$$x E_t y \iff \exists n \exists m \forall k \ (x(n + k) = y(m + k)).$$

Show that E_t is hyperfinite but not smooth. Thus $E \sim_B E_0$.

Exercise 8.1.3 Show that a hypersmooth equivalence relation is essentially countable iff it is idealistic. (*Hint*: Use Exercise 7.5.1.)

8.2   Borel orbit equivalence relations

In Section 3.4 we already analyzed some conditions guaranteeing Borelness of orbit equivalence relations. For instance, if a Borel action of a Polish group is free then the orbit equivalence relation is Borel. We now give some more characterizations of Borelness of orbit equivalence relations.

Recall the following notation about group actions. Let G be a Polish group and X a Borel G-space. For $x \in X$ the stabilizer of x is the closed subgroup $G_x = \{g \in G : g \cdot x = x\}$. For $x, y \in X$ we denote

$$G_{x,y} = \{g \in G : g \cdot x = y\}.$$

Then $G_{x,y}$ is a left coset of G_x since for any $g, h \in G_{x,y}$, $h^{-1}g \in G_x$. We view G_x and $G_{x,y}$ as elements of the Effros Borel space $F(G)$ of all closed subsets of G.

Theorem 8.2.1 (Becker–Kechris)
Let G be a Polish group and X a Borel G-space. Then the following are equivalent:

(i) *E_G^X is Borel.*

(ii) *The map $x \mapsto G_x$ from X into $F(G)$ is Borel.*

(iii) *The map $(x, y) \mapsto G_{x,y}$ from $X \times X$ into $F(G)$ is Borel.*

Proof. We show that (i)$\Rightarrow$(ii)$\Rightarrow$(iii)$\Rightarrow$(i). (iii)$\Rightarrow$(i) is obvious since $x E_G^X y$ iff $G_{x,y} \neq \emptyset$. We next show (ii)$\Rightarrow$(iii). For this let $U \subseteq G$ be nonempty open. Then

$$G_{x,y} \cap U \neq \emptyset \iff \exists g \in G \ (g \cdot x = y \wedge G_x \cap g^{-1}U \neq \emptyset)$$
$$\iff G_{x,y} \neq \emptyset \wedge \forall g \in G \ (g \cdot x = y \Rightarrow G_x \cap g^{-1}U \neq \emptyset)$$

The computation shows that the condition is both $\boldsymbol{\Sigma}_1^1$ and $\boldsymbol{\Pi}_1^1$, and is therefore Borel.

It remains to show (i)$\Rightarrow$(ii). For this we assume E_G^X is Borel. We use the uniform version of Theorem 4.4.6, Exercise 4.4.4, for $E_G^X \subseteq X \times X$, to obtain a Borel set $B \subseteq X \times X \times \omega$ with the following properties. For $y \in X$ and $n \in \omega$ let $B_{y,n} = \{x \in X : (x, y, n) \in B\}$ and let σ_y be the topology generated by the countable collection $\{B_{y,n} : n \in \omega\}$. Then for each $y \in X$, $B_{y,0} = [y]_G$, and (X, σ_y) is a Polish G-space. We may assume that $\{B_{y,n} : n \in \omega\}$ is a countable base for σ_y.

We are now ready to show that the map $x \mapsto G_x$ from X into $F(G)$ is Borel. For this we fix a countable base $\mathcal{U} = \{U_m\}$ of nonempty open sets for the topology of G. It is easy to see that the countable collection $\{U_m^{-1} U_k : m, k \in \omega\}$ is still a base for the topology of G. Moreover,

$$G_x \cap U_m^{-1} U_k \neq \emptyset \iff U_m G_x \cap U_k \neq \emptyset.$$

Thus, in order to finish the proof, it suffices to show that for any nonempty open $U, V \subseteq G$, the set $\{x \in X : UG_x \cap V \neq \emptyset\}$ is Borel. For this we claim that

$$UG_x \cap V \neq \emptyset \iff \exists m \, \forall n \, (U_m \subseteq U \wedge x \in B_{x,n}^{\triangle U_m} \rightarrow x \in B_{x,n}^{\triangle V}).$$

Suppose $UG_x \cap V \neq \emptyset$ and let $g_0 \in G_x$ and $h_0 \in U$ with $h_0 g_0 \in V$. Let $U_m \subseteq U$ with $h_0 \in U_m$ and $U_m g_0 \subseteq V$. Now for any $A \subseteq X$,

$$\begin{aligned} x \in A^{\triangle U_m} &\iff \exists^* h \in U_m \ (h \cdot x \in A) \\ &\iff \exists^* h' \in U_m g_0 \subseteq V \ (h' \cdot x \in A) \\ &\implies x \in A^{\triangle V}. \end{aligned}$$

Thus the direction ($\Rightarrow$) follows with $A = B_{n,x}$ for any $n \in \omega$. For the direction ($\Leftarrow$) assume $UG_x \cap V = \emptyset$. Then $U \cdot x \cap V \cdot x = \emptyset$. Consider the topology σ_x on X. (X, σ_x) is a Polish G-space and $[x]_G$ is σ_x-open. By Effros' Theorem 3.2.4 the map $gG_x \mapsto g \cdot x$ is a homeomorphism from G/G_x onto $[x]_G$. It follows that $U \cdot x = UG_x \cdot x$ is σ_x-open. Since $\{B_{x,n} : n \in \omega\}$ is a countable base for σ_x, there exists $n \in \omega$ such that $x \in B_{x,n} \subseteq U \cdot x$. It follows that $x \notin B_{n,x}^{\triangle V}$. This completes the proof of the claim. □

The following is another useful characterization of Borelness of orbit equivalence relations.

Theorem 8.2.2 (Sami)
Let G be a Polish group and X a Polish G-space. Then E_G^X is Borel iff there is $\alpha < \omega_1$ such that every G-orbit is $\boldsymbol{\Pi}_\alpha^0$.

Proof. If E_G^X is Borel then there is $\alpha < \omega_1$ such that E_G^X is a $\boldsymbol{\Pi}_\alpha^0$ subset of $X \times X$. Then every G-orbit is a section of E_G^X, and hence is $\boldsymbol{\Pi}_\alpha^0$. Conversely, consider for any $\alpha < \omega_1$ the equivalence relation E_α defined by

$$x E_\alpha y \iff \text{ for all } \mathbf{\Pi}_\alpha^0 \text{ invariant subsets } A \subseteq X,\ x \in A \text{ iff } y \in A.$$

If every G-orbit is $\mathbf{\Pi}_\alpha^0$ then $E_\alpha = E_G^X$. Hence it suffices to show that E_α is $\mathbf{\Pi}_1^1$ for all $\alpha < \omega_1$. For this we use the fact that there is a $\mathbf{\Pi}_\alpha^0$ subset U of $\omega^\omega \times X$ which is universal for $\mathbf{\Pi}_\alpha^0$ subsets of X (see Exercise 1.5.6). Consider the action of G on $\omega^\omega \times X$ defined by

$$g \cdot (z, x) = (z, g \cdot x).$$

Then the action is still continuous. Now let $T = U^{*G} \subseteq \omega^\omega \times X$. Then for any $z \in \omega^\omega$ and $x \in X$,

$$x \in T_z \iff \forall^* g \in G\ g \cdot (z, x) \in U \iff \forall^* g \in G\ g \cdot x \in U_z \iff x \in (U_z)^{*G}.$$

We claim that $\{T_z : z \in \omega^\omega\}$ consists of all $\mathbf{\Pi}_\alpha^0$ invariant subsets of X. Since $T_z = (U_z)^{*G}$ and $U_z \in \mathbf{\Pi}_\alpha^0$, T_z is $\mathbf{\Pi}_\alpha^0$ invariant. Conversely, if A is $\mathbf{\Pi}_\alpha^0$ invariant, then there is $z \in \omega^\omega$ with $U_z = A$, and furthermore, $T_z = (U_z)^{*G} = A^{*G} = A$. This finishes the proof of the claim. We thus have that

$$x E_\alpha y \iff \forall z \in \omega^\omega\ (x \in T_z \leftrightarrow y \in T_z),$$

and hence E_α is $\mathbf{\Pi}_1^1$. □

The assumption of a Polish group action is crucial in the theorem. It is easy to construct non-Borel equivalence relations with all equivalence classes closed (see Exercise 7.1.2).

We give some examples of Borel orbit equivalence relations. By Corollary 7.5.3 any locally compact Polish group action generates an essentially countable equivalence relation, and thus a Borel one. In particular, the following scheme will yield many examples of Borel orbit equivalence relations with potentially distinct complexity in the Borel reducibility hierarchy. Consider an arbitrary uncountable Polish group H. Let G be an arbitrary countable subgroup of H. Then the coset equivalence relation H/G is a countable equivalence relation. Note that G is regarded to have the discrete topology, rather than the subspace topology inherited from H. However, the Borel structures on G generated by these topologies are the same. The following concept is an abstraction of this circumstance.

Definition 8.2.3

Let H be a Polish group. A **Polishable subgroup** G of H is a Borel subgroup of H which admits a Polish topology τ generating the Borel structure on G inherited from that of H, and such that (G, τ) is a topological group.

If H is a Polish group and G is a Polishable subgroup of H, then the translation action of G on H is continuous and the coset equivalence relation

H/G is obviously a Borel orbit equivalence relation. The seemingly technical requirement of Polishability is actually fulfilled by many examples other than countable subgroups. For example, the classical Banach spaces ℓ_p ($1 \leq p < \infty$) and c_0 are Polishable subgroups of the additive group $\mathbb{R}^\omega$.

The Borel reducibility relation among these examples of Borel orbit equivalence relations are extremely complicated. In fact many questions about them remain open. We will discuss them again in the next few sections and the rest of the book.

Exercise 8.2.1 Let G be a Polish group. Let $S(G)$ be the set of all closed subgroups of G and $C(G)$ be the set of all (left) cosets of closed subgroups of G. Show that both $S(G)$ and $C(G)$ are Borel subsets of $F(G)$, and hence are standard Borel spaces.

Exercise 8.2.2 (Dixmier) Show that for any Polish group G there is a Borel set $T \subseteq S(G) \times G$ such that for any $H \in S(G)$, T_H is a transversal for the left cosets of H.

Exercise 8.2.3 Let G be a Polish group and X a Borel G-space. Show that E_G^X is Borel iff there is a Borel function $g : X \times X \to G$ such that for all $x E_G^X y$, $g(x,y) \cdot x = y$.

Exercise 8.2.4 Show that ℓ_p ($1 \leq p < \infty$) and c_0 are Polishable subgroups of $\mathbb{R}^\omega$.

8.3 A jump operator for Borel equivalence relations

In this section we discuss an operation that will increase the Borel complexity of any nontrivial Borel equivalence relation.

Definition 8.3.1
Let E be a Borel equivalence relation on a standard Borel space X. The **Friedman–Stanley jump** (or simply **jump**) of E, denoted by E^+, is the equivalence relation on X^ω defined by

$$(x_n) E^+ (y_n) \iff \{[x_n]_E : n \in \omega\} = \{[y_n]_E : n \in \omega\}.$$

E^+ is obviously an equivalence relation. To see that it is Borel, just note that

$$(x_n) E^+ (y_n) \iff \forall n \exists m (x_n E y_m) \wedge \forall m \exists n (x_n E y_m).$$

Thus in general if E is $\mathbf{\Sigma}_\alpha^0$ for $\alpha < \omega_1$, then E^+ is $\mathbf{\Pi}_{\alpha+1}^0$. It is easy to see that $E \leq_B E^+$ for any equivalence relation E. We will show later in this section that $E <_B E^+$ for any Borel equivalence relation E with more than one equivalence class. To do this we need first to prove a theorem of Friedman on Borel diagonalizations. Friedman's original proof uses forcing and Borel determinacy; here we give a proof by Harrington using Borel determinacy and some facts of effective descriptive set theory.

We first give a brief review of Borel determinacy. For any subset A of ω^ω we consider a two-person game G_A played as follows:

$$
\begin{array}{c c c c c}
\text{I} & n_0 & & n_2 & \cdots \\
\text{II} & & n_1 & & n_3 & \cdots
\end{array}
$$

where $n_i \in \omega$ for all $i \in \omega$. Let $x \in \omega^\omega$ be such that $x(i) = n_i$ for all $i \in \omega$. Then Player I wins iff $x \in A$. The set A is called the **payoff set** for the game G_A. The game G_A is **determined** if either Player I or Player II has a winning strategy in the game. Borel determinacy is the statement that G_A is determined for all Borel subsets A of ω^ω. This is a well-known theorem of Martin.

Theorem 8.3.2 (Martin)
If $A \subseteq \omega^\omega$ is Borel then G_A is determined.

We will also use the facts about Π_1^1 singletons reviewed in Section 1.7, especially Corollary 1.7.8.

We are now ready to prove a nonseparation theorem. Recall that LO $\subseteq$ $2^{\omega \times \omega}$ denotes the set of linear orderings of ω and WO $\subseteq$ LO denotes the set of well-orderings of ω. Here we fix a computable bijection $\langle \cdot, \cdot \rangle : \omega \times \omega \to \omega$, and via coding regard LO and WO as subsets of 2^ω. For $x \in$ LO let $<_x$ denote the linear ordering of ω coded by x, that is,

$$
n <_x m \iff \langle n, m \rangle \in x.
$$

Theorem 8.3.3
Let A be a Σ_1^1 set such that WO $\subseteq A \subseteq$ LO. Then there is no Δ_1^1 set $B \subseteq 2^\omega \times 2^\omega$ such that WO $\times (A - \text{WO}) \subseteq B$ and $(A - \text{WO}) \times \text{WO} \cap B = \emptyset$.

Proof. Let $C \subseteq 2^\omega \times \omega^\omega$ be Π_1^0 such that

$$
x \in A \iff \exists y \in \omega^\omega (x, y) \in C.
$$

Let $\pi : 2^\omega \times \omega^\omega \to 2^\omega$ be the projection to the first coordinate. Then for any $z \in C$, $\pi(z) \in A$, and in particular, $\pi(z) \in$ LO. Assume toward a contradiction that the described Δ_1^1 set B exists. Then for any $u, v \in C$, if $\pi(u) \in$ WO and $\pi(v) \notin$ WO then $(\pi(u), \pi(v)) \in B$, and if $\pi(u) \notin$ WO and $\pi(v) \in$ WO then $(\pi(u), \pi(v)) \notin B$.

For each Π_1^1 set P we consider a game G^P as follows. First let $C^P \subseteq \omega^\omega \times \omega^\omega$ be Π_1^0 such that

$$
x \notin P \iff \exists y \in \omega^\omega (x, y) \in C^P.
$$

Then let T^P be a tree on $\omega \times \omega$ such that $[T^P] = C^P$. For each $x \in \omega^\omega$, let T_x^P be the tree on ω defined by

$$s \in T_x^P \iff (x \restriction \mathrm{lh}(s), s) \in T^P.$$

Then we have that

$$x \in P \iff \forall y \in \omega^\omega (x, y) \notin [T^P] \iff T_x^P \text{ is well-founded.}$$

The game G^P is now played as follows:

$$
\begin{array}{llll}
\text{I} & n & & a_1, u_1, \varphi_1 \\
\text{II} & & \epsilon & a_2, u_2, \varphi_2
\end{array}
$$

where

(i) $n \in \omega$, $\epsilon \in \{0, 1\}$, $a_1, a_2 \in 2^\omega$;

(ii) if $\epsilon = 0$ then $n \in a_1$ but $n \notin a_2$; if $\epsilon = 1$ then $n \in a_2$ but $n \notin a_1$;

(iii) $u_1, u_2 \in C$ (thus $\pi(u_1), \pi(u_2) \in A \subseteq \mathrm{LO}$);

(iv) for $i = 1, 2$, $\varphi_i : T_{a_i}^P \to \omega$ is strictly (reverse) order-preserving: if $s, t \in T_{a_i}^P$ and $s \subsetneq t$ then $\varphi_i(t) <_{\pi(u_i)} \varphi_i(s)$.

Then elements $a_1, u_1, \varphi_1, a_2, u_2, \varphi_2$ are regarded as coded by subsets of ω and are understood to be played alternatively by the two players just as in the standard game. Then Player I wins the play iff $(\pi(u_1), \pi(u_2)) \in B$. For each fixed Π_1^1 set P the above game can be equivalently played on ω as a standard game where the payoff set is Borel. Thus by Borel determinacy the game G^P is determined.

We make some observations about the game G^P where P is a singleton. For this we fix an arbitrary Π_1^1 singleton a and replace all notation with superscript P by superscript a. We claim that Player II has a winning strategy in the game G^a. Since G^a is determined, it is enough to show that Player I does not have a winning strategy in G^a. Toward a contradiction assume he does. Consider the following play of the game. At the beginning Player I chooses $n \in \omega$. Let Player II play ϵ according to whether $n \in a$. If $n \in a$ let $\epsilon = 1$; otherwise let $\epsilon = 0$. Then let Player II play $a_2 = a$. Now T_a^a is well-founded, hence there exist well-ordering x of ω and strictly order-preserving map $\varphi : T_a^a \to \omega$ with $\varphi(t) <_x \varphi(s)$ for all $s \subsetneq t$ in T_a^a. Let $u_2 \in C$ be such that $\pi(u_2) = x$ and $\varphi_2 = \varphi$. Then the rules (i) through (iv) of the game are obeyed. Let Player I play according to his winning strategy. Then (i) through (iv) hold and $(\pi(u_1), \pi(u_2)) \in B$. Since $\pi(u_2) \in \mathrm{WO}$ the separation property of B requires that $\pi(u_1) \in \mathrm{WO}$. Thus we have that $\varphi_1 : T_{a_1}^a \to \omega$ is a strictly order-preserving map from $T_{a_1}^a$ into a well-order coded by $\pi(u_1)$. This implies that $T_{a_1}^a$ is well-founded, and thus $a_1 \in \{a\}$ and in fact $a_1 = a = a_2$. However, by (ii) $n \in a_2$ iff $n \notin a_1$, and thus $a_1 \neq a_2$, a contradiction.

Let σ be a winning strategy of Player II in G^a. σ is coded by a real in ω^ω. Let $\sigma^* : \omega \to \{0,1\}$ be such that $\sigma^*(n) = \epsilon$ iff Player II responses with ϵ according to σ when Player I's first move is n. Apparently σ^* is recursive in σ. We note that $\sigma^*(n) = 1$ iff $n \in a$. To see this suppose $n \in a$ but $\sigma^*(n) = 0$. Then let Player I plays $a_1 = a$ (note that (ii) is satisfied), $u_1 \in C$ with $\pi(u_1) \in \mathrm{WO}$, $\varphi_1 : T_a^a \to \omega$ strictly order-preserving into $<_{\pi(u_1)}$, where the choice of u_1 and φ_1 are similar to the argument in the preceding paragraph. Then since Player II wins we have $(\pi(u_1), \pi(u_2)) \notin B$, and by the separation property of B this requires that $\pi(u_2) \in \mathrm{WO}$. By the above argument again we have that $a_2 = a$, a contradiction to (ii). The argument for the situation $n \notin a$ but $\sigma^*(n) = 1$ is similar.

Finally note that for all Π_1^1 sets P the payoff set for the game G^P is Δ_1^1, hence Σ_α^0 for some $\alpha < \omega_1$. Since there are only countably many Π_1^1 sets there exists $\beta < \omega_1$ such that all payoff sets are Σ_β^0. Let $U \subseteq \omega \times \omega^\omega$ be universal for all Σ_β^0 subsets of ω^ω. Consider the set Q of all $\tau \in (\omega^\omega)^\omega$ such that for any $k \in \omega$, $\tau(k) \in \omega^\omega$ is a winning strategy for either Player I or Player II in the game G_{U_k}. Then Q is nonempty Π_1^1. By Theorem 1.7.7 Q contains a Π_1^1 singleton τ_0. We arrive at a final contradiction to Corollary 1.7.8 by showing that every Π_1^1 singleton is recursive in τ_0.

Let a be an arbitrary Π_1^1 singleton. Let $k \in \omega$ be such that U_k is the payoff set of the game G^a. Then $\tau_0(k)$ is a winning strategy for Player II in G^a. Therefore $n \in a$ iff $\tau_0(k)^*(n) = 1$. This shows that a is recursive in $\tau_0(k)^*$, which is in turn recursive in τ_0, and hence a is recursive in τ_0. □

Recall that $D \subseteq \omega^\omega$ is $\mathbf{\Pi_1^1}$-complete if D is $\mathbf{\Pi_1^1}$ and for all $\mathbf{\Pi_1^1}$ sets $P \subseteq \omega^\omega$ there is a continuous function $f : \omega^\omega \to \omega^\omega$ such that $f^{-1}(D) = P$. The set WO is $\mathbf{\Pi_1^1}$-complete, and it is easy to see that a set D is $\mathbf{\Pi_1^1}$-complete iff there is a continuous function $f : \mathrm{LO} \to \omega^\omega$ such that $f^{-1}(D) = \mathrm{WO}$. With these concepts the above theorem can be generalized and relativized.

Theorem 8.3.4
Let $D \subseteq \omega^\omega$ be $\mathbf{\Pi_1^1}$-complete and $A \subseteq \omega^\omega$ be $\mathbf{\Sigma_1^1}$ with $D \subseteq A$. Then there is no Borel set $B \subseteq \omega^\omega \times \omega^\omega$ such that $D \times (A - D) \subseteq B$ and $(A - D) \times D \cap B = \emptyset$.

Proof. First note that the proof of the previous theorem can be relativized. Let $f : \mathrm{LO} \to \omega^\omega$ be continuous such that $f^{-1}(D) = \mathrm{WO}$. Let $A' = f^{-1}(A)$. Then $A' \subseteq \mathrm{LO}$ is $\mathbf{\Sigma_1^1}$ and $\mathrm{WO} \subseteq A'$ since $D \subseteq A$. Assume that there is a Borel set B with the separation property. Let $B' = \{(u, v) \in \mathrm{LO} \times \mathrm{LO} : (f(u), f(v)) \in B\}$. Then B' is Borel in $\mathrm{LO} \times \mathrm{LO}$, and $\mathrm{WO} \times (A' - \mathrm{WO}) \subseteq B'$ and $(A' - \mathrm{WO}) \times \mathrm{WO} \cap B' = \emptyset$. Let $x \in \omega^\omega$ be such that $A' \in \Sigma_1^1(x)$ and $B' \in \Delta_1^1(x)$. This contradicts the previous theorem relativized to x. □

We are now ready to give Harrington's proof of Friedman's **Borel diagonalization theorem** for Borel equivalence relations.

Theorem 8.3.5 (Friedman)

Let E be a Borel equivalence relation on a standard Borel space X. Then there is no Borel function $F : X^\omega \to X$ such that, for all $x = (x_n)$ and $y = (y_n)$ in X^ω,

(a) if xE^+y then $F(x)EF(y)$, and

(b) $(F(x), x_n) \notin E$ for all $n \in \omega$.

Proof. Assume that the Borel function F exists with properties (a) and (b). Without loss of generality we may assume that $X = \omega^\omega$. Let $z_0 \in \omega^\omega$ be an arbitrarily fixed element. Let $A \subseteq 2^\omega \times (\omega^\omega)^\omega$ be the set of all elements $u = (x, f)$ where $x \in \mathrm{LO}$ and $f : \omega \to \omega^\omega$ such that, for all $n \in \omega$,

(i) if n is $<_x$-least then $f(n) = z_0$; and

(ii) if n is not $<_x$-least, then letting $m_0, m_1, \ldots$ enumerate the set $\{m \in \omega : m <_x n\}$ in the increasing order of natural numbers (possibly with repetitions), we have that $f(n) = F(f(m_0), f(m_1), \ldots)$.

Because of property (a) we can write $f(n) = F(f \restriction \{m : m <_x n\})$ without specifying the order in which the elements are enumerated. Note that A is Borel. Let $\pi : 2^\omega \times (\omega^\omega)^\omega \to 2^\omega$ be the projection to the first coordinate. Then $\mathrm{WO} \subseteq \pi(A)$ since for any $x \in \mathrm{WO}$ a function f satisfying (i) and (ii) can be defined by a transfinite induction on the ordinal length of $<_x$. Define $D \subseteq A$ by

$$(x, f) \in D \iff (x, f) \in A \wedge x \in \mathrm{WO}.$$

Then it is clear that D is a $\mathbf{\Pi}_1^1$-complete subset of $2^\omega \times (\omega^\omega)^\omega$. We are going to apply Theorem 8.3.4 to the sets D and A, regarding the underlying space as a homeomorphic copy of ω^ω.

For any $u = (x, f)$ and $v = (y, g)$ in A, we define a binary relation $R_{u,v} \subseteq \omega \times \omega$ by

$$n R_{u,v} m \iff f(n) E g(m).$$

Because of property (b) we have that if $n R_{u,v} m_1$ and $n R_{u,v} m_2$ then $m_1 = m_2$. Similarly, if $n_1 R_{u,v} m$ and $n_2 R_{u,v} m$ then $n_1 = n_2$. Thus $R_{u,v}$ is in fact a partial bijection between two subsets of ω. Let $\varphi_{u,v}$ denote the partial bijection given by $R_{u,v}$. Define $I_{u,v} \subseteq \mathrm{dom}(\varphi_{u,v})$ and $J_{u,v} \subseteq \mathrm{range}(\varphi_{u,v})$ by letting

$$\begin{aligned} n \in I_{u,v} \iff & \{k : k \leq_x n\} \subseteq \mathrm{dom}(\varphi_{u,v}) \\ & \wedge \varphi_{u,v}(\{k : k \leq_x n\}) = \{l : l \leq_y \varphi_{u,v}(n)\} \\ & \wedge \forall k, k' \leq_x n \, (k <_x k' \to \varphi_{u,v}(k) <_y \varphi_{u,v}(k')) \end{aligned}$$

and $J_{u,v} = \varphi_{u,v}(I_{u,v})$.

Now define $B \subseteq A \times A$ as follows. For $u = (x, f)$ and $v = (y, g)$, let $(u, v) \in B$ iff either $I_{u,v} = \omega$ or (there is n with $I_{u,v} = \{k : k <_x n\}$ and

$J_{u,v} \neq \omega$). Note that B is Borel. We will arrive at a contradiction by checking that $D \times (A - D) \subseteq B$ and $(A - D) \times D \cap B = \emptyset$. For this let $u = (x, f)$ and $v = (y, g)$ be in A. First suppose $x \in \mathrm{WO}$ and $y \notin \mathrm{WO}$. If $I_{u,v} \neq \omega$, then since $(I_{u,v}, <_x)$ is a well-order and $\varphi_{u,v}$ is order-preserving, we get that $(J_{u,v}, <_y)$ is a well-order and hence $J_{u,v} \neq \omega$ since $y \notin \mathrm{WO}$. This shows that $(u, v) \in B$. Next suppose $x \notin \mathrm{WO}$ and $y \in \mathrm{WO}$. Since $(I_{u,v}, <_x)$ is a well-order but $x \notin \mathrm{WO}$, we have $I_{u,v} \neq \omega$. We claim that $(u, v) \notin B$. Otherwise, there is n with $I_{u,v} = \{k : k <_x n\}$ and $J_{u,v} \neq \emptyset$. Since $(J_{u,v}, <_y)$ is an initial segment of the well-order $<_y$ there is m such that $J_{u,v} = \{l : l <_y m\}$. By our definition, for $k \in I_{u,v}$, $f(k) E g(\varphi_{u,v}(k))$. Thus $f \upharpoonright \{k : k <_x n\} E^+ g \upharpoonright \{l : l <_y m\}$, and therefore $f(n) = F(f \upharpoonright \{k : k <_x n\}) E g(m) = F(g \upharpoonright \{l : l <_y m\})$. This means that $m R_{u,v} n$ or $\varphi_{u,v}(n) = m$, and further $n \in I_{u,v}$, contradicting our assumption that $n \notin I_{u,v}$. $\qquad\square$

Theorem 8.3.6 (Friedman–Stanley)

Let E be a Borel equivalence relation on a standard Borel space X. Suppose that E has more than one equivalence class. Then $E <_B E^+$.

Proof. It suffices to show that $E^+ \not\leq_B E$. Assume toward a contradiction that $f : X^\omega \to X$ is a Borel reduction from E^+ to E. We will consider the space $(X^\omega)^\omega$ and the equivalence relation $(E^+)^+$ on it. For notational clarity we denote a typical element of X^ω by $x = (x_n)$, where each $x_n \in X$, and a typical element of $(X^\omega)^\omega$ by $\overline{x} = (x^k)$, where $x^k \in X^\omega$. We will also omit the subscripts when we denote the equivalence class of an element. This will not cause any confusion.

Let $C = \{x = (x_n) \in X^\omega : \forall n, m \ x_n = x_m\}$. Thus elements of C are constant sequences in X^ω. We note that there is $x \in C$ such that $f(x) \neq x_0$. Otherwise we would have that $f(C) = X$. This implies that, for every $y \in X^\omega$, there is some $x \in C$ with $f(x) = f(y)$, and thus $x E^+ y$ since f is a reduction, and in fact $y_n E x_0$ for all n. It follows that there is only one E-equivalence class on X, contradicting our assumption. We fix a $z \in C$ with $f(z) \neq z_0$.

For each $\overline{x} = (x^k) \in (X^\omega)^\omega$ let $F(\overline{x}) \in X^\omega$ be a canonical enumeration (possibly with repetitions) of the set

$$\{f(z), f(x^k) : \forall n \ [f(x^k)] \neq [x^k_n]\}.$$

Note that $F : (X^\omega)^\omega \to X^\omega$ is Borel. We claim that F is invariant, that is, if $\overline{x}(E^+)^+\overline{y}$ then $F(\overline{x}) E^+ F(\overline{y})$. For this suppose $\overline{x}(E^+)^+\overline{y}$. Let k be such that $[f(x^k)] \neq [x^k_n]$ for all n. Let l be such that $x^k E^+ y^l$. Since f is a reduction $[f(y^l)] = [f(x^k)]$ whereas $\{[x^k_n] : n \in \omega\} = \{[y^l_n] : n \in \omega\}$. Hence $[f(y^l)] \neq [y^l_n]$ for all n. By symmetry we have that $F(\overline{x}) E^+ F(\overline{y})$.

Next we claim that for any $\overline{x}$, $(F(\overline{x}), x^k) \notin E^+$ for all k. Assume that for some $\overline{x}$ and k, $F(\overline{x}) E^+ x^k$. Let $A = \{z, x^p : \forall n \ [f(x^p)] \neq [x^p_n]\}$. We consider two cases. Case 1: there is $y \in A$ with $f(x^k) E f(y)$. Then in fact $x^k E^+ y$,

$[f(x^k)] = [f(y)]$, and $\{[y_n] : n \in \omega\} = \{[x_n^k] : n \in \omega\}$. This implies that $[f(x^k)] \neq [x_n^k]$ for all n, and therefore $x^k \in A$. However since $F(\overline{x})E^+x^k$, $[f(x^k)] \in \{[f(y)] : y \in A\} = \{[x_n^k] : n \in \omega\}$, a contradiction. Case 2: there is no $y \in A$ with $f(x^k)Ef(y)$. In particular $x^k \notin A$. Thus for some n, $[f(x^k)] = [x_n^k]$. Again by $F(\overline{x})E^+x^k$ it follows that $[f(x^k)] \in \{[f(y)] : y \in A\}$, contradicting our case assumption.

We have thus constructed a Borel diagonalizer of E^+, contradicting Theorem 8.3.5. $\Box$

Exercise 8.3.1 Show that $E \leq_B E^+$ for any equivalence relation E on a standard Borel space X.

Exercise 8.3.2 Show that if $E \leq_B F$ then $E^+ \leq_B F^+$.

Exercise 8.3.3 Let $=^+$ denote $\mathrm{id}(\omega^\omega)^+$. Show that $E_\infty \leq_B =^+$ and in fact $E_\infty^+ \sim_B =^+$.

Let E be an equivalence relation on a standard Borel space X. Define an equivalence relation E^* on X^ω by

$$xE^*y \iff xE^+y \wedge \forall z \in X \; |\{n \in \omega : x_n = z\}| = |\{n \in \omega : y_n = z\}|.$$

Exercise 8.3.4 Show that $E^* \sim_B E^+$.

Exercise 8.3.5 Let G be a Polish group and X a Polish G-space. Show that $(E_G^X)^*$ is the orbit equivalence relation induced by an action of $G \times S_\infty$ on X^ω, where S_∞ is the infinite permutation group.

8.4 Examples of F_σ equivalence relations

Recall from Theorem 6.4.4 that every G_δ equivalence relation is smooth. Thus from the point of view of descriptive complexity the simplest kind of Borel equivalence relations are F_σ ones. However, this is not to say that the Borel complexity of these equivalence relations in the Borel reducibility hierarchy is easy to describe. In this section we give an overview of some canonical examples of F_σ equivalence relations.

We have introduced at least three concrete nonsmooth F_σ equivalence relations: E_0, E_∞, and E_1. Of course, any countable Borel equivalence relation is F_σ. Consider the following additional examples arising from classical Banach spaces.

Recall that for a real number $1 \leq p < \infty$,

$$\ell_p = \left\{ x = (x_n) \in \mathbb{R}^\omega : \sum_{n=0}^\infty |x_n|^p < \infty \right\},$$

and

$$\ell_\infty = \left\{ x = (x_n) \in \mathbb{R}^\omega : \sup_{n \in \omega} |x_n| < \infty \right\}.$$

All of them are F_σ subgroups of the additive group $\mathbb{R}^\omega$. Thus for $1 \leq p \leq \infty$ the coset equivalence relation $\mathbb{R}^\omega / \ell_p$ defined by

$$x \; \mathbb{R}^\omega / \ell_p \; y \iff x - y \in \ell_p$$

is an F_σ equivalence relation on the Polish space $\mathbb{R}^\omega$. For notational simplicity we also denote the equivalence relation by ℓ_p.

We first note that the equivalence relations ℓ_p all have nontrivial Borel complexity.

Lemma 8.4.1
For any $1 \leq p \leq \infty$, $E_0 \leq_B \ell_p$.

Proof. We use Theorem 6.2.1. For this note that each ℓ_p is a group of homeomorphisms on $\mathbb{R}^\omega$. It is also easy to see that ℓ_p is dense, and thus the equivalence relation has a dense orbit. It remains to show that the equivalence relation is meager. By Kuratowski–Ulam it suffices to show that every orbit is meager. Since each orbit is a homeomorphic copy of ℓ_p, it suffices to show that ℓ_p is meager in $\mathbb{R}^\omega$. We do this for $p = \infty$ first. For each $N \in \omega$ let $A_N = \{x \in \mathbb{R}^\omega : \sup_n |x_n| \leq N\}$. Then A_N is closed in $\mathbb{R}^\omega$ and $\ell_\infty = \bigcup_{N \in \omega} A_N$. We claim that A_N is nowhere dense. Otherwise there is a basic open $U \subseteq \mathbb{R}^\omega$ with $U \subseteq A_N$. Without loss of generality assume $U = (a_0, b_0) \times (a_1, b_1) \times \dots (a_n, b_n) \times \mathbb{R}^\omega$. Consider $V \subseteq U$ defined by

$$V = (a_0, b_0) \times (a_1, b_1) \times \dots (a_n, b_n) \times (N, N+1) \times \mathbb{R}^\omega.$$

Then $V \subseteq A_N$. However, for any $x \in V$ we have that $|x_{n+1}| > N$ and therefore $\sup_n |x_n| > N$, contradicting $x \in A_N$. A similar argument shows that ℓ_p is also meager for $1 \leq p < \infty$. ◻

Our next goal is to show that ℓ_∞ is the most complex among these equivalence relations. In fact, we show that ℓ_∞ is universal among all K_σ equivalence relations on Polish spaces. Recall that a subset of a topological space is K_σ if it is the union of countably many compact sets. If X is a Polish space, E a K_σ equivalence relation on X, and $\Delta = \{(x, x) : x \in X\}$, then Δ is a closed subset of E which is homeomorphic to X, and it follows that X is K_σ.

Theorem 8.4.2 (Rosendal)
Let E be a K_σ equivalence relation on a Polish space X. Then $E \leq_B \ell_\infty$.

Proof. Let (X_n) be an increasing sequence of compact subsets of X with $X = \bigcup_n X_n$. Let $\Delta = \{(x,x) : x \in X\}$. Let (F_n) be an increasing sequence of compact subsets of $X \times X$ with $E = \bigcup_n F_n$. Without loss of generality we may assume $\Delta \cap X_n^2 \subseteq F_n$. Define an increasing sequence (R_n) of compact relations by induction as follows. Let $R_0 = F_0 \cap X_0^2$. Assuming R_n has been defined, let

$$R_{n+1} = R_n \cup (R_n \circ R_n) \cup (F_n \cap X_n^2).$$

It is easy to see that each R_n is compact and that $E = \bigcup_n R_n$.

Let $\{U_m\}_{m \in \omega}$ be a countable base for the topology of X. Fix a bijection $\langle \cdot, \cdot \rangle : \omega \times \omega \to \omega$ such that for all $n, m \in \omega$, $\langle n, m \rangle \geq n$. We define a function $\varphi : X \to \mathbb{R}^\omega$ by

$$\varphi(x)(\langle n, m \rangle) = \begin{cases} k, & \text{if } k \leq n \text{ is the least such that } \exists y \in U_m \ (x,y) \in R_k, \\ n, & \text{if } \forall y \in U_m \ (x,y) \notin R_n. \end{cases}$$

Note that φ is a Borel map. To see this it suffices to check that for any $m, n \in \omega$, the set $\{x \in X : \exists y \in U_m \ (x,y) \in R_n\}$ is Borel. We have that

$$\exists y \in U_m \ (x,y) \in R_n \iff \exists l (\overline{U_l} \subseteq U_m \wedge \exists y \in \overline{U_l} \cap X_n \ (x,y) \in R_n).$$

The set $\{x : \exists y \in \overline{U_l} \cap X_n \ (x,y) \in R_n\}$ is compact, and thus the desired set is in fact F_σ.

We claim that φ is Borel reduction from E to ℓ_∞. Suppose $x, y \in X$ and xEy. Fix the least $n_0 \in \omega$ with $(x,y) \in R_{n_0}$ and $(y,x) \in R_{n_0}$. Then for any $z \in X$ and $k \in \omega$, if $(x,z) \in R_k$ then $(y,z) \in R_{n_0} \circ R_k \subseteq R_{\max\{n_0, k\}+1}$, and by symmetry, if $(y,z) \in R_k$ then $(y,z) \in R_{\max\{n_0,k\}+1}$. It follows that if $\varphi(x)(\langle n, m \rangle) = k < n$, then $\varphi(y)(\langle n, m \rangle) \leq \max\{n_0, k\} + 1$, and therefore $|\varphi(x) - \varphi(y)| \leq n_0 + 1$. By symmetry if $\varphi(x)(\langle n, m \rangle) < n$ then $|\varphi(x)(\langle n, m \rangle) - \varphi(y)(\langle n, m \rangle)| \leq n_0 + 1$. If both $\varphi(x)(\langle n, m \rangle) = n = \varphi(y)(\langle n, m \rangle)$, then trivially $|\varphi(x)(\langle n, m \rangle) - \varphi(y)(\langle n, m \rangle)| \leq n_0 + 1$. We have shown that $\varphi(x) - \varphi(y) \in \ell_\infty$.

Conversely, suppose $\varphi(x) - \varphi(y) \in \ell_\infty$. Let k be large enough such that $y \in X_k$ and $|\varphi(x)(N) - \varphi(y)(N)| \leq k$ for all N. Fix an arbitrary m with $y \in U_m$. Since $\Delta \cap X_k^2 \subseteq F_k$, $(y,y) \in R_{k+1}$ and then $\varphi(y)(\langle 2k + 2, m \rangle) \leq k + 1$. It follows that $\varphi(x)(\langle 2k + 2, m \rangle) \leq 2k + 1$, and thus there is $z \in U_m$ with $(x,z) \in R_{k+1}$. We have shown that, for any m with $y \in U_m$ there is $z \in U_m$ with $(x,z) \in R_{k+1}$. Thus we have a sequence (z_m) with $z_m \to y$ and $(x, z_m) \in R_{k+1}$. Since R_{k+1} is compact, we have that $(x,y) \in R_{k+1}$, and therefore $(x,y) \in E$. ∎

It is easy to see that all of the equivalence relations E_0, E_∞, E_1, and ℓ_p ($1 \leq p \leq \infty$) are K_σ (see Exercise 8.4.3). From this we immediately get that all of them are Borel reducible to ℓ_∞. It is not known if ℓ_∞ is universal among all F_σ equivalence relations.

Before closing this section we mention some results without proofs. Dougherty and Hjorth [29] have shown that the Borel reducibility order $<_B$ for the equivalence relations ℓ_p $(1 \leq p < \infty)$ respects the order of real numbers.

Theorem 8.4.3 (Dougherty–Hjorth)
For $1 \leq p < q < \infty$, $\ell_p <_B \ell_q$.

For $1 \leq p < \infty$ the space ℓ_p is a separable Banach space with the norm

$$\|x\|_p = \left(\sum_{n=0}^{\infty} |x_n|^p \right)^{1/p}.$$

This makes the equivalence relation ℓ_p an orbit equivalence relation of a Polish group action. In particular, all ℓ_p $(1 \leq p < \infty)$ are idealistic. Since E_1 is not Borel reducible to any orbit equivalence relations (Theorem 10.6.1), $E_1 \not\leq_B \ell_p$ for $1 \leq p < \infty$.

On the other hand we just showed that $E_1 \leq_B \ell_\infty$, and so the equivalence relation ℓ_∞ is not an orbit equivalence relation of a Polish group action. This implies that topological group ℓ_∞ is not a Polishable subgroup of $\mathbb{R}^\omega$, which can also be seen directly (see Lemma 9.3.3). For now we note that, with the usual norm

$$\|x\|_\infty = \sup_{n \in \omega} |x_n|,$$

the space ℓ_∞ is not separable.

Exercise 8.4.1 Show that for any $1 \leq p \leq \infty$, ℓ_p is an F_σ subset of $\mathbb{R}^\omega$.

Exercise 8.4.2 Show that ℓ_p is meager in $\mathbb{R}^\omega$ for $1 \leq p < \infty$.

Exercise 8.4.3 Show that all equivalence relations E_0, E_∞, E_1, and ℓ_p $(1 \leq p \leq \infty)$ are K_σ.

8.5   Examples of Π_3^0 equivalence relations

We define the power operator on equivalence relations.

Definition 8.5.1
Let E be an equivalence relation on a standard Borel space X. The **power** of E is the equivalence relation E^ω on X^ω defined by

$$x E^\omega y \iff \forall n \in \omega \; x_n E y_n.$$

Apparently $E \leq_B E^\omega$ and $E^\omega \leq_B F^\omega$ if $E \leq_B F$. However, $(E^\omega)^\omega \sim_B E^\omega$. Thus the power operator is not a jump operator. If E is $\mathbf{\Pi}^0_\alpha$ for $\alpha < \omega_1$, then E^ω is also $\mathbf{\Pi}^0_\alpha$. Thus the powers of all F_σ equivalence relations we considered in the last section are examples of $\mathbf{\Pi}^0_3$ equivalence relations, among which E^ω_0 has the simplest Borel complexity. It is easy to see that if E is an orbit equivalence relation induced by an action of a Polish group G, then E^ω is induced by an action of G^ω.

We show that E^ω_0 is not Borel reducible to any $\mathbf{\Sigma}^0_3$ equivalence relation. It follows that the powers of all other equivalence relations have the same property.

Theorem 8.5.2
Let E be a $\mathbf{\Sigma}^0_3$ equivalence relation on a Polish space X. Then $E^\omega_0 \not\leq_B E$.

Proof. Assume toward a contradiction that $f \colon (2^\omega)^\omega \to X$ is a Borel reduction of E^ω_0 to E. Since f is Borel, and hence Baire measurable, there is a comeager set $C \subseteq (2^\omega)^\omega$ such that $f \restriction C$ is continuous. Without loss of generality we may assume C is G_δ. Then $E^\omega_0 \cap C^2$ is $\mathbf{\Sigma}^0_3$, since

$$(x, y) \in E^\omega_0 \cap C^2 \iff x, y \in C \land (f(x), f(y)) \in E.$$

To arrive at a contradiction it suffices to establish the following claim: For every dense G_δ set $C \subseteq (2^\omega)^\omega$, $E^\omega_0 \cap C^2$ is $\mathbf{\Pi}^0_3$-complete.
Fix any bijection $\langle \cdot, \cdot \rangle \colon \omega \times \omega \to \omega$. Let

$$B = \{x \in 2^\omega : \forall i \, \exists j \, \forall k \geq j \, x(\langle i, k \rangle) = 1\}.$$

Then B is $\mathbf{\Pi}^0_3$-complete. To see this, let $A \subseteq 2^\omega$ be $\mathbf{\Pi}^0_3$, say

$$A = \bigcap_i \bigcup_j \bigcap_k A_{i,j,k},$$

where each $A_{i,j,k}$ is clopen. Define $\rho \colon 2^\omega \to 2^\omega$ as follows. Given $x \in 2^\omega$, for each $i, k \in \omega$ let $a_{x,i,k} \in \omega$ be the least integer $j \leq k$, if one exists, such that $\forall k' \leq k \, (x \in A_{i,j,k'})$. Let

$$\rho(x)(\langle i, k \rangle) = \begin{cases} 1, & \text{if } a_{x,i,k} \text{ and } a_{x,i,k-1} \text{ are both defined and are equal,} \\ 0, & \text{otherwise.} \end{cases}$$

It is easy to see that ρ is continuous and $x \in A \iff \rho(x) \in B$. This shows that B is $\mathbf{\Pi}^0_3$-complete.
Write $C = \bigcap_n D_n$ where each D_n is open dense in $(2^\omega)^\omega$. We next define two continuous functions $\pi_1, \pi_2 \colon 2^\omega \to 2^{\omega \times \omega}$ so that $x \mapsto (\pi_1(x), \pi_2(x))$ reduces the set B to $E^\omega_0 \restriction C^2$. For this we define two sequences $(u_s)_{s \in 2^{<\omega}}$, $(v_s)_{s \in 2^{<\omega}}$ and a function $p \colon \omega \to \omega$ so that the following hold:

(i) for any $n \in \omega$, if $n = \langle i, k \rangle$ then $i, p(n) + k < p(n+1)$;

(ii) for any $n \in \omega$ and $\mathrm{lh}(s) = n$, $u_s, v_s \in 2^{p(n) \times p(n)}$;

(iii) for $s \subseteq t$, $u_s \subseteq u_t$ and $v_s \subseteq v_t$;

(iv) if $\mathrm{lh}(s) = n$ then $N_{u_s}, N_{v_s} \subseteq D_n$;

(v) for any $n \in \omega$, if $s(n) = 1$ then $u_s(j, l) = v_s(j, l)$ for all $(j, l) \in p(n + 1) \times p(n + 1) - p(n) \times p(n)$;

(vi) for $n = \langle i, k \rangle$, if $s(n) = 0$ then $u_s(i, p(n) + k) \neq v_s(i, p(n) + k)$ and $u_s(j, l) = v_s(j, l)$ for $(j, l) \in p(n+1) \times p(n+1) - p(n) \times p(n) - \{(i, p(n) + k)\}$.

It follows easily from the density of D_n that such sequences can be constructed.

We let $\pi_1(x) = \bigcup_n u_{x \restriction n}$ and $\pi_2(x) = \bigcup_n v_{x \restriction n}$. Clearly π_1, π_2 are continuous. By (iv) $\pi_1(x), \pi_2(x) \in C$ for any $x \in 2^\omega$. If $x \in B$, then for each i let $k(i)$ be such that $\forall k \geq k(i)\ x(\langle i, k \rangle) = 1$. Fix $i \in \omega$. For any $n \geq \langle i, k(i) \rangle$, if $x(n) = 1$ then by (v) for any $p(n) \leq j < p(n + 1)$ we have $\pi_1(x)(i, j) = \pi_2(x)(i, j)$; if $x(n) = 0$ then $n = \langle i', k \rangle$ for some $i' \neq i$, since otherwise $k \geq k(i)$ and we have $x(\langle i, k \rangle) = 1$ by assumption. In this latter case by (vi) we still have that $\pi_1(x)(i, j) = \pi_2(x)(i, j)$ if $i < p(n + 1)$ and $p(n) \leq j < p(n + 1)$. Thus $\pi_1(x)(i, k) = \pi_2(x)(i, k)$ for all $k \geq k_0$. This shows that $\pi_1(x)\, E_0^\omega\, \pi_2(x)$.

If $x \notin B$, then for some i we have that $x(\langle i, k \rangle) = 0$ for infinitely many k. Fix such an i. Then $\pi_1(x)(i, p(\langle i, k \rangle) + k) \neq \pi_2(x)(i, p(\langle i, k \rangle) + k)$ for infinitely many k. This shows that $(\pi_1(x), \pi_2(x)) \notin E_0^\omega$. □

We introduce some other $\mathbf{\Pi}_3^0$ equivalence relations. First consider the classical Banach space

$$c_0 = \{x \in \mathbb{R}^\omega \ : \ \lim_n |x_n| = 0\}$$

with the norm $\| \cdot \|_\infty$. c_0 is a $\mathbf{\Pi}_3^0$ subgroup of $\mathbb{R}^\omega$. We consider the coset equivalence relation $\mathbb{R}^\omega/c_0$ defined by

$$x\ \mathbb{R}^\omega/c_0\ y \iff x - y \in c_0$$

and still denote it by c_0. Then the equivalence relation c_0 is $\mathbf{\Pi}_3^0$.

Lemma 8.5.3
$E_0^\omega \leq_B c_0$.

Proof. Let $\langle \cdot, \cdot \rangle : \omega \times \omega \to \omega$ be a bijection such that $\langle n, m \rangle \leq \langle n', m' \rangle$ if $n \leq n'$ and $m \leq m'$. Define $\varphi : 2^{\omega \times \omega} \to \mathbb{R}^\omega$ by

$$\varphi(x)(\langle n, m \rangle) = 2^{-n} x(n, m).$$

Clearly φ is a Borel function. We check that it is a reduction from E_0^ω to c_0. Let $x E_0^\omega y$, that is, $\forall n \exists m(n) \forall m \geq m(n)\ x(n, m) = y(n, m)$. Let $\epsilon > 0$. Fix

n_0 large enough such that $2^{-n_0} < \epsilon$. Let $m_0 = \max\{m(i) : i < n_0\}$. We show that for all $N \geq \langle n_0, m_0 \rangle$, $|\varphi(x)(N) - \varphi(y)(N)| < \epsilon$. Let $N \geq \langle n_0, m_0 \rangle$. Say $N = \langle n, m \rangle$. If $n \geq n_0$ then $|\varphi(x)(N) - \varphi(y)(N)| \leq 2^{-n} \leq 2^{-n_0} < \epsilon$. If $n < n_0$ then $m \geq m(n)$ since $N = \langle n, m \rangle \geq \langle n_0, m_0 \rangle \geq \langle n, m_0 \rangle \geq \langle n, m(n) \rangle$. Thus $x(n, m) = y(n, m)$ and $\varphi(x)(N) - \varphi(y)(N) = 0$. This shows that $\lim_N \varphi(x)(N) - \varphi(y)(N) = 0$ and so $\varphi(x) - \varphi(y) \in c_0$.

Conversely let $(x, y) \notin E_0^\omega$. Fix n such that $(x_n, y_n) \notin E_0$, that is, $x(n, m) \neq y(n, m)$ for infinitely many m. Then $|\varphi(x)(\langle n, m \rangle) - \varphi(y)(\langle n, m \rangle)| = 2^{-n}$ for infinitely many m, and thus $\lim_N [\varphi(x)(N) - \varphi(y)(N)] \neq 0$. This shows that $\varphi(x) - \varphi(y) \notin c_0$. $\quad\square$

Another class of examples of $\mathbf{\Pi}_3^0$ equivalence relations are the Friedman–Stanley jumps of F_σ equivalence relations. Among these the simplest is the jump of the identity relation on a Polish space, for instance, $\mathrm{id}(2^\omega)^+$. For notational simplicity we denote this equivalence relation by $=^+$.

Lemma 8.5.4
$E_0^\omega \leq_B =^+$.

Proof. It is easy to see that $E_0^+ \sim_B =^+$ (see Exercise 8.3.3). Thus it suffices to show that $E_0^\omega \leq E_0^+$. Fix a bijection $\langle \cdot, \cdot \rangle : \omega \times \omega \to \omega$. Let $\tau : 2^\omega \to 2^\omega$ be defined by

$$\tau(z)(\langle i, j \rangle) = \begin{cases} z(j), & \text{if } i = 0, \\ 1, & \text{otherwise.} \end{cases}$$

For each $n \in \omega$ define a function $\varphi_n : 2^\omega \to 2^\omega$ by

$$\varphi_n(z)(\langle i, j \rangle) = \begin{cases} \tau(z)(j), & \text{if } i = n, \\ 0, & \text{otherwise.} \end{cases}$$

Then for each $n \in \omega$ and $z_1, z_2 \in 2^\omega$, we have that $z_1 E_0 z_2$ iff $\varphi_n(z_1) E_0 \varphi_n(z_2)$. Moreover, for $n_1 \neq n_2$ and any $z_1, z_2 \in 2^\omega$, $(\varphi_{n_1}(z_1), \varphi_{n_2}(z_2)) \in E_0$ since for infinitely many m, $\varphi_{n_1}(z_1)(\langle n_1, m \rangle) = \tau(z_1)(m) = 1$, but for all m, $\varphi_{n_2}(z_2)(\langle n_1, m \rangle) = 0$.

We then define $\rho : (2^\omega)^\omega \to (2^\omega)^\omega$ by letting

$$(\rho(x))_n = \varphi_n(x_n).$$

Suppose $x E_0^\omega y$. Then for all $n \in \omega$, $x_n E y_n$, and therefore $\varphi_n(x_n) E_0 \varphi_n(y_n)$. It follows that $\{[\varphi_n(x_n)]_{E_0} : n \in \omega\} = \{[\varphi_n(y_n)]_{E_0} : n \in \omega\}$, and thus $\rho(x) E_0^+ \rho(y)$. Conversely, suppose $(x, y) \notin E_0^\omega$. Let n be such that $(x_n, y_n) \notin E_0$. Then $(\varphi_n(x_n), \varphi_m(y)) \notin E_0$ for all $m \in \omega$. Thus $[\varphi_n(x_n)]_{E_0} \notin \{[\varphi_n(y_n)]_{E_0} : n \in \omega\}$ and $(\rho(x), \rho(y)) \notin E_0^+$. $\quad\square$

The last equivalence relation we consider in this section is measure equivalence. Let M be the space of Borel probability measures on 2^ω. Every $\mu \in M$

is completely determined by the sequence $(\mu(N_s))_{s \in 2^{<\omega}}$. Hence M can be viewed as a subspace of $[0,1]^\omega$, and therefore is a Polish space. For $\mu, \nu \in M$, we say that μ is **absolutely continuous** with respect to ν, denoted $\mu \ll \nu$, if for any Borel set $A \subseteq 2^\omega$, $\nu(A) = 0 \Rightarrow \mu(A) = 0$. Two measures μ and ν are **equivalent** if both $\mu \ll \nu$ and $\nu \ll \mu$. We use $\equiv_m$ to denote the measure equivalence relation.

It is easy to see that

$$\mu \ll \nu \iff \forall \epsilon > 0 \exists \delta > 0 \; \forall \text{ basic open } U \; (\nu(U) < \delta \Rightarrow \mu(U) < \epsilon).$$

It follows that $\ll$ is a $\mathbf{\Pi}^0_3$ relation and $\equiv_m$ is a $\mathbf{\Pi}^0_3$ equivalence relation.

Lemma 8.5.5
$=^+ \;\leq_B\; \equiv_m$.

Proof. We give an informal sketch of the proof. The exact details of the definitions are left to the reader. Intuitively we will assign to each countable subset S of 2^ω a Borel probability measure μ_S so that $S = S'$ iff $\mu_S \equiv_m \mu_{S'}$. This is easy to do: let μ_S be any Borel probability measure supported on S. Then for any Borel set $A \subseteq 2^\omega$, $\mu(A) = 0$ iff $A \cap S = \emptyset$ iff $\nu(A) = 0$. $\qquad\square$

There are many important facts omitted in our discussion so far. Before closing this section we mention some known facts without proof.

The Spectral Theorem in functional analysis essentially established that the measure equivalence relation is Borel bireducible to the orbit equivalence relation of the conjugacy action of the unitary group on itself. In particular this establishes that the measure equivalence is Borel bireducible with an orbit equivalence relation of a Polish group action. This implies (by Theorem 10.6.1) that E_1 is not reducible to any of the equivalence relations we discussed in this section.

Kakutani studied the measure equivalence relation on product measures. He essentially showed that $\ell_2 \leq_B \equiv_m$ (see Reference [104]). It follows from Theorem 8.4.3 that $\ell_p \leq_B \equiv_m$ if $1 \leq p \leq 2$. It is not known if $\ell_p \leq_B \equiv_m$ for any $p > 2$.

Hjorth and Kechris [78] proved a dichotomy theorem that for any $E \leq E_0^\omega$, either $E \sim_B E_0^\omega$ or E is hyperfinite.

Exercise 8.5.1 Let $C \subseteq \mathbb{R}^\omega$ be comeager. Show that $c_0 \cap C^2$ is $\mathbf{\Pi}^0_3$-complete.

Exercise 8.5.2 Show that $(c_0)^\omega \sim_B c_0$.

Exercise 8.5.3 Give an example of an equivalence relation E such that $E^+ <_B E^\omega$.

Exercise 8.5.4 Show that $(=^+)^\omega \sim_B =^+$ and $(\equiv_m)^\omega \sim_B \equiv_m$.

Chapter 9

Analytic Equivalence Relations

The study of general analytic equivalence relations is a very important and yet challenging subject. In this chapter we will discuss general dichotomy theorems for analytic equivalence relations, universal analytic equivalence relations, and other examples of analytic but non-Borel equivalence relations. Orbit equivalence relations of Polish group actions are in general analytic, and often non-Borel, and therefore provide many interesting examples and opportunities of applications of general theorems. The Borel reducibility hierarchy for analytic equivalence relations is more complicated than the one for Borel equivalence relations.

9.1 The Burgess trichotomy theorem

The Silver dichotomy theorem for $\mathbf{\Pi}^1_1$ equivalence relations fails for $\mathbf{\Sigma}^1_1$ equivalence relations. For an example, consider the equivalence relation E_{ω_1} defined on the standard Borel space LO by

$$x E_{\omega_1} y \iff (x \notin \mathrm{WO} \wedge y \notin \mathrm{WO}) \vee \exists \varphi : (\omega, <_x) \cong (\omega, <_y).$$

Then E_{ω_1} is a $\mathbf{\Sigma}^1_1$ equivalence relation, and there are apparently ω_1 many E_{ω_1}-equivalence classes. To elaborate more on the equivalence classes of E_{ω_1}, there is a special E_{ω_1}-equivalence class consisting of all nonwell-orderings of ω, and then every other E_{ω_1}-equivalence class consists of well-orderings of ω, and two elements $x, y \in \mathrm{WO}$ are E_{ω_1}-equivalent iff $<_x$ and $<_y$ have the same order type. In particular, E_{ω_1} has uncountably many equivalence classes. However, E_{ω_1} does not have perfectly many classes. In fact, if P is a perfect set of E-inequivalent elements in some compatible Polish topology of LO, then P is in particular a Borel subset of LO. Note that the cardinalities of $P \cap \mathrm{WO}$ and of P differ by at most one. But since $P \cap \mathrm{WO}$ is again Borel, by the boundedness principle there is $\alpha < \omega_1$ such that $P \cap \mathrm{WO} \subseteq \mathrm{WO}_\alpha$, and therefore $P \cap \mathrm{WO}$ is in fact countable. It follows that P is countable, a contradiction.

In this section we prove a trichotomy theorem of Burgess [13] stating that a $\mathbf{\Sigma}^1_1$ equivalence relation has either perfectly many, countably many, or ω_1 many equivalence classes. Note that if a $\mathbf{\Sigma}^1_1$ equivalence relation E has only countably many equivalence classes it is in fact Borel and smooth. Thus having ω_1 many equivalence classes is typical of all counterexamples of the Silver dichotomy for Σ_1^1 equivalence relations.

To prove the Burgess trichotomy theorem we need to show another theorem of Burguess representing every Σ_1^1 equivalence relation as a intersection of ω_1 many Borel equivalence relations. We need also recall some basic notions about ordinals. The notation ω_1 is used to represent the least uncountable ordinal as well as the set of all countable ordinals. We also use it to denote the least uncountable cardinal number. The space ω_1 of all countable ordinals is endowed with the **order topology**, where subbasic open sets are of the form $\{\alpha < \omega_1 : \alpha < \alpha_0\}$ and $\{\alpha < \omega_1 : \alpha > \alpha_0\}$ for $\alpha_0 < \omega_1$. A subset C of ω_1 is **closed** if C is closed in the order topology of ω_1, and C is a **club** if it is closed and unbounded above. The intersection of countably many clubs in ω_1 is still a club in ω_1. If $f : \omega_1 \to \omega_1$ is a continuous function satisfying $f(\alpha) \geq \alpha$ for all $\alpha < \omega_1$, then there is a club $C \subseteq \omega_1$ such that $f(\alpha) = \alpha$ for all $\alpha \in C$.

The following lemma is an immediate corollary of the boundedness principle Theorem 1.6.10.

Lemma 9.1.1
Let X be a standard Borel space and $f_1, \ldots, f_n, g_1, \ldots, g_m : X \to \mathrm{LO}$ be Borel functions. Suppose $f_1^{-1}(\mathrm{WO}) \cap \cdots \cap f_n^{-1}(\mathrm{WO}) \subseteq g_1^{-1}(\mathrm{WO}) \cup \cdots \cup g_m^{-1}(\mathrm{WO})$. Then there is a club $C \subseteq \omega_1$ such that for all $\alpha \in C$,

$$f_1^{-1}(\mathrm{WO}_\alpha) \cap \cdots \cap f_n^{-1}(\mathrm{WO}_\alpha) \subseteq g_1^{-1}(\mathrm{WO}_\alpha) \cup \cdots \cup g_m^{-1}(\mathrm{WO}_\alpha).$$

Proof. We define a continuous function $\alpha \mapsto \beta(\alpha)$ as follows. For each $\alpha < \omega_1$, let $\beta(\alpha)$ be the least $\beta \geq \alpha$ such that

$$f_1^{-1}(\mathrm{WO}_\alpha) \cap \cdots \cap f_n^{-1}(\mathrm{WO}_\alpha) \subseteq g_1^{-1}(\mathrm{WO}_\beta) \cup \cdots \cup g_m^{-1}(\mathrm{WO}_\beta).$$

To see that this $\beta(\alpha)$ is well defined, note that by our assumption

$$f_1^{-1}(\mathrm{WO}_\alpha) \cap \cdots \cap f_n^{-1}(\mathrm{WO}_\alpha) \subseteq g_1^{-1}(\mathrm{WO}) \cup \cdots \cup g_m^{-1}(\mathrm{WO}).$$

Let A denote the set on the left-hand side. Then A is a Borel subset of X. It follows that $A - (g_2^{-1}(\mathrm{WO}) \cup \cdots \cup g_m^{-1}(\mathrm{WO}))$ is Σ_1^1. Since

$$A - (g_2^{-1}(\mathrm{WO}) \cup \cdots \cup g_m^{-1}(\mathrm{WO})) \subseteq g_1^{-1}(\mathrm{WO}),$$

we have that

$$g_1(A - (g_2^{-1}(\mathrm{WO}) \cup \cdots \cup g_m^{-1}(\mathrm{WO}))) \subseteq \mathrm{WO}.$$

The set on the left-hand side is Σ_1^1. Hence by the boundedness principle there is some $\beta_1 \geq \alpha$ such that

$$A - (g_2^{-1}(\mathrm{WO}) \cup \cdots \cup g_m^{-1}(\mathrm{WO})) \subseteq g_1^{-1}(\mathrm{WO}_{\beta_1}).$$

Repeating the argument for $g_2, \ldots, g_m$, we obtain $\beta_2, \ldots, \beta_m \geq \alpha$ similarly. Let $\beta = \max\{\beta_1, \ldots, \beta_m\}$. Then we have that $A \subseteq g_1^{-1}(\mathrm{WO}_\beta) \cup \cdots \cup g_m^{-1}(\mathrm{WO}_\beta)$. This shows that $\beta(\alpha)$ is well defined. It is easy to see that $\alpha \mapsto \beta(\alpha)$ is continuous. We can thus get a club $C \subseteq \omega_1$ such that for all $\alpha \in C$, $\beta(\alpha) = \alpha$. ☐

Theorem 9.1.2 (Burgess)

Let E be a Σ_1^1 equivalence relation on a standard Borel X. Then there are Borel equivalence relations E_α on X for $\alpha < \omega_1$ such that $E = \bigcap_{\alpha < \omega_1} E_\alpha$.

Proof. Let $f : X^2 \to \mathrm{LO}$ be a Borel function such that $X^2 - E = f^{-1}(\mathrm{WO})$. For $\alpha < \omega_1$ let $A_\alpha = X^2 - f^{-1}(\mathrm{WO}_\alpha)$. Then each A_α is a Borel binary relation on X and $E = \bigcap_{\alpha < \omega_1} A_\alpha$. We claim that there is a club $C \subseteq \omega_1$ such that for all $\alpha \in C$, A_α is an equivalence relation. Clearly the theorem follows from this claim. It is easy to see that every A_α is reflexive. To prove the claim we first show that there is a club $C \subseteq \omega_1$ such that for all $\alpha \in C$, A_α is symmetric. For this let $f_1(x, y) = f(x, y)$ and $g_1(x, y) = f(y, x)$ for all $(x, y) \in X^2$. Then since E is symmetric, $f_1^{-1}(\mathrm{WO}) = X^2 - E = g_1^{-1}(\mathrm{WO})$. By applying Lemma 9.1.1 for both $f_1^{-1}(\mathrm{WO}) \subseteq g_1^{-1}(\mathrm{WO})$ and $g_1^{-1}(\mathrm{WO}) \subseteq f_1^{-1}(\mathrm{WO})$, we get a club $C \subseteq \omega_1$ such that for all $\alpha \in C$, $f_1^{-1}(\mathrm{WO}_\alpha) = g_1^{-1}(\mathrm{WO}_\alpha)$. It follows that A_α is symmetric for all $\alpha \in C$. Next we show that there is a club $C \subseteq \omega_1$ such that for all $\alpha \in C$, A_α is transitive. The claim then follows since the intersection of two clubs is still a club. For this define $f_1, g_1, g_2 : X^3 \to \mathrm{LO}$ by

$$f_1(x, y, z) = f(x, z), \quad g_1(x, y, z) = f(x, y), \text{ and } g_2(x, y, z) = f(y, z).$$

Then for any $(x, y, z) \in X^3$,

$$
\begin{aligned}
(x, y, z) \in f_1^{-1}(\mathrm{WO}) &\iff (x, z) \notin E \\
&\implies (x, y) \notin E \vee (y, z) \notin E \\
&\iff (x, y, z) \in g_1^{-1}(\mathrm{WO}) \cup g_2^{-1}(\mathrm{WO}).
\end{aligned}
$$

This means that $f_1^{-1}(\mathrm{WO}) \subseteq g_1^{-1}(\mathrm{WO}) \cup g_2^{-1}(\mathrm{WO})$. Thus by Lemma 9.1.1 there is a club $C \subseteq \omega_1$ such that for all $\alpha \in C$, $f_1^{-1}(\mathrm{WO}_\alpha) \subseteq g_1^{-1}(\mathrm{WO}_\alpha) \cup g_2^{-1}(\mathrm{WO}_\alpha)$. It follows that for $\alpha \in C$, A_α is transitive, since if $(x, y), (y, z) \in A_\alpha$ then $(x, y, z) \notin g_1^{-1}(\mathrm{WO}_\alpha) \cup g_2^{-1}(\mathrm{WO}_\alpha)$ and therefore $(x, y, z) \notin f_1^{-1}(\mathrm{WO}_\alpha)$ and so $(x, z) \in A_\alpha$. ☐

Recall from Definition 5.4.3 that a collection $\mathcal{S}$ of subsets of a standard Borel space X generates the equivalence relation

$$xEy \iff \forall S \in \mathcal{S} \ (x \in S \leftrightarrow y \in S).$$

Definition 9.1.3

An equivalence relation E on a standard Borel space X is ω_1-**smooth** if E is generated by a set S of ω_1 many Borel subsets of X.

Theorem 9.1.4

Let E be an ω_1-smooth equivalence relation on a Polish space X. Then either E has $\leq \omega_1$ many equivalence classes or else E has perfectly many classes.

Proof. We fix a complete metric for X for the entire proof. Let B_α, $\alpha < \omega_1$, witness that E is ω_1-smooth. Since both B_α and $X - B_\alpha$ are Suslin, we have sequences $(C_s^{\alpha,0})_{s\in\omega^{<\omega}}$, $(C_s^{\alpha,1})_{s\in\omega^{<\omega}}$ of closed sets with $\text{diam}(C_s^{\alpha,i}) < 2^{-\text{lh}(s)}$ for $i = 0, 1$ and all $\alpha < \omega_1$, such that

$$B_\alpha = \bigcup_{x\in\omega^\omega} \bigcap_{n\in\omega} C_{x\restriction n}^{\alpha,0}, \text{ and } X - B_\alpha = \bigcup_{x\in\omega^\omega} \bigcap_{n\in\omega} C_{x\restriction n}^{\alpha,1}.$$

As usual we may assume that $C_t^{\alpha,i} \subseteq C_s^{\alpha,i}$ if $s \subseteq t$. For $s \in \omega^{<\omega}$ we let

$$B_s^{\alpha,i} = \bigcup_{x\in\omega^\omega, s\subseteq x} \bigcap_{n\in\omega} C_{x\restriction n}^{\alpha,i}.$$

Then $B_s^{\alpha,i} \subseteq C_s^{\alpha,i}$.

Assume that E has more than ω_1 many equivalence classes. Let $Z \subseteq X$ be a set of ω_2 pairwise E-inequivalent elements. By induction on $n \in \omega$, and for each $s \in 2^n$, we construct a set D_s, an ordinal $\alpha_s < \omega_1$, and $\sigma(s, m) \in \omega^n$ for $m < n$ so that the following conditions are satisfied:

(i) $\sigma(s, m) \subseteq \sigma(t, m)$ if $s \subseteq t$ and $m < \text{lh}(s)$;

(ii) $D_s = \bigcap_{m<n} B_{\sigma(s,m)}^{\alpha_{s\restriction m}, s(m)}$ for $n > 0$;

(iii) $|Z \cap D_s| = \omega_2$.

To begin, let $\sigma(s, 0) = \emptyset$ for any $s \in 2^{<\omega}$, $\alpha_\emptyset = 0$, and $D_\emptyset = X$. In general suppose $n \in \omega$, $s \in 2^n$, and for all $m < n$, $\alpha_{s\restriction m}$ and $\sigma(s, m)$ have been defined so that (i) through (iii) are satisfied. In the induction step we define α_s and $\sigma(s\,\hat{}\,i, m)$ for $i = 0, 1$ and $m \leq n$. Then by (ii) $D_{s\,\hat{}\,i}$ is defined for $i = 0, 1$.

We first note that there exists $\alpha < \omega_1$ such that $Z \cap D_s \cap B_\alpha$ and $(Z \cap D_s) - B_\alpha$ both have cardinality ω_2. Otherwise, for each $\alpha < \omega_1$ let Y_α be either $Z \cap D_s \cap B_\alpha$ or $(Z \cap D_s) - B_\alpha$, whichever has cardinality ω_1 or less. Then the set $X - \bigcup_{\alpha<\omega_1} Y_\alpha$ has at most one element of Z. But since $\bigcup_{\alpha<\omega_1} Y_\alpha$ has ω_1 many elements of Z, it follows that Z has cardinality ω_1, a contradiction.

To continue our construction, we let α_s be the least ordinal $\alpha < \omega_1$ such that both $Z \cap D_s \cap B_\alpha$ and $(Z \cap D_s) - B_\alpha$ have cardinality ω_2. Note that $Z \cap D_s \cap B_{\alpha_s}$ is now a subset of

$$\bigcap_{m<n} B_{\sigma(s,m)}^{\alpha_{s\restriction m}, s(m)} \cap B_{\alpha_s} = \bigcap_{m<n} \bigcup_{k\in\omega} B_{\sigma(s,m)\,\hat{}\,k}^{\alpha_{s\restriction m}, s(m)} \cap \bigcup_{t\in\omega^{n+1}} B_t^{\alpha_s,0},$$

which is a finite union of sets of the form

$$\bigcap_{m<n} B^{\alpha_s\restriction m,\, s(m)}_{\sigma(s,m)\hat{}\, k_m} \cap B^{\alpha_s,0}_t$$

for $k_m \in \omega$ and $t \in \omega^{n+1}$. At least one of these sets contains w_2 many elements of Z. Thus there exists k_m for $m < n$ and $t \in \omega^{n+1}$ such that if we define $\sigma(s\hat{}\,0, m) = \sigma(s,m)\hat{}\, k_m$ for $m < n$ and $\sigma(s\hat{}\,0, n) = t$, then the corresponding $D_{s\hat{}\,0}$ contains w_2 many elements of Z. It is clear that all of (i) through (iii) are satisfied. The definitions of $\sigma(s\hat{}\,1, m)$ for $m \le n$ is similar by considering the set $(Z \cap D_s) - B_{\alpha_s}$. This finishes our construction.

Now the sequence $(D_s)_{s \in 2^\omega}$ gives rise to a Suslin set

$$A = \bigcup_{x \in 2^\omega} \bigcap_{n \in \omega} \overline{D_{x\restriction n}}$$

via the Suslin operation. Since $\operatorname{diam}(D_s) < 2^{-\mathrm{lh}(s)}$ we get that for any $x \in 2^\omega$, the set $\bigcap_{n \in \omega} \overline{D_{x\restriction n}}$ is a singleton, say with element z_x. Moreover, if $x(m) = 0$, then by our construction, for any $n > m$,

$$D_{x\restriction n} \subseteq C^{\alpha_x\restriction(m+1),0}_{\sigma(x\restriction n, m)}.$$

It follows from (i) that

$$z_x \in \bigcap_{n>m} \overline{D_{x\restriction n}} \subseteq \bigcap_{n>m} C^{\alpha_x\restriction(m+1),0}_{\sigma(x\restriction n, m)} \subseteq B_{\alpha_x\restriction(m+1)}.$$

Thus $z_x \in B_{\alpha_x\restriction(m+1)}$. Similarly, if $y(m) = 1$, then $z_y \notin B_{\alpha_y\restriction(m+1)}$. These imply that if $x, y \in 2^\omega$ with $x \ne y$, then there is $\alpha < w_1$ such that either $x \in B_\alpha$ and $y \notin B_\alpha$, or $x \notin B_\alpha$ and $y \in B_\alpha$. In either case we have that x and y are E-inequivalent. Thus we have produced a Suslin set A of E-inequivalent elements containing uncountably many elements. By the perfect set theorem (Theorem 1.6.6) A contains a perfect set of E-inequivalent elements. □

Theorem 9.1.5 (Burgess)
Let E be a Σ^1_1 equivalence relation on a standard Borel space X. Then E has either perfectly many, countably many, or w_1 many equivalence classes.

Proof. By Theorem 9.1.2 E can be written as the intersection of w_1 many Borel equivalence relations. Let E_α, $\alpha < w_1$, be Borel equivalence relations such that $E = \bigcap_{\alpha < w_1} E_\alpha$. Each E_α has either countably many or perfectly many equivalence classes by the Silver dichotomy theorem. If any of E_α has perfectly many equivalence classes, then so does E. So assume each E_α has only countably many equivalence classes, and let them be enumerated as $C_{\alpha,n}$ for $n \in \omega$. Then we have that

$$xEy \iff \forall\alpha\forall n\ (x \in C_{\alpha,n} \leftrightarrow y \in C_{\alpha,n}).$$

This means that E is ω_1-smooth, and hence by Theorem 9.1.4, E has $\leq$ ω_1 many equivalence classes. It follows that E has either perfectly many, countably many, or ω_1 many equivalence classes. □

Exercise 9.1.1 Write E_{ω_1} as an intersection of ω_1 many Borel equivalence relations.

Exercise 9.1.2 Recall that a quasiorder is a reflexive and transitive relation. Show that any $\mathbf{\Sigma}^1_1$ quasiorder on a standard Borel space X is the intersection of ω_1 many Borel quasiorders.

Exercise 9.1.3 Let X be a standard Borel space and $E \subseteq F$ equivalence relations on X, where E is $\mathbf{\Sigma}^1_1$ and F is $\mathbf{\Pi}^1_1$. Show that there is a Borel equivalence relation R such that $E \subseteq R \subseteq F$.

Exercise 9.1.4 Let X be a standard Borel space and $E = \bigcap_{\alpha<\omega_1} E_\alpha$, where E_α are $\mathbf{\Sigma}^1_1$ equivalence relations on X. Show that E has either perfectly many, countably many, or ω_1 many equivalence classes.

Exercise 9.1.5 Let X be a standard Borel space and $E = \bigcap_{\alpha<\omega_1} E_\alpha$, where E_α are $\mathbf{\Pi}^1_1$ equivalence relations on X. Show that E has either perfectly many, countably many, or ω_1 many equivalence classes.

Exercise 9.1.6 For a standard Borel space X and a collection $\mathcal{S}$ of subsets of X, let $E(\mathcal{S})$ denote the equivalence relation generated by $\mathcal{S}$. Show the following corollary of the proof of Theorem 9.1.4: If $\mathcal{S}$ is a set of ω_1 many Borel subsets of a standard Borel space X, then either $E(\mathcal{S})$ has $\leq \omega_1$ many equivalence classes or there is a countable $\mathcal{T} \subseteq \mathcal{S}$ such that $E(\mathcal{T})$ has perfectly many equivalence classes.

9.2 Definable reductions among analytic equivalence relations

In this section we discuss definable reductions among $\mathbf{\Sigma}^1_1$ equivalence relations. We start with a discussion of Borel reducibility and show that it is sometimes inadequate to capture the intuitive reductions among general $\mathbf{\Sigma}^1_1$ equivalence relations. We will then introduce a more general notion of definable reducibility. The downside, however, of considering this broader notion is that a significant amount of metamathematics is needed to prove statements about it. Since these metamathematical techniques are beyond the scope of this book, we will only mention some results without proving them. On the other hand, we will introduce the very useful notion of Ulm classifiability and also briefly discuss universal analytic equivalence relations.

Recall that the conclusion of the Silver dichotomy can be reformulated in terms of Borel reducibility as either $E \leq_B \mathrm{id}(\omega)$ or $\mathrm{id}(2^\omega) \leq_B E$. For the Burgess trichotomy theorem, these Borel reducibility statements still correspond to the cases when E has either countably many or perfectly many equivalence classes. However, when E has ω_1 many equivalence classes, there is no obvious way to characterize E in the Borel reducibility hierarchy. We elaborate some more below.

Consider the equivalence relation F_{ω_1} defined on 2^ω by

$$x F_{\omega_1} y \iff \omega_1^{\mathrm{CK}(x)} = \omega_1^{\mathrm{CK}(y)}.$$

Since $\{\omega_1^{\mathrm{CK}(x)} : x \in 2^\omega\}$ is unbounded in ω_1, clearly F_{ω_1} has ω_1 many equivalence classes.

Lemma 9.2.1
F_{ω_1} is Σ_1^1 equivalence relation with every equivalence class Borel.

Proof. Note that $\omega_1^{\mathrm{CK}(x)} \leq \omega_1^{\mathrm{CK}(y)}$ iff

$$\forall m \exists n \; (\{m\}^x \in \mathrm{WO} \to \exists \varphi : (\omega, \{m\}^x) \cong (\omega, \{n\}^y)),$$

which implies that F_{ω_1} is Σ_1^1. Now fix $x \in 2^\omega$ and $m, n \in \omega$ with $\{m\}^x \in \mathrm{WO}$. Then we also have that $(\omega, \{m\}^x) \cong (\omega, \{n\}^y)$ iff

(a) $\{n\}^y \in \mathrm{WO}$,

(b) for all strictly order-preserving φ from $(\omega, \{n\}^y)$ into $(\omega, \{m\}^x)$, φ is onto, and

(c) for all strictly order-preserving ψ from $(\omega, \{m\}^x)$ into $(\omega, \{n\}^y)$, ψ is onto.

Thus if we let $B_{m,n} = \{y : (\omega, \{m\}^x) \cong (\omega, \{n\}^y)\}$, then $B_{m,n}$ is Δ_1^1, in particular Borel, when $\{m\}^x \in \mathrm{WO}$. Now

$$\left\{ y : \omega_1^{\mathrm{CK}(x)} = \omega_1^{\mathrm{CK}(y)} \right\} = \bigcap_{\{m\}^x \in \mathrm{WO}} \bigcup_n B_{m,n},$$

hence the F_{ω_1}-equivalence class of x is Borel. □

Now F_{ω_1} is indeed a Σ_1^1 equivalence relation and apparently has ω_1 many equivalence classes. However, if we compare it to E_{ω_1} defined in the preceding section via Borel reducibility, they turn out to be incomparable.

Lemma 9.2.2
$E_{\omega_1} \not\leq_B F_{\omega_1}$ and $F_{\omega_1} \not\leq_B E_{\omega_1}$.

Proof. Note that if $x \notin$ WO then the E_{ω_1}-equivalence class of x is not Borel. Now every F_{ω_1}-equivalence class is Borel, which implies that any equivalence relation Borel reducible to F_{ω_1} has the same property. Thus $E_{\omega_1} \not\leq_B F_{\omega_1}$. On the other hand, assume $f : 2^\omega \to$ LO is a Borel reduction from F_{ω_1} to E_{ω_1}. There is at most one F_{ω_1}-equivalence class C such that $f(C) \subseteq$ LO $-$ WO. Since C is Borel, it follows that $f(2^\omega - C)$ is Σ_1^1 in LO. Since $f(2^\omega - C) \subseteq$ WO, by the boundedness principle there is $\alpha < \omega_1$ such that $f(2^\omega - C) \subseteq$ WO$_\alpha$. This is a contradiction to the assumption that f is a reduction since $E_{\omega_1} \upharpoonright$ WO$_\alpha$ has only countably many equivalence classes. □

Nevertheless on an intuitive level we should be able to define a reduction from F_{ω_1} to E_{ω_1}, that is, given $x \in 2^\omega$ to define $\Phi(x) \in$ WO so that its order type is $\omega_1^{\text{CK}(x)}$. A straightforward strategy to achieve this is the following. Consider the countable set $\{e \in \omega : \{e\}^x \in \text{WO}\}$ and enumerate it as $\{e_n : n \in \omega\}$. Then define

$$\Phi(x)(\langle n, k\rangle, \langle m, l\rangle) = \begin{cases} 1, & \text{if } n < m, \\ 0, & \text{if } m < n, \\ \{e_n\}(k, l), & \text{if } m = n. \end{cases}$$

It is easy to see that Φ is as required. From the discussions above Φ is not a Borel function. To allow reductions like Φ we need to consider classes of functions larger than that of Borel functions. This takes us to the realm of the projective hierarchy, and one natural class of functions to use is that of Δ_2^1 functions. One can check that Φ is Δ_2^1.

Two basic issues arise with the consideration of Δ_2^1 reductions. The first is that the usual axioms of set theory no longer seem sufficient to determine the answers of many questions. Thus for many results axioms beyond ZFC are assumed and it is not clear that they can be removed from the assumptions. The second, a more subtle, issue is that, even for results which turn out to be provable within ZFC, the proof techniques go beyond the classical or effective descriptive set theory we have developed and been using in this book. The most powerful method employed in these proofs is that of forcing. Because of these issues we will not prove any result involving Δ_2^1 reductions, even if some of the proofs are classical. Instead, we will mention some results without proof so as to provide the reader a brief overview of the subject.

One of the common axioms beyond ZFC used in the results about Δ_2^1 reductions is Σ_1^1 determinacy. This is the statement that every game G_A is determined if the payoff set $A \subseteq \omega^\omega$ is Σ_1^1. Burgess has shown under Σ_1^1 determinacy that if an analytic equivalence relation E has $\leq \omega_1$-many equivalence classes, then E is Δ_2^1 reducible to E_{ω_1}. For this reason it is customary to denote E_{ω_1} by id(ω_1), so that the Burgess trichotomy theorem becomes a full analog to the Silver dichotomy.

Hjorth and Kechris [76] have considered a generalization of the Glimm–Effros dichotomy for Σ_1^1 equivalence relations. Here the analog of the smooth clause, or the concrete classifiability, is the following notion of Ulm classifiability, inspired by Ulm's classification of countable abelian p-groups (which we will review more in Section 13.4).

For any $\alpha < \omega$ let 2^α be the space of all transfinite sequences $(x_\beta)_{\beta<\alpha}$ where $x_\beta \in \{0,1\}$ for all $\beta < \alpha$. Elements of 2^α code subsets of α. If $\alpha < \beta$ then we view 2^α as a subset of 2^β in the natural sense: for any $x \in 2^\omega$, regard $x(\xi) = 0$ for all $\alpha \le \xi < \beta$. Let $2^{<\omega_1}$ be the increasing union of all 2^α for $\alpha < \omega_1$. Elements of $2^{<\omega_1}$ are called **Ulm invariants**, since Ulm used such elements to classify countable abelian p-groups up to isomorphism (see Section 13.4). We provide a way to code Ulm invariants by reals, as follows. Given any $(x,y) \in \mathrm{WO} \times 2^\omega$, let $\alpha = \mathrm{ot}(<_x)$. For each $n \in \omega$ with $y(n) = 1$, let $<_{x,n}$ be the restriction of $<_x$ on the set $\{m \in \omega : m <_x n\}$, and then let $\beta_n = \mathrm{ot}(<_{x,n})$. We then associate $U(x,y) \in 2^\alpha$ by letting it code the set $\{\beta_n : y(n) = 1\}$. Consider the equivalence relation E_{Ulm} on $\mathrm{LO} \times 2^\omega$ defined by $(x,y)E_{\mathrm{Ulm}}(x',y')$ if $U(x,y)$ and $U(x',y')$ code the same set of ordinals. A somewhat tedious but straightforward calculation shows that E_{Ulm} is a Σ_1^1 equivalence relation (Exercise 9.2.2).

Definition 9.2.3
An equivalence relation E on a standard Borel space X is **Ulm classifiable** if it is Δ_2^1 reducible to E_{Ulm}.

The equivalence relation E_{Ulm} is customarily denoted by $\mathrm{id}(2^{<\omega_1})$ for the obvious reason that it can be viewed as the identity relation on $2^{<\omega_1}$. In this notation an equivalence relation E is Ulm classifiable iff E is Δ_2^1 reducible to $\mathrm{id}(2^{<\omega_1})$. Hjorth and Kechris [76] proved under Σ_1^1 determinacy that if E is a Σ_1^1 equivalence relation, then either E is Ulm classifiable or $E_0 \le_B E$.

Finally we note that there are equivalence relations universal for all analytic equivalence relations with respect to Borel reducibility.

Proposition 9.2.4
There exists a universal Σ_1^1 equivalence relation, that is, a Σ_1^1 equivalence relation E such that for any Σ_1^1 equivalence relation F, $F \le_B E$.

Proof. Let $U \subseteq \omega^\omega \times (\omega^\omega)^2$ be a universal set for all Σ_1^1 subsets of $(\omega^\omega)^2$, that is, for any Σ_1^1 $A \subseteq (\omega^\omega)^2$ there is $x \in \omega^\omega$ such that $A = U_x$. Define U^* by letting

$$(x,y,z) \in U^* \iff (x,y,z) \in U \lor (x,z,y) \in U.$$

Then U^* is still Σ_1^1. Let $\langle \cdot, \cdot \rangle : \omega^\omega \times \omega^\omega \to \omega^\omega$ be a homeomorphism. Then define an equivalence relation E on ω^ω by

$$\langle x,y \rangle E \langle x',y' \rangle \iff x = x' \land (y = y' \lor (x,y,y') \in U^* \lor$$
$$\exists y_0 = y, y_1, \ldots, y_n = y' \ \forall i < n \ (x, y_i, y_{i+1}) \in U^*).$$

Informally E is a direct sum of transitive closures of U_x^* for all $x \in \omega^\omega$. It is clear that E is a Σ_1^1 equivalence relation. Suppose F is any Σ_1^1 equivalence relation on ω^ω. Then $F \subseteq (\omega^\omega)^2$ and so there is $x \in \omega^\omega$ such that $F = U_x$. We have that $U_x^* = U_x = F$ is a transitive relation. Define $f : \omega^\omega \to \omega^\omega$ by letting $f(y) = \langle x, y \rangle$. Then for any $y, y' \in \omega^\omega$, yFy' iff $(y, y') \in U_x^* = F$ iff $(x, y, y') \in U^*$ iff $\langle x, y \rangle E \langle x, y' \rangle$. This shows that, in fact, $F \sqsubseteq_c E$. □

Exercise 9.2.1 Show that the reduction Φ defined in this section is $\sigma(\Sigma_1^1)$-measurable.

Exercise 9.2.2 Show that E_{Ulm} is a Σ_1^1 equivalence relation.

Exercise 9.2.3 Show that there is a universal Σ_1^1 equivalence relation.

9.3 Actions of standard Borel groups

Orbit equivalence relations of Polish group actions form an important class of analytic equivalence relations. More generally, one can consider actions in which the acting group is standard Borel and still get that the resulting orbit equivalence relation is analytic.

Definition 9.3.1
A **standard Borel group** is a group G with a standard Borel structure $\mathcal{B}$ on G such that the group operations are Borel with respect to $\mathcal{B}$.

Borel subgroups of Polish groups are standard Borel groups with their induced Borel structure. In particular, all Polish groups are standard Borel groups with their Borel structure. The converse, as we will see below, is not true.

Definition 9.3.2
A standard Borel group G is **Polishable** if there is a Polish topology τ inducing the Borel structure on G so that (G, τ) is a topological group, that is, the group operations are continuous with respect to τ.

Recall that in Definition 8.2.3 we defined Polishable subgroups of Polish groups. The above definition is a generalization of Definition 8.2.3. By Theorem 2.3.3, and more explicitly by Exercise 2.3.7, the topological group topology witnessing the Polishability of a standard Borel group is unique.

Lemma 9.3.3
ℓ_∞ is a standard Borel group that is not Polishable.

Proof. ℓ_∞ is an F_σ subgroup of the Polish group $\mathbb{R}^\omega$, hence is a standard Borel group. Assume τ is a Polish group topology on ℓ_∞ witnessing its Polishability. For each $N \in \omega$ define

$$A_N = \left\{ (x_n) \in \mathbb{R}^\omega \ : \ \|(x_n)\|_\infty = \sup_n |x_n| \le N \right\}.$$

Then A_N is a compact, hence closed subset of $\mathbb{R}^\omega$ and $\ell_\infty = \bigcup_N A_N$. It follows that A_N is a Borel subset of (ℓ_∞, τ), and in particular has the Baire property. By the Baire category theorem there is $N \in \omega$ such that A_N is nonmeager in (ℓ_∞, τ). By Theorem 2.3.2 $A_N - A_N$ contains a τ-open nbhd of the identity element of ℓ_∞. Note that $A_N - A_N \subseteq A_{2N}$, and from the separability of τ there is a countable set $C \subseteq \ell_\infty$ such that

$$\ell_\infty = \bigcup_{x \in C} (x + A_{2N}).$$

Let $C = \{x^n \ : \ n \in \omega\}$. We construct a $y \in A_{5N}$ such that $y \notin x^n + A_{2N}$ for any $n \in \omega$, arriving at a contradiction. The construction of $y = (y_n)$ is by diagonalization:

$$y_n = \begin{cases} 5N, & \text{if } |(x^n)_n| \le 2N, \\ 0, & \text{if } |(x^n)_n| > 2N. \end{cases}$$

It is easy to see that $|y_n - (x^n)_n| > 2N$, hence $y \notin x^n + A_{2N}$ for any $n \in \omega$. $\Box$

There are many examples of non-Polishable standard Borel groups. Another such example can be found in Exercise 9.3.1. Here we mention a third one which is also useful in our discussion later this section.

Recall from Section 2.6 that for any metric space (X, d) the Graev metric δ can be defined on the free group $\mathbb{F}(X)$ making it a topological group. Moreover if (X, d) is separable then $\mathbb{F}(X)$ is separable and its completion $\overline{\mathbb{F}}(X)$ becomes a Polish group. If in addition (X, d) is compact, then by Exercise 2.6.3 $\mathbb{F}(X)$ becomes a K_σ subset of $\overline{\mathbb{F}}(X)$. We summarize these simple observations in the following lemma.

Lemma 9.3.4
If X is a compact Polish space then there is a Borel structure on $\mathbb{F}(X)$ making it a standard Borel group.

Proof. Let d be any compatible metric on X and let δ be the Graev metric on $\mathbb{F}(X)$. Then $\mathbb{F}(X)$ becomes a Borel subgroup of the Polish group $\overline{\mathbb{F}}(X)$ and therefore a standard Borel group. $\Box$

If X is not finite then it has an accumulation point, and from the definition of the Graev metric we can see that δ on $\mathbb{F}(X)$ is not complete. In fact by a similar argument as in the proof of Lemma 9.3.3 one can show that $\mathbb{F}(X)$ is not Polishable if X is any uncountable compact Polish space. We leave the details of these proofs to the reader (Exercises 9.3.2 and 9.3.3).

We now turn to actions of standard Borel groups.

Definition 9.3.5
If G is a standard Borel group, X a standard Borel space and $a : G \times X \to X$ a Borel action, then we call X a **standard Borel G-space**.

If G is a standard Borel group and X a standard Borel G-space, then the orbit equivalence relation E_G^X is obviously Σ_1^1. The following result of Shelah shows that every Σ_1^1 equivalence relation arises this way.

Theorem 9.3.6 (Shelah)
Let X be a standard Borel space and E a Σ_1^1 equivalence relation on X. Then there is a standard Borel group G and a Borel action of G on X such that $E = E_G^X$.

Proof. Let τ be a Polish topology on X giving rise to the standard Borel structure. Since E is Σ_1^1 as a subset of $(X, \tau)^2$, there is a closed set $F \subseteq (X, \tau)^2 \times \omega^\omega$ such that $xEy \iff \exists z \in \omega^\omega \ (x, y, z) \in F$. As a closed subset of $(X, \tau)^2 \times \omega^\omega$, F is a Polish space. For each $(x, y, z) \in F$ we define a Borel isomorphism $\pi_{(x,y,z)} : X \to X$ by

$$\pi_{(x,y,z)}(w) = \begin{cases} y, & \text{if } w = x, \\ x, & \text{if } w = y, \\ u, & \text{otherwise.} \end{cases}$$

Now let Y be a compact Polish space with the same cardinality as F. For example, if F is uncountable then let $Y = 2^\omega$; if F is countable then let Y be a closed countable subset of 2^ω with $|F| = |Y|$. Then there is a Borel isomorphism φ between Y and F. We can therefore associate for any $a \in Y$ a Borel isomorphism $\pi_{\varphi(a)}$.

Let $G = \mathbb{F}(Y)$ be the standard Borel group as given in Lemma 9.3.4. Then there is a natural action of G on X induced by

$$a \cdot w = \pi_{\varphi(a)}(w)$$

for $a \in Y$ and $w \in X$. It is straightforward to check that this action is Borel and that $E = E_G^X$. $\quad\square$

Exercise 9.3.1 Let $\mathbb{T}$ be the multiplicative group of the unit circle $\{e^{i\theta} : 0 \leq \theta < 2\pi\}$. Let $H = \mathbb{T}^\omega$ and

$$G = \{(x_n) \in H : \exists m \ \forall n \geq m \ (x_n = 1)\}.$$

Show that

(i) $\mathbb{T}$ and H are compact Polish groups;

(ii) G is an F_σ subgroup of the Polish group H, hence is a standard Borel group;

(iii) G is not Polishable;

(iv) E_1 is Borel reducible to the coset equivalence relation H/G.

Exercise 9.3.2 Let X be a compact Polish space, d a compatible metric on X, and x an accumulation point in X. Show that there is a sequence (y_n) in X with $y_n \to x$ as $n \to \infty$ such that the following sequence (u_n) in $\mathbb{F}(X)$ is Cauchy in the Graev metric δ but does not converge in $\mathbb{F}(X)$:

$$u_0 = e, \quad u_{n+1} = u_n y_n x^{-1}.$$

Thus δ is not complete on $\mathbb{F}(X)$.

Exercise 9.3.3 Show that for any uncountable compact Polish space X the standard Borel group $\mathbb{F}(X)$ is not Polishable.

9.4 Wild Polish groups

For a general $\mathbf{\Sigma}_1^1$ equivalence relation the very first attempt to classify it is to decide whether it is Borel or non-Borel. In practice a good place to start is to examine its equivalence classes. Of course, if there is one non-Borel class then we conclude that the equivalence relation is non-Borel. However, for an orbit equivalence relation induced by a Polish group action this strategy will not work since every one of its orbits is Borel. Thus it is somewhat more challenging to decide if an orbit equivalence relation, being $\mathbf{\Sigma}_1^1$ in general, is Borel or not. In this section we investigate some non-Borel orbit equivalence relations.

Theorem 8.2.2 provides a criterion: an orbit equivalence relation on a Polish space is non-Borel iff it has orbits of arbitrarily high complexity. This is useful in some context. For instance, if a locally compact Polish group acts continuously on a Polish space, then every orbit is F_σ and hence the equivalence relation is Borel (in fact F_σ). However, it is not clear how to directly apply this theorem when the acting group is not locally compact. For instance, one of the simplest nonlocally compact Polish groups is $\mathbb{Z}^\omega$. More work is needed to decide if $\mathbb{Z}^\omega$ can induce a non-Borel orbit equivalence relation. In fact, it was a documented question asked by Sami [134] whether any abelian group can induce a non-Borel orbit equivalence relation. This was answered positively by Solecki [140].

We introduce some terminology.

Definition 9.4.1

A Polish group G is **tame** if every G-orbit equivalence relation is Borel; it is **wild** if it is not tame.

In this section we present a proof by Solecki that $\mathbb{Z}^\omega$ is wild. In fact, Solecki has considered groups of the form $\prod_n G_n$, where G_n are abelian, and has completely determined which of them are tame.

Let X be the set of all trees on $\mathbb{Z}$, that is, $T \in X$ iff $T \subseteq \mathbb{Z}^{<\omega}$ is closed under taking initial segments. X is a Borel subset of $2^{\mathbb{Z}^{<\omega}}$, hence is itself a standard Borel space. We define an action of $\mathbb{Z}^\omega$ on X as follows. For $T \in X$ and $\sigma \in \mathbb{Z}^\omega$, let

$$\sigma \cdot T = \bigcup_{n \in \omega} (\sigma \restriction n) + (T \cap \mathbb{Z}^n).$$

Clearly $\sigma \cdot T \in X$ and the action is Borel. Throughout this section we denote the orbit equivalence relation by $E(\mathbb{Z}^\omega)$, or E for simplicity. Our objective is to show that E is non-Borel.

Recall that a tree is well-founded if it does not have any infinite branch, and is ill-founded otherwise. If $T_1, T_2 \in X$, we let

$$\Phi(T_1, T_2) = \{t \in \mathbb{Z}^{<\omega} : \forall m \leq \mathrm{lh}(t)\ (t \restriction m) + (T_1 \cap \mathbb{Z}^m) = T_2 \cap \mathbb{Z}^m\}.$$

Then it is easy to see that $\Phi(T_1, T_2) \in X$, and $T_1 E T_2$ iff $[\Phi(T_1, T_2)] \neq \emptyset$ iff $\Phi(T_1, T_2)$ is ill-founded. Note that for any $n \in \omega$, $\Phi(T_1, T_2) \cap \mathbb{Z}^n$ is a coset of a subgroup of $\mathbb{Z}^n$. This motivates the following concepts.

Definition 9.4.2

Let $T \subseteq \mathbb{Z}^{<\omega}$ be a tree on $\mathbb{Z}$. T is a **coset tree** if for all $n \in \omega$, $T \cap \mathbb{Z}^n$ is a coset of a subgroup of $\mathbb{Z}^n$. T is a **group tree** if for all $n \in \omega$, $T \cap \mathbb{Z}^n$ is a subgroup of $\mathbb{Z}^n$.

Note that in general if G is a group and $C \subseteq G$ is nonempty, then C is a left-coset of a subgroup of G iff for all $g_1, g_2, g_3 \in C$, $g_1 g_2^{-1} g_3 \in C$. The same condition also characterizes the right-cosets in G. Thus even without commutativity we may address them simply as cosets. If C is a coset in G and $g \in C$, then $g^{-1}C$ is a subgroup of G. In fact $g^{-1}C$ is independent of the choice of $g \in C$, and it is the unique subgroup of G for which C is a left-coset. For an arbitrary $C \subseteq G$, we define

$$\gamma C = \begin{cases} g^{-1}C, & \text{if } C \text{ is a nonempty coset in } G \text{ and } g \in C, \\ \emptyset, & \text{otherwise.} \end{cases}$$

Then for any $C \subseteq G$, γC is a subgroup of G.

Now if T is a coset tree on $\mathbb{Z}$, we let

$$\gamma T = \bigcup_{n \in \omega} \gamma(T \cap \mathbb{Z}^n).$$

Then γT is a group tree on $\mathbb{Z}$. Moreover, $\Phi(\gamma T, T) = T$.

When a tree T on $\mathbb{Z}$ is well-founded we may speak of its rank, denoted by $\operatorname{rk}(T)$, which is a countable ordinal. Here in order to study ill-founded trees, we define the following more general notion. For any tree T on $\mathbb{Z}$, let $T' = \{t \in T : \exists s \in T \; t \subsetneq s\}$. Then $T' \subseteq T$. By transfinite induction define T^α for $\alpha < \omega_1$ as follows: $T^0 = T$, $T^{\alpha+1} = (T^\alpha)'$, and $T^\lambda = \bigcap_{\alpha<\lambda} T^\alpha$, if λ is limit. Since T is countable there is a countable ordinal $\beta < \omega_1$ such that $T^\beta = T^{\beta+1}$. Define the **height** of T, denoted $\operatorname{ht}(T)$, to be the least ordinal β such that $T^\beta = T^{\beta+1}$. For a well-founded tree T the height coincides with the rank of T, and if $\beta = \operatorname{ht}(T) = \operatorname{rk}(T)$, then $T^\beta = \emptyset$.

Putting all these concepts together we can now reduce the problem to the construction of well-founded coset trees on $\mathbb{Z}$ of arbitrarily high rank.

Lemma 9.4.3
If there are well-founded coset trees on $\mathbb{Z}$ of arbitrarily high rank then $E(\mathbb{Z}^\omega)$ is non-Borel.

Proof. Assume toward a contradiction that E is Borel. Note that Φ is a Borel map from X^2 into X, and

$$(T_1, T_2) \notin E \iff \Phi(T_1, T_2) \text{ is well-founded.}$$

Thus $\Phi(X^2 - E)$ is a Σ_1^1 set of well-founded trees, and by the boundedness principle for Σ_1^1 sets of well-founded trees (Theorem 1.6.11) there is $\alpha < \omega_1$ such that $\operatorname{rk}(\Phi(T_1, T_2)) < \alpha$ for all $(T_1, T_2) \notin E$. Now if T is any well-founded coset tree with $\operatorname{rk}(T) \geq \omega$, γT is ill-founded, and since $\Phi(\gamma T, T) = T$ is well-founded, $(\gamma T, T) \notin E$. It follows that $\operatorname{rk}(T) = \operatorname{rk}(\Phi(\gamma T, T)) < \alpha$. This shows that all well-founded coset trees on $\mathbb{Z}$ have rank $< \max\{\omega, \alpha\}$, contradicting our assumption. ⬚

Next we further reduce the problem to the construction of group trees on $\mathbb{Z}$ of arbitrarily high height. Note that there are no well-founded group trees of rank $\geq \omega$, since the element $(0, 0, 0, \ldots)$ is always a branch of such a tree. So we do need to consider the concept of height when dealing with complex group trees. We also need to examine closely the tree nodes in the transfinite pruning process. For this define for any $t \in T$ the **rank** of t in T by

$$\operatorname{rk}_T(t) = \begin{cases} \text{the least } \beta \text{ such that } t \notin T^\beta, & \text{if such } \beta \text{ exists,} \\ \omega_1, & \text{otherwise.} \end{cases}$$

For any $T \in X$, if $\alpha \leq \operatorname{ht}(T)$ is a successor ordinal, then there is $t \in T$ with $\operatorname{rk}_T(t) = \alpha$.

Lemma 9.4.4
Let T be a group tree with $\operatorname{ht}(T) > \omega$. Then there exist $t_n \in \mathbb{Z}^n$ such that for all $n \in \omega$, $(t_{n+1} \upharpoonright n) - t_n \in T$, and $\bigcup_n t_n + (T \cap \mathbb{Z}^n)$ is a well-founded coset tree.

Proof. First we note a general fact about countable groups. If G is a countable group and (G_n) is a strictly decreasing sequence of subgroups of G, then there exist $g_n \in G$ such that for all $n \in \omega$, $g_{n+1} \in g_n G_n$, and $\bigcap_n g_n G_n = \emptyset$. To see this, enumerate G as $\{h_n : n \in \omega\}$ and inductively define $g_{n+1} \in G$ such that $g_{n+1} \in g_n G_n$ and $h_n \notin g_{n+1} G_{n+1}$.

Now let T be a group tree with $\mathrm{ht}(T) > \omega$. Let $s_0 \in T$ be such that $\mathrm{rk}_T(s_0) = \omega + 1$. Let $k_0 = \mathrm{lh}(s_0) + 1$. Then there exists a sequence $(u_n)_{n \in \omega}$ such that for all $n \in \omega$, $\mathrm{lh}(u_n) = k_0$, $s \subseteq u_n$, and $\mathrm{rk}_T(u_n) < \mathrm{rk}_T(u_{n+1}) < \omega$. Let $l_n = k_0 + \mathrm{rk}_T(u_n) + 1$ and π_n be the projection of $\mathbb{Z}^{l_n}$ onto the first k_0 coordinates. Then π_n is a group homomorphism from $\mathbb{Z}^{l_n}$ to $\mathbb{Z}^{k_0}$, and therefore $\pi_n(T \cap \mathbb{Z}^{l_n})$ is a subgroup of $T \cap \mathbb{Z}^{k_0}$. Note that $u_n \notin \pi_n(T \cap \mathbb{Z}^{l_n})$ but $u_n \in \pi_m(T \cap \mathbb{Z}^{l_m})$ for all $m > n$. Thus if we let $G_n = \pi_n(T \cap \mathbb{Z}^{l_n})$, then (G_n) is a strictly decreasing sequence of subgroups of $T \cap \mathbb{Z}^{k_0}$. By the general fact above we may let $v_n \in T \cap \mathbb{Z}^{k_0}$ be such that $v_{n+1} \in v_n + G_n$ and $\bigcap_n (v_n + G_n) = \emptyset$.

Next we inductively define a sequence (w_n) such that $w_n \in \mathbb{Z}^{l_n}$, $\pi_n(w_n) = v_n$, and $(w_{n+1} \restriction l_n) - w_n \in T$. To begin with, let $w_0 \in \mathbb{Z}^{l_0}$ be any extension of v_0. In general, assume w_n has been defined to satisfy the inductive hypothesis. Since $v_{n+1} \in v_n + \pi_n(T \cap \mathbb{Z}^{l_n})$, there is $w \in T \cap \mathbb{Z}^{l_n}$ such that $\pi_n(w) = v_{n+1} - v_n = v_{n+1} - \pi_n(w_n)$. Thus $v_{n+1} = \pi_n(w + w_n)$. Let $w_{n+1} \in \mathbb{Z}^{l_{n+1}}$ be any extension of $w + w_n$. Then $\pi_{n+1}(w_{n+1}) = \pi_n(w + w_n) = v_{n+1}$ and $(w_{n+1} \restriction l_n) - w_n = w \in T$.

Finally we define $t_n \in \mathbb{Z}^n$. For any $n \in \omega$ let $m \in \omega$ be the least such that $n \leq l_m$. Then let $t_n = w_m \restriction n$. We verify that the t_n have the required properties. To see that $(t_{n+1} \restriction n) - t_n \in T$, it is without loss of generality to assume $n = l_m$ for some $m \in \omega$. Then $t_n = w_m$ but $t_{n+1} = w_{m+1} \restriction (n+1)$. Therefore, $(t_{n+1} \restriction n) - t_n = (w_{m+1} \restriction l_m) - w_m \in T$. Now let $S = \bigcup_n t_n + (T \cap \mathbb{Z}^\omega)$. It follows that S is a coset tree. It remains to verify that S is well-founded.

We claim that for any $s \in S \cap \mathbb{Z}^{k_0}$, $\mathrm{rk}_S(s) < \omega$. For this let $s \in S \cap \mathbb{Z}^{k_0}$. Since $\bigcap_n v_n + \pi_n(T \cap \mathbb{Z}^{l_n}) = \emptyset$, we have that $s \notin v_n + \pi_n(T \cap \mathbb{Z}^{l_n})$ for some $n \in \omega$. Assume $s' \in S \cap \mathbb{Z}^{l_n}$ with $s \subseteq s'$. Then

$$s = \pi_n(s') \in \pi_n(t_{l_n}) + \pi_n(T \cap \mathbb{Z}^{l_n}) = \pi_n(w_n) + \pi_n(T \cap \mathbb{Z}^{l_n}) = v_n + \pi_n(T \cap \mathbb{Z}^{l_n}),$$

a contradiction. This shows that $\mathrm{rk}_S(s) < l_n$, and therefore $\mathrm{rk}_S(s) < \omega$. It follows that $\mathrm{rk}(S) < \omega + k_0$, and hence S is well-founded. $\qquad\square$

Lemma 9.4.5

Let T be a group tree, $\alpha < \omega_1$, and $t_n \in \mathbb{Z}^n$ satisfy $(t_{n+1} \restriction n) - t_n \in T^\alpha$. Let $S = \bigcup_n t_n + (T \cap \mathbb{Z}^n)$. Then S is a coset tree, and for all $\beta \leq \alpha$, $S^\beta = \bigcup_n t_n + (T^\alpha \cap \mathbb{Z}^n)$.

Proof. It is easy to see that if T is a coset (group) tree then T' is a coset (respectively group) tree. It follows that for any $\beta < \omega_1$, T^β is a coset (group) tree if T is. We only check that $S' = \bigcup_n t_n + (T' \cap \mathbb{Z}^n)$. The lemma then follows by an easy induction.

Let $s \in S' \cap \mathbb{Z}^n$. Then there are $t \in T \cap \mathbb{Z}^n$, $s' \in S \cap \mathbb{Z}^{n+1}$, and $t' \in T \cap \mathbb{Z}^{n+1}$ such that $s = t_n + t$, $s \subseteq s'$, and $s' = t_{n+1} + t'$. Since $(t_{n+1} \restriction n) - t_n \in T'$ there is also $u \in T \cap \mathbb{Z}^{n+1}$ with $(t_{n+1} \restriction n) \subseteq t_n \subseteq u$. Then

$$t = s - t_n = (s' \restriction n) - t_n = (t_{n+1} \restriction n) + (t' \restriction n) - t_n = u \restriction n + t' \restriction n = (u + t') \restriction n.$$

Since T is a group tree $u + t' \in T$, and so $S' \subseteq \bigcup_n t_n + (T' \cap \mathbb{Z}^n)$. The converse is obvious. □

Lemma 9.4.6

For any $\alpha < \omega_1$ if there is a group tree of height $> \alpha + \omega$ then there is a well-founded coset tree S of rank $\geq \alpha$.

Proof. Let T be a group tree of height $> \alpha + \omega$. Then $T^\alpha \neq \emptyset$. Apply Lemma 9.4.4 to obtain $t_n \in T^\alpha \cap \mathbb{Z}^n$ for $n \in \omega$ such that $(t_{n+1} \restriction n) - t_n \in T^\alpha$ and $\bigcup_n t_n + (T^\alpha \cap \mathbb{Z}^\omega)$ is well-founded. Let $S = \bigcup_n t_n + (T \cap \mathbb{Z}^n)$. Then by Lemma 9.4.5 S is a coset tree with $S^\alpha \neq \emptyset$ well-founded. Thus S is a well-founded coset tree with $\mathrm{rk}(S) \geq \alpha$. □

Theorem 9.4.7 (Makkai–Solecki)

There are group trees on $\mathbb{Z}$ of arbitrarily high height.

Proof. Let $\phi : \mathbb{Z}^2 \to \mathbb{Z}$ be the homomorphism $\phi(a, b) = a + b$. Let $(G_n^0), (G_n^1)$ be strictly decreasing sequences of subgroups of $\mathbb{Z}$ such that $\bigcap_n G_n^0 = \bigcap_n G_n^1 = \{0\}$ and $\phi(G_n^0 \times G_n^1) = \mathbb{Z}$ for all $n \in \omega$. For example, we may take $G_n^0 = 2^n \mathbb{Z}$ and $G_n^1 = 3^n \mathbb{Z}$.

By transfinite induction on $\alpha < \omega_1$ we define a group tree T_α such that

(a) if $\alpha = \beta + 1$ is a successor, then $(T_\alpha)^\beta \cap \mathbb{Z} = \mathbb{Z}$ and $(T_\alpha)_a$ is well-founded if $a \neq 0$;

(b) if α is a limit, then for any $\beta < \alpha$ there is $n \in \omega$ such that

$$G_n^0 \times G_n^1 \subseteq (T_\alpha)^\beta \cap \mathbb{Z}^2,$$

and if $t \in \mathbb{Z}^2$ and $t \neq 0^\smallfrown 0$, then $(T_\alpha)_t$ is well-founded.

Granting this construction it is clear that $\mathrm{ht}(T_\alpha) \geq \alpha$.

Let $T_0 = \{0^n : n \in \omega\}$ and $T_1 = T_0 \cup \mathbb{Z}$. If $\alpha = \beta + 1$ and β is a successor, let

$$T_\alpha = \{\emptyset\} \cup \mathbb{Z} \cup \{t(0)^\smallfrown t : t \in T_\beta, \mathrm{lh}(t) \geq 1\}.$$

If $\alpha = \beta + 1$ and β is a limit, let

$$T_\alpha = \text{the tree generated by } \{\phi(t(0), t(1))^\smallfrown t : t \in T_\beta, \mathrm{lh}(t) \geq 2\}.$$

With these definitions (a) is satisfied.

Suppose α is a limit. To satisfy (b) we construct two group trees S_0 and S_1 and an increasing sequence of successor ordinals $\beta_n \to \alpha$ such that

(i) $G_n^0 \subseteq (S_0)^{\beta_n}$ and $G_n^1 \subseteq (S_1)^{\beta_n}$ for all $n \in \omega$;

(ii) $(S_0)_a$ and $(S_1)_a$ are well-founded if $a \neq 0$.

If S_0 and S_1 are defined as above, then let

$$T_\alpha = \{t : t \upharpoonright \{2k : k \in \omega\} \in S_0 \text{ and } t \upharpoonright \{2k+1 : k \in \omega\} \in S_1 \}.$$

Then it is easy to see that (b) is satisfied. Fix any increasing sequence of successor ordinals $\beta_n \to \alpha$. Apparently the constructions of S_0 and S_1 are parallel. We only construct S_0 and then the construction of S_1 will be similar.

Note that T_{β_n} have already been constructed to satisfy (a). S_0 will be an amalgamation of the T_{β_n} as follows. Let $\{I_n\}_{n\in\omega}$ be a partition of $\omega - \{0\}$ into infinitely many infinite subsets. For each $n \in \omega$ define a tree U_n by

$$s \in U_n \iff s = \emptyset \lor [s(0) \in G_n^0 \land s \upharpoonright I_n \in T_{\beta_n} \land$$
$$\forall i < \mathrm{lh}(s)(i = \min I_n \Rightarrow s(i) = s(0) \land i \notin I_n \Rightarrow s(i) = 0)].$$

Then U_n is a group tree. Let

$$S_0 = \bigcup_k \langle \bigcup_n U_n \cap \mathbb{Z}^k \rangle.$$

Then S_0 is a group tree by definition. To see that $G_n^0 \subseteq (S_0)^{\beta_n}$ let $a \in G_n^0$. Let $s \in \mathbb{Z}^{\min I_n + 1}$ be such that $s(0) = s(\min I_n) = a$ and $s(i) = 0$ for all $0 < i < \min I_n$. Then $s \in U_n \subseteq S_0$. Since $(T_{\beta_n})^{\beta_n - 1} \cap \mathbb{Z} = \mathbb{Z}$, $s \in (U_n)^{\beta_n - 1} \subseteq (S_0)^{\beta_n - 1}$. Thus $a \in (S_0)^{\beta_n}$ and (i) is satisfied. To show (ii) let $a \neq 0$ and toward a contradiction let $\sigma \in \mathbb{Z}^\omega$ be an infinite branch of $(S_0)_a$. Then $\sigma(0) = a$. Let $n \in \omega$ be large enough such that $a \notin G_n^0$. We have that $\sigma(\min I_i) = 0$ for all $i < n$. Otherwise if $\sigma(\min I_i) \neq 0$ for $i < n$, then by our construction of S_0, $\sigma \upharpoonright I_i$ is an infinite branch of T_{β_i} with $(\sigma \upharpoonright I_i)(0) = \sigma(\min I_i) \neq 0$, violating our inductive assumption (a) for T_{β_i}. Let $k > \max\{\min I_i : i < n\}$ and $s = \sigma \upharpoonright k \in S_0$. Then by the definition of S_0 we have that $s \in \langle \bigcup_m U_m \cap \mathbb{Z}^k \rangle$. Since U_m are group trees we may assume $s = s_{m_1} + s_{m_2} + \cdots + s_{m_l} + s_{m_{l+1}} + \cdots + s_{m_p}$ with $m_1 < m_2 < m_l < n \leq m_{l+1} < \cdots < m_p$, where $s_{m_j} \in U_{m_j}$ for $j = 1, \ldots, p$. It follows that

$$a = s_{m_1}(\min I_{m_1}) + \cdots + s_{m_l}(\min I_{m_l}) + s_{m_{l+1}}(\min I_{m_{l+1}}) + \cdots + s_{m_p}(\min I_{m_p}).$$

From the above argument we get that $s_{m_j}(\min I_{m_j}) = 0$ for $j = 1, \ldots, l$. However, $s_{m_j}(\min I_{m_j}) \in G_n^0$ for $j = l+1, \ldots, p$. Hence $a \in G_n^0$, a contradiction. This shows that S_0 has the required properties. □

Corollary 9.4.8 (Solecki)
$E(\mathbb{Z}^\omega)$ is non-Borel and $\mathbb{Z}^\omega$ is wild.

Exercise 9.4.1 Let G be a group and $C \subseteq G$ is nonempty. Show that the following are equivalent:

(i) C is a left-coset of some subgroup of G.

(ii) C is a right-coset of some subgroup of G.

(iii) For all $g_1, g_2, g_3 \in C$, $g_1 g_2^{-1} g_3 \in C$.

Exercise 9.4.2 Show that if T is a coset tree then $\Phi(\gamma T, T) = T$. Give a counterexample where T is not a coset tree.

Exercise 9.4.3 Show that if T is a coset tree then $(\gamma T)' = \gamma(T')$.

Exercise 9.4.4 Let T be a group tree and $t_n \in \mathbb{Z}^n$ such that $(t_{n+1} \restriction n) - t_n \in T \cap \mathbb{Z}^n$. Show that $\bigcup_n t_n + (T \cap \mathbb{Z}^n)$ is a coset tree.

9.5 The topological Vaught conjecture

It is not known if the Silver dichotomy holds for orbit equivalence relations of Polish group actions. This is one of the central open problems in the entire invariant descriptive set theory.

Proposition 9.5.1
Let G be a Polish group. Then the following statements are equivalent:

(a) For any Polish G-space X, either there are countably many orbits or there are perfectly many orbits.

(b) For any Polish G-space X and invariant Borel set $B \subseteq X$, either B contains countably many orbits or B contains perfectly many orbits.

(c) For any Borel G-space X, either there are countably many orbits or there are perfectly many orbits.

Proof. The implications (c)$\Rightarrow$(b)$\Rightarrow$(a) are obvious. (a)$\Rightarrow$(c) follows from Theorem 4.4.6. ☐

For any Polish group G, the **topological Vaught conjecture for** G refers to any of the above equivalent statements. We denote it as $\mathrm{TVC}(G)$. Then the **topological Vaught conjecture** is the statement

TVC(G) holds for any Polish group G.

The topological Vaught conjecture was first formulated by Miller. But its counterpart in countable model theory goes back to Vaught. We will elaborate on the connections in Chapter 11.

The topological Vaught conjecture for any tame Polish group is true by the Silver dichotomy. In particular, TVC(G) holds for any locally compact Polish group G. Sami [134] proved the topological Vaught conjecture for all abelian Polish groups. Hjorth and Solecki [83] proved it for all nilpotent Polish groups. Solecki [83] proved it for all Polish groups with two-sided invariant metrics. The strongest result up to date is the following theorem of Becker.

Theorem 9.5.2 (Becker)
TVC(G) *holds for all* cli *Polish groups, that is, Polish groups with complete left-invariant metrics.*

Proof.   This is a corollary of Theorem 6.5.1.   ⬚

By Theorem 2.2.11 the class of cli Polish groups is closed under closed subgroups, quotient groups, and group extensions. Also it is easy to see that it is closed under taking countable products. We next consider the abstract closure properties of the class of all Polish groups satisfying the topological Vaught conjecture.

Theorem 9.5.3
Let $\mathcal{C}$ be the class of Polish groups G for which TVC(G) *holds. Then*

(i) *if $G \in \mathcal{C}$ and H is a closed subgroup of G, then $H \in \mathcal{C}$;*

(ii) *if $G \in \mathcal{C}$ and H is a closed normal subgroup of G, then $G/H \in \mathcal{C}$;*

(iii) *if $G \in \mathcal{C}$ and G is a closed subgroup of H where H/G is countable, then $H \in \mathcal{C}$.*

Proof.   For (i) we let X be a Polish H-space. By Theorem 3.5.2 there is a Polish G-space Y such that X is a closed subset of Y and that every G-orbit contains a unique H-orbit. It follows that if Y has only countably many G-orbits then X has only countably many G-orbits. Assume Y has perfectly many G-orbits, and let $Z \subseteq Y$ be a perfect set of pairwise E_G^Y-inequivalent elements. Then the relation $R = E_G^Y \upharpoonright (Z \times X)$ is $\mathbf{\Sigma}_1^1$ and for any $y \in Z$ there is $x \in X$ such that $yE_G^Y x$ by Theorem 3.5.2. By the Jankov–von Neumann uniformization theorem R has a $\sigma(\mathbf{\Sigma}_1^1)$-measurable uniformization, that is, there is a $\sigma(\mathbf{\Sigma}_1^1)$-measurable function $f : Z \to X$ such that for any $y \in Z$, $yRf(y)$. Since $\sigma(\mathbf{\Sigma}_1^1)$-measurable functions are Baire measurable (by the remarks preceding Proposition 2.3.1), it follows that there is a comeager set $C \subseteq Z$ such that $f \upharpoonright C$ is continuous. Without loss of generality assume

C is G_δ, hence an uncountable Polish space. Then $f : C \to X$ is a continuous reduction from $\mathrm{id}(C)$ to E_H^X, since for any $y_1 \neq y_2 \in C$, $(y_1, y_2) \notin E_G^Y$, and thus $(f(y_1), f(y_2)) \notin E_G^X$ and therefore $(f(y_1), f(y_2)) \notin E_H^X$. This shows that X has perfectly many G-orbits.

(ii) is obvious since every Polish G/H-space is also a G-space.

For (iii) we consider a Polish H-space X. X is also a Polish G-space with the restricted action of G, and so either there are only countably many G-orbits or there are perfectly many G-orbits. The assumption implies that every H-orbit contains countably many G-orbits. Thus in the case there are only countably many G-orbits there are only countably many H-orbits as well. In the second case, assume $Y \subseteq X$ is a perfect subset of E_G^X-inequivalent elements. Then $E = E_G^X \restriction Y$ is a $\mathbf{\Sigma}_1^1$ equivalence relation on Y with every equivalence class countable. Thus E has the Baire property. By Theorem 5.3.1, and more explicitly Exercise 5.3.1, there are perfectly many E-equivalence classes in Y, which gives a perfect set of E_H^X-inequivalent elements in X. ☐

It follows that the topological Vaught conjecture is equivalent to $\mathrm{TVC}(G)$ for any universal Polish group G. By the results of Section 2.5 it is equivalent to $\mathrm{TVC}(\mathrm{Iso}(\mathbb{U}))$ or to $\mathrm{TVC}(H(\mathbb{I}^{\mathbb{N}}))$.

In the rest of this section we consider a more general notion.

Definition 9.5.4
Let X be a standard Borel space and E an equivalence relation on X. The **topological Vaught conjecture for** (X, E), denoted $\mathrm{TVC}(X, E)$, is the statement

> every invariant Borel subset of X either contains only countably many E-equivalence classes or perfectly many E-equivalence classes.

Recall from Section 5.2 that a faithful Borel reduction from an equivalence relation E on X to F on Y is a Borel reduction $f : X \to Y$ witnessing $E \leq_B F$ and such that for any invariant Borel subset $A \subseteq X$, $[f(A)]_F$ is Borel. The following proposition shows that TVC is closed under faithful Borel reductions.

Theorem 9.5.5
Let X, Y be standard Borel spaces and E, F be $\mathbf{\Sigma}_1^1$ equivalence relations on X, Y, respectively. If $E \leq_{fB} F$, then $\mathrm{TVC}(Y, F)$ implies $\mathrm{TVC}(X, E)$.

Proof. Without loss of generality assume that X and Y are Polish spaces. Let f be a faithful Borel reduction from E to F. Let $A \subseteq X$ be invariant Borel. Then $[f(A)]_F$ is invariant Borel. Suppose $\mathrm{TVC}(Y, F)$. If $[f(A)]_F$ contains only countably many F-classes, then A contains only countably many E-classes. Otherwise, suppose $[f(A)]_F$ contains perfectly many F-classes, and let $P \subseteq [f(A)]_F$ be a perfect subset of F-inequivalent elements. Consider the

set
$$D = \{(y, x) \in P \times X : f(x)Fy\}.$$

Then D is $\mathbf{\Sigma}_1^1$ and for all $y \in P$ there is $x \in X$ with $(y, x) \in D$. By the Jankov–von Neumann uniformization theorem D has a $\sigma(\mathbf{\Sigma}_1^1)$-measurable uniformization g. Thus $g : P \to X$ is Baire measurable and for $y_1 \neq y_2 \in P$, $(g(y_1), g(y_2)) \notin E$. Let now $C \subseteq P$ be a dense G_δ subset such that $g \restriction C$ is continuous. Then $g \restriction C : C \to X$ and $g(C)$ is a perfect subset of X with E-inequivalent elements. $\qquad\Box$

Thus to prove TVC(G) for a Polish group it suffices to consider a Polish G-space to which every other Polish G-space is faithfully Borel reducible.

Exercise 9.5.1 Show that if H is a countable group then TVC(G) holds iff TVC$(G \times H)$ holds.

For any standard Borel space X and equivalence relation E on X the **Glimm–Effros dichotomy for** (X, E), denoted GED(X, E), is the statement

for any invariant Borel subset $A \subseteq X$, either $E \restriction A$ is smooth or $E_0 \sqsubseteq_c E \restriction A$.

Exercise 9.5.2 Let X, Y be standard Borel spaces and E, F be $\mathbf{\Sigma}_1^1$ equivalence relations on X, Y, respectively. Show that if $E \leq_{fB} F$, then GED(Y, F) implies GED(X, E).

The following exercise problems outline a proof of the topological Vaught conjecture for abelian Polish groups by Sami. Assume G is an abelian Polish group and X a Polish G-space. Define an equivalence relation E on X by

$$xEy \iff G_x = G_y,$$

where $G_x = \{g \in G : g \cdot x = x\}$ is the stabilizer group of $x \in X$.

Exercise 9.5.3 Show that E is a $\mathbf{\Pi}_1^1$ equivalence relation and $E_G^X \subseteq E$. Deduce that if there are perfectly many E-equivalence classes then there are perfectly many G-orbits.

Exercise 9.5.4 Assume there are only countably many E-equivalence classes. Show that each E-equivalence class is Borel and that for each E-equivalence class C the relation $E_G^X \restriction C$ is Borel. (*Hint*: Use Theorem 8.2.1.)

Exercise 9.5.5 Show that either there are only countably many G-orbits in X or there are perfectly many G-orbits.

Chapter 10

Turbulent Actions of Polish Groups

In this chapter we turn back to Polish group actions and orbit equivalence relations. The Borel reducibility hierarchy of analytic equivalence relations has taken shape with the results we have established in the previous chapters. However, it is also noticeable that the results were obtained with a variety of machineries ranging from Baire category methods, measure theoretic methods, group theory, and descriptive set theory. There are still many pairs of equivalence relations we have introduced so far but have not mentioned their reducibility relation, either because it is still an open problem or because its known proof goes beyond the limitation of length or scope of this book. It is thus clear that a sweeping method to prove reducibility or, more challengingly, nonreducibility would be much desirable. The greatest success up to date is Hjorth's theory of turbulence, which we present in this chapter. It will turn out that orbit equivalence relations from turbulent actions are not reducible to any S_∞-orbit equivalence relation. This powerful theorem has many applications in the classification problems of mathematics, and its potential is still being discovered in current research.

The results in this entire chapter are due to Hjorth [70], unless otherwise indicated. Expositions of further results of the theory of turbulence can be found in References [70] and [101].

10.1 Homomorphisms and generic ergodicity

In this section we review the Baire category method for showing nonsmoothness of orbit equivalence relations and introduce some new concepts.

Let G be a Polish group and X a Polish G-space. Recall from Section 6.1 that E_G^X is generically ergodic iff every G-invariant Borel set is either meager or comeager. We showed in Proposition 6.1.9 that E_G^X is generically ergodic iff it has a dense orbit, and in Proposition 6.1.10 that if E_G^X is generically ergodic and has no comeager orbits, then E_G^X is not smooth. The proof of Proposition 6.1.10 (similar to that of Proposition 6.1.6) is by contradiction. We now turn this proof into a positive statement by working a little harder.

Proposition 10.1.1
Let G be a Polish group, X a Polish G-space, and $f : X \to 2^\omega$ a Baire measurable G-invariant map, that is, $f(x) = f(y)$ whenever $x E_G^X y$. Suppose E_G^X is generically ergodic. Then there is a comeager set $C \subseteq X$ on which f is constant.

Proof. First by Proposition 6.1.9 (iii) there is an invariant dense G_δ set $Y \subseteq X$ such that every orbit in Y is dense. Now Y is still a Polish G-space. Thus it suffices to find a comeager set $C \subseteq Y$ on which f is constant. Let $C \subseteq Y$ be comeager such that $f \upharpoonright C$ is continuous. By the continuity of the action the map $x \mapsto g \cdot x$ is a homeomorphism of X onto X for all $g \in G$. Thus we have that $\forall g \in G \ \forall^* x \in Y \ g \cdot x \in C$. Then by the Kuratowski–Ulam theorem, $\forall^* x \in Y \forall^* g \in G \ g \cdot x \in C$. Therefore we may find and fix an $x \in Y$ such that $\forall^* g \in G \ g \cdot x \in C$. We claim that $[x]_G \cap C$ is dense. For this let $y \in Y$ be arbitrary and $U \subseteq Y$ be open with $y \in U$. Since $[x]_G$ is dense the set $\{g \in G : g \cdot x \in U\}$ is nonempty. But it is also open, so there is $g \in G$ such that $g \cdot x \in U \cap C$. This shows that $[x]_G \cap C \cap U \neq \emptyset$.

Now $f \upharpoonright C$ takes the constant value $f(x)$. To see this, let $z \in C$. By the density of $[x]_G \cap C$ there is a sequence $(x_n)_{n \in \omega}$ in $[x]_G \cap C$ such that $x_n \to z$ as $n \to \infty$. However, f is continuous on C, and therefore $f(x_n) \to f(z)$ as $n \to \infty$. But $f(x_n) = f(x)$ by the invariance assumption, so $f(z) = f(x)$.   □

In view of its proof the hypotheses of the proposition can be weakened to assume a Baire measurable G-invariant map f defined on a comeager set.

Proposition 10.1.2
Let G be a Polish group, X a Polish G-space, $C \subseteq X$ a comeager set, and $f : C \to 2^\omega$ a Baire measurable G-invariant map. Suppose E_G^X is generically ergodic. Then there is a comeager set $C' \subseteq X$ on which f is constant.

Proof. As in the above proof there is an invariant comeager set C_0 such that every orbit in C_0 is dense. Now $C \cap C_0$ is still comeager, we may let $C' \subseteq C \cap C_0$ be comeager such that $f \upharpoonright C'$ is continuous. The same proof as above gives that f is constant on C'.   □

A number of things can be extracted from these propositions and their proofs. We start with some definitions.

Definition 10.1.3
Let X, Y be sets and E, F equivalence relations on X, Y, respectively.

(1) A map $f : X \to Y$ is a **homomorphism** from E to F if for any $x_1, x_2 \in X$, $x_1 E x_2 \Rightarrow f(x_1) F f(x_2)$.

(2) If X is a Polish space, then a **generic homomorphism** from E to F is a map f from a comeager subset C_0 of X to Y so that for some comeager set $C_1 \subseteq C_0$ f is a homomorphism from $E \restriction C_1$ to F.

Definition 10.1.4
Let X, Y be Polish spaces and E, F equivalence relations on X, Y, respectively. We say that E is **generically F-ergodic** if for every Baire measurable generic homomorphism f from E to F there is a comeager set $C \subseteq X$ such that $f(x_1) F f(x_2)$ for any $x_1, x_2 \in C$.

With the new terminology Proposition 10.1.2 can be succinctly stated as

> Any generically ergodic orbit equivalence relation is generically $\mathrm{id}(2^\omega)$-ergodic.

It turns out that these notions are actually equivalent.

Lemma 10.1.5
Let G be a Polish group and X a Polish G-space. Then E_G^X is generically ergodic iff it is generically $\mathrm{id}(2^\omega)$-ergodic.

Proof. Suppose E_G^X is generically $\mathrm{id}(2^\omega)$-ergodic. Let $\{U_n\}_{n \in \omega}$ be a countable base for X consisting of nonempty open sets. Define $f : X \to 2^\omega$ by

$$f(x)(n) = 1 \iff \exists g \in G \; g \cdot x \in U_n \iff x \in [U_n]_G.$$

Since $[U_n]_G$ is open for each $n \in \omega$, f is a Borel map. It is clear that f is a homomorphism from E_G^X to $\mathrm{id}(2^\omega)$. It follows that there is a comeager set $C \subseteq X$ on which f is constant. We claim that for any $x \in C$ and $n \in \omega$, $f(x)(n) = 1$. Assume not and let $x \in C$ and $n \in \omega$ be such that $f(x)(n) = 0$. Then $x \notin [U_n]_G$. Since C is comeager and $[U_n]_G$ is open, $C \cap [U_n]_G \neq \emptyset$. Let $y \in C \cap [U_n]_G$. Then $f(y)(n) = 1$. Now $x, y \in C$ and $f(x) \neq f(y)$, contradiction. Now by the claim for any $x \in C$ and $n \in \omega$, $[x]_G \cap U_n \neq \emptyset$ since $x \in [U_n]_G$. This means that for any $x \in C$, $[x]_G$ is dense. $\quad\square$

Hence the notion of generic F-ergodicity is a generalization of generic ergodicity. This is why we kept using the terminology even if generic ergodicity means no more than the existence of a dense orbit.

Of course generic ergodicity is ubiquitous, since every orbit is dense in its closure, which is an invariant subset and hence a Polish G-space. To infer nonsmoothness as we did in Proposition 6.1.10 we need nonexistence of comeager orbits. Under the assumption of generic ergodicity every invariant Borel set is either meager or comeager, and in particular so is every orbit. Thus the nonexistence of comeager orbits is equivalent to the assumption that every orbit is meager. To summarize, if a Polish G-space is generically ergodic and every orbit is meager, then E_G^X is nonsmooth. But we already proved more in Theorem 6.2.1. There it was shown that if a Polish G-space has only meager orbits and is generically ergodic, then $E_0 \sqsubseteq_c E_G^X$. This discussion motivates the following definition.

Definition 10.1.6
Let X be a Polish space and E an equivalence relation on X. We say that E is **properly generically ergodic** if every E-equivalence class is meager and E is generically ergodic. If F is an equivalence relation on a Polish space Y, then E is **properly generically F-ergodic** if every E-equivalence class is meager and E is generically F-ergodic.

With this terminology Theorem 6.2.1 can be stated as

> If an orbit equivalence relation E is properly generically ergodic then $E_0 \sqsubseteq_c E$; in particular, E is not smooth.

In general proper generic F-ergodicity will be the condition we are after because it is enough to guarantee nonreducibility.

Lemma 10.1.7
Let X, Y be Polish spaces and E, F be equivalence relations on X, Y, respectively. Suppose E is properly generically F-ergodic. Then there is no Baire measurable reduction from E to F. In particular $E \not\leq_B F$. Moreover, for any comeager subset $C \subseteq X$, $E{\restriction}C \not\leq_B F$.

Proof. If f is a Baire measurable reduction from $E \restriction C$ to F for some comeager $C \subseteq X$ then in particular it is a generic homomorphism from E to F, and by generic F-ergodicity there is a comeager set $C_0 \subseteq C$ such that $f \restriction C_0 \subseteq [y]_F$ for some $y \in Y$. But since f is a reduction on C this implies that $C_0 \subseteq [x]_E$ for some $x \in X$. This contradicts the assumption that every E-class is meager. □

In later sections of this chapter we will consider, as we did in Proposition 10.1.2, structural properties of the actions that can guarantee proper generic F-ergodicity. In doing this we will assume that the orbit equivalence relations we consider are all properly generically ergodic.

Exercise 10.1.1 Let X, Y be Polish spaces, E, F be equivalence relations on X, Y, respectively, and $f : C \to Y$ for some comeager subset $C \subseteq X$. Show that f is a generic homomorphism iff the set $\{(x, y) \in E : (f(x), f(y)) \notin F\}$ is meager in $C \times C$.

Exercise 10.1.2 Let E be an equivalence relation on a Polish space. Show that the following are equivalent:

(i) E is generically id(2)-ergodic;

(ii) E is generically id(ω)-ergodic;

(iii) Every E-invariant set with the Baire property is either meager or comeager.

Moreover, if $E = E_G$ where G is a group of homeomorphisms on X, then E is generically id(ω)-ergodic iff E is generically ergodic.

Exercise 10.1.3 Let G be a countably infinite group acting by shift on 2^G. Show that the orbit equivalence relation E_G is properly generically ergodic.

Exercise 10.1.4 Show that if E is (properly) generically F_1-ergodic and $F_2 \leq_B F_1$, then E is (properly) generically F_2-ergodic. In particular, if F has perfectly many equivalence classes then any (properly) generically F-ergodic equivalence relation is also (properly) generically ergodic.

10.2 Local orbits of Polish group actions

The key idea of the Hjorth theory of turbulence is to consider local orbits of Polish group actions. The analysis of the local orbits reveals a crucial difference between the actions of S_∞ and those inducing orbit equivalence relations generically ergodic with respect to certain S_∞-orbit equivalence relations.

Throughout this section we let G be a Polish group, X a Polish G-space, $\{U_n\}_{n\in\omega}$ a countable base for X, and $\{V_n\}_{n\in\omega}$ a countable open nbhd base for 1_G. We also let G_0 be a countable dense subgroup of G and enumerate it as $\{\gamma_n\}_{n\in\omega}$.

Definition 10.2.1
For $x \in X$, $U \subseteq X$ open with $x \in U$ and $V \subseteq G$ with $1_G \in V$, the **local U-V-orbit of x**, denoted $\mathcal{O}(x, U, V)$, is the set of $y \in U$ for which there exist $l \in \omega$, $x = x_0, x_1, \ldots, x_l = y \in U$, and $g_0, \ldots, g_{l-1} \in V$, such that $x_{i+1} = g_i \cdot x_i$ for all $i < l$.

Note that the local orbits are in general Σ_1^1. The following lemma collects some basic facts about local orbits. They are easy to check and we leave the proof to the reader.

Lemma 10.2.2
Let $U \subseteq X$ be open, $x, y, z \in U$, and $V \subseteq G$ open with $1_G \in V$. Then the following hold:

(i) $x \in \mathcal{O}(x, U, V)$.

(ii) If $U' \subseteq U$ and $1_G \in V' \subseteq V$ are open, then $\mathcal{O}(x, U', V') \subseteq \mathcal{O}(x, U, V)$.

(iii) If $y \in \mathcal{O}(x, U, V)$ and $z \in \mathcal{O}(y, U, V)$ then $z \in \mathcal{O}(x, U, V)$.

(iv) If $y \in \mathcal{O}(x, U, V)$ then $\mathcal{O}(y, U, V) \subseteq \mathcal{O}(x, U, V)$.

(v) If $V = V^{-1}$ then $y \in \mathcal{O}(x, U, V)$ iff $x \in \mathcal{O}(y, U, V)$.

(vi) If $y \in \mathcal{O}(x, U, V)$ then there is an open V' with $1_G \in V'$ such that $V' \cdot y \in \mathcal{O}(x, U, V)$.

Example 10.2.3

Let Y be a perfect Polish space. Consider the action of S_∞ on Y^ω by shift:

$$(g \cdot x)(n) = x(g^{-1}(n)).$$

Now let $U \subseteq Y^\omega$ be open and $x \in U$. Then there is a basic open set $U_0 \subseteq U$ such that $x \in U_0$. We may assume

$$U_0 = W_0 \times W_1 \times \cdots \times W_m \times Y^\omega$$

for open sets $W_0, W_1, \ldots, W_m \subseteq Y$. Now let $V \subseteq S_\infty$ be the basic open set $V = \{g \in S_\infty : g(i) = i \; \forall i \le m\}$. Then $V \cdot U_0 = U_0$ and

$$\mathcal{O}(x, U, V) \subseteq \{x(0)\} \times \cdots \times \{x(m)\} \times Y^\omega.$$

Note that $\mathcal{O}(x, U, V)$ is nowhere dense, and in fact so is $\overline{\mathcal{O}(x, U, V)}$.

The equivalence relation induced by the above action was considered in Exercises 8.3.4 and 8.3.5, where it was denoted as $\mathrm{id}(Y)^*$ and was shown to be Borel bireducible with $\mathrm{id}(Y)^+$, or $=^+$.

It will turn out that generically $=^+$-ergodic orbit equivalence relations behave differently: their local orbits will be somewhere dense. It is clear that a set is nowhere dense iff its closure is. So for our purposes we are going to investigate the closure of local orbits. Next we define codes for them and study their properties.

Notation 10.2.4

For a local orbit $\mathcal{O}(x, U, V)$, let $\varphi_0(x, U, V) = \{k \in \omega : U_k \cap \mathcal{O}(x, U, V) \neq \emptyset\}$.

Note that $\varphi_0(x, U, V)$ is an element of 2^ω coding the closure of $\mathcal{O}(x, U, V)$ in the sense that $\varphi_0(x, U, V) = \varphi_0(x', U', V')$ iff $\overline{\mathcal{O}(x, U, V)} = \overline{\mathcal{O}(x', U', V')}$.

Lemma 10.2.5

For any open $U \subseteq X$ and $1_G \in V \subseteq G$, the map $\varphi_0(\cdot, U, V) : X \to 2^\omega$ is continuous. In particular, $\varphi_0 : X \times \omega \times \omega \to 2^\omega$ is continuous.

Proof. Let $\{W_m\}_{m \in \omega}$ be a countable base for G. Assume without loss of generality that $U_k \subseteq U$. We show that $\{x \in U : \mathcal{O}(x, U, V) \cap U_k \neq \emptyset\}$ is open. For this let $\mathcal{O}_l(x, U, V)$ be the set of $y \in U$ for which there are $x =$

$x_0, x_1, \ldots, x_l = y \in U$ and $g_0, \ldots, g_{l-1} \in V$ such that $x_{i+1} = g_i \cdot x_i$ for all $i < l$. Then by continuity of the action we have that if $y \in \mathcal{O}_l(x, U, V) \cap U_k$ then there are basic open U_n with $x \in U_n \subseteq U$ and basic open $W_{m_0}, W_{m_1}, \ldots, W_{m_{l-1}}$ with $g_i \in W_{m_i}$ for all $i < l$, such that

$$(W_{m_{l-1}} \cdots W_{m_0}) \cdot U_n \subseteq U_k, \quad \text{and for all } i < l, \ (W_{m_i} \cdots W_{m_0}) \cdot U_n \subseteq U.$$

This implies that

$$\mathcal{O}(x, U, V) \cap U_k \neq \emptyset$$

$$\Longleftrightarrow \exists l \in \omega \ \mathcal{O}_l(x, U, V) \cap U_k \neq \emptyset$$

$$\Longleftrightarrow \exists l \in \omega \ \exists n \in \omega \ \exists m_0, \ldots, m_{l-1} \in \omega \ [x \in U_n \wedge$$
$$(W_{m_{l-1}} \cdots W_{m_0}) \cdot U_n \subseteq U_k \wedge \forall i < l \ (W_{m_i} \cdots W_{m_0}) \cdot U_n \subseteq U].$$

It is now clear that the set $\{x \in U : \mathcal{O}(x, U, V) \cap U_k \neq \emptyset\}$ is a countable union of basic open sets. □

Lemma 10.2.6
For any open $U \subseteq X$, $x \in U$, and open $1_G \in V \subseteq G$,

$$\{\varphi_0(x_0, U, V) : x_0 \in [x]_G \cap U\} = \{\varphi_0(\gamma_n \cdot x, U, V) : n \in \omega, \gamma_n \cdot x \in U\}.$$

In particular, the set $\{\varphi_0(x_0, U, V) : x_0 \in [x]_G \cap U\}$ is countable.

Proof. Let $g_0 \in G$ and $x_0 = g_0 \cdot x \in [x]_G \cap U$. By continuity of the action there is an open set $V' = V'^{-1} \subseteq V$ such that $V' \cdot x_0 \subseteq U$. Thus for all $g \in V'$, $g \cdot x_0 \in \mathcal{O}(x_0, U, V)$ and $x_0 = g^{-1} \cdot (g \cdot x_0) \in \mathcal{O}(g \cdot x_0, U, V)$. By Lemma 10.2.2 (iv) $\mathcal{O}(x_0, U, V) = \mathcal{O}(g \cdot x_0, U, V)$ for all $g \in V'$. Now let $\gamma_n \in V' g_0 \cap G_0$. Then $\gamma_n \cdot x = g g_0 \cdot x = g \cdot x_0$ for some $g \in V'$. Hence $\mathcal{O}(x, U, V) = \mathcal{O}(\gamma_n \cdot x, U, V)$. □

Theorem 10.2.7
Let G be a Polish group and X a Polish G-space. Suppose E_G^X is generically $=^+$-ergodic. Then for all open $U \subseteq X$ and $1_G \in V \subseteq G$ open there is a comeager set $C \subseteq X$ such that for all $x \in C$ and $x_0 \in [x]_G \cap U$, $\mathcal{O}(x_0, U, V)$ is somewhere dense.

Proof. Fix U and V and consider the map $\varphi : X \to (2^\omega)^\omega$ defined by

$$\varphi(x)(n) = \begin{cases} 1^\frown \varphi_0(\gamma_n \cdot x, U, V), & \text{if } g_n \cdot x \in U, \\ (0, 0, \ldots), & \text{otherwise.} \end{cases}$$

φ is Borel. We claim that there is a comeager set C_0 such that $\varphi \upharpoonright C_0$ is a homomorphism from $E_G^X \upharpoonright C_0$ to $=^+$. Let

$$C_0 = X - G_0 \cdot (\overline{U} - U).$$

Note that $\overline{U} - U$ is nowhere dense, and by the continuity of the action, for each $n \in \omega$, $g_n \cdot (\overline{U} - U)$ is also nowhere dense. Thus $G_0 \cdot (\overline{U} - U)$ is meager. This shows that C_0 is comeager. To see that $\varphi \upharpoonright C_0$ is a homomorphism let $x E_G^X x'$. If $[x]_G \cap U = \emptyset$ then $\varphi(x)(n) = (0, 0, \dots) = \varphi(x')(n)$ for all $n \in \omega$, so $\varphi(x) = \varphi(x')$. Therefore we assume $[x]_G \cap U \neq \emptyset$. By Lemma 10.2.6

$$\{\varphi_0(\gamma_n \cdot x, U, V) : n \in \omega, \gamma_n \cdot x \in U\} = \{\varphi_0(\gamma_n \cdot x', U, V) : n \in \omega, \gamma_n \cdot x' \in U\}.$$

If both $G_0 \cdot x \subseteq U$ and $G_0 \cdot x' \subseteq U$ then we already have that $\varphi(x) =^+ \varphi(x')$. Otherwise if both $G_0 \cdot x \not\subseteq U$ and $G_0 \cdot x' \not\subseteq U$ then we also have that $\varphi(x) =^+ \varphi(x')$. Assume we are in the remaining cases, and by symmetry assume $G_0 \cdot x \subseteq U$ and $G_0 \cdot x' \not\subseteq U$. Then by the continuity of the action $[x]_G = G \cdot x \subseteq \overline{G_0 \cdot x} \subseteq \overline{U}$ but for some $\gamma_n \in G_0$, $\gamma_n \cdot x' \notin U$. This shows that $\gamma_n \cdot x' \in \overline{U} - U$ and hence $x' \in \gamma_n^{-1} \cdot (\overline{U} - U) \subseteq G_0 \cdot (\overline{U} - U)$, contradicting our assumption that $x' \in C_0$.

Applying generic $=^+$-ergodicity of E_G^X we obtain a comeager set $C \subseteq C_0$ such that for any $x, y \in C$, $\varphi(x) =^+ \varphi(y)$. Let $x \in C$ and $x_0 \in [x]_G \cap U$. We verify that $\mathcal{O}(x_0, U, V)$ is somewhere dense. Assume not. Let $A = \mathcal{O}(x_0, U, V)$. Then A is nowhere dense and $C' = X - G_0 \cdot A$ is comeager. Let $y \in C \cap C'$. Then by $\varphi(x) =^+ \varphi(y)$ there is $\gamma_n \in G_0$ such that $\varphi_0(\gamma_n \cdot y, U, V) = \varphi_0(x_0, U, V)$. Hence $\gamma_n \cdot y \in \mathcal{O}(\gamma_n \cdot y, U, V) = \mathcal{O}(x_0, U, V) = A$. This implies that $y \in G_0 \cdot A$, contradicting our assumption that $y \in C'$. □

Exercise 10.2.1 Prove Lemma 10.2.2.

Exercise 10.2.2 Show that for any open $U \subseteq G$, $x \in U$, open $1_G \in V \subseteq G$, and $g \in G$, $g \cdot \mathcal{O}(x, U, V) = \mathcal{O}(g \cdot x, g \cdot U, gVg^{-1})$.

Exercise 10.2.3 Let $U \subseteq X$ be open and $V \subseteq G$ be open with $1_G \in V = V^{-1}$. Define $\sim_{U,V}$ on U by

$$x \sim_{U,V} y \iff \mathcal{O}(x, U, V) = \mathcal{O}(y, U, V).$$

Show that $\sim_{U,V}$ is an equivalence relation with countably many equivalence classes.

10.3   Turbulent and generically turbulent actions

In Theorem 10.2.7 a purely topological condition is identified from the consideration of generic $=^+$-ergodicity. This condition was the main part of Hjorth's definition of turbulence, which we now give.

Definition 10.3.1

Let G be a Polish group and X a Polish G-space. The action of G on X is **turbulent** if

(T1) every orbit is meager,

(T2) every orbit is dense, and

(T3) every local orbit is somewhere dense, that is, for any open $U \subseteq G$, $x \in U$, and open $1_G \in V \subseteq G$, $\mathcal{O}(x, U, V)$ is somewhere dense.

To facilitate our discussions we introduce some more definitions.

Definition 10.3.2

Let G be a Polish group and X a Polish G-space. The action of G on X is **generically turbulent** if it is turbulent on a comeager subset of X.

Definition 10.3.3

Let G be a Polish group and X a Polish G-space.

(1) The action of G on X is **preturbulent** if for all $x, y \in X$, $U \subseteq X$ open with $x \in U$ and open $1_G \in V \subseteq G$, $\overline{\mathcal{O}(x, U, V)} \cap [y]_G \neq \emptyset$.

(2) The action of G on X is **generically preturbulent** if it is preturbulent on a comeager subset of X.

(3) The action of G on X is **properly generically preturbulent** if it is generically preturbulent and every orbit is meager.

It is easy to see that preturbulence is weaker than the condition (T2)+(T3) but implies (T2). It follows that generic turbulence implies proper generic preturbulence. The main objective of this section is to show that the converse is true.

Lemma 10.3.4

Let $Y = \{(x_n) \in (2^\omega)^\omega : \forall n \neq m \in \omega \; x_n \neq x_m\}$. Then $=^+ \upharpoonright Y \sim_B =^+$.

Proof. For any $z \in 2^\omega$ and $k \in \omega$, let $\zeta(z, k) = 0^{k\frown}1^\frown z$, that is, $\zeta(z, k) \in 2^\omega$ is such that

$$\zeta(z, k)(i) = 1 \iff i = k \; \vee \; (i \geq k+1 \; \wedge \; z(i - k - 1) = 1).$$

It is easy to check that for all $z, z' \in 2^\omega$, $\zeta(z, k) = \zeta(z', k')$ iff $z = z'$ and $k = k'$.

Now for any $x = (x_n) \in (2^\omega)^\omega$ let $S_x = \{\zeta(x_n, k) : k \in \omega\}$. Then S_x is a countably infinite set, and for $x, x' \in (2^\omega)^\omega$, $S_x = S_{x'}$ iff $x =^+ x'$. Let $\eta(x) \in (2^\omega)^\omega$ be an element encoding S_x without repetitions. Then $\eta(x) =^+ \eta(x')$ iff $x =^+ x'$. This η is a Borel reduction of $=^+$ to $=^+ \upharpoonright Y$. □

Theorem 10.3.5
Let G be a Polish group and X a Polish G-space. If the action of G on X is preturbulent, then E_G^X is generically $=^+$-ergodic.

Proof. Let $\{V_n\}_{n \in \omega}$ a countable open nbhd base for 1_G.

Let f be a Baire measurable generic homomorphism from E_G^X to $=^+ \restriction Y$, where Y is as in Lemma 10.3.4. Let $C \subseteq X$ be a comeager Borel set so that $f \restriction C$ is a continuous homomorphism from $E_G^X \restriction C$ to $=^+ \restriction Y$ and that $C \subseteq C^{*G}$, that is, for any $x \in C$, $\forall^* g \in G$ $g \cdot x \in C$. To see that such a C exists, let C_0 be a dense G_δ set so that $f \restriction C_0$ is a continuous homomorphism from $E_G^X \restriction C_0$ to $=^+ \restriction Y$. Note that C_0^{*G} is comeager and Borel. Let $C = C_0 \cap C_0^{*G}$. Then $C \subseteq C_0$ is comeager Borel, and $C \subseteq C^{*G}$.

Since $f \restriction C$ is a homomorphism from $E_G^X \restriction C$ to $=^+ \restriction Y$, for any $x \in C$, $k \in \omega$, and $x' \in [x]_G \cap C$, there is a unique $l \in \omega$ such that $(f(x))_k = (f(x'))_l$. Fix $x \in C$ and $k \in \omega$. We claim that

> for some $l \in \omega$ there is a nonempty open $W \subseteq G$ so that for all $g \in W$ with $g \cdot x \in C$, $(f(g \cdot x))_l = (f(x))_k$.

To see this, let $H_l = \{g \in G : g \cdot x \in C \wedge (f(g \cdot x))_l = (f(x))_k\}$ for each $l \in \omega$. Then $\bigcup_l H_l = \{g \in G : g \cdot x \in C\}$ is a comeager set in G. Hence for some $l \in \omega$ H_l is nonmeager and therefore comeager in some nonempty open $W \subseteq G$. Fix such an $l \in \omega$ and $W \subseteq G$. Now suppose $g \in W$ with $g \cdot x \in C$. Then there is a sequence (g_n) of elements of $W \cap H_l$ such that $g_n \to g$ as $n \to \infty$. Then $g_n \cdot x \in C$ and $(f(g_n \cdot x))_l = (f(x))_k$ for all $n \in \omega$. By continuity of f on C, and of the action, $(f(g \cdot x))_l = (f(x))_k$. This proves the claim.

It follows from the claim that for any $x \in C$ there are $l \in \omega$, $g \in G$ and an open $1_G \in V \subseteq G$ such that for any $h, h' \in V$ so that $hg \cdot x, h'g \cdot x \in C$, we have $(f(hg \cdot x))_l = (f(h'g \cdot x))_l$. In view of this fact, define, for each $x \in C$ and $l \in \omega$, $N(x, l) = n + 1$ where n is the least such that for any $h \in V_n$ with $h \cdot x \in C$, we have $(f(h \cdot x))_l = (f(x))_l$; if no such n exists then let $N(x, l) = 0$. Note that $N(x, l)$ is a Borel function from $C \times \omega$ to ω, since by continuity of $f \restriction C$,

$$\forall h \in V_n \ (h \cdot x \in C \Rightarrow (f(h \cdot x))_l = (f(x))_l)$$
$$\iff \forall^* h \in V_n \ (h \cdot x \in C \Rightarrow (f(h \cdot x))_l = (f(x))_l)$$

is Borel.

Now let $D \subseteq C$ be a comeager Borel set such that $N \restriction (D \times \omega)$ is continuous and that $D \subseteq D^{*G}$. Such a D exists by a similar argument as that for the existence of C. The above claim implies that for any $x \in D$ and $k \in \omega$ there are $x' \in [x]_G \cap D$ and $l \in \omega$ such that $N(x', l) > 0$ and $(f(x'))_l = (f(x))_k$.

Let $x, y \in D$. We show that $f(x) =^+ f(y)$. By symmetry it suffices to show that $(f(x))_k \in \{(f(y))_n : n \in \omega\}$ for any $k \in \omega$. For this fix $k \in \omega$. There are $x' \in [x]_G \cap D$ and $l \in \omega$ such that $N(x', l) > 0$ and $((f(x'))_l = (f(x))_k$. Hence without loss of generality we may assume that

$x = x'$ and $k = l$. Thus $N(x, k) > 0$. Suppose $N(x, k) = m + 1$. Since $N \restriction (D \times \omega)$ is continuous, there is an open set $U \subseteq X$ with $x \in U$ such that for all $z \in U \cap D$, $N(z, k) = m + 1$. This means that for all $z \in U \cap D$ and $g \in V_m$ with $g \cdot z \in D$, $(f(g \cdot z))_k = (f(z))_k$. Let $W \subseteq G$ be open with $1_G \in W^3 \subseteq V_m$, and let $V = W \cap W^{-1}$.

Now by preturbulence $\overline{\mathcal{O}(x, U, V)} \cap [y]_G \neq \emptyset$. This means that there is $y_0 \in [y]_G \cap U$ and sequence $(x_i)_{i \in \omega}$ of elements in U and $(g_i)_{i \in \omega}$ of elements of V such that $x_0 = x$, $x_{i+1} = g_i \cdot x_i$ for all $i \in \omega$, and $y_0 \in \overline{\{x_i : i \in \omega\}}$. Let d be a compatible complete metric on X. We define sequences $(x_i')_{i \in \omega}$ of elements of U and $(\epsilon_i)_{i \in \omega}$ of elements of V such that for all $i \in \omega$, $\epsilon_i \cdot x_i = x_i'$, $d(x_i, x_i') < 2^{-i}$, and $x_i' \in D \cap U$. For this fix $i \in \omega$. If $i = 0$ let $\epsilon_0 = 1_G$ and $x_0' = x_0 = x$. Assume $i > 0$. Let $h_{i-1} = g_{i-1} g_{i-2} \cdots g_0$. Then $x_i = h_{i-1} \cdot x$. Since $x \in D$, we have that for comeager many $g \in G$, $g \cdot x \in D$. It follows that there are comeager many $g \in V h_{i-1}$ such that $g \cdot x \in D$. Let $\epsilon_i \in V$ be such that $\epsilon_i h_{i-1} \cdot x \in D \cap U$ and that $d(\epsilon_i h_{i-1} \cdot x, h_{i-1} \cdot x) < 2^{-i}$. This means that $d(\epsilon_i \cdot x_i, x_i) < 2^{-i}$. So if we let $x_i' = \epsilon_i \cdot x_i$ the required properties are fulfilled.

Now note that $y_0 \in \overline{\{x_i' : i \in \omega\}}$. Also since $x_i' \in D \cap U$, $N(x_i', k) = m + 1$. Thus for all $g \in V_m$ with $g \cdot x_i' \in D$, $(f(g \cdot x_i'))_k = (f(x_i'))_k$. For each $i \in \omega$, $x_{i+1}' = \epsilon_{i+1} g_i \epsilon_i^{-1} \cdot x_i'$. But since $\epsilon_{i+1} g_i \epsilon_i^{-1} \in V^3 \subseteq W^3 \subseteq V_m$, we have that $(f(x_{i+1}'))_k = (f(x_i'))_k$. This means that for all $i \in \omega$, $(f(x_i'))_k = (f(x))_k$.

In order to apply the continuity of $f \restriction D$ we need $\delta \in V_m$ with the following properties:

(i) $\delta \cdot y_0 \in D$,

(ii) for all $i \in \omega$, $\delta \cdot x_i \in D$, and

(iii) for all $i \in \omega$, $(f(\delta \cdot x_i))_k = (f(x_i))_k = (f(x))_k$.

Granting such a δ we have that $\delta \cdot y \in \overline{\{\delta \cdot x_i : i \in \omega\}}$, and then by the continuity of $f \restriction D$, $(f(\delta \cdot y_0))_k = (f(x))_k$, which finishes the proof. However, since $y \in D \subseteq D^{*G}$, the set of δ such that $\delta \cdot y_0 \in D$ is comeager. Similarly, since each $x_i \in D$, the set of δ such that $\delta \cdot x_i \in D$ is comeager for each $i \in \omega$. Finally, since for any $z \in U \cap D$, $N(z, k) = m+1$, we have $(f(\delta \cdot x_i))_k = (f(x_i))_k$ whenever $\delta \in V_m$ and $\delta \cdot x_i \in D$. The set of such δ is comeager in V_m. By taking δ to be in the intersection of all these sets, we obtain the required properties (i) through (iii). ∎

Corollary 10.3.6
Let G be a Polish group and X a Polish G-space. Then the action of G on X is generically preturbulent iff E_G^X is generically $=^+$-ergodic.

Proof. Suppose the action is preturbulent on a dense G_δ set $C \subseteq X$. Then it is also preturbulent on the invariant dense G_δ set C^{*G}. It follows from Theorem 10.3.5 that E_G^X is generically $=^+$-ergodic. Conversely, if E_G^X is generically $=^+$-ergodic then by Theorem 10.2.7 there is a comeager set $C \subseteq X$ such that for all $x \in C$, open $U \subseteq X$ with $x \in U$, and open $1_G \in V \subseteq G$, $\overline{\mathcal{O}(x, U, V)}$ contains some nonempty open set. Note that E_G^X is generically ergodic (Exercise 10.1.4), thus there is a dense orbit. It follows that there is a comeager set $D \subseteq C$ all of whose orbits are dense. Then for any $y \in D$, $\overline{\mathcal{O}(x, U, V)} \cap [y]_G \neq \emptyset$. ▯

Corollary 10.3.7
Let G be a Polish group and X a Polish G-space. Then the following are equivalent:

(a) *The action of G on X is generically turbulent.*

(b) *The action of G on X is properly generically preturbulent.*

(c) *E_G^X is properly generically $=^+$-ergodic.*

Proof. It is obvious that (a) implies (b). The equivalence of (b) and (c) is immediate from the preceding theorem. Then (c) implies (a) because of Theorem 10.2.7. ▯

Exercise 10.3.1 Show that if an action satisfies (T2) and (T3) of Definition 10.3.1 then it is preturbulent. Show that in a preturbulent action every orbit is dense.

Let G be a Polish group and X a Polish G-space. Call a point $x \in X$ **turbulent** if for all open $U \subseteq X$ with $x \in U$ and open $1_G \in V \subseteq G$, $\mathcal{O}(x, U, V)$ is somewhere dense. Call an orbit $[x]_G$ **turbulent** if every $y \in [x]_G$ is turbulent.

Exercise 10.3.2 Show that x is turbulent iff for all open $U \subseteq X$ with $x \in U$ and open $1_G \in V \subseteq G$, there is an open $U' \subseteq U$ with $x \in U' \subseteq \overline{\mathcal{O}(x, U', V)}$.

Exercise 10.3.3 Let bases for X and G be given. Show that x is turbulent iff for all basic open $U \subseteq X$ with $x \in U$ and basic open $1_G \in V \subseteq G$, $\mathcal{O}(x, U, V)$ is somewhere dense.

Exercise 10.3.4 Show that the set of all turbulent points is G_δ.

Exercise 10.3.5 Show that if x is turbulent and $y \in [x]_G$ then y is turbulent. Thus an orbit is turbulent iff it contains a turbulent point.

Exercise 10.3.6 Show that if there is a dense turbulent orbit then there is a comeager set of points all of whose orbits are dense turbulent.

Exercise 10.3.7 Let G be a Polish group and X a Polish G-space. Show that the following are equivalent:

(i) The action of G on X is generically turbulent.

(ii) There is a dense turbulent orbit in X and every orbit is meager.

(iii) There is a comeager set of points all of whose orbits are dense, meager, and turbulent.

10.4 The Hjorth turbulence theorem

In this section we prove the Hjorth turbulence theorem stating that any orbit equivalence relation of a turbulent action is properly generically E-turbulent for any S_∞-orbit equivalence relation E. This is the main theorem of this chapter.

We need the following lemma in the main proof. The lemma gives a generic uniform property for homomorphisms of orbit equivalence relations.

Lemma 10.4.1
Let G, H be Polish groups, X a Polish G-space, Y a Polish H-space, and f a Baire measurable generic homomorphism from E_G^X to E_H^Y. Then for any open $1_H \in W \subseteq H$ there is a comeager set $C \subseteq X$ such that for any $x \in C$ there is an open $1_G \in V \subseteq G$ such that

$$\forall^* g \in V \; \exists h \in W \; (f(g \cdot x) = h \cdot f(x)).$$

Proof. Fix an open $1_H \in W \subseteq H$. Let $A \subseteq X$ be defined as

$$x \in A \iff x \in (f^{-1}(W \cdot f(x)))^{\triangle G}.$$

It is straightforward to check that the lemma states that A is comeager. Let $C_0 \subseteq X$ be a dense G_δ such that f is a continuous homomorphism from $E_G^X \upharpoonright C_0$ to E_H^Y. Then C_0^{*G} is comeager and Borel. We claim that $C_0 \cap C_0^{*G} \subseteq A^{*G}$.

For this fix also $x \in C_0 \cap C_0^{*G}$. Let $W_0 \subseteq H$ be open such that $1_H \in W_0 = W_0^{-1} \subseteq W_0^2 \subseteq W$. Then there is a countable set $R \subseteq H$ such that $W_0 R = H$. Since $f \upharpoonright C_0$ is a homomorphism from $E_G^X \upharpoonright C_0$ to E_H^Y, we have that

$$[x]_G \cap C_0 \subseteq f^{-1}(H \cdot f(x)) = f^{-1}(W_0 R \cdot f(x)) = \bigcup_{r \in R} f^{-1}(W_0 r \cdot f(x)).$$

For each $r \in R$ let $G_r = \{g \in G : g \cdot x \in C_0 \cap f^{-1}(W_0 r \cdot f(x))\}$. Note that $W_0 r \cdot f(x)$ is Σ_1^1 in Y, from which it follows that $C_0 \cap f^{-1}(W_0 r \cdot f(x))$ is Σ_1^1 in C_0 and G_r is Σ_1^1 in G, and they all have the Baire property. Then for each $r \in R$ let $O_r \subseteq G$ be open such that $G_r \triangle O_r$ is meager. Then $\bigcup_{r \in R} O_r$ is dense open in G and each G_r is comeager in O_r. Let $D = \bigcup_{r \in R}(G_r \cap O_r)$. Then D is comeager in G. We next verify that for all $g \in D$, $g \cdot x \in A$.

For this fix $g \in D$. Say $g \in G_r \cap O_r$. Then $g \cdot x \in C_0 \cap f^{-1}(W_0 r \cdot f(x))$. Let $V \subseteq G$ be open with $1_G \in V$ and $Vg \subseteq O_r$. Then for comeager many $g' \in V$, $g'g \in G_r \cap O_r$, and so $g'g \cdot x \in C_0 \cap f^{-1}(W_0 r \cdot f(x))$. This means that there are $h_0, h'_0 \in W_0$ such that $f(g \cdot x) = h_0 r \cdot f(x)$ and $f(g'g \cdot x) = h'_0 r \cdot f(x)$. Thus $f(g'g \cdot x) = h_0 h_0'^{-1} \cdot f(g \cdot x)$. Since $h_0 h'_0 \in W_0^2 \subseteq W$, we have that $f(g'g \cdot x) \in W \cdot f(g \cdot x)$, or $g' \cdot (g \cdot x) \in f^{-1}(W \cdot f(g \cdot x))$. This gives that $g \cdot x \in A$, hence the claim is proved.

By the claim A^{*G} is comeager, and hence $\forall^* x \in X \; \forall^* g \in G \; g \cdot x \in A$. By Kuratowski–Ulam, $\forall^* g \in G \; \forall^* x \in X \; g \cdot A$. Let g be any such element. Then $\forall^* x \; g \cdot x \in A$. This states that $g^{-1} \cdot A$ is comeager. And by continuity of the action, A is comeager. $\qquad\Box$

Theorem 10.4.2
Let G be a Polish group, X a Polish G-space, and Y a Polish S_∞-space. If the action of G on X is preturbulent, then E_G^X is generically $E_{S_\infty}^Y$-ergodic.

Proof. Fix a countable base $\{U_n\}_{n \in \omega}$ for X with $U_0 = X$ and a countable nbhd base $\{V_n\}_{n \in \omega}$ for 1_G in G with $V_0 = G$. Let $d_Y < 1$ be a compatible complete metric on Y. Let D be the complete metric on S_∞ given by $D(h_1, h_2) = d(h_1, h_2) + d(h_1^{-1}, h_2^{-1})$, where d is the usual metric on S_∞ inherited from the Baire space ω^ω, that is,

$$d(h_1, h_2) = \begin{cases} 2^{-n}, & \text{if } n \in \omega \text{ is the least such that } h_1(n) \neq h_2(n), \\ 0, & \text{otherwise.} \end{cases}$$

Let $e = 1_{S_\infty}$. For $k \in \omega$ let $N_k = \{h \in S_\infty : d(h, e) < 2^{-k}\} = \{h \in S_\infty : h(i) = i \; \forall i \leq k\}$. Then $\{N_k\}_{k \in \omega}$ is a nbhd base for e in S_∞.

Let f be a Baire measurable generic homomorphism from E_G^X to $E_{S_\infty}^Y$. By Lemma 10.4.1 for any $N_k \subseteq S_\infty$ there are comeager many $x \in X$ such that there is some basic open $V_m \subseteq G$ with

$$\forall^* g \in V_m \; \exists h \in N_k \; f(g \cdot x) = h \cdot f(x).$$

Then there is a comeager set of $x \in X$ such that for any N_k there is V_m so that the above displayed property holds. Let $C_0 \subseteq X$ be such a comeager set that is Borel and such that $f \restriction C_0$ is a continuous homomorphism from $E_G^X \restriction C_0$ to E_H^Y. In view of this define, for $x \in X$ and $k \in \omega$, $N(x, k) = m+1$ if m is the least so that the above property holds; and $N(x, k) = 0$ if there is no such m. Then for any $x \in C_0$ and $k \in \omega$, $N(x, k) > 0$. Note that $N \restriction (C_0 \times \omega)$ is Baire measurable since the defining condition is Σ_1^1.

Since the action of G on X is preturbulent, by Theorems 10.3.5 and 10.2.7, for all $U_n \subseteq X$ and $V_m \subseteq G$ there is a comeager set of $x \in X$ such that for all $x_0 \in [x]_G \cap U_n$, $\mathcal{O}(x, U_n, V_m)$ is somewhere dense. It follows that there is a comeager set of $x \in X$ such that for all U_n and V_m, for all $x_0 \in [x]_G \cap U_n$, $\mathcal{O}(x, U_n, V_m)$ is somewhere dense.

Also by Theorem 10.3.5 E_G^X is generically $=^+$-ergodic, and therefore it is generically ergodic. It follows that there is a comeager set of $x \in X$ in which every orbit is dense.

Taking the intersections of these comeager sets, we obtain a comeager $C_1 \subseteq C_0$ such that

   (i) for any $x \in C_1$, $[x]_G \cap C_1$ is dense;

   (ii) $f \upharpoonright C_1$ is a continuous homomorphism from $E_G^X \upharpoonright C_1$ to E_H^Y;

   (iii) $N \upharpoonright (C_1 \times \omega) \to (\omega - \{0\})$ is continuous; and

   (iv) for all $U_n \subseteq X$, $V_m \subseteq G$, and $x \in C_1 \cap U_n$, $\mathcal{O}(x, U_n, V_m)$ is somewhere dense.

Finally let $C = C_1 \cap C_1^{*G}$. Then C is comeager, and we claim that for all $x, y \in C$, $f(x) E_{S_\infty}^Y f(y)$. To prove this, fix $x, y \in C$. We construct $g, h \in S_\infty$ such that $g \cdot f(x) = h \cdot f(y)$. The elements g, h will be approximated by two D-Cauchy sequences (g_i) and (h_i) in S_∞. Each g_i or h_i arises from finding x_i or y_i respectively so that $g_i \cdot f(x) = f(x_i)$ or $h_i \cdot f(y) = f(y_i)$. To guarantee that $(f(x_i))_{i \in \omega}$ and $(f(y_i))_{i \in \omega}$ converge to the same point in Y we will construct sequences of basic open sets $(U_{n_x(i)})_{i \in \omega}$ and $(U_{n_y(i)})_{i \in \omega}$ so that $U_{n_x(i+1)} \subseteq U_{n_y(i)} \subseteq U_{n_x(i)}$ for all $i \in \omega$ and so that $\operatorname{diam}_Y f(U_{n_x(i)}) \to 0$ as $i \to \infty$. Finally these basic open sets $U_{n_x(i)}$ and $U_{n_y(i)}$ will be found using the turbulence condition (iv); in doing this we also construct sequences of basic open nbhds $V_{m_x(i)}$ and $V_{m_y(i)}$.

In summary we construct sequences of elements indexed by $i \in \omega$,

$$n_x(i), n_y(i), m_x(i), m_y(i) \in \omega, \quad x_i, y_i \in C, \quad g_i, h_i \in S_\infty,$$

so that the following are satisfied for all $i \in \omega$:

   (1) $x_0 = x$, $y_0 = y$; $g_0 = h_0 = e$;

   (2) $U_{n_x(i+1)} \subseteq U_{n_y(i)} \subseteq U_{n_x(i)}$;

   (3_x) $\mathcal{O}(x_i, U_{n_x(i)}, V_{m_x(i)})$ is dense in $U_{n_y(i)}$;

   (3_y) $\mathcal{O}(y_i, U_{n_y(i)}, V_{m_y(i)})$ is dense in $U_{n_x(i+1)}$;

   (4_x) $x_{i+1} \in U_{n_x(i+1)} \cap C \cap \mathcal{O}(x_i, U_{n_x(i)}, V_{m_x(i)})$;

   (4_y) $y_{i+1} \in U_{n_y(i+1)} \cap C \cap \mathcal{O}(y_i, U_{n_y(i)}, V_{m_y(i)})$;

   (5) $\operatorname{diam}_Y f(U_{n_x(i)} \cap C) \leq 2^{-i}$;

   (6) $g_i \cdot f(x) = f(x_i)$; $h_i \cdot f(y) = f(y_i)$;

   (7) $D(g_i, g_{i+1}) \leq 2^{-i}$; $D(h_i, h_{i+1}) \leq 2^{-i}$;

   (8_x) for $i > 0$, $k_x(i) = \max\{g_i(l), g_i^{-1}(l) : l \leq i\}$, for all $z \in U_{n_x(i)} \cap C$, $N(z, k_x(i)) = m_x(i) + 1$;

(8_y) for $k_y(i) = \max\{h_i(l), h_i^{-1}(l) : l \le i\}$, for all $z \in U_{n_y(i)} \cap C$,
$N(z, k_y(i)) = m_y(i) + 1$.

Granting the construction, (g_i) and (h_i) are D-Cauchy by (7); and $d_Y(g_i \cdot f(x), h_i \cdot f(y)) < 2^{-i}$ by (2), (4), (5), and (6). Let $g_i \to g$ and $h_i \to h$ as $i \to \infty$. Then $g \cdot f(x) = h \cdot f(y)$, as required.

To begin the construction, we let $n_x(0) = m_x(0) = 0$, $g_0 = e$, $x_0 = x$. Then $U_{n_x(0)} = X$ and $V_{m_x(0)} = G$. Note that $\mathcal{O}(x, X, G) = [x]_G$ is dense in X. Also let $y_0 = y$ and $h_0 = e$. Then $k_y(0) = 0$. By the continuity of N on $C \times \omega$ there is some U_n such that $y \in U_n$ and for all $z \in U_n \cap C$, $N(z, 0) = N(y, 0) = m + 1$ for some $m \in \omega$. Let $n_y(0), m_y(0)$ be such m, n, respectively. Then (3_x) and (8_y) are fulfilled. This finishes the definition for $i = 0$.

By induction assume we have completed the definition for the index i. We indicate how to define $n_x(i + 1), m_x(i + 1), x_{i+1}, g_{i+1}$. The definition for $n_y(i + 1), m_y(i + 1), y_{i+1}, h_{i+1}$ is similar.

Since $y_i \in C$ we have by (iv) that $\mathcal{O}(y_i, U_{n_y(i)}, V_{m_y(i)})$ is somewhere dense. Let $W \subseteq U_{n_y(i)}$ be nonempty open such that $\mathcal{O}(y_i, U_{n_y(i)}, V_{m_y(i)})$ is dense in W and such that $\text{diam}_Y f(W \cap C) \le 2^{-(i+1)}$. This guarantees requirements (2), (3_y), and (5). We now choose x_{i+1} to satisfy (4_x).

For notational simplicity let $U = U_{n_x(i)}$, $V = V_{m_x(i)}$, and $z_0 = x_i$. Note first that $\mathcal{O}(z_0, U, V)$ is dense in W since it is dense in $U_{n_y(i)}$ and $W \subseteq U_{n_y(i)}$. Let $z_1, \ldots, z_l \in U$, $\gamma_0, \ldots, \gamma_{l-1} \in V$ be such that $z_{j+1} = \gamma_j \cdot z_j$ for all $j < l$ and $z_l \in W$. In particular the sequence $z_0, z_1, \ldots, z_l$ witnesses that $z_l \in \mathcal{O}(x, U, V) \cap W$. We will find $\epsilon_0 = 1_G, \ldots, \epsilon_l \in G$ such that, letting $z_j' = \epsilon_j \cdot z_j$ for all $j \le l$, we have that $z_0', z_1', \ldots, z_l' \in C \cap U$, $z_l' \in W$, and $\epsilon_{i+1} \gamma_i \epsilon_i^{-1} \in V$. This guarantees that the sequence $z_0', z_1', \ldots, z_l'$ of elements of C witnesses that $z_l' \in \mathcal{O}(x, U, V) \cap W \cap C$. For this let $1_G \in V' = V'^{-1}$ be open such that for all $j \le l$, $V' \gamma_j V' \subseteq V$. It suffices to choose $\epsilon_1, \ldots, \epsilon_l \in V'$ so that $\epsilon_j \cdot z_j \in C \cap U$ and $z_l \in W$. But for each $j \le l$ the set of such ϵ_j is comeager in a nonempty open subset of V', hence the required ϵ_j can be chosen as required. This construction will eventually guarantee (4_x).

Again for notational simplicity we suppress the $'$ and assume that $z_0 = x_i, z_1, \ldots, z_l \in U \cap C$, $\gamma_0, \ldots, \gamma_{l-1} \in V$ are defined such that $z_{j+1} = \gamma_j \cdot z_j$ for all $j < l$ and that $z_l \in W$. We indicate how to fulfill (7). Since $z_j \in U \cap C$ for all $j \le l$, by (8_x) of index i, for $k = k_x(i) = \max\{g_i(\lambda), g_i^{-1}(\lambda) : \lambda \le i\}$, $N(z_j, k) = m_x(i) + 1$. This means that

$$\forall^* \gamma \in V = V_{m_x(i)} \; \exists p \in N_k \; f(\gamma \cdot z_j) = p \cdot f(z_j).$$

By repeating the construction in the preceding paragraph we may assume that $\gamma_0, \ldots, \gamma_{l-1}$ are chosen from such comeager sets that there are $f(\gamma_j \cdot z_j) = p_j \cdot f(z_j)$ for some $p_j \in N_k$ for all $j < l$. This gives that $f(z_{j+1}) = p_j \cdot z_j$ and therefore

$$f(z_l) = p_{l-1} \ldots p_0 \cdot f(z_0) = (p_{l-1} \ldots p_0) \cdot f(x_i).$$

Note that $p_{l-1} \ldots p_0 \in N_k$.

In view of the above argument, we may let $x_{i+1} = z_l$ as above. Then $g_{i+1} = (p_{l-1} \ldots p_0)g_i$, which implies that $d(g_{i+1}, g_i), d(g_{i+1}, g_i) \leq 2^{-(i+1)}$, and so $D(g_{i+1}, g_i) \leq 2^{-i}$. This fulfills (6) and (7).

Finally, let $k_x(i+1) = \max\{g_{i+1}(\lambda), g_{i+1}^{-1}(\lambda) : \lambda \leq i+1\}$ and $m_x(i+1) = N(x_{i+1}, k_x(i+1)) - 1$. By the continuity of $N \restriction (C \times \omega)$, let $n_x(i+1)$ be such that $x_{i+1} \subseteq V_{n_x(i+1)} \subseteq W$ and for all $z \in V_{n_x(i+1)} \cap C$, $N(z, k_x(i+1)) = m_x(i+1)$. Then (4_x) and (8_x) are fulfilled. ∎

We refer to the following corollary as the **Hjorth turbulence theorem**.

Corollary 10.4.3
Let G be a Polish group and X a Polish G-space. If the action of G on X is turbulent, then E_G^X is properly generically $E_{S_\infty}^Y$-ergodic for any Polish S_∞-space Y; in particular, $E_G^X \not\leq_B E_{S_\infty}^Y$.

The following corollary is a continuation of Corollary 10.3.7.

Corollary 10.4.4
Let G be a Polish group and X a Polish G-space. Then the following are equivalent:

(a) The action of G on X is generically turbulent.

(d) E_G^X is properly generically $E_{S_\infty}^Y$-ergodic for all Polish S_∞-space Y.

Exercise 10.4.1 Let G be a Polish group, X a Polish G-space, H a closed subgroup of S_∞, and Y a Polish H-space. Suppose $=^+ \leq_B F$. Show that the action of G on X is generically turbulent iff E_G^X is properly generically F-ergodic.

10.5   Examples of turbulence

In this section we will show that the equivalence relations c_0, ℓ_p ($1 \leq p < +\infty$) we introduced in Sections 8.4 and 8.5 are not Borel reducible to any S_∞-orbit equivalence relation. This is by showing that the actions are actually turbulent.

Definition 10.5.1
A subset S of $\mathbb{R}^\omega$ is **strongly dense** if $S \neq \mathbb{R}^\omega$ but for every $n \in \omega$ and $(x_0, \ldots, x_{n-1}) \in \mathbb{R}^n$, there is $y \in S$ such that $y_i = x_i$ for all $i < n$.

Theorem 10.5.2
Let $G \subseteq \mathbb{R}^\omega$ be a strongly dense Polishable subgroup of $\mathbb{R}^\omega$. Then the translation action of G on $\mathbb{R}^\omega$ is turbulent.

Proof. First it is easy to check that every orbit is dense. Thus the action is generically ergodic, and every invariant Borel set, in particular, every orbit, is either meager or comeager. However, every orbit is a homeomorphic copy of G; thus if any orbit is comeager then every orbit is comeager, contradicting the assumption that G is a proper subset of $\mathbb{R}^\omega$ (and therefore there is more than one orbit). This shows that every orbit is meager.

It remains to check the condition (T3) from the definition of turbulence. It suffices to show that for all $x \in \mathbb{R}^\omega$, basic open $U \subseteq \mathbb{R}^\omega$ with $x \in U$ and open $1_G \in V \subseteq G$, $\mathcal{O}(x, U, V)$ is somewhere dense. For this let $x \in U$, where

$$U = \{y \in \mathbb{R}^\omega \ : \ |y_i - x_i| < \epsilon \ \forall i < n\}$$

for some $n \in \omega$ and $\epsilon > 0$. Let τ be the Polish group topology on G. Let $1_G \in V \subseteq G$ be τ-open. We claim that $\mathcal{O}(x, U, V)$ is indeed dense in the entire U.

For this let $y \in U$ be arbitrary, and let U_0 be a basic open set containing y. Without loss of generality we may assume that for some $\delta < \epsilon$ and $m \geq n$,

$$U_0 = \{z \in \mathbb{R}^\omega \ : \ |z_i - y_i| < \delta \ \forall i < m\}.$$

Consider the projection $\pi_m : G \to \mathbb{R}^m$ where for all $g \in G$, $\pi_n(g) = (g_0, \ldots, g_{m-1})$. By Polishability of G, π_m is a Borel homomorphism from the Polish group (G, τ) into $\mathbb{R}^m$. Since G is strongly dense, π_m is onto. Thus by Theorem 2.3.3 π_m is both continuous and open.

Let $W = \pi_m(V)$. Then $(0, \ldots, 0) \in W$ and $W \subseteq \mathbb{R}^m$ is open. Thus for some $\eta > 0$, we have

$$W_0 = \{w \in \mathbb{R}^m \ : \ |w_i| < \eta \ \forall i < m\} \subseteq W.$$

Now let N be large enough such that $|x_i - y_i| < N\eta$ for all $i < m$. Let $w \in \mathbb{R}^m$ be such that $w_i = (y_i - x_i)/N$ for all $i < m$. Then $w \in W_0$. Let $g \in V$ be such that $\pi_m(g) = w$. Then the sequence

$$x, g + x, 2g + x, \ldots, Ng + x$$

witnesses that $Ng + x \in \mathcal{O}(x, U, V)$. Finally $Ng + x \in U_0$, and thus $\mathcal{O}(x, U, V) \cap U_0 \neq \emptyset$. □

Corollary 10.5.3
For any S_∞-orbit equivalence relation E and any $1 \leq p < +\infty$,

$$c_0, \ell_p \not\leq_B E.$$

Proof. The Polishable groups c_0, ℓ_p ($1 \leq p < +\infty$) are strongly dense in $\mathbb{R}^\omega$. □

In fact a large collection of Polish group actions have been examined and proved to be turbulent. The details of these proofs very often depend on the nature of the group and the action involved, and offer little insight into further development of the theory of turbulence. The theory of turbulence establishes an important criterion of classifiability for classification problems in mathematics.

10.6 Orbit equivalence relations and E_1

In this section we apply the techniques developed in this chapter and prove a theorem of Kechris and Louveau [102] about the hypersmooth equivalence relation E_1. The proof we present is due to Hjorth [70].

Recall from Section 8.1 that the equivalence relation E_1 is defined on $2^{\omega \times \omega}$. By the Borel isomorphism between 2^ω and $\mathbb{R}$, E_1 is obviously Borel isomorphic to the following equivalence relation on $\mathbb{R}^\omega$, which we still denote by E_1:

$$x E_1 y \iff \exists n \in \omega \; \forall m \geq n \; x_m = y_m.$$

Note that E_1 can be written as an increasing sequence of smooth equivalence relations F_n where

$$x F_n y \iff \forall m \geq n \; x_m = y_m.$$

Each F_n can be viewed as an orbit equivalence relation of an action of $\mathbb{R}^n$ on $\mathbb{R}^\omega$ given by

$$(g \cdot x)_m = \begin{cases} g_m + x_m, & \text{if } m < n, \\ x_m, & \text{otherwise.} \end{cases}$$

The action is apparently continuous. The increasing union of $\mathbb{R}^n$ for all $n \in \omega$, which we would denote by $\mathbb{R}^{<\omega}$, is customarily denoted as c_{00} in Banach space theory. c_{00} is an F_σ subgroup of $\mathbb{R}^\omega$, and therefore is a standard Borel group. Similar to Lemma 9.3.3 one can show that c_{00} is not Polishable (see Exercise 10.6.1).

Theorem 10.6.1 (Kechris–Louveau)
Let G be a Polish group and X a Borel G-space. Then $E_1 \not\leq_B E_G^X$.

Proof. As usual we assume without loss of generality that X is a Polish G-space. Assume $f : \mathbb{R}^\omega \to X$ is a Borel reduction from E_1 to E_G^X. Then for each $n \in \omega$, f is a Borel homomorphism from F_n to E_G^X. By applying Lemma 10.4.1 to f for all $n \in \omega$, we can obtain a dense G_δ subset C of $\mathbb{R}^\omega$ such that

(i) $f \restriction C : C \to X$ is continuous,

(ii) for all $n \in \omega$, all open nbhd W of 1_G in G, and $x \in C$, there is an open nbhd V of the identity of $\mathbb{R}^n$ such that

$$\forall^* h \in V \,\exists g \in W \,(\,f(h \cdot x) = g \cdot f(x)\,).$$

Let $\{O_i\}_{i \in \omega}$ be a sequence of dense open subsets of $\mathbb{R}^\omega$ such that $C = \bigcap_{i \in \omega} O_i$. For each $n \in \omega$ let $C_n = C^{*\mathbb{R}^n}$. Then each C_n is comeager in $\mathbb{R}^\omega$ and $\mathbb{R}^n$-invariant. It follows that for any $n \in \omega$, $\bigcap_{m \geq n} C_m$ is comeager in $\mathbb{R}^\omega$ and $\mathbb{R}^n$-invariant.

Let d_G be a compatible complete metric on G. We are ready to construct sequences of elements indexed by $i \in \omega$,

$$x^i \in \mathbb{R}^\omega, \ h^i \in H_i, \ U_i \text{ open in } \mathbb{R}^\omega, \ g_i \in G,$$

such that the following conditions hold:

(1) $x^{i+1} = h^i \cdot x^i$;

(2) $x^i \in U_i \cap C \cap \bigcap_{m \geq i} C_m$;

(3) $h^i_j > 0$ for all $j < i$;

(4) $h^i_j < 2^{-i}$ for all $j < i$;

(5) $\overline{U_{i+1}} \subseteq U_i \subseteq O_i$;

(6) $g_i \cdot f(x^0) = f(x^i)$;

(7) $d_G(g_i, g_{i+1}) < 2^{-i}$.

Granting the construction, we will arrive at a contradiction as follows. By (1) and (4) $\lim_i x^i$ exists in $\mathbb{R}^\omega$, which we denote by x. By (3), x^0 and x are E_1-inequivalent. By (5) $x \in C$. However, by (7) there is $g_\infty \in G$ with $g_i \to g_\infty$ as $i \to \infty$. Thus by (2), (6), and the continuity of $f \restriction C$,

$$g_\infty \cdot f(x^0) = \lim_i g_i \cdot f(x^0) = \lim_i f(x^i) = f(x).$$

This shows that $f(x^0) E_G^X f(x)$, and by our assumption that f is a reduction, $x^0 E_1 x$, a contradiction.

It remains to construct the sequences by induction on i. To begin with, let $x^0 \in C \cap \bigcap_{m \geq 0} C_m$ be arbitrary. Since $x^0 \in O_0$, let U_0 be any open set in $\mathbb{R}^\omega$ with $x^0 \in U_0 \subseteq O_0$. Then let $g_0 = 1_G$. In the inductive step we assume that x^i, U_i, and g_i have been defined. We define h^i, U_{i+1} and g_{i+1} to satisfy the conditions (1) through (7) for appropriate indices.

Let W be an open nbhd of 1_G such that for all $g \in W$, $d_G(gg_i, g_i) < 2^{-i}$. By (ii) let V_0 be an open nbhd of the identity in $\mathbb{R}^i$ such that

$$\forall^* h \in V_0 \,\exists g \in W \,(\,f(h \cdot x^i) = g \cdot f(x^i)\,).$$

Let $D \subseteq V_0$ be a comeager set in V_0 witnessing the displayed property. Let $V \subseteq V_0$ be also an open nbhd of the identity of $\mathbb{R}^n$ such that $V \cdot x^i \subseteq U_i$ and for all $h \in V$ and $j < i$, $h_j < 2^{-i}$. Now note that the set $D \cap \{h \in V : h \cdot x^i \in C \cap \bigcap_{m \geq i} C_m\}$ is comeager in V. Since $\{h \in V : \forall j < i\ h_j > 0\}$ is an open subset of V, we have that

$$D \cap \{h \in V : h \cdot x^i \in C \cap \bigcap_{m \geq i} C_m \ \wedge\ \forall j < i\ h_j > 0\} \neq \emptyset.$$

Let h^i be an arbitrary element of this nonempty set. Then we have that $h^i \cdot x^i \in U_i \cap C \cap \bigcap_{m \geq i} C_m$. Let $x^{i+1} = h^i \cdot x^i$. Conditions (1), (3), and (4) are satisfied. Since $h^i \in D$, there is $g \in W$ such that $f(h^i \cdot x^i) = g \cdot f(x^i)$. Fix such a g, and let $g_{i+1} = g g_i$. Then $g \cdot f(x^i) = g g_i \cdot f(x^0) = g_{i+1} \cdot f(x^0)$, and since $g \in W$, $d_G(g_{i+1}, g_i) = d_G(g g_i, g_i) < 2^{-i}$. Thus (6) and (7) are fulfilled. Finally since $x_{i+1} \in U_i \cap C = U_i \cap \bigcap_i O_i$, we may find open set U_{i+1} such that $x_{i+1} \in U_{i+1} \subseteq \overline{U_{i+1}} \subseteq U_i \cap O_i$. This gives (2) and (5). We have thus finished the construction as required. □

Of course it follows from the theorem that c_{00} is not Polishable, since otherwise E_1 would be itself an orbit equivalence relation of a Polish group action.

Exercise 10.6.1 Give a direct proof that c_{00} is not Polishable.

Part III

Countable Model Theory

Chapter 11

Polish Topologies of Infinitary Logic

In this part of the book we study S_∞-orbit equivalence relations. In Sections 2.4 and 3.6 we have already seen connections between S_∞-orbit equivalence relations and isomorphism relations for countable models. In this part we will establish a full duality between the two subjects. Historically many of the model theoretic results predated and in fact motivated their group action counterpart. On the other hand, the study of isomorphism relation of countable structures also benefited from the perspective of Polish group actions. Since the realization that they are the same subject our understanding of countable model theory has advanced significantly.

11.1   A review of first-order logic

Recall that some notion of logic has been reviewed in Sections 2.4 and 3.6, where we gave a characterization of closed subgroups of S_∞ by automorphism groups of countable structures and discussed the logic actions of S_∞ on $\mathrm{Mod}(L)$ for countable relational languages L. In this section we review more concepts of logic.

Let $L = \{R_i\}_{i \in I}$ be a countable relational language, where R_i is an n_i-ary relation symbol. Fix once and for all **variable** symbols

$$v_0, v_1, \ldots, v_n, \ldots$$

and let $\mathfrak{V} = \{v_i : i \in \omega\}$. An **atomic** L-**formula** is an expression of the form $R_i(x_1, \ldots, x_{n_i})$, where $i \in I$ and $x_1, \ldots, x_{n_i} \in \mathfrak{V}$.

We will use the logical connectives

$$\neg,\ \wedge,\ \vee,\ \exists,\ \forall$$

and necessary parentheses to obtain more complicated formal expressions. A **negated atomic** L-**formula** is an expression of the form $\neg\varphi$, where φ is an atomic L-formula. The set of L-**formulas** is the smallest set $\mathfrak{F}$ of expressions satisfying the following closure properties:

(i) if φ is an atomic L-formula then $\varphi \in \mathfrak{F}$;

(ii) if $\varphi, \psi \in \mathfrak{F}$ then $\neg\varphi, \varphi \wedge \psi, \varphi \vee \psi \in \mathfrak{F}$;

(iii) if $\varphi \in \mathfrak{F}$ and $x \in \mathfrak{V}$, then $\exists x\varphi, \forall x\varphi \in \mathfrak{F}$.

It is clear that any formula is of finite length and the set $\mathfrak{F}$ is countable. As in the usual practice of formal mathematics, when forming more complicated L-formulas we use parentheses to eliminate any possible ambiguity in parsing the formula, and omit them when there is no danger of confusion. Thus any L-formula can be uniquely classified as one of the six forms mentioned in (i) through (iii). This allows us to prove statements about L-formulas by induction on their forms as well as make inductive definitions. For example, the concept of a **subformula** of an L-formula φ is inductively defined on the form of φ as follows:

(i) φ is a subformula of φ;

(ii) if $\varphi = \neg\psi$ then ψ is a subformula of φ; if $\varphi = \psi \wedge \psi'$ or $\varphi = \psi \vee \psi'$ then both ψ and ψ' are subformulas of φ;

(iii) if $\varphi = \exists x\psi$ or $\varphi = \forall x\psi$ then ψ is a subformula of φ;

(iv) if ϕ is a subformula of φ and ψ is a subformula of ϕ, then ψ is a subformula of φ.

Example 11.1.1
Let R be a binary relation symbol. The following formula φ attempts to describe R as an equivalence relation:

$$\forall v_0 R(v_0, v_0) \wedge \forall v_0 \forall v_1 (\neg R(v_0, v_1) \vee R(v_1, v_0))$$
$$\wedge \forall v_0 \forall v_1 \forall v_2 (\neg R(v_0, v_1) \vee \neg R(v_1, v_2) \vee R(v_0, v_2)).$$

The clause on symmetry $\neg R(v_0, v_1) \vee R(v_1, v_0)$ is a subformula of φ.

The example shows that it is desirable to introduce logical connectives $\rightarrow$ and $\leftrightarrow$ as informal substitutes of their formal counterparts. With these the symmetry clause can be replaced by a more readable expression $R(v_0, v_1) \rightarrow R(v_1, v_0)$.

A variable can occur in different subformulas of an L-formula. The occurrences sometimes have different properties. An occurrence in φ of a variable x is **bound** if it happens in a subformula of the form $\exists x\psi$ or $\forall x\psi$; otherwise the occurrence is **free**. A variable x is a **free variable** of φ if there is a free occurrence of x in φ. An L-formula is an L-**sentence** if it does not have any free variables.

Example 11.1.2
The formula φ defined in Example 11.1.1 is in fact a sentence. Consider the following formula ψ:

$$\varphi \wedge \exists v_1 \neg R(v_0, v_1).$$

The explicit occurrence of v_0 in ψ is free, making v_0 a free variable of ψ, despite the fact that v_0 also occurs bound in the subformula φ. The formula ψ attempts to address a hypothetical element whose equivalence class is not the whole domain.

In the above example the variable v_0 occurs both free and bound in the formula ψ, which is legitimate but confusing. It is clear that we may avoid this by replacing all bound occurrences of v_0 by a new variable symbol, say v_3, without changing the intended meaning of the formula. In general, it is not hard to show that any formula is logically equivalent to one with none of the variables occurring both free and bound.

A formula is called **quantifier-free** if it does not contain any quantifiers $\exists$ or $\forall$. A formula is in **prenex normal form** if it is of the form

$$Q_1 x_1 \ldots Q_k x_k \varphi$$

where $Q_1, \ldots, Q_k \in \{\exists, \forall\}$ are quantifiers and φ is quantifier-free. It is well known that any formula is logically equivalent to one in prenex normal form.

Before continuing we need to deal with a nuisance. Suppose in the above examples we need to describe a property of the equivalence relation that every equivalence class contains exactly one element. Then the natural formula to employ is

$$\forall v_0 \forall v_1 (R(v_0, v_1) \to v_0 = v_1).$$

However, under our definitions this is not a legitimate L-formula, since the equality symbol is not specified to be in L. Even if the symbol $=$ is added to the language, it is impossible to describe the desired properties of the equality relation with a formula. Therefore, our approach is to add $=$ as a default symbol beyond those in L and make sure that it is always interpreted as the true equality relation. To emphasize this distinction, we call an $(L \cup \{=\})$-formula an $L_{\omega\omega}$-**formula** and similarly handle other concepts defined above. Note that this is not the standard practice in the literature. Some authors choose to specify that L is a language with equality and continue to call our $L_{\omega\omega}$-formulas L-formulas.

What we have just defined is the syntax of first-order logic $L_{\omega\omega}$. In this notation the first subscript ω stands for the fact that only finitary (formally binary) conjunctions $\wedge$ and disjunctions $\vee$ are allowed in any formula, and the second subscript ω stands for the fact that only finitely many quantifiers $\exists$ or $\forall$ are allowed in any formula.

We now turn to the semantics of first-order logic. Consider an L-structure $M = (|M|, \{R_i^M\}_{i \in I})$ as defined in Section 2.4. An **assignment** is a function $\epsilon : \mathfrak{V} \to |M|$. For a variable $x \in \mathfrak{V}$ and an element $a \in |M|$, we define a new assignment $\epsilon[a/x]$ by letting

$$\epsilon[a/x](y) = \begin{cases} a, & \text{if } y = x, \\ \epsilon(y), & \text{otherwise.} \end{cases}$$

Given an $L_{\omega\omega}$-formula φ, we define a **satisfaction relation** $(M, \epsilon) \models \varphi$ by induction on the form of φ as follows:

(i) If φ is $x = y$ for $x, y \in \mathfrak{V}$, then $(M, \epsilon) \models \varphi$ iff $\epsilon(x) = \epsilon(y)$.

(ii) If φ is $R_i(x_1, \ldots, x_{n_i})$ for $i \in I$ and $x_1, \ldots, x_{n_i} \in \mathfrak{V}$, then $(M, \epsilon) \models \varphi$ iff $R_i^M(\epsilon(x_1), \ldots, \epsilon(x_{n_i}))$ holds.

(iii) If φ is $\neg\psi$, then $(M, \epsilon) \models \varphi$ iff $(M, \epsilon) \not\models \psi$.

(iv) If φ is $\phi \wedge \psi$, then $(M, \epsilon) \models \varphi$ iff both $(M, \epsilon) \models \phi$ and $(M, \epsilon) \models \psi$.

(v) If φ is $\phi \vee \psi$, then $(M, \epsilon) \models \varphi$ iff either $(M, \epsilon) \models \phi$ or $(M, \epsilon) \models \psi$.

(vi) If φ is $\exists x \psi$ where $x \in \mathfrak{V}$, then $(M, \epsilon) \models \varphi$ iff there is $a \in |M|$ such that $(M, \epsilon[a/x]) \models \psi$.

(vii) If φ is $\forall x \psi$ where $x \in \mathfrak{V}$, then $(M, \epsilon) \models \varphi$ iff for all $a \in |M|$, $(M, \epsilon[a/x]) \models \psi$.

We say that M **satisfies** φ, or M **models** φ, and denote it by $M \models \varphi$, if $(M, \epsilon) \models \varphi$ for all assignments ϵ. These definitions make precise our usual interpretation of formulas by statements about the model.

It is easy to check that if φ is a sentence, that is, every variable in φ is bound, then for any assignment ϵ, $(M, \epsilon) \models \varphi$ iff $M \models \varphi$. The following notation will be handy. If ϵ is an assignment, $x_1, \ldots, x_k \in \mathfrak{V}$ are distinct, and $a_1, \ldots, a_k \in |M|$ arbitrary, we define

$$\epsilon[a_1/x_1, \ldots, a_k/x_k] = \epsilon[a_1/x_1] \ldots [a_k/x_k].$$

If φ is a formula with its free variables among $x_1, \ldots, x_k$, we will write φ as $\varphi(x_1, \ldots, x_k)$ to emphasize it. For $a_1, \ldots, a_k \in |M|$, we write

$$M \models \varphi[a_1/x_1, \ldots, a_k/x_k], \text{ or simply } M \models \varphi[a_1, \ldots, a_k],$$

if for any assignment ϵ, $(M, \epsilon[a_1/x_1, \ldots, a_k/x_k]) \models \varphi$. It is not hard to check that the definition makes sense since the satisfaction relation $(M, \epsilon) \models \varphi(x_1, \ldots, x_k)$ does not depend on the values of ϵ outside $\{x_1, \ldots, x_k\}$.

An $L_{\omega\omega}$-**theory**, or simply a **theory**, T is a set of $L_{\omega\omega}$-sentences. A nonempty L-structure M is a **model** of T, and we denote $M \models T$, if $M \models \varphi$ for all $\varphi \in T$. The theory T is **consistent** if there exists a model of T, and **inconsistent** otherwise. The famous compactness theorem of Gödel states that a theory T is consistent iff every finite subset of T is consistent. On the other hand, the **theory of an L-structure** M, denoted $\mathrm{Th}(M)$, is defined by

$$\mathrm{Th}(M) = \{\varphi : \varphi \text{ is an } L_{\omega\omega}\text{-sentence and } M \models \varphi\}.$$

A theory T is **complete** if $T = \mathrm{Th}(M)$ for some L-structure M. Note that by our definition for any $L_{\omega\omega}$-sentence φ and complete theory T, either $\varphi \in T$ or $\neg\varphi \in T$.

In a refined analysis we define $L_{\omega\omega}$-types. For $n \in \omega$, an $L_{\omega\omega}$-n-**type**, or simply an n-**type**, S is a set of $L_{\omega\omega}$-formulas such that for some distinct variables $x_1, \ldots, x_n \in \mathfrak{V}$ all formulas in S have free variables among $x_1, \ldots, x_n$. In this situation we write $S(x_1, \ldots, x_n)$ to emphasize the hypothesis that all formulas in S have free variables among $x_1, \ldots, x_n$. A **type** is an n-type for some $n \in \omega$. If $S(x_1, \ldots, x_n)$ is an n-type, M an L-structure, and $a_1, \ldots, a_n \in M$, we say that $(a_1, \ldots, a_n)$ **realizes** S, and write $M \models S[a_1, \ldots, a_n]$, if for all $\varphi(x_1, \ldots, x_n) \in S$, $M \models \varphi[a_1, \ldots, a_n]$. An n-type $S(x_1, \ldots, x_n)$ is **consistent with a theory** T if there is a model M of T and $a_1, \ldots, a_n \in |M|$ such that $M \models S[a_1, \ldots, a_n]$, and **inconsistent with** T otherwise. In this case we say that S is a **type of** T. A type is **consistent** (**inconsistent**) if it is consistent (inconsistent, respectively) with the empty theory. A type $S(x_1, \ldots, x_n)$ is a **complete type of a theory** T if for all formulas $\varphi(x_1, \ldots, x_n)$, either $S \cup \{\varphi\}$ or $S \cup \{\neg\varphi\}$ is inconsistent with T. A type is **complete** if it is a complete type of the empty theory. For an L-structure and $a_1, \ldots, a_n \in |M|$, the **type of the tuple** $(a_1, \ldots, a_n)$ is defined as

$$\mathrm{tp}(a_1, \ldots, a_n) = \{\varphi(v_0, \ldots, v_{n-1}) : M \models \varphi[a_1, \ldots, a_n]\}.$$

If M is a model of a theory T and $a_1, \ldots, a_n \in |M|$, then the type of the tuple $(a_1, \ldots, a_n)$ is obviously a complete type consistent with T.

Exercise 11.1.1 Let $L = \{<\}$. Find $L_{\omega\omega}$-formulas whose intended interpretations are the following:

(1) $<$ is a linear order.

(2) $<$ is dense linear order without endpoints.

(3) $<$ is a linear order of cardinality n for some fixed $n \in \omega$.

Exercise 11.1.2 Find a relational language L and an $L_{\omega\omega}$-sentence φ such that the models of φ are exactly Boolean algebras.

Exercise 11.1.3 Let R be a 3-ary relation symbol and $L = \{R\}$. Find an $L_{\omega\omega}$-sentence φ such that all models M of φ are groups with the multiplication defined by $a \cdot b = c$ iff $M \models R[a, b, c]$.

Exercise 11.1.4 Show that if $\varphi(x_1, \ldots, x_k)$ is a formula whose free variables are among $x_1, \ldots, x_k$ and ϵ_1, ϵ_2 are assignments such that $\epsilon_1(x_j) = \epsilon_2(x_j)$ for $j = 1, \ldots, k$, then $(M, \epsilon_1) \models \varphi$ iff $(M, \epsilon_2) \models \varphi$.

Exercise 11.1.5 Let L be the empty language.

(a) Find an $L_{\omega\omega}$-theory T such that the models of T are exactly infinite sets.

(b) Use the compactness theorem to show that there is no $L_{\omega\omega}$-sentence φ such that the models of φ are exactly infinite sets. (*Hint:* Assume such φ exists. Let T be given by (a) and consider $T \cup \{\neg\varphi\}$.)

11.2   Model theory of infinitary logic

In this section we define the infinitary logic $L_{\omega_1\omega}$ and review more concepts that eventually will lead to connections with invariant descriptive set theory. For a more comprehensive treatment of infinitary logic the reader should consult Reference [3] or [105].

We continue to consider a countable relational language L. To define the infinitary logic we use the new logical connectives $\bigwedge$ and $\bigvee$.

Definition 11.2.1
Let L be a countable relational language. The set $\mathfrak{F}_{\omega_1}$ of $L_{\omega_1\omega}$-**formulas** is the smallest set of expressions satisfying the following closure properties:

(i) if φ is an atomic $L_{\omega\omega}$-formula then $\varphi \in \mathfrak{F}_{\omega_1}$;

(ii) if $\varphi \in \mathfrak{F}_{\omega_1}$ then $\neg\varphi \in \mathfrak{F}_{\omega_1}$;

(iii) if $\varphi \in \mathfrak{F}_{\omega_1}$ and $x \in \mathfrak{V}$ then $\exists x\varphi, \forall x\varphi \in \mathfrak{F}_{\omega_1}$;

(iv) if Φ is a countable subset of $\mathfrak{F}_{\omega_1}$ then $\bigwedge \Phi, \bigvee \Phi \in \mathfrak{F}_{\omega_1}$.

Similar to the finitary case we can define the notion of a **subformula** by induction on the form of a formula φ as follows:

(i) φ is a subformula of φ;

(ii) if $\varphi = \neg\psi$, or $\varphi = \exists x\psi$, or $\varphi = \exists x\psi$ for some $x \in \mathfrak{V}$, then ψ is a subformula of φ;

(iii) if $\varphi = \bigwedge \Phi$ or $\varphi = \bigwedge \Phi$ for some countable $\Phi \subseteq \mathfrak{F}_{\omega_1}$, and $\psi \in \Phi$, then ψ is a subformula of φ;

(iv) if ϕ is a subformula of φ and ψ is a subformula of ϕ then ψ is a subformula of φ.

Also in a similar fashion one can define the notion of free and bound variables. Then an $L_{\omega_1\omega}$-formula is an $L_{\omega_1\omega}$-**sentence** if it does not have any free variables. The following simple lemma can be proved using induction.

Lemma 11.2.2
If φ is an $L_{\omega_1\omega}$-formula with only finitely many free variables, then so does any subformula of φ. In particular, if ψ is a subformula of an $L_{\omega_1\omega}$-sentence then ψ has only finitely many free variables.

The semantics of $L_{\omega_1\omega}$ is also similar to the finitary case, with the only obvious modification being the following:

If M is an L-structure, ϵ an assignment, and Φ a countable subset of $L_{\omega_1\omega}$-formulas, then

$$(M, \epsilon) \models \bigwedge \Phi \text{ iff for all } \varphi \in \Phi, (M, \epsilon) \models \varphi,$$

and

$$(M, \epsilon) \models \bigvee \Phi \text{ iff for some } \varphi \in \Phi, (M, \epsilon) \models \varphi.$$

We also use other concepts, notation, and conventions of first-order logic whenever they are applicable to $L_{\omega_1\omega}$. Obviously the infinitary logic has more expressive power. Note that any $L_{\omega\omega}$-theory T is countable, and therefore is logically equivalent to the infinitary sentence $\bigwedge T$. The set $\mathfrak{F}_{\omega_1}$ has the size of the continuum, but every $L_{\omega_1\omega}$-formula has only countably many subformulas.

Definition 11.2.3
A set F of $L_{\omega_1\omega}$-formulas is a **fragment** if it satisfies the following closure properties:

(i) all $L_{\omega\omega}$-formulas are in F;

(ii) if $\varphi \in F$ and ψ is a subformula of φ, then $\psi \in F$;

(iii) if $\varphi, \psi \in F$ and $x \in \mathfrak{V}$, then $\neg\varphi, \varphi \wedge \psi, \varphi \vee \psi, \exists x \varphi, \forall x \varphi \in F$.

(iv) if $\varphi \in F$ and ψ is obtained from φ by a change of variables, then $\psi \in F$.

If $A \subseteq \mathfrak{F}_{\omega_1}$ then the smallest fragment containing A is called the fragment **generated by** A and is denoted $F(A)$.

If $A \subseteq \mathfrak{F}_{\omega_1}$ is countable, then so is $F(A)$. In particular, if A contains a single formula φ, then we write $F(\varphi)$ for $F(\{\varphi\})$. Apparently the set of all $L_{\omega\omega}$-formulas is the smallest fragment.

We now turn to some concepts in the model theory of $L_{\omega_1\omega}$.

Definition 11.2.4
Let M, N be L-structures and F a fragment.

(1) An $L_{\omega_1\omega}$-**theory** is a set of $L_{\omega_1\omega}$-sentences. An F-**theory** is a set of sentences in F.

(2) The F-**theory** of M, denoted $\text{Th}_F(M)$, is the set of all sentences $\varphi \in F$ such that $M \models \varphi$.

(3) M and N are F-**elementarily equivalent**, denoted $M \equiv_F N$, if $\text{Th}_F(M) = \text{Th}_F(N)$, that is, for any sentence $\varphi \in F$, $M \models \varphi$ iff $N \models \varphi$.

(4) An injection $f : |M| \to |N|$ is an F-**elementary embedding** if for any formula $\varphi(x_1, \ldots, x_n) \in F$ all of whose free variables are among $x_1, \ldots, x_n \in \mathfrak{V}$, and for any $a_1, \ldots, a_n \in M$,

$$M \models \varphi[a_1, \ldots, a_n] \iff N \models \varphi[f(a_1), \ldots, f(a_n)].$$

(5) M is a **substructure**, or a **submodel**, of N if $|M| \subseteq |N|$ and for any $R_i \in L$ and $a_1, \ldots, a_{n_i} \in M$,

$$M \models R_i[a_1, \ldots, a_{n_i}] \iff N \models R_i[a_1, \ldots, a_{n_i}].$$

(6) M is an F-**elementary substructure** of N, denoted $M \prec_F N$, if M is a substructure of N and the identity embedding is an F-elementary embedding.

All concepts defined here generalize some similar ones in the first-order logic. We denote these first-order concepts by dropping the subscript F in the corresponding notation.

Since L is a relational language, any subset in an L-structure gives rise to a substructure. Thus in our context, substructures correspond uniquely to subsets of structures. We therefore adopt the following notation. If M is a substructure of N, we write $M \subseteq N$. Conversely, if N is an L-structure and S is a subset of $|N|$ then we also use S to denote the unique substructure of N determined by S. The following lemma collects some easy facts about the concepts defined above.

Lemma 11.2.5
Let L be a countable relational language, F a fragment, and M, N, P L-structures. Then the following hold:

(a) *If f is an F-elementary embedding from M into N, then $f : M \cong f(M)$ and $f(M) \prec_F N$.*

(b) *If $M \prec_F N$, then $M \equiv_F N$.*

(c) *If $M \prec_F P$ and $P \prec_F N$, then $M \prec_F N$.*

Proof. (a) It follows from the F-elementarity for atomic formulas that $f : M \cong f(M)$. Now suppose $\varphi(x_1, \ldots, x_n) \in F$ and $b_1, \ldots, b_n \in f(M)$. Let $a_1 = f^{-1}(b_1), \ldots, a_n = f^{-1}(b_n)$. Then $f(M) \models \varphi[b_1, \ldots, b_n]$ iff $M \models \varphi[a_1, \ldots, a_n]$ since f is an isomorphism. But $M \models \varphi[a_1, \ldots, a_n]$ iff $N \models \varphi[b_1, \ldots, b_n]$ by the assumption that f is an F-elementary embedding. Thus we have $f(M) \models \varphi[b_1, \ldots, b_n]$ iff $N \models \varphi[b_1, \ldots, b_n]$. This shows that $f(M) \prec_F N$. (b) and (c) are obvious. $\Box$

The following lemma is usually referred to as the **Tarski–Vaught criterion**.

Lemma 11.2.6 (Tarski–Vaught)
Let L be a countable relational language, F a fragment, N an L-structure, and $M \subseteq N$. Then $M \prec_F N$ iff for all formulas $\varphi(x_1, \ldots, x_n, y) \in F$ and $a_1, \ldots, a_n \in M$, if there is $b \in N$ such that $N \models \varphi[a_1, \ldots, a_n, b]$ then there is $b_0 \in M$ such that $N \models \varphi[a_1, \ldots, a_n, b_0]$.

Proof. For ($\Rightarrow$) suppose there is $b \in N$ such that $N \models \varphi[a_1, \ldots, a_n, b]$. Then $\exists y \varphi \in F$ and $N \models \exists y \varphi[a_1, \ldots, a_n]$, and therefore $M \models \exists y \varphi[a_1, \ldots, a_n]$ by elementarity. This gives some $b_0 \in M$ such that $M \models \varphi[a_1, \ldots, a_n, b_0]$. But by elementarity again we have $N \models \varphi[a_1, \ldots, a_n, b_0]$ as desired. For ($\Leftarrow$), we show by induction on the form of a formula $\psi(x_1, \ldots, x_n) \in F$ that for any $a_1, \ldots, a_n \in M$, $M \models \psi[a_1, \ldots, a_n]$ iff $N \models \psi[a_1, \ldots, a_n]$. The case when ψ is atomic is immediate, since $M \subseteq N$. The case when φ is $\neg \psi$ is trivial. If ψ is $\bigwedge \Phi$ for a countable set $\Phi \subseteq F$, then every formula in Φ has free variables among $x_1, \ldots, x_n$, and by the inductive hypotheses

$$\begin{aligned} M \models \psi[a_1, \ldots, a_n] &\iff \text{for all } \phi \in \Phi, \ M \models \phi[a_1, \ldots, a_n] \\ &\iff \text{for all } \phi \in \Phi, \ N \models \phi[a_1, \ldots, a_n] \\ &\iff N \models \psi[a_1, \ldots, a_n]. \end{aligned}$$

The case when ψ is $\bigvee \Phi$ is similar. Now if ψ is $\exists y \varphi$, then the free variables of φ are among $x_1, \ldots, x_n, y$. Suppose $M \models \psi[a_1, \ldots, a_n]$. Then there is $b_0 \in M$ with $M \models \varphi[a_1, \ldots, a_n, b_0]$, and by the inductive hypothesis $N \models \varphi[a_1, \ldots, a_n b_0]$, which gives that $N \models \psi[a_1, \ldots, a_n]$. Conversely, suppose $N \models \psi[a_1, \ldots, a_n]$. Then there is $b \in N$ such that $N \models \varphi[a_1, \ldots, a_n, b]$. By our assumption there is $b_0 \in M$ such that $N \models \varphi[a_1, \ldots, a_n, b]$ and by the inductive hypothesis $M \models \varphi[a_1, \ldots, a_n, b_0]$, which gives that $M \models \psi[a_1, \ldots, a_n]$. This shows that $M \models \psi[a_1, \ldots, a_n]$ iff $N \models \psi[a_1, \ldots, a_n]$. Finally if ψ is $\forall y \varphi$, then we apply the above argument to $\neg \varphi(x_1, \ldots, x_n, y)$ and get $M \models \exists y \neg \varphi[a_1, \ldots, a_n]$ iff $N \models \exists y \neg \varphi[a_1, \ldots, a_n]$. Thus

$$\begin{aligned} M \models \psi[a_1, \ldots, a_n] &\iff M \not\models \exists y \neg \varphi[a_1, \ldots, a_n] \\ &\iff N \not\models \exists y \neg \varphi[a_1, \ldots, a_n] \\ &\iff N \models \psi[a_1, \ldots, a_n]. \end{aligned}$$

☐

Thus the elementarity of a submodel lies on the existence of witnesses for existential formulas.

Exercise 11.2.1 Show that every $L_{\omega_1 \omega}$-formula has only countably many subformulas.

Exercise 11.2.2 Prove Lemma 11.2.2.

Exercise 11.2.3 Recall that for a prime number p a p-group is a group in which every element has order p^n for some $n \in \omega$. Find a countable relational language L and an $L_{\omega_1 \omega}$-sentence φ such that the models of φ are exactly the abelian p-groups. (Compare Exercise 11.1.3.)

Exercise 11.2.4 Let $L = \{<\}$ and $\alpha < \omega_1$. Find an $L_{\omega_1 \omega}$-sentence φ_α such that for any L-structure M, $M \models \varphi_\alpha$ iff $\mathrm{ot}(M) = \alpha$.

Exercise 11.2.5 Show that if M is a substructure of N then for any quantifier-free formula $\varphi(x_1, \ldots, x_n)$ and $a_1, \ldots, a_n \in M$, $M \models \varphi[a_1, \ldots, a_n]$ iff $N \models \varphi[a_1, \ldots, a_n]$.

Exercise 11.2.6 Let M be an L-structure and $(M_n)_{n \in \omega}$ a sequence of substructures of M. Suppose $M = \bigcup_n M_n$ and $M_n \prec_F M_{n+1}$ for all $n \in \omega$. Show that $M_n \prec_F M$ for all $n \in \omega$.

Exercise 11.2.7 Let L be a countable relational language, F a fragment, N an L-structure, and $M \subseteq N$. Show that $M \prec_F N$ iff for all formulas $\varphi(x_1, \ldots, x_n, y) \in F$ and $a_1, \ldots, a_n \in M$, $M \models \exists y \varphi[a_1, \ldots, a_n]$ iff $N \models \exists y \varphi[a_1, \ldots, a_n]$.

11.3 Invariant Borel classes of countable models

In this section we consider the canonical topology and invariant Borel subsets of $\mathrm{Mod}(L)$ following the approach of Vaught [161].

Let $L = \{R_i\}_{i \in I}$ be a countable relational language, with each R_i an n_i-ary relation symbol. Recall from Section 3.5 that the bijection

$$\prod_{i \in I} 2^{\omega^{n_i}} = X_L \longrightarrow \mathrm{Mod}(L)$$

$$x \longmapsto M_x$$

associates with each element of the product space X_L a countable L-structure. With this natural correspondence $\mathrm{Mod}(L)$ becomes a compact Polish space, and we call this topology on $\mathrm{Mod}(L)$ the **canonical topology**. For a refined analysis we use the following notation.

Definition 11.3.1

For any $L_{\omega_1\omega}$-formula $\varphi(x_1, \ldots, x_n)$ and tuple $\vec{a} = (a_1, \ldots, a_n) \in \omega^n$, let

$$\mathrm{Mod}(\varphi, \vec{a}) = \{M \in \mathrm{Mod}(L) : M \models \varphi[a_1, \ldots, a_n]\}.$$

In particular, if φ is a sentence, we let

$$\mathrm{Mod}(\varphi) = \{M \in \mathrm{Mod}(L) : M \models \varphi\}.$$

For a set A of $L_{\omega_1\omega}$-formulas, the **topology generated by** A, denoted t_A, is the topology on $\mathrm{Mod}(L)$ generated by the subbase

$$\mathcal{B}_A = \{\mathrm{Mod}(\varphi, \vec{a}) : \varphi \in A, \ \vec{a} \in \omega^{<\omega}\}.$$

We denote by $\mathcal{B}_{\text{atom}}$ the subbase given by all atomic and negated atomic $L_{\omega\omega}$-formulas, and t_{atom} the topology generated by $\mathcal{B}_{\text{atom}}$. With this notation the canonical topology on $\text{Mod}(L)$ given by X_L is just t_{atom}. Similarly, denote by $\mathcal{B}_{\text{qf}}$ the subbase given by all quantifier-free $L_{\omega\omega}$-formulas, and t_{qf} the topology generated by $\mathcal{B}_{\text{qf}}$. Since quantifier-free formulas are exactly the Boolean combinations of atomic formulas, and in particular atomic and negated atomic formulas are quantifier-free, the topologies t_{atom} and t_{qf} are the same. In the following lemma we give another alternative subbase. For $n \in \omega$ we let $\vec{n}$ denote the tuple $(0, 1, \ldots, n-1) \in \omega^n$. Note that $\vec{0}$ is the empty tuple $\emptyset$.

Lemma 11.3.2

Let

$$\mathcal{B}_0 = \{\, \text{Mod}(\varphi, \vec{n}) \ : \ n \in \omega, \ \varphi \text{ is a quantifier-free } L_{\omega\omega}\text{-formula} \,\}.$$

Then $\mathcal{B}_0$ is a subbase for the canonical topology on $\text{Mod}(L)$.

Proof. We show that for any quantifier-free $L_{\omega\omega}$-formula $\psi(x_1, \ldots, x_k)$ and any tuple $\vec{a} = (a_1, \ldots, a_k) \in \omega^k$, there is a quantifier-free formula φ and $n \in \omega$ such that

$$\text{Mod}(\varphi, \vec{n}) = \text{Mod}(\psi, \vec{a}).$$

First note that we may assume without loss of generality that $a_1, \ldots, a_k$ are distinct. In fact, suppose $a_i = a_j$ for $i < j$. Then we let $\vec{a}'$ be obtained from the tuple $\vec{a}$ by removing a_j, that is, $\vec{a}' = (a_1, \ldots, a_{j-1}, a_{j+1}, \ldots, a_k)$. Also let ψ' be obtained from ψ by replacing all occurrences of x_j by x_i. Then for any $N \in \text{Mod}(L)$, $N \models \psi[\vec{a}]$ iff $N \models \psi'[\vec{a}']$. Therefore repetitions in the tuple $\vec{a}$ can be eliminated by iteratively applying this procedure.

Next we may assume that $a_1 < \cdots < a_k$. In fact, if $a_{j_1} < \cdots < a_{j_k}$ is an enumeration of the tuple $\vec{a}$ in the increasing order, then we may let $\vec{a}'' = (a_{j_1}, \ldots, a_{j_k})$ and ψ'' be obtained from ψ by replacing all occurrences of x_i by x_{j_i} for all $i = 1, \ldots, k$. For any $N \in \text{Mod}(L)$, we still have that $N \models \psi[\vec{a}]$ iff $N \models \psi''[\vec{a}'']$.

Finally let $n = a_k + 1$ and φ be obtained from ψ by replacing all occurrences of x_i by v_{a_i} for all $i = 1, \ldots, k$. Note that all (free) variables of ψ are among $x_1, \ldots, x_k$, and it follows that all variables of φ are among $v_{a_1}, \ldots, v_{a_k}$. In particular, the free variables of φ are among $v_0, \ldots, v_{n-1}$. It is again obvious that $\text{Mod}(\psi, \vec{a}) = \text{Mod}(\varphi, \vec{n})$. $\square$

Note that we actually showed that $\mathcal{B}_0 = \mathcal{B}_{\text{qf}}$.

Lemma 11.3.3

For any $L_{\omega_1\omega}$-formula $\varphi(x_1, \ldots, x_n)$ and $\vec{a} = (a_1, \ldots, a_n) \in \omega^n$, $\text{Mod}(\varphi, \vec{a})$ is a Borel subset of $\text{Mod}(L)$.

Proof. This is proved by induction on all $L_{\omega_1\omega}$-formulas with finitely many free variables. If φ is atomic then $\mathrm{Mod}(\varphi, \bar{a})$ is open for any $\bar{a}$. The inductive cases follow from the following computations:

$$\mathrm{Mod}(\bigwedge \Phi, \bar{a}) = \bigcap_{\psi \in \Phi} \mathrm{Mod}(\psi, \bar{a}), \qquad \mathrm{Mod}(\bigvee \Phi, \bar{a}) = \bigcup_{\psi \in \Phi} \mathrm{Mod}(\psi, \bar{a}),$$

$$\mathrm{Mod}(\exists x\psi, \bar{a}) = \bigcup_{b \in \omega} \mathrm{Mod}(\psi, \bar{a}\,\hat{}\,b), \qquad \mathrm{Mod}(\forall x\psi, \bar{a}) = \bigcap_{b \in \omega} \mathrm{Mod}(\psi, \bar{a}\,\hat{}\,b).$$

$\square$

Also recall from Section 3.5 that the isomorphism relation among L-structures is an orbit equivalence relation induced by the logic action of S_∞. Specifically, for $g \in S_\infty$, $M \in \mathrm{Mod}(L)$, and any n-ary $R \in L$,

$$R^{g \cdot M}(a_1, \ldots, a_n) \iff R^M(g^{-1}(a_1), \ldots, g^{-1}(a_n))$$

for any $\bar{a} = (a_1, \ldots, a_n) \in \omega^n$. In the notation of logic, the action is determined by specifying that

$$g \cdot M \models R[\bar{a}] \iff M \models R[g^{-1}(\bar{a})].$$

The logic action is clearly continuous.

Definition 11.3.4
Let L be a countable relational language. An **invariant Borel class** of countable L-structures is an S_∞-invariant Borel subset of $\mathrm{Mod}(L)$.

If φ is an $L_{\omega_1\omega}$-sentence then $\mathrm{Mod}(\varphi)$ is Borel by Lemma 11.3.3, and isomorphism invariant since if $M \cong N$ then $M \models \varphi$ iff $N \models \varphi$. Thus for any sentence φ, $\mathrm{Mod}(\varphi)$ is an invariant Borel class. Next we show that the converse is true. For this we use the Vaught transforms and the following notation.

Let $[\omega]^n$ denote the set of tuples $(a_0, \ldots, a_{n-1}) \in \omega^n$ where $a_0, \ldots, a_{n-1}$ are distinct. Note that $[\omega]^0 = \{\emptyset\}$. For each $\bar{a} = (a_0, \ldots, a_{n-1}) \in [\omega]^n$, let $N_{\bar{a}} \subseteq S_\infty$ be the set of all elements $f \in S_\infty$ such that $f(a_i) = i$ for $i < n$. Note that $N_\emptyset = S_\infty$. For any set $B \subseteq \mathrm{Mod}(L)$ and $n \in \omega$, define $B^{[*n]}, B^{[\triangle n]} \subseteq \mathrm{Mod}(L) \times [\omega]^n$ by

$$B^{[\triangle n]} = \left\{ (M, \bar{a}) : M \in B^{\triangle N_{\bar{a}}} \right\}, \quad B^{[*n]} = \left\{ (M, \bar{a}) : M \in B^{* N_{\bar{a}}} \right\}.$$

Lemma 11.3.5
If $B \subseteq \mathrm{Mod}(L)$ is Borel, then for any $n \in \omega$, there are $L_{\omega_1\omega}$-formulas $\varphi_n(x_1, \ldots, x_n)$ and $\psi_n(x_1, \ldots, x_n)$ such that

$$(M, \bar{a}) \in B^{[\triangle n]} \iff M \in \mathrm{Mod}(\varphi_n, \bar{a}),$$

$$(M, \bar{a}) \in B^{[*n]} \iff M \in \mathrm{Mod}(\psi_n, \bar{a}).$$

Proof. Let $\mathcal{B}_0$ be the subbase defined in Lemma 11.3.2. Let $\mathcal{B}$ be the collection of all sets $B \subseteq \mathrm{Mod}(L)$ such that the conclusion of the lemma holds. We show that $\mathcal{B}_0 \subseteq \mathcal{B}$ and that $\mathcal{B}$ is a σ-algebra.

In the definition of the formulas we use the abbreviation $(\forall y_1 \ldots y_k)^{\neq} \psi$ to stand for the formula

$$\forall y_1 \ldots \forall y_k \left(\bigwedge_{1 \le i < j \le k} y_i \ne y_j \rightarrow \psi \right).$$

For $\mathcal{B}_0 \subseteq \mathcal{B}$ we let φ be quantifier-free and $n \in \omega$, and check that $\mathrm{Mod}(\varphi, \vec{n}) \in \mathcal{B}$. Note that for any $M \in \mathrm{Mod}(L)$, $\vec{a} \in [\omega]^n$ and $g \in N_{\vec{a}}$,

$$g \cdot M \models \varphi[\vec{n}] \iff M \models \varphi[g^{-1}(\vec{n})] \iff M \models \varphi[\vec{a}],$$

where the right-hand side is independent of g. Thus

$$(M, \vec{a}) \in \mathrm{Mod}(\varphi, \vec{n})^{[\triangle n]} \iff \exists^* g \in N_{\vec{a}} \; g \cdot M \in \mathrm{Mod}(\varphi, \vec{n}) \iff M \models \varphi[\vec{a}].$$

Therefore we can take $\varphi_n = \varphi$. Similarly we can also take $\psi_n = \varphi$. This shows that $\mathrm{Mod}(\varphi, \vec{a}) \in \mathcal{B}$.

Next suppose $B_m \in \mathcal{B}$ for $m \in \omega$ and let φ_n^m and ψ_n^m be witnesses. Suppose the free variable of φ_n^m and ψ_n^m are among $x_1, \ldots, x_n$. We show that $\bigcup_{m \in \omega} B_m \in \mathcal{B}$. Let $\varphi_n = \bigvee_{m \in \omega} \varphi_n^m$. Then

$$(M, \vec{a}) \in \left(\bigcup_{m \in \omega} B_m \right)^{[\triangle n]} \iff \exists^* g \in N_{\vec{a}} \; g \cdot M \in \bigcup_{m \in \omega} B_m$$

$$\iff \exists m \in \omega \; \exists^* g \in N_{\vec{a}} \; g \cdot M \in B_m$$

$$\iff \exists m \in \omega \; M \models \varphi_n^m[\vec{a}]$$

$$\iff M \models \varphi_n[\vec{a}].$$

The $*$-transform is a bit more complicated. Fix $n \in \omega$. For each $k \in \omega$ we let

$$\psi_{n,k} = (\forall y_1 \ldots y_k)^{\neq} \left(\bigwedge_{\substack{1 \le i \le n \\ 1 \le j \le k}} x_i \ne y_k \rightarrow \bigvee_{m \in \omega} \varphi_{n+k}^m(x_1, \ldots, x_n, y_1, \ldots, y_k) \right),$$

and put $\psi_n = \bigwedge_{k \in \omega} \psi_{n,k}$. Note that the free variables of $\psi_{n,k}$ and ψ_n are still among $x_1, \ldots, x_n$. Now we have

$$(M, \vec{a}) \in \left(\bigcup_{m \in \omega} B_m \right)^{[*n]}$$

$$\Longleftrightarrow \forall^* g \in N_{\vec{a}} \, \exists m \in \omega \, g \cdot M \in B_m$$

$$\Longleftrightarrow \forall \vec{b} \left(\vec{a}{}^{\frown}\vec{b} \in [\omega]^{n+k} \rightarrow \exists m \in \omega \, \exists^* g \in N_{\vec{a}{}^{\frown}\vec{b}} \, g \cdot M \in B_m \right)$$

$$\Longleftrightarrow \forall \vec{b} \left(\vec{a}{}^{\frown}\vec{b} \in [\omega]^{n+k} \rightarrow M \models \bigvee_{m \in \omega} \varphi_{n+k}^m[\vec{a}{}^{\frown}\vec{b}] \right)$$

$$\Longleftrightarrow M \models \psi_n[\vec{a}].$$

It remains to show that if $B \in \mathcal{B}$ then $\mathrm{Mod}(L) - B \in \mathcal{B}$. For this let φ_n and ψ_n be the witnesses of $B \in \mathcal{B}$. Then it is easy to see that

$$(M, \vec{a}) \in (\mathrm{Mod}(L) - B)^{[\triangle n]} \iff M \models \neg \psi_n[\vec{a}]$$

and

$$(M, \vec{a}) \in (\mathrm{Mod}(L) - B)^{[*n]} \iff M \models \neg \varphi_n[\vec{a}].$$

▯

Note that the lemma also applies in the case $n = 0$.

Theorem 11.3.6
Let L be a countable relational language and $\mathcal{C} \subseteq \mathrm{Mod}(L)$. Then $\mathcal{C}$ is an invariant Borel class iff there is an $L_{\omega_1\omega}$-sentence φ such that $\mathcal{C} = \mathrm{Mod}(\varphi)$.

Proof. The implication ($\Leftarrow$) is clear. For ($\Rightarrow$) let $\mathcal{C}$ be an invariant Borel class. Then $\mathcal{C} = \mathcal{C}^{\triangle S_\infty}$ and $\mathcal{C}^{[\triangle 0]} = \mathcal{C} \times \{\emptyset\}$. By Lemma 11.3.5 for $n = 0$, there is a sentence φ such that $(M, \emptyset) \in \mathcal{C}^{[\triangle 0]}$ iff $M \in \mathrm{Mod}(\varphi)$. Thus $\mathcal{C} = \mathrm{Mod}(\varphi)$.
▯

Theorem 11.3.6 has far-reaching consequences on the isomorphism relations of invariant Borel classes.

Notation 11.3.7
Let L be a countable relational language. For any $L_{\omega_1\omega}$-sentence φ we let $\cong_\varphi$ denote the isomorphism relation $\cong \restriction \mathrm{Mod}(\varphi)$.

Since $\mathrm{Mod}(\varphi)$ is an invariant Borel subset of $\mathrm{Mod}(L)$, it is a Borel S_∞-space with the action inherited from the logic action. Thus $\cong_\varphi$ is an S_∞-orbit equivalence relation. The following theorem establishes the converse.

Theorem 11.3.8
For any Borel S_∞-space X there is a countable relational language L and an $L_{\omega_1\omega}$-sentence φ such that $E^X_{S_\infty}$ is Borel isomorphic to $\cong_\varphi$. In particular, $E^X_{S_\infty} \sim_B \cong_\varphi$.

Proof. By Theorem 3.6.1 there is a countable relational language L and a Borel S_∞-embedding $j : X \to \mathrm{Mod}(L)$. It follows that $j(X)$ is an invariant Borel class and that j is a Borel S_∞-isomorphism between X and $j(X)$. Now by Theorem 11.3.6 there is an $L_{\omega_1\omega}$-sentence φ such that $j(X) = \mathrm{Mod}(\varphi)$. Then j is a Borel isomorphism between $E^X_{S_\infty}$ and $\cong\restriction j(X)$, or $\cong_\varphi$. □

The following theorem also shows that being an isomorphism relation is closed under Borel reductions for Borel orbit equivalence relations.

Theorem 11.3.9
Let G be a Polish group, X a Borel G-space, L a countable relational language, and φ an $L_{\omega_1\omega}$-sentence. Suppose that $\cong_\varphi$ is Borel and $E^X_G \leq_B \cong_\varphi$. Then there is an $L_{\omega_1\omega}$-sentence σ such that $\mathrm{Mod}(\sigma) \subseteq \mathrm{Mod}(\varphi)$ and $E^X_G \sim_B \cong_\sigma$.

Proof. Let $f : X \to \mathrm{Mod}(\varphi)$ be a Borel reduction from E^X_G to $\cong_\varphi$. By Theorem 5.2.3 f is a faithful Borel reduction. And thus $Y = [f(X)]_\cong$ is a Borel subset of $\mathrm{Mod}(\varphi)$. By Theorem 11.3.6 there is an $L_{\omega_1\omega}$-sentence σ such that $\mathrm{Mod}(\sigma) = Y$. Then by Corollary 5.2.4 $E^X_G \sim_B \cong_\sigma$. □

The most important open problems on isomorphism relations are about the number of nonisomorphic countable models in invariant Borel classes. The **Vaught conjecture** is the statement

Any complete first-order theory either has only countably many nonisomorphic countable models or has perfectly many nonisomorphic countable models.

And the $L_{\omega_1\omega}$-**Vaught conjecture** states that

For any countable relational language L and $L_{\omega_1\omega}$-sentence φ, $\cong_\varphi$ either has only countably many classes or has perfectly many classes.

The Vaught conjecture is a special case of the $L_{\omega_1\omega}$-Vaught conjecture since any complete first-order theory (that is, $L_{\omega\omega}$-theory) T is logically equivalent to $\varphi = \bigwedge T$. In both statements "perfectly many" refers to a perfect subset of $\mathrm{Mod}(\varphi)$.

Recall that $\mathrm{TVC}(G)$, the topological Vaught conjecture for G, is the statement (among several equivalent formulations)

For any Borel G-space E^X_G has either only countably many orbits or perfectly many orbits.

It is thus immediate from Theorem 11.3.8 that the $L_{\omega_1\omega}$-Vaught conjecture is equivalent to TVC(S_∞).

Exercise 11.3.1 Let

$$\mathcal{B}_{00} = \{\, \mathrm{Mod}(\varphi, \vec{n}) \;:\; n \in \omega, \quad \varphi \text{ is atomic or negated atomic} \,\}.$$

Show that $\mathcal{B}_{00} = \mathcal{B}_{\mathrm{atom}}$.

Exercise 11.3.2 In the proof of Lemma 11.3.5 show directly that $\mathcal{B}$ is closed under countable intersection.

Exercise 11.3.3 For each of the following equivalence relations E find a countable relational language L and an $L_{\omega_1\omega}$-sentence φ such that $\cong_\varphi \sim_B E$:

(a) $\mathrm{id}(\omega)$,

(b) $\mathrm{id}(2^\omega)$,

(c) E_0.

11.4   Polish topologies generated by countable fragments

The technique of change of topology in the context of Polish group actions, as we presented in Chapter 4 following the approaches of References [8] and [5], has been a powerful tool in invariant descriptive set theory. This technique has its natural motivation in the model theoretic context (for example, see References [152], [9], and [124]), and was used to obtain stronger results for countable models than for general orbit equivalence relations (for example, see References [77] and [82]).

In the preceding section we have mostly considered the canonical topology on $\mathrm{Mod}(L)$ and its various (sub)bases. From Chapter 4 we know abstractly that there are many other Polish topologies we can put on $\mathrm{Mod}(L)$. In particular, for any invariant Borel class $\mathcal{C} \subseteq \mathrm{Mod}(L)$ there exists a Polish topology finer than the canonical topology such that $\mathcal{C}$ becomes clopen but the logic action is still continuous. In this section we consider some natural bases for such topologies.

First note that by Definition 11.3.1 for any set of $L_{\omega_1\omega}$-formulas A we can alway associate a subbase $\mathcal{B}_A$ and therefore a topology t_A generated by $\mathcal{B}_A$. However, the definition only uses formulas in A with only finitely many free variables. That is to say, if we let A' be the set of formulas in A with only finitely many free variables, then $\mathcal{B}_{A'} = \mathcal{B}_A$ and $t_{A'} = t_A$. Thus in our future discussions we may always assume that the set A consists of only formulas with only finitely many free variables.

Theorem 11.4.1
Let L be a countable relational language and F a countable fragment of $L_{\omega_1\omega}$. Then $\mathcal{B}_F$ is a clopen base for the topology t_F and $(\mathrm{Mod}(L), t_F)$ is a Polish S_∞-space.

Proof. For any formula $\varphi(x_1, \ldots, x_n)$, let $S(\varphi)$ be the set of all quantifier-free $L_{\omega\omega}$-formulas, all subformulas of φ, and their negations. We first prove by induction on the form of φ that $\mathcal{B}_{S(\varphi)}$ is a clopen base for a Polish topology.

If φ is quantifier-free then $\mathcal{B}_{S(\varphi)} = \mathcal{B}_{\mathrm{qf}}$ is a clopen base for t_{qf}. Also $S(\neg\varphi) = S(\varphi)$. We next consider $\varphi = \bigwedge \Phi$ for a countable set Φ, and note that $S(\varphi) = \{\varphi, \neg\varphi\} \cup \bigcup_{\psi \in \Phi} S(\psi)$. By the inductive hypotheses each $\mathcal{B}_{S(\psi)}$ is a clopen base for a Polish topology finer than the canonical topology. It follows from Lemma 4.2.2 that $\bigcup_{\psi \in \Phi} S(\psi)$ is also a clopen base for a Polish topology τ. Now fix $\vec{a} = (a_1, \ldots, a_n) \in \omega^n$. Note that $\mathrm{Mod}(\varphi, \vec{a}) = \bigcap_{\psi \in \Phi} \mathrm{Mod}(\psi, \vec{a})$ and is therefore τ-closed. By Lemma 4.2.1 the topology $\tau \cup \{\mathrm{Mod}(\varphi, \vec{a})\}$ is Polish. It follows from Lemma 4.2.2 again that

$$\tau \cup \{\mathrm{Mod}(\varphi, \vec{a}) : \vec{a} \in \omega^n\}$$

generates a Polish topology, for which $\mathcal{B}_{S(\varphi)}$ is a clopen base. This proves the inductive case for $\varphi = \bigwedge \Phi$. The argument for $\varphi = \bigvee \Phi$ is similar. Next we consider $\varphi = \exists y \psi(x_1, \ldots, x_n, y)$. For each $\vec{a} \in \omega^n$, $\mathrm{Mod}(\varphi, \vec{a}) = \bigcup_{b \in \omega} \mathrm{Mod}(\psi, \vec{a}^\frown b)$, and is therefore $t_{S(\psi)}$-open. However,

$$\mathrm{Mod}(\neg\varphi, \vec{a}) = \bigcap_{b \in \omega} \mathrm{Mod}(\neg\psi, \vec{a}^\frown b)$$

is $t_{S(\psi)}$-closed. By a similar argument as above using Lemmas 4.2.1 and 4.2.2 we get that $\mathcal{B}_{S(\varphi)}$ is a clopen base for a Polish topology. The case $\varphi = \forall y \psi(x_1, \ldots, x_n, y)$ is similar.

Now that F is a countable fragment, we have that

$$\mathcal{B}_F = \bigcup_{\varphi \in F} S(\varphi)$$

is a countable union. Thus by Lemma 4.2.2 again $\mathcal{B}_F$ generates a Polish topology. It is a clopen base since F is closed under negation and finitary logical connectives $\wedge$ and $\vee$.

To show the continuity of the logic action with respect to t_F, let $g \in S_\infty$, $M \in \mathrm{Mod}(L)$, $\mathrm{Mod}(\varphi, \vec{a}) \in \mathcal{B}_F$, and suppose $g \cdot M \in \mathrm{Mod}(\varphi, \vec{a})$. Note that $g^{-1} \cdot \mathrm{Mod}(\varphi, \vec{a}) = \mathrm{Mod}(\varphi, g^{-1}(\vec{a}))$. Thus if we let $\vec{b} = g^{-1}(\vec{a})$ and $N_{\vec{a}, \vec{b}} = \{f \in S_\infty : f(\vec{b}) = \vec{a}\}$, then $g \in N_{\vec{a}, \vec{b}}$ is a basic open set in S_∞, $M \in \mathrm{Mod}(\varphi, \vec{b})$, and for any $f \in N_{\vec{a}, \vec{b}}$ and $N \in \mathrm{Mod}(\varphi, \vec{b})$, $f \cdot N \in \mathrm{Mod}(\varphi, f(\vec{b})) = \mathrm{Mod}(\varphi, \vec{a})$. $\quad\square$

Thus topologies generated by countable fragments are concrete examples of finer Polish topologies on $\mathrm{Mod}(L)$. If $\mathcal{C}$ is an invariant Borel class and φ is an

$L_{\omega_1\omega}$-sentence such that $\mathcal{C} = \text{Mod}(\varphi)$, then the fragment $F(\varphi)$ generated by φ is countable and generates a Polish topology $t_{F(\varphi)}$ in which $\mathcal{C}$ is clopen and such that $(\text{Mod}(L), t_{F(\varphi)})$ is a Polish S_∞-space. In particular, $(\text{Mod}(\varphi), t_{F(\varphi)})$ is a Polish S_∞-space whose orbit equivalence relation is $\cong_\varphi$.

As we will see, model theoretic properties of countable structures in a countable fragment F correspond nicely to topological properties in the topology t_F. For simplicity we use $[M]$, rather than $[M]_\cong$, to denote the isomorphism class of M.

Theorem 11.4.2

Let L be a countable relational language and F a countable fragment of $L_{\omega_1\omega}$. Then for any $M, N \in \text{Mod}(L)$, $M \equiv_F N$ iff $\overline{[M]}^{t_F} = \overline{[N]}^{t_F}$.

Proof. For each sentence $\varphi \in F$, $\text{Mod}(\varphi)$ is a basic open set in t_F. Therefore, if $\overline{[M]}^{t_F} = \overline{[N]}^{t_F}$, then for any sentence $\varphi \in F$, $[M] \cap \text{Mod}(\varphi) \neq \emptyset$ iff $[N] \cap \text{Mod}(\varphi) \neq \emptyset$. Since $\text{Mod}(\varphi)$ is invariant, this implies that $M \in \text{Mod}(\varphi)$ iff $N \in \text{Mod}(\varphi)$. Hence $M \equiv_F N$. Conversely, suppose $M \equiv_F N$. We show that for any $\text{Mod}(\varphi, \vec{a}) \in \mathcal{B}_F$, $[M] \cap \text{Mod}(\varphi, \vec{a}) \neq \emptyset$ iff $[N] \cap \text{Mod}(\varphi, \vec{a}) \neq \emptyset$. This of course implies that $\overline{[M]}^{t_F} = \overline{[N]}^{t_F}$. For this suppose $\varphi(x_1, \ldots, x_n) \in F$, $\vec{a} \in \omega^n$, and assume $[M] \cap \text{Mod}(\varphi, \vec{a}) \neq \emptyset$. By symmetry it suffices to show that $[N] \cap \text{Mod}(\varphi, \vec{a}) \neq \emptyset$. First note that we may assume without loss of generality that $\vec{a} \in [\omega]^n$ by a standard argument of change of variables as in the proof of Lemma 11.3.2. Without loss of generality we may also assume $M \in \text{Mod}(\varphi, \vec{a})$. Thus $M \models \varphi[\vec{a}]$, and in particular $M \models (\exists x_1 \ldots x_n)^{\neq} \varphi(x_1, \ldots, x_n)$, where $(\exists x_1 \ldots x_n)^{\neq} \varphi$ is an abbreviation of

$$\exists x_1 \ldots \exists x_n \left(\bigwedge_{1 \leq i < j \leq n} x_i \neq x_j \wedge \varphi \right).$$

Now the sentence $(\exists x_1 \ldots x_n)^{\neq} \varphi(x_1, \ldots, x_n) \in F$, and by $M \equiv_F N$ we get that $N \models (\exists x_1 \ldots x_n)^{\neq} \varphi(x_1, \ldots, x_n)$. Hence there is $\vec{b} \in [\omega]^n$ such that $N \models \varphi[\vec{b}]$. Let $g \in S_\infty$ be such that $g(\vec{b}) = \vec{a}$. Then $g \cdot N \models \varphi[\vec{a}]$, or $g \cdot N \in \text{Mod}(\varphi, \vec{a})$. This shows that $[N] \cap \text{Mod}(\varphi, \vec{a}) \neq \emptyset$ as desired. $\square$

By this theorem F-theories of countable models correspond to the t_F-closure of their isomorphism classes. This has an immediate corollary on ω-categoricity, which is defined below.

Definition 11.4.3

*Let L be a countable relational language. An $L_{\omega_1\omega}$-theory T is ω-**categorical** if $M \cong N$ whenever $M, N \in \bigcap_{\varphi \in T} \text{Mod}(\varphi)$.*

Thus an ω-categorical theory has a unique countable model up to isomorphism.

Corollary 11.4.4
*Let L be a countable relational language and F a countable fragment of $L_{\omega_1\omega}$.
Then $[M]$ is closed in t_F iff $\mathrm{Th}_F(M)$ is ω-categorical.*

Proof. First suppose $[M]$ is closed in t_F and let $N \equiv_F M$. Then by Theorem 11.4.2, $[N] \subseteq \overline{[N]}^{t_F} = \overline{[M]}^{t_F} = [M]$. This shows that $N \in [M]$ and hence $N \cong M$. Conversely, if $\mathrm{Th}_F(M)$ is ω-categorical then $[M] = \bigcap_{\varphi \in \mathrm{Th}_F(M)} \mathrm{Mod}(\varphi)$, which is t_F-closed. $\square$

Next we consider F-elementary embeddability. For this we recall that the canonical metric on S_∞ (the one inherited from the Baire space ω^ω) is

$$d(f,g) = \begin{cases} 2^{-n}, & \text{if } n \in \omega \text{ is the least such that } f(n) \neq g(n), \\ 0, & \text{if } f = g. \end{cases}$$

This metric is left-invariant but not complete.

Theorem 11.4.5 (Becker)
Let L be a countable relational language and F a countable fragment of $L_{\omega_1\omega}$. Let d be the canonical metric on S_∞. Then for any $M, N \in \mathrm{Mod}(L)$, the following are equivalent:

(a) *There is an F-elementary embedding from M into N.*

(b) *There is a d-Cauchy sequence $(g_n)_{n \in \omega}$ in S_∞ such that $g_n \cdot M \to N$ in t_F.*

Proof. (a)$\Rightarrow$(b): Let $j : M \to N$ be an F-elementary embedding. We show that for any $\varphi(x_1, \ldots, x_n) \in F$, $\vec{a} \in \omega^n$, and $m \in \omega$, if $N \in \mathrm{Mod}(\varphi, \vec{a})$ then there is $f_m \in S_\infty$ such that $f_m(i) = j(i)$ for all $i < m$ and $f_m \cdot M \in \mathrm{Mod}(\varphi, \vec{a})$. As before we may assume that $\vec{a} \in [\omega]^n$. We may also assume that $k \leq n$ and $a_1, \ldots, a_k \in \{j(0), \ldots, j(m-1)\}$ and $a_{k+1}, \ldots, a_n \notin \{j(0), \ldots, j(m-1)\}$. Then by our assumption $N \models \psi[a_1, \ldots, a_k]$, where $\psi(x_1, \ldots, x_k)$ is the formula

$$(\exists x_{k+1} \ldots x_n)^{\neq} \Big(\bigwedge_{\substack{1 \leq i \leq k \\ k+1 \leq i' \leq n}} x_i \neq x_{i'} \wedge \varphi \Big).$$

Since $\psi \in F$ it follows from F-elementarity that $M \models \psi[j^{-1}(a_1), \ldots, j^{-1}(a_k)]$. Let $b_i = j^{-1}(a_i)$ for $1 \leq i \leq k$. Then there are $b_{k+1}, \ldots, b_n \in \omega$ such that $\vec{b} \in [\omega]^n$ and $M \models \varphi[\vec{b}]$. Now let $f_m \in S_\infty$ by such that $f_m(i) = j(i)$ for all $i < m$, and

$$f_m(b_{k+1}) = a_{k+1}, \ldots, f_m(b_n) = a_n.$$

Then $f_m(b_i) = a_i$ for all $1 \leq i \leq n$. Therefore $f_m \cdot M \models \varphi[\vec{a}]$, or $f_m \cdot M \in \mathrm{Mod}(\varphi, \vec{a})$.

Now since t_F is first countable we may let U_n be a decreasing sequence of basic open sets of the form $\text{Mod}(\varphi, \bar{a}) \in \mathcal{B}_F$ such that $\bigcap_n U_n = \{N\}$. By the above claim there is $g_n \in S_\infty$ such that $g_n(i) = j(i)$ for $i < n$ and $g_n \cdot M \in U_n$. Clearly $g_n \cdot M \to N$ in t_F. However, for any $n < n'$, $d(g_n, g_{n'}) \leq 2^{-(n+1)}$, hence $(g_n)_{n \in \omega}$ is d-Cauchy.

(b)$\Rightarrow$(a): Let $(g_n)_{n \in \omega}$ be a d-Cauchy sequence in S_∞ such that $g_n \cdot M \to N$ in t_F. Then for any $k \in \omega$ there is n_k such that for all $n, m \geq n_k$, $d(g_n, g_m) < 2^{-k}$. This implies in particular that for all $n, m \geq n_k$, $g_n(k) = g_m(k)$. Thus we may let $j(k) = \lim_n g_n(k)$, and j is well defined. It is easy to see that j is an injection. We verify that j is an F-elementary embedding from M into N.

Let $\varphi(x_1, \ldots, x_n) \in F$ and $\bar{a} \in \omega^n$. Assume $M \models \varphi[\bar{a}]$ but $N \not\models \varphi[j(\bar{a})]$, that is, $N \notin \text{Mod}(\varphi, j(\bar{a}))$. Since $g_n \cdot M \to N$ in t_F and $\text{Mod}(\varphi, j(\bar{a}))$ is clopen, there is $N \in \omega$ such that for all $m \geq N$, $g_m \cdot M \notin \text{Mod}(\varphi, j(\bar{a}))$. But let $m > N$ be large enough such that $j(a_i) = g_m(a_i)$ for all $1 \leq i \leq n$. Then we have $g_m \cdot M \models \varphi[g_m(\bar{a})]$, and since $j(\bar{a}) = g_m(\bar{a})$, $g_m \cdot M \models \varphi[j(\bar{a})]$, a contradiction. This shows that $M \models \varphi[\bar{a}]$ implies $N \models \varphi[j(\bar{a})]$. For the converse note that the same argument gives that $M \models \neg\varphi[\bar{a}]$ implies $N \models \neg\varphi[j(\bar{a})]$, and therefore $N \models \varphi[j(\bar{a})]$ implies $M \models \varphi[\bar{a}]$. Thus j is an F-elementary embedding. □

Clause (b) of this theorem obviously implies that $N \in \overline{[M]}^{t_F}$, and thus $\overline{[N]}^{t_F} \subseteq \overline{[M]}^{t_F}$. It is curious to observe that clause (b) does not trivially imply that $\overline{[M]}^{t_F} = \overline{[N]}^{t_F}$, since given d-Cauchy sequence (g_n) such that $g_n \cdot M \to N$ it is not clear how to construct another d-Cauchy sequence (h_n) so that $h_n \cdot N \to M$ (note that (g_n^{-1}) is not necessarily d-Cauchy). However, by combining the above two theorems this implication is true, since it follows from the existence of an F-elementary embedding from M into N that $M \equiv_F N$.

In the following exercises let L be a countable relational language and F a countable fragment of $L_{\omega_1 \omega}$.

Exercise 11.4.1 Let $\mathcal{B}_{0F} = \{\text{Mod}(\varphi, \bar{n}) : \varphi \in F, n \in \omega\}$. Show that $\mathcal{B}_{0F} = \mathcal{B}_F$.

Exercise 11.4.2 Show that for any $U \in \mathcal{B}_F$, $[U]_\cong = S_\infty \cdot U$ is clopen in t_F.

Exercise 11.4.3 Show that $\equiv_F$ is a closed equivalence relation on $(\text{Mod}(L), t_F)$.

11.5 Atomic models and G_δ orbits

By now it should have been clear that for the study of countable models the connection between methods of logic and topology is natural and intrinsic. In this section we continue to explore this connection and establish some deeper results. To do this we will have to rely on some model theoretic results we state without proof. We start with some more definitions.

Definition 11.5.1
Let L be a countable relational language and F a countable fragment of $L_{\omega_1\omega}$.

(1) For any $M \in \mathrm{Mod}(L)$ and $\vec{a} \in \omega^n$, the **F-type of $\vec{a}$ over** M is

$$\mathrm{tp}_F(\vec{a}/M) = \{\varphi(x_1,\dots,x_n) \in F : M \models \varphi[\vec{a}]\}.$$

An **F-type** is a set of formulas Φ such that for some $M \in \mathrm{Mod}(L)$ and $\vec{a} \in \omega^n$, $\Phi \subseteq \mathrm{tp}_F(\vec{a}/M)$. An F-type Φ is **complete** if $\Phi = \mathrm{tp}_F(\vec{a}/M)$ for some M and $\vec{a} \in \omega^n$.

(2) An F-type Φ is **realized** in M if $\Phi \subseteq \mathrm{tp}_F(\vec{a}/M)$ for some $\vec{a} \in \omega^n$; otherwise it is **omitted** in M.

The following definitions are specific about a complete F-theory T.

Definition 11.5.2
Let L be a countable relational language and F a countable fragment of $L_{\omega_1\omega}$.

(1) An F-theory T is **complete** if $T = \mathrm{Th}_F(M)$ for some $M \in \mathrm{Mod}(L)$.

(2) If T is a complete theory and $M \in \mathrm{Mod}(L)$, then M is a **model of T**, denoted $M \models T$, if $M \models \varphi$ for all $\varphi \in T$. The set of all models of T in $\mathrm{Mod}(L)$ is denoted by $\mathrm{Mod}(T)$.

(3) Let T be a complete F-theory. An F-type is an **F-type of T** if it is realized in some countable model of T. It is a **complete F-type of T** if it is a complete type realized in a countable model of T.

(4) Let T be a complete F-theory and Φ an F-type of T. Φ is **principal** if there is a complete F-type Ψ of T such that $\Phi \subseteq \Psi$ and there is a formula $\varphi \in \Psi$ such that for any $M \in \mathrm{Mod}(T)$ and $\psi \in \Phi$, $M \models \forall x_1 \dots \forall x_n\,(\varphi \to \psi)$; otherwise Φ is **nonprincipal**.

Note that if F is a countable fragment so is any complete F-theory or F-type. For any F-type Φ if we let $\varphi = \bigwedge \Phi$, then $M \models \forall x_1 \dots \forall x_n\,(\varphi \to \psi)$ for all $M \in \mathrm{Mod}(T)$; however $\varphi \notin \Phi$ and in general $\varphi \notin F$. Thus in the definition of principal types it is essential to refer to the theory T and the fragment F.

The following lemma is a standard fact in model theory and is the main reason to consider principal types.

Lemma 11.5.3
Let L be a countable relational language, F a countable fragment of $L_{\omega_1\omega}$, and T a complete F-theory. Then every principal F-type of T is realized in any $M \in \mathrm{Mod}(T)$.

Proof. Let Φ be a principal F-type of T, witnessed by complete F-type Ψ of T and formula $\varphi \in \Psi$. Let $M \in \mathrm{Mod}(T)$ and $\vec{a} \in \omega^n$ be such that $\Phi \subseteq \Psi = \mathrm{tp}_F(\vec{a}/M)$. Then $M \models \exists x_1 \ldots \exists x_n \ \varphi$. Now $\exists x_1 \ldots \exists x_n \varphi$ is a sentence in F, and thus in $\mathrm{Th}_F(M) = T$ since T is a complete F-theory. It follows that for any $N \in \mathrm{Mod}(T)$, $N \models \exists x_1 \ldots \exists x_n \varphi$. Let $\vec{b} \in \omega^n$ be such that $N \models \varphi[\vec{b}]$. But then by principality $N \models \forall x_1 \ldots \forall x_n (\varphi \to \psi)$ for all $\psi \in \Phi$. Hence $N \models \psi[\vec{b}]$ for all $\psi \in \Phi$. This means that Φ is realized in N as witnessed by $\vec{b}$. $\qquad\qquad\qquad\qquad\qquad\qquad\qquad\qquad\qquad\qquad\qquad\quad$ $\Box$

It turns out that the converse of the lemma is also true. This is implied by the following theorem known as the **omitting types theorem**, which we state without proof. A proof for the first-order case of the omitting types theorem can be found in most textbooks in model theory, for example, Reference [84]; for the case of infinitary logic the proof is similar (see References [3] and [105]).

Theorem 11.5.4
Let L be a countable relational language, F a countable fragment of $L_{\omega_1\omega}$, and T a complete F-theory. For any countable set $\{\Phi_n\}_{n\in\omega}$ of nonprincipal F-types of T there exists $M \in \mathrm{Mod}(T)$ such that M omits all types Φ_n for $n \in \omega$.

We now turn to atomic models.

Definition 11.5.5
*Let L be a countable relational language, F a countable fragment of $L_{\omega_1\omega}$, and T a complete F-theory. A model $M \in \mathrm{Mod}(T)$ is F-**atomic** if all F-types of T realized in M are principal.*

Note that when referring to F-atomic models it is unnecessary, although sometimes only convenient, to specify the complete F-theory since a model M can only be F-atomic with respect to $\mathrm{Th}_F(M)$.

Theorem 11.5.6
Let L be a countable relational language, F a countable fragment of $L_{\omega_1\omega}$, and T a complete F-theory.

(i) *If $M, N \in \mathrm{Mod}(T)$ and M is F-atomic, then there is an F-elementary embedding from M into N.*

(ii) *If $M, N \in \mathrm{Mod}(T)$ are both F-atomic, then $M \cong N$.*

Proof. We only prove (i). The proof of (ii) is by a back-and-forth argument similar to (i) and is left as an exercise.

Suppose M is F-atomic. For each $n \in \omega$ we let $\Phi_n(v_0, \ldots, v_{n-1}) = \mathrm{tp}_F(\vec{n}/M)$ and $\varphi_n(v_0, \ldots, v_{n-1}) \in F$ be the witness for the principality of Φ_n. Let

$N \in \mathrm{Mod}(T)$. We define a distinct sequence $(b_n)_{n\in\omega}$ in $|N|$ such that for all $n \in \omega$, $N \models \varphi_n[b_0,\ldots,b_{n-1}]$. To begin with we note that $N \models \Phi_0$ and that by Lemma 11.5.3 Φ_1 is realized in N, hence there is $b_0 \in \omega$ such that $N \models \varphi_1[b_0]$. In general suppose $b_0,\ldots,b_{n-1}$ have been defined so that $N \models \varphi_n[b_0,\ldots,b_{n-1}]$. Since $M \models \varphi_n[0,\ldots,n-1]$ and $M \models \varphi_{n+1}[0,\ldots,n-1,n]$, we have that $M \models \psi[0,\ldots,n-1]$, where ψ is the formula

$$\exists v_n \left(\bigwedge_{i<n} v_n \neq v_i \ \wedge\ \varphi_n \wedge \varphi_{n+1} \right).$$

Since $\psi \in F$, we have $\psi \in \Phi_n$, and by principality of Φ_n, $N \models \forall v_0 \ldots \forall v_{n-1}(\varphi \to \psi)$. This implies that

$$N \models \exists v_n \left(\bigwedge_{i<n} v_n \neq v_i \ \wedge\ \varphi_n \wedge \varphi_{n+1} \right)[b_0,\ldots,b_{n-1}]$$

since $N \models \varphi[b_0,\ldots,b_{n-1}]$. Therefore there is $b_n \in \omega$ distinct from $b_0,\ldots,b_{n-1}$ such that $N \models \varphi_{n+1}[b_0,\ldots,b_{n-1},b_n]$. This finishes the definition of the sequence (b_n).

We claim that the map $j(n) = b_n$, $n \in \omega$, is an F-elementary embedding from M into N. To show this it suffices to show that for any $n \in \omega$ and formula $\phi(v_0,\ldots,v_{n-1}) \in F$, $M \models \phi[\bar{n}]$ iff $N \models \phi[j(\bar{n})]$. For this fix $n \in \omega$ and $\phi \in F$. It suffices to show that $M \models \phi[\bar{n}]$ implies $N \models \phi[j(\bar{n})]$. So we suppose $M \models \phi[\bar{n}]$. This means that $\phi \in \Phi_n$ and therefore by principality of Φ_n witnessed by φ_n, $N \models \forall v_0 \ldots \forall v_{n-1}(\varphi_n \to \phi)$. Now $N \models \varphi_n[j(\bar{n})]$, and it follows that $N \models \phi[j(\bar{n})]$, as required. □

Theorem 11.5.7 (Miller–Suzuki)
Let L be a countable relational language and F a countable fragment of $L_{\omega_1\omega}$. Then M is F-atomic iff $[M]$ is G_δ in t_F.

Proof. Suppose M is F-atomic, and for each $n \in \omega$, let $\{\Phi^n_m\}_{m\in\omega}$ enumerate the set $\{\mathrm{tp}_F(\bar{a}/M) : \bar{a} \in \omega^n\}$. Note that $\{\Phi^n_m : n,m \in \omega\}$ in fact enumerate all complete principal F-types of $T = \mathrm{Th}_F(M)$. For each $n,m \in \omega$, let $\varphi^n_m \in F$ be a formula witnessing that Φ^n_m is principal. Let φ be the $L_{\omega_1\omega}$-sentence

$$\bigwedge_{n\in\omega} \forall x_1 \ldots \forall x_n \bigvee_{m\in\omega} \varphi^n_m.$$

Then any model $N \in \mathrm{Mod}(T)$ of φ is F-atomic. To see this let $N \models \varphi$. Then for any $\bar{a} \in \omega^n$ there is $m \in \omega$ such that $N \models \varphi^n_m[\bar{a}]$. It follows that for any $\psi \in \Phi^n_m$, $N \models \psi[\bar{a}]$, and therefore $\mathrm{tp}_F(\bar{a}/N) = \Phi^n_m$ is principal. Since N realizes only principal types, it is F-atomic.

By Theorem 11.5.6 if $N \in \mathrm{Mod}(T)$ and $N \models \varphi$ then $N \cong M$. Thus for $N \in \mathrm{Mod}(T)$, $N \models \varphi$ iff $N \cong M$. This implies that

$$N \in [M] \iff N \models \varphi \iff N \in \bigcap_{n\in\omega} \bigcap_{\bar{a}\in\omega^n} \bigcup_{m\in\omega} \mathrm{Mod}(\varphi^n_m, \bar{a}),$$

and thus $[M]$ is G_δ in t_F.

Conversely, assume that $[M]$ is G_δ in t_F. Let $[M] = \bigcap_{n \in \omega} \bigcup_{m \in \omega} U_{n,m}$, where each $U_{n,m} = \text{Mod}(\varphi_{n,m}, \vec{m})$ for $\varphi_{n,m} \in F$ (see Exercise 11.4.1), and such that $U_{n,m} \subseteq U_{n,m'}$ whenever $m < m'$, and $U_{n,m} \supseteq U_{n',m}$ whenever $n < n'$. This can be achieved by a proof similar to that of Lemma 11.3.2 and by using finitary conjunctions (as n increases) and disjunctions (as m increases). Then $[M] = [M]^{*S_\infty}$, and by a computation similar to that in the proof of Lemma 11.3.5 we have that

$$N \in [M]^{*S_\infty}$$

$$\Longleftrightarrow \forall^* g \in S_\infty \ \forall n \in \omega \ \exists m \in \omega \ g \cdot N \in U_{n,m}$$

$$\Longleftrightarrow \forall n \in \omega \ \forall^* g \in S_\infty \ \exists m \in \omega \ g \cdot N \in U_{n,m}$$

$$\Longleftrightarrow \forall n \in \omega \ \forall k \in \omega \ \forall^* g \in S_\infty \ \exists m \in \omega \ g \cdot N \in U_{\max\{n,k\},m}$$

$$\Longleftrightarrow \forall n \in \omega \ \forall \vec{b} \in [\omega]^k \ \exists m \geq \max\{n,k\} \ \exists^* g \in N_{\vec{b}} \ g \cdot N \in U_{\max\{n,k\},m}$$

$$\Longleftrightarrow \forall n \in \omega \ \forall \vec{b} \in [\omega]^n \ \exists m \geq n \ \exists^* g \in N_{\vec{b}} \ g \cdot N \in U_{n,m}$$

$$\Longleftrightarrow \forall n \in \omega \ \forall \vec{b} \in [\omega]^n \ \exists m \geq n$$
$$\exists \vec{c} \ (\vec{b}^\frown \vec{c} \in [\omega]^m \ \wedge \exists^* g \in N_{\vec{b}^\frown \vec{c}} \ g \cdot N \in \text{Mod}(\varphi_{n,m}, \vec{m}))$$

$$\Longleftrightarrow \forall n \in \omega \ \forall \vec{b} \in [\omega]^n \ \exists m \geq n$$
$$\exists \vec{c} \ (\vec{b}^\frown \vec{c} \in [\omega]^m \ \wedge M \in \text{Mod}(\varphi_{n,m}, \vec{b}^\frown \vec{c})).$$

Thus for $m \geq n$ if we let $\psi_{n,m}(x_1, \ldots, x_n) \in F$ be the formula

$$\exists x_{n+1} \ldots \exists x_m \ (\bigwedge_{1 \leq i < j \leq m} x_i \neq x_j \ \wedge \varphi_{n,m}),$$

then $N \in [M]$ iff $N \models \bigwedge_{n \in \omega} \forall x_1 \ldots \forall x_n \bigvee_{m \geq n} \psi_{n,m}$.

Now let $\Phi_n = \{\neg \psi_{n,m} : m \in \omega\}$ for $n \in \omega$. Then $\Phi_n \subseteq F$. If Φ_n is an F-type of T then by Lemma 11.5.3 it is nonprincipal since it is not realized in M. Let $S \subseteq \omega$ be such that Φ_n is an F-type iff $n \in S$. We are ready to verify that M is F-atomic. Assume not, and let Ψ be a nonprincipal F-type realized in M. Then the collection

$$\{\Phi_n : n \in S\} \cup \{\Psi\}$$

is a countable set of nonprincipal types of $T = \text{Th}_F(M)$. By the Omitting Types Theorem 11.5.4 there is an $N \in \text{Mod}(T)$ omitting all types in the collection. For $n \notin S$ there are also no $\vec{a} \in \omega^n$ such that $\Phi_n \subseteq \text{tp}_F(\vec{a}/N)$ since Φ_n is not an F-type of T. Therefore $N \models \bigwedge_{n \in \omega} \forall x_1 \ldots \forall x_n \bigvee_{m \geq n} \psi_{n,m}$. This shows that $N \in [M]$, or $N \cong M$. However, N omits the F-type Ψ, whereas M realizes it, hence $N \ncong M$, a contradiction. $\qquad\square$

As a corollary we give a model theoretic characterizations for smoothness of isomorphism relations.

Theorem 11.5.8
Let L be a countable relational language and φ an $L_{\omega_1\omega}$-sentence. Then the following are equivalent:

(i) *$\cong_\varphi$ is smooth.*

(ii) *There is a countable fragment F (containing φ) such that for any $M \in \mathrm{Mod}(\varphi)$, $\mathrm{Th}_F(M)$ is ω-categorical.*

(iii) *There is a countable fragment F (containing φ) such that every $M \in \mathrm{Mod}(L)$ is F-atomic.*

Proof. (i)$\Rightarrow$(ii): Let X be a Polish space and $f : \mathrm{Mod}(\varphi) \to X$ be Borel such that $M \cong N$ iff $f(M) = f(N)$. Let $\{U_n\}_{n\in\omega}$ be a countable base for X. Then

$$M \cong N \iff \forall n \in \omega \; M \in f^{-1}(U_n) \leftrightarrow N \in f^{-1}(U_n).$$

Now each $f^{-1}(U_n)$ is an invariant Borel class, so there is an $L_{\omega_1\omega}$-sentence ψ_n such that $f^{-1}(U_n) = \mathrm{Mod}(\psi_n)$. Let F be a countable fragment generated by $\{\varphi, \psi_0, \psi_1, \ldots\}$. Then in the topology τ_F on $\mathrm{Mod}(\varphi)$, each isomorphism class $[M]$ is closed. Then by Corollary 11.4.4, $\mathrm{Th}_F(M)$ is ω-categorical.

(ii)$\Rightarrow$(iii) follows immediately from Corollary 11.4.4 and Theorem 11.5.7.

(iii)$\Rightarrow$(i): By Theorem 11.5.7 for any $M \in \mathrm{Mod}(L)$, $[M]$ is G_δ in τ_F on $\mathrm{Mod}(\varphi)$. By Theorem 6.4.4 $\cong_\varphi$ is smooth. $\qquad\square$

Exercise 11.5.1 Prove Theorem 11.5.6 (ii).

Exercise 11.5.2 Assuming the statement of Theorem 11.5.7, deduce that if M and N are both F-atomic models of the same complete F-theory, then $M \cong N$.

Chapter 12

The Scott Analysis

The Scott analysis is a framework to determine the isomorphism type of any countable structure through descriptions of the structure in infinitary logic. As an abstract solution to the isomorphism problem, it provides useful information about invariant Borel classes of countable models. A generalized Scott analysis on arbitrary Polish G-spaces provides a scenario of how the orbit equivalence relations might be reduced to isomorphism relations.

12.1 Elements of the Scott analysis

In this section we present the original Scott analysis for countable models. In the setup we have a countable relational language L and a countable L-structure M. The objective is to obtain an $L_{\omega_1 \omega}$-sentence φ_M, known as the canonical Scott sentence for M, with the property that for any countable L-structure N, if $N \models \varphi_M$, then $N \cong M$. This sentence is built up by a transfinite process. At each stage of the process the formulas obtained are considered approximations of the final product. By the Scott analysis we refer to this whole process and the partial information these approximations provide.

Definition 12.1.1
Let L be a countable relational language. The **quantifier rank** of an $L_{\omega_1 \omega}$-formula is inductively defined as follows:

$$\begin{aligned}
&\mathrm{qr}(\varphi) = 0 \text{ if } \varphi \text{ is atomic,} \\
&\mathrm{qr}(\neg \varphi) = \mathrm{qr}(\varphi), \\
&\mathrm{qr}(\bigwedge \Phi) = \mathrm{qr}(\bigvee \Phi) = \sup\{\mathrm{qr}(\varphi) : \varphi \in \Phi\}, \\
&\mathrm{qr}(\exists x \varphi) = \mathrm{qr}(\forall x \varphi) = \mathrm{qr}(\varphi) + 1.
\end{aligned}$$

Definition 12.1.2
Let L be a countable relational language, $M, N \in \mathrm{Mod}(L)$, $\vec{a}, \vec{b} \in \omega^n$ for some $n \in \omega$, and $\alpha < \omega_1$. We say that $(M, \vec{a})$ and $(N, \vec{b})$ are α-**equivalent**, denoted $(M, \vec{a}) \equiv_\alpha (N, \vec{b})$, if for all $L_{\omega_1 \omega}$-formula $\varphi(x_1, \ldots, x_n)$ with $\mathrm{qr}(\varphi) \leq \alpha$, $M \models$

$\varphi[a]$ iff $N \models \varphi[\vec{b}]$. We also say that M and N are α-**equivalent**, denoted $M \equiv_\alpha N$, if for all $L_{\omega_1\omega}$-sentence φ with $\mathrm{qr}(\varphi) \leq \alpha$, $M \models \varphi$ iff $N \models \varphi$.

Thus $M \equiv_\alpha N$ can be interpreted as the special case $(M, \emptyset) \equiv_\alpha (N, \emptyset)$.

Definition 12.1.3

Let L be a countable relational language and $M \in \mathrm{Mod}(L)$. For any tuple $\vec{a} \in \omega^n$ and $\alpha < \omega_1$, the **canonical Scott formula** of rank α for $\vec{a}$ in M, denoted $\varphi_\alpha^{M,\vec{a}}(\vec{v})$ where $\vec{v} = (v_0, \ldots, v_{n-1}) \in \mathfrak{V}^n$, is inductively defined as follows:

$$\varphi_0^{M,\vec{a}}(\vec{v}) = \bigwedge\{\,\theta(\vec{v}) : \theta \text{ is atomic or negated atomic and } M \models \theta[\vec{a}]\,\},$$

$$\varphi_{\alpha+1}^{M,\vec{a}}(\vec{v}) = \varphi_\alpha^{M,\vec{a}}(\vec{v}) \wedge \bigwedge_{b\in\omega} (\exists v_n)\varphi_\alpha^{M,\vec{a}\,^\frown b}(\vec{v}\,^\frown v_n) \wedge (\forall v_n) \bigvee_{b\in\omega} \varphi_\alpha^{M,\vec{a}\,^\frown b}(\vec{v}\,^\frown v_n),$$

$$\varphi_\lambda^{M,\vec{a}}(\vec{v}) = \bigwedge_{\alpha<\lambda} \varphi_\alpha^{M,\vec{a}}(\vec{v}), \text{ if } \lambda \text{ is a limit.}$$

It is easy to check that $\mathrm{qr}(\varphi_\alpha^{M,\vec{a}}) = \alpha$ and $M \models \varphi_\alpha^{M,\vec{a}}[\vec{a}]$.

Lemma 12.1.4

Let L be a countable relational language, $M, N \in \mathrm{Mod}(L)$, $\vec{a}, \vec{b} \in \omega^n$ for some $n \in \omega$, and $\alpha < \omega_1$. Then the following are equivalent:

(i) $(M, \vec{a}) \equiv_\alpha (N, \vec{b})$.

(ii) $N \models \varphi_\alpha^{M,\vec{a}}[\vec{b}]$.

(iii) $\varphi_\alpha^{M,\vec{a}} = \varphi_\alpha^{N,\vec{b}}$.

Proof. The implication (i)$\Rightarrow$(ii) easily follows from the definitions. We show (ii)$\Rightarrow$(iii) and (iii)$\Rightarrow$(i). First we show (ii)$\Rightarrow$(iii) by induction on α. The cases of $\alpha = 0$ and α being a limit are both straightforward from the definitions. For the successor case, assume $N \models \varphi_{\alpha+1}^{M,\vec{a}}[\vec{b}]$. Thus

$$N \models \varphi_\alpha^{M,\vec{a}}[\vec{b}] \text{ and } N \models \left(\bigwedge_{c\in\omega} (\exists v_n)\varphi_\alpha^{M,\vec{a}\,^\frown c}\right)[\vec{b}] \text{ and } N \models (\forall v_n) \bigvee_{c\in\omega} \varphi_\alpha^{M,\vec{a}\,^\frown c}[\vec{b}].$$

By $N \models \varphi_\alpha^{M,\vec{a}}[\vec{b}]$ and the inductive hypothesis $\varphi_\alpha^{M,\vec{a}} = \varphi_\alpha^{N,\vec{b}}$. To show $\varphi_{\alpha+1}^{M,\vec{a}} = \varphi_{\alpha+1}^{N,\vec{b}}$ it suffices to prove that

$$\{\varphi_\alpha^{M,\vec{a}\,^\frown c} : c \in \omega\} = \{\varphi_\alpha^{N,\vec{b}\,^\frown d} : d \in \omega\}.$$

To see this let first $c \in \omega$. Then $N \models (\exists v_n)\varphi_\alpha^{M,\vec{a}^\frown c}[\vec{b}]$, and therefore there is $d \in \omega$ such that $N \models \varphi_\alpha^{M,\vec{a}^\frown c}[\vec{b}^\frown d]$. By the inductive hypothesis, $\varphi_\alpha^{M,\vec{a}^\frown c} = \varphi_\alpha^{N,\vec{b}^\frown d}$. Conversely, let $d \in \omega$ be arbitrary. Then

$$N \models \bigvee_{c \in \omega} \varphi_\alpha^{M,\vec{a}^\frown c}[\vec{b}^\frown d].$$

Thus there is $c \in \omega$ such that $N \models \varphi_\alpha^{M,\vec{a}^\frown c}[\vec{b}^\frown d]$, and by the inductive hypothesis $\varphi_\alpha^{M,\vec{a}^\frown c} = \varphi_\alpha^{N,\vec{b}^\frown d}$. This finishes the proof of (ii)$\Rightarrow$(iii).

Next we show (iii)$\Rightarrow$(i) by induction on α. For $\alpha = 0$, assume $\varphi_0^{M,\vec{a}} = \varphi_0^{N,\vec{b}}$. Then for any atomic or negated atomic formula θ, $M \models \theta[\vec{a}]$ iff $N \models \theta[\vec{b}]$. Let $\psi(x_1,\ldots,x_n)$ be a quantifier-free formula. We need to verify that $M \models \psi[\vec{a}]$ iff $N \models \psi[\vec{b}]$ by induction on the form of ψ. If ψ is atomic then there is nothing to prove. The case $\psi = \neg\psi'$ is easy. Suppose $\psi = \bigwedge \Phi$. Then for any $\psi' \in \Phi$, $M \models \psi'[\vec{a}]$ iff $N \models \psi'[\vec{b}]$ by the inductive hypothesis, and it follows that $M \models \bigwedge \Phi$ iff $N \models \bigwedge \Phi$. The case $\psi = \bigvee \Phi$ follows similarly. This completes the case $\alpha = 0$.

Suppose next α is a limit, and $\varphi_\alpha^{M,\vec{a}} = \varphi_\alpha^{N,\vec{b}}$. Since $\varphi_\alpha^{M,\vec{a}} = \bigwedge_{\beta < \alpha} \varphi_\beta^{M,\vec{a}}$ and for each $\beta < \alpha$, $\mathrm{qr}(\varphi_\beta^{M,\vec{a}}) = \beta$, we must have that for each $\beta < \alpha$, $\varphi_\beta^{M,\vec{a}} = \varphi_\beta^{N,\vec{b}}$. By the inductive hypotheses, it follows that $(M,\vec{a}) \equiv_\beta (N,\vec{b})$ for all $\beta < \alpha$. We check that $(M,\vec{a}) \equiv_\alpha (N,\vec{b})$. For this let $\varphi(x_1,\ldots,x_n)$ be a formula with $\mathrm{qr}(\varphi) \le \alpha$. We again use induction on the form of α. The cases that φ is atomic and that $\varphi = \neg\varphi'$ are easy. Suppose $\varphi = \bigwedge \Phi$. Then for every $\varphi' \in \Phi$, $\mathrm{qr}(\varphi') \le \alpha$. By the inductive hypotheses, for any $\phi' \in \Phi$, $M \models \phi'[\vec{a}]$ iff $N \models \phi'[\vec{b}]$. Hence $M \models \varphi[\vec{a}]$ iff $N \models \varphi[\vec{b}]$. The case $\varphi = \bigvee \Phi$ is similar. This completes the case α is a limit.

Finally suppose $\varphi_{\alpha+1}^{M,\vec{a}} = \varphi_{\alpha+1}^{N,\vec{b}}$. It follows in particular that $N \models \varphi_\alpha^{M,\vec{a}}[\vec{b}]$, and since (ii)$\Rightarrow$(iii), we have that $\varphi_\alpha^{M,\vec{a}} = \varphi_\alpha^{N,\vec{b}}$. By the inductive hypothesis, $(M,\vec{a}) \equiv_\alpha (N,\vec{b})$. Also by the proof of (ii)$\Rightarrow$(iii), we have that

$$\{\varphi_\alpha^{M,\vec{a}^\frown c} : c \in \omega\} = \{\varphi_\alpha^{N,\vec{b}^\frown d} : d \in \omega\}.$$

Let φ be a formula with $\mathrm{qr}(\varphi) = \alpha + 1$. Suppose $\varphi = \exists x\psi$. If $M \models \varphi[\vec{a}]$, then for some $c \in \omega$ we have $M \models \psi[\vec{a}^\frown c]$. By the above equality there is $d \in \omega$ such that $\varphi_\alpha^{M,\vec{a}^\frown c} = \varphi_\alpha^{N,\vec{b}^\frown d}$, and by the inductive hypothesis, $(M,\vec{a}^\frown c) \equiv_\alpha (N,\vec{b}^\frown d)$. Since $\mathrm{qr}(\psi) = \alpha$, it follows that $N \models \psi[\vec{b}^\frown d]$ and $N \models \varphi[\vec{b}]$. By symmetry we have shown that $M \models \varphi[\vec{a}]$ iff $N \models \varphi[\vec{b}]$. The case $\alpha = \forall x\psi$ is similar. We have thus completed the successor case and the proof of (iii)$\Rightarrow$(i). $\square$

The proof of Lemma 12.1.4 is routine and uneventful, but the content of the lemma is rather surprising. Note that there are uncountably many $L_{\omega_1\omega}$-formulas of any rank; however the lemma says that for every rank it takes just one formula to summarize all the truth of a tuple on that rank. It is obvious from the definition of α-equivalence that $\equiv_\beta \subseteq \equiv_\alpha$ if $\alpha \leq \beta$. By the lemma it follows that if $\varphi_\alpha^{M,\vec{a}} \neq \varphi_\alpha^{N,\vec{b}}$ then $\varphi_\beta^{M,\vec{a}} \neq \varphi_\beta^{N,\vec{b}}$ for all $\beta \geq \alpha$.

Definition 12.1.5

Let L be a countable relational language and $M, N \in \mathrm{Mod}(L)$. For any $n \in \omega$ and $\vec{a}, \vec{b} \in \omega^n$, we say that $(M, \vec{a})$ and $(N, \vec{b})$ are ∞-**equivalent** or $L_{\omega_1\omega}$-**equivalent**, denoted by $(M, \vec{a}) \equiv_\infty (M, \vec{b})$, if $(M, \vec{a}) \equiv_\alpha (N, \vec{b})$ for all $\alpha < \omega_1$. We write $(M, \vec{a}) \cong (N, \vec{b})$ if there is an isomorphism π from M onto N such that $\pi(\vec{a}) = \vec{b}$.

We write $M \equiv_\infty N$ for $(M, \emptyset) \equiv_\infty (N, \emptyset)$.

Lemma 12.1.6 (Karp)

Let L be a countable relational language and $M, N \in \mathrm{Mod}(L)$. For any $n \in \omega$ and $\vec{a}, \vec{b} \in \omega^n$, then $(M, \vec{a}) \equiv_\infty (N, \vec{b})$ iff $(M, \vec{a}) \cong (N, \vec{b})$. In particular, $M \equiv_\infty N$ iff $M \cong N$.

Proof. By an easy induction on α one can show that if $(M, \vec{a}) \cong (N, \vec{b})$ then $(M, \vec{a}) \equiv_\alpha (N, \vec{b})$. Then ($\Leftarrow$) follows. To show ($\Rightarrow$), assume $(M, \vec{a}) \equiv_\infty (N, \vec{b})$. Without loss of generality we may assume $\vec{a} = (a_0, \ldots, a_{n-1}), \vec{b} = (b_0, \ldots, b_{n-1}) \in [\omega]^n$, since for any $i \neq j < n$, $a_i = a_j$ iff $b_i = b_j$ by $(M, \vec{a}) \equiv_0 (N, \vec{b})$. By induction on $m \in \omega$ we define $a_{n+m}, b_{n+m} \in \omega$ such that for all $m \in \omega$,

(a) if m is even, a_{n+m} is the least element of $\omega - \{a_0, \ldots, a_{n+m-1}\}$; if m is odd, b_{n+m} is the least element of $\omega - \{b_0, \ldots, b_{n+m-1}\}$;

(b) letting $\vec{a}^m = (a_0, \ldots, a_{n+m-1})$ and $\vec{b}^m = (b_0, \ldots, b_{n+m-1})$, we have $(M, \vec{a}^m) \equiv_\infty (N, \vec{b}^m)$.

Before carrying out the construction we note the following facts. For any $k \in \omega$ and tuples $\vec{c}, \vec{d} \in \omega^k$, we let $\alpha(\vec{c}, \vec{d})$ be the least ordinal $\alpha < \omega_1$ such that $(M, \vec{c}) \not\equiv_\alpha (M, \vec{d})$, if such an α exists; and let $\alpha(\vec{c}, \vec{d}) = 0$ if $(M, \vec{c}) \equiv_\infty (N, \vec{d})$. Let

$$\alpha(M, N) = \sup\{\alpha(\vec{c}, \vec{d}) : k \in \omega, \ \vec{c}, \vec{d} \in \omega^k\}.$$

Then $\alpha(M, N) < \omega_1$. It is straightforward from the definition that if $(M, \vec{c}) \equiv_\beta (N, \vec{d})$ for some $\beta \geq \alpha(M, N)$, then $(M, \vec{c}) \equiv_\infty (N, \vec{d})$. Thus to satisfy (b) we only need to make sure that $(M, \vec{a}^m) \equiv_{\alpha(M,N)} (N, \vec{b}^m)$.

Now for $m = 0$ let a_n be the least element of $\omega - \{a_0, \ldots, a_{n-1}\}$. Since $(M, \vec{a}) \equiv_\infty (N, \vec{b})$, and in particular, $(M, \vec{a}) \equiv_{\alpha(M,N)+1} (N, \vec{b})$, we have that $N \models \varphi_{\alpha(M,N)+1}^{M,\vec{a}}[\vec{b}]$. It follows that

$$N \models \bigwedge_{c \in \omega} (\exists v_n) \varphi_{\alpha(M,N)}^{M, \vec{a}^\frown c}[\vec{b}],$$

and in particular

$$N \models (\exists v_n)\varphi^{M,\vec{a}^\frown a_n}_{\alpha(M,N)}[\vec{b}].$$

Let $b_n \in \omega$ be such that

$$N \models \varphi^{M,\vec{a}^1}_{\alpha(M,N)}[\vec{b}^\frown b_n].$$

Then by Lemma 12.1.4 $(M, \vec{a}^1) \equiv_{\alpha(M,N)} (N, \vec{b}^1)$, and therefore $(M, \vec{a}^1) \equiv_\infty$ $(N, \vec{b}^1)$ as required by (c).

The case $m = 1$ is similar with the roles of a's and b's reversed. The general even case is similar to the case $m = 0$ and the odd case is similar to that of $m = 1$. This completes the construction of the sequences $(a_{n+m})_{m\in\omega}$ and $(b_{n+m})_{m\in\omega}$.

Define $\pi(a_i) = b_i$ for all $i \in \omega$. We check that π is an isomorphism from M onto N. It follows from (a) that π is a bijection from ω onto ω. Suppose $R \in L$ is an l-ary relation symbol and $M \models R[a_{n_1}, \ldots, a_{n_l}]$. Let $m \in \omega$ be large enough such that $\{n_1, \ldots, n_l\} \subseteq \{0, \ldots, n + m\}$. Then $N \models R[b_{n_1}, \ldots, b_{n_l}]$ since $(M, \vec{a}^m) \equiv_0 (N, \vec{b}^m)$. By symmetry we have that $M \models R[a_{n_1}, \ldots, a_{n_l}]$ iff $N \models R[b_{n_1}, \ldots, b_{n_l}]$. ⬜

Definition 12.1.7

Let L be a countable relational language and $M \in \mathrm{Mod}(L)$. The **Scott rank** of M, denoted by $\mathrm{sr}(M)$, is the least ordinal α such that for any $n \in \omega$ and tuples $\vec{a}, \vec{b} \in \omega^n$, if $\varphi^{M,\vec{a}}_\alpha = \varphi^{M,\vec{b}}_\alpha$ then $\varphi^{M,\vec{a}}_{\alpha+1} = \varphi^{M,\vec{b}}_{\alpha+1}$. The **canonical Scott sentence** of M, denoted by φ_M, is

$$\varphi^{M,\emptyset}_{\mathrm{sr}(M)} \wedge \bigwedge_{\substack{n\in\omega \\ \vec{a}\in\omega^n}} \forall v_0 \ldots \forall v_{n-1} \left(\varphi^{M,\vec{a}}_{\mathrm{sr}(M)}(v_0, \ldots, v_{n-1}) \rightarrow \varphi^{M,\vec{a}}_{\mathrm{sr}(M)+1}(v_0, \ldots, v_{n-1}) \right).$$

Note that $\mathrm{qr}(\varphi_M) = \mathrm{sr}(M) + \omega$.

Theorem 12.1.8 (Scott)

Let L be a countable relational language and $M, N \in \mathrm{Mod}(L)$. Then the following are equivalent:

(i) $M \cong N$.

(ii) $N \models \varphi_M$.

(iii) $\varphi_M = \varphi_N$.

Proof. We first verify that $M \models \varphi_M$. Of course $M \models \varphi^{M,\emptyset}_{\mathrm{sr}(M)}$. Let $\vec{a} \in \omega^n$. By the definition of the Scott rank, for any $\vec{b} \in \omega^n$, if $\varphi^{M,\vec{a}}_{\mathrm{sr}(M)} = \varphi^{M,\vec{b}}_{\mathrm{sr}(M)}$ then

$\varphi^{M,\vec{a}}_{\text{sr}(M)+1} = \varphi^{M,\vec{b}}_{\text{sr}(M)+1}$. By Lemma 12.1.4 it is equivalent to stating that if $M \models \varphi^{M,\vec{a}}_{\text{sr}(M)}[\vec{b}]$ then $M \models \varphi^{M,\vec{a}}_{\text{sr}(M)+1}[\vec{b}]$. Thus we have that

$$M \models \forall \vec{v} \, (\varphi^{M,\vec{a}}_{\text{sr}(M)} \rightarrow \varphi^{M,\vec{a}}_{\text{sr}(M)+1}).$$

Hence $M \models \varphi_M$.

Now the implications (i)$\Rightarrow$(iii)$\Rightarrow$(ii) are obvious. It remains to show (ii)$\Rightarrow$(i). For this assume $N \models \varphi_M$.

We claim that for any $\alpha \geq \text{sr}(M) + 1$,

$$N \models \bigwedge_{\substack{n \in \omega \\ \vec{a} \in \omega^n}} \forall v_0 \ldots \forall v_{n-1} \, (\varphi^{M,\vec{a}}_{\text{sr}(M)}(v_0, \ldots, v_{n-1}) \rightarrow \varphi^{M,\vec{a}}_{\alpha}(v_0, \ldots, v_{n-1})).$$

This is proved by induction on $\alpha \geq \text{sr}(M) + 1$. The case $\alpha = \text{sr}(M) + 1$ is immediate from the assumption $N \models \varphi_M$. The case α is a limit follows easily from the inductive hypotheses. We only consider the case $\alpha = \beta + 1$ where $\beta \geq \text{sr}(M) + 1$. For this let $n \in \omega$ and $\vec{a}, \vec{b} \in \omega^n$ such that $N \models \varphi^{M,\vec{a}}_{\text{sr}(M)}[\vec{b}]$. We need to show that $N \models \varphi^{M,\vec{a}}_{\beta+1}[\vec{b}]$. By the inductive hypothesis $N \models \varphi^{M,\vec{a}}_{\beta}[\vec{b}]$. Let $c \in \omega$. Since $N \models \varphi^{M,\vec{a}}_{\text{sr}(M)+1}[\vec{b}]$ by $N \models \varphi_M$, we can find $d \in \omega$ such that $N \models \varphi^{M,\vec{a}^\frown c}_{\text{sr}(M)}[\vec{b}^\frown d]$. By the inductive hypothesis it follows that $N \models \varphi^{M,\vec{a}^\frown c}_{\beta}[\vec{b}^\frown d]$. A similar, symmetric argument gives that for any $d \in \omega$ there is $c \in \omega$ such that $N \models \varphi^{M,\vec{a}^\frown c}_{\beta}[\vec{b}^\frown d]$. This finishes the proof for the successor case.

By the $n = 0$ case of the claim we obtain that $N \models \varphi^{M,\emptyset}_{\alpha}$ for all $\alpha \geq \text{sr}(M) + 1$. By Lemma 12.1.4 $N \equiv_{\infty} M$, and hence by Lemma 12.1.6 $N \cong M$.
□

Exercise 12.1.1 Show that if $(M, \vec{a}) \equiv_{\text{sr}(M)} (M, \vec{b})$ then $(M, \vec{a}) \equiv_{\infty} (M, \vec{b})$.

A tuple $\vec{a}$ is $L_{\omega_1 \omega}$-**definable** in M if there is an $L_{\omega_1 \omega}$-formula $\theta(\vec{x})$ such that $M \models \varphi[\vec{a}]$ and $M \models \neg\varphi[\vec{b}]$ for all $\vec{b} \neq \vec{a}$. We say that $(M, \vec{a})$ is **rigid** if there is no nontrivial automorphism π of M with $\pi(\vec{a}) = \vec{a}$. Similarly, M is **rigid** if it has no nontrivial automorphism.

Exercise 12.1.2 Show that $\vec{a}$ is $L_{\omega_1 \omega}$-definable in M iff for all $\pi \in \text{Aut}(M)$, $\pi(\vec{a}) = \vec{a}$.

Exercise 12.1.3 Show that M is **rigid**, that is, has no nontrivial automorphism, iff every element of M is $L_{\omega_1 \omega}$-definable in M.

Exercise 12.1.4 Show that if $(M, \vec{a})$ is rigid for some $\vec{a}$ then M has only countably many automorphisms.

12.2 Borel approximations of isomorphism relations

Throughout this section we fix a countable relational language L. We consider $\mathrm{Mod}(L)$ with the canonical topology and investigate the descriptive complexity of the sets and relations arising in the Scott analysis.

Lemma 12.2.1
For any $\alpha < \omega_1$ the relation $\equiv_\alpha$ is a Borel equivalence relation with every equivalence class $\mathbf{\Pi}^0_{1+2\cdot\alpha}$.

Proof. By Lemma 12.1.4 and the definition of canonical Scott formulas we can inductively characterize $\equiv_\alpha$ as follows:

$$(M,\vec{a}) \equiv_0 (N,\vec{b}) \iff \text{ for all atomic formula } \theta,\ M \models \theta[\vec{a}] \text{ iff } N \models \theta[\vec{b}].$$

$$(M,\vec{a}) \equiv_{\alpha+1} (N,\vec{b}) \iff \forall c \in \omega\ \exists d \in \omega\ (M, \vec{a}^\frown c) \equiv_\alpha (N, \vec{b}^\frown d)$$
$$\wedge \forall d \in \omega\ \exists c \in \omega\ (M, \vec{a}^\frown c) \equiv_\alpha (N, \vec{b}^\frown d).$$

$$(M,\vec{a}) \equiv_\lambda (N,\vec{b}) \iff \forall \alpha < \lambda\ (M,\vec{a}) \equiv_\alpha (N,\vec{b}), \text{ if } \lambda \text{ is a limit.}$$

The lemma follows from this characterization by an easy induction. □

Again since there are uncountably many formulas of any given quantifier rank, the lemma is not clear from the definition of $\equiv_\alpha$.

By Lemma 12.1.6 the isomorphism relation on $\mathrm{Mod}(L)$ is $\bigcap_{\alpha<\omega} \equiv_\alpha$. Thus $\equiv_\alpha$ can be viewed as Borel approximations of the isomorphism relation. We will show that these approximations are canonical in the sense that for any invariant Borel class $\mathrm{Mod}(\varphi)$ if $\cong_\varphi$ is Borel then it coincides with $\equiv_\alpha \upharpoonright \mathrm{Mod}(\varphi)$ for some $\alpha < \omega_1$. Before proving this we need the following observation.

Lemma 12.2.2
If an invariant Borel class $\mathcal{C}$ is a $\mathbf{\Pi}^0_\alpha$ subset of $\mathrm{Mod}(L)$, then there is a formula φ with $\mathrm{qr}(\varphi) \leq \omega \cdot \alpha$ such that $\mathcal{C} = \mathrm{Mod}(\varphi)$.

Proof. Examine the proof of Lemma 11.3.5 and count the quantifier rank of the formulas involved. We get that if $B \subseteq \mathrm{Mod}(L)$ is $\mathbf{\Pi}^0_\alpha$ then for each $n \in \omega$ there are formulas φ_n and ψ_n with $\mathrm{qr}(\varphi_n), \mathrm{qr}(\psi_n) \leq \omega \cdot \alpha$ such that

$$(M,\vec{a}) \in B^{[\triangle n]} \iff M \in \mathrm{Mod}(\varphi_n, \vec{a}),$$
$$(M,\vec{a}) \in B^{[*n]} \iff M \in \mathrm{Mod}(\psi_n, \vec{a}).$$

Now if $\mathcal{C}$ is an invariant Borel class then $\mathcal{C}^{[*0]} = \mathcal{C} \times \{\emptyset\}$ and $\mathcal{C} = \mathrm{Mod}(\varphi_0)$, where $\mathrm{qr}(\varphi_0) \leq \omega \cdot \alpha$. □

In particular if a single isomorphism class $[M]_\cong$ is $\mathbf{\Pi}^0_\alpha$ then there is φ with $\mathrm{qr}(\varphi) \le \omega \cdot \alpha$ so that for any $N \in \mathrm{Mod}(L)$, if $N \models \varphi$ then $N \cong M$. Thus the sentence has the similar property as the canonical Scott sentence. We next prove that the quantifier rank of such a sentence is a good approximation of the Scott rank.

Lemma 12.2.3
Let $M \in \mathrm{Mod}(L)$ and $\alpha < \omega_1$. Suppose for all $N \in \mathrm{Mod}(L)$, if $M \equiv_\alpha N$ then $M \cong N$. Then $\mathrm{sr}(M) \le \alpha + \omega$.

Proof. Let F be the countable fragment of $L_{\omega_1\omega}$ generated by $\{\varphi_\beta^{M,\emptyset} : \beta < \alpha + \omega\}$. Then for every $n \in \omega$, $\vec{a} \in \omega^n$ and $k \in \omega$, $\varphi_{\alpha+k}^{M,\vec{a}} \in F$ since it is a subformula of $\varphi_{\alpha+k+n}^{M,\emptyset}$. Let $T = \mathrm{Th}_F(M)$. We note that for any $N \in \mathrm{Mod}(T)$, $N \cong M$. This is because, if $N \in \mathrm{Mod}(T)$, then $N \models \varphi_\alpha^{M,\emptyset}$ since $\varphi_\alpha^{M,\emptyset} \in F$, and by Lemma 12.1.4 $N \equiv_\alpha M$; therefore $N \cong M$ by our assumption. Thus M is the unique model of T up to isomorphism. We claim that M is an F-atomic model of T. To see this, assume otherwise, that there is a nonprincipal F-type Φ of T realized in M. Then by the omitting types theorem there is a model $N \in \mathrm{Mod}(T)$ omitting the type Φ, and thus $N \not\cong M$, contradiction.

Let $n \in \omega$ and $\vec{a} \in \omega^n$ be arbitrary. Then $\mathrm{tp}_F(\vec{a}/M)$ is principal. Hence there is an F-formula $\psi_{\vec{a}} \in \mathrm{tp}_F(\vec{a}/M)$ such that for all $\phi \in \mathrm{tp}_F(\vec{a}/M)$, $N \in \mathrm{Mod}(T)$ and $\vec{b} \in \omega^n$, if $N \models \psi_{\vec{a}}[\vec{b}]$ then $N \models \phi[\vec{b}]$. By our construction $\mathrm{qr}(\psi_{\vec{a}}) < \alpha + \omega$. Let $\beta_{\vec{a}} = \mathrm{qr}(\psi_{\vec{a}})$. We now have that for all $\vec{b} \in \omega^n$, if $(M, \vec{a}) \equiv_{\beta_{\vec{a}}} (M, \vec{b})$, then $(M, \vec{a}) \equiv_F (M, \vec{b})$, and therefore $(M, \vec{a}) \equiv_{\alpha+\omega} (M, \vec{b})$.

We now claim that for all $n \in \omega$ and $\vec{a}, \vec{b} \in \omega^n$, if $(M, \vec{a}) \equiv_{\alpha+\omega} (M, \vec{b})$ then $(M, \vec{a}) \cong (M, \vec{b})$. This implies that $\mathrm{sr}(M) \le \alpha + \omega$. To prove the claim we assume $(M, \vec{a}) \equiv_{\alpha+\omega} (M, \vec{b})$ and use a back-and-forth construction. We indicate one step of the construction. For any $c \in \omega$, we have that $(M, \vec{a}) \equiv_{\beta_{\vec{a}^\frown c}+1} (M, \vec{b})$, and hence there is $d \in \omega$ such that $(M, \vec{a}^\frown c) \equiv_{\beta_{\vec{a}^\frown c}} (M, \vec{b}^\frown d)$. This implies that $(M, \vec{a}^\frown c) \equiv_{\alpha+\omega} (M, \vec{b}^\frown d)$. Now it is clear that the back-and-forth construction can sustain to give an automorphism π of M so that $\pi(\vec{a}) = \vec{b}$. ∎

Theorem 12.2.4
Let L be a countable relational language and φ an $L_{\omega_1\omega}$-sentence. Then the following are equivalent:

(i) *$\cong_\varphi$ is Borel.*

(ii) *There is $\alpha < \omega_1$ such that $\cong_\varphi$ coincides with $\equiv_\alpha \upharpoonright \mathrm{Mod}(\varphi)$.*

(iii) *There is $\alpha < \omega_1$ such that $\mathrm{sr}(M) < \alpha$ for all $M \in \mathrm{Mod}(\varphi)$.*

Proof. The implication (ii)⇒(i) is obvious by Lemma 12.2.1. Next we show (iii)⇒(ii). Note that (iii) implies that there is $\alpha < \omega_1$ such that $\mathrm{sr}(M) + \omega < \alpha$

for all $M \in \text{Mod}(\varphi)$. Now for $M, N \in \text{Mod}(\varphi)$, if $M \equiv_\alpha N$, then $N \models \varphi_M$ and therefore $N \cong M$ by Scott's Theorem 12.1.8. This shows that $\equiv_\alpha \restriction \text{Mod}(\varphi)$ coincides with $\cong_\varphi$.

Finally we show (i)$\Rightarrow$(iii). Assume $\cong_\varphi$ is $\mathbf{\Pi}^0_\beta$ for some $\beta < \omega_1$. Then in particular every orbit is $\mathbf{\Pi}^0_\beta$. By Lemma 12.2.2 for every $M \in \text{Mod}(\varphi)$ there is a sentence ψ_M with $\text{qr}(\psi_M) \le \omega \cdot \beta$ such that $[M]_\cong = \text{Mod}(\psi_M)$. Let $\alpha = \omega \cdot \beta + \omega + 1 = \omega \cdot (\beta + 1) + 1$. By Lemma 12.2.3 $\text{sr}(M) < \alpha$. □

We next define a cofinal family of Borel isomorphism relations known as the **Friedman–Stanley tower**. The basic idea is to start with the simplest equivalence relation, the identity relation on a standard Borel space, and iterate the Friedman–Stanley jump operation transfinitely. Recall that for equivalence relation E on a standard Borel space X, the Friedman–Stanley jump E^+ is an equivalence relation on X^ω defined as

$$x E^+ y \iff \{[x_n]_E : n \in \omega\} = \{[y_n]_E : n \in \omega\}.$$

To define the Friedman–Stanley tower we need first to fix some notation for exponentiation of countable ordinals. We note that the notation we introduce is not standard, and the purpose to introduce it is to avoid confusion with product spaces.

Notation 12.2.5
For any $\alpha < \omega_1$ we define the ordinal $\epsilon(\alpha)$ inductively as follows:

$$\epsilon(0) = \omega,$$
$$\epsilon(\alpha + 1) = \epsilon(\alpha) \cdot \omega,$$
$$\epsilon(\lambda) = \sup_{\alpha < \lambda} \epsilon(\alpha), \text{ if } \lambda \text{ is a limit.}$$

The usual notation for $\epsilon(\alpha)$ is $\omega^{1+\alpha}$; there is obviously a danger of confusion with the product space such as $\omega^\alpha = \prod_{i<\alpha} \omega$. The equivalence relations we are considering are on product spaces $\omega^{\epsilon(\alpha)}$. We fix a natural homeomorphism between the space $\omega^{\epsilon(\alpha+1)}$ and $\left(\omega^{\epsilon(\alpha)}\right)^\omega$.

Each of the ordinals $\epsilon(\alpha)$ is a limit itself, and if $\alpha \le \beta$ then $\epsilon(\alpha) + \epsilon(\beta) = \epsilon(\beta)$. If λ is a limit, then we also have $\epsilon(\lambda) = \sum_{\alpha < \lambda} \epsilon(\alpha)$. This induces a homeomorphism between the space $\omega^{\epsilon(\lambda)}$ and the product space $\prod_{\alpha < \lambda} \omega^{\epsilon(\alpha)}$.

Notation 12.2.6
For any $\alpha < \omega_1$ we define the equivalence relation $=^{\alpha+}$ on $\omega^{\epsilon(\alpha)}$ inductively as follows:

$$=^{0+} = \text{id}(\omega^\omega),$$
$$=^{(\alpha+1)+} = \left(=^{\alpha+}\right)^+,$$
$$=^{\lambda+} = \prod_{\alpha < \lambda} =^{\alpha+}, \text{ if } \lambda \text{ is a limit.}$$

By induction on $\alpha < \omega_1$ it is easy to verify that $=^{\alpha+}$ is Borel bireducible to some S_∞-orbit equivalence relation and that each $=^{\alpha+}$ is a $\mathbf{\Pi}^0_{1+2\cdot\alpha}$ equivalence relation on $\omega^{\epsilon(\alpha)}$.

Lemma 12.2.7
Let L be a countable relational language. For any $\alpha < \omega_1$, $\equiv_\alpha$ is Borel reducible to $=^{\alpha+}$.

Proof. By induction on $\alpha < \omega_1$ we show that the equivalence relation $\equiv_\alpha$ on $\mathrm{Mod}(L) \times \omega^{<\omega}$ is Borel reducible to $=^{\alpha+}$. For $\alpha = 0$, $\equiv_0$ is closed and hence smooth; therefore $\equiv_0$ is Borel reducible to $\mathrm{id}(\omega^\omega)$. Assume f_α is a Borel reduction of $\equiv_\alpha$ on $\mathrm{Mod}(L) \times \omega^{<\omega}$ to $=^{\alpha+}$ on $\omega^{\epsilon(\alpha)}$. Then note that

$$(M, \vec{a}) \equiv_{\alpha+1} (N, \vec{b}) \iff \forall c \in \omega \, \exists d \in \omega \, (M, \vec{a}^\frown c) \equiv_\alpha (N, \vec{b}^\frown d)$$
$$\forall d \in \omega \, \exists c \in \omega \, (M, \vec{a}^\frown c) \equiv_\alpha (N, \vec{b}^\frown d).$$

Thus for any $(M, \vec{a}) \in \mathrm{Mod}(L) \times \omega^{<\omega}$ we let $f_{\alpha+1}(M, \vec{a})$ be an element of $\left(\omega^{\epsilon(\alpha)}\right)^\omega$ such that

$$f_{\alpha+1}(M, \vec{a})(c) = f_\alpha(M, \vec{a}^\frown c).$$

Then $(M, \vec{a}) \equiv_{\alpha+1} (N, \vec{b})$ iff $f_{\alpha+1}(M, \vec{a}) =^{(\alpha+1)+} f_{\alpha+1}(N, \vec{b})$. This finishes the proof of the successor case. Assume λ is a limit and for all $\alpha < \lambda$ let f_α be a Borel reduction from $\equiv_\alpha$ to $=^{\alpha+}$. Then define

$$f_\lambda(M, \vec{a}) = (f_\alpha(M, \vec{a}))_{\alpha < \lambda},$$

and we have

$$(M, \vec{a}) \equiv_\lambda (N, \vec{b})$$
$$\iff \forall \alpha < \lambda \, (M, \vec{a}) \equiv_\alpha (N, \vec{b})$$
$$\iff \forall \alpha < \lambda \, f_\alpha(M, \vec{a}) =^{\alpha+} f_\alpha(N, \vec{b})$$
$$\iff f_\lambda(M, \vec{a}) =^{\lambda+} f_\lambda(N, \vec{b}).$$

$\square$

It follows immediately from this lemma and Theorem 12.2.4 that in the Borel reducibility hierarchy the Friedman–Stanley tower is cofinal for all Borel S_∞-orbit equivalence relations or Borel isomorphism relations.

Corollary 12.2.8
Let L be a countable relational language and φ an $L_{\omega_1\omega}$-sentence. Then $\cong_\varphi$ is Borel iff there is $\alpha < \omega_1$ such that $\cong_\varphi$ is Borel reducible to $=^{\alpha+}$.

Corollary 12.2.9
Let G be a closed subgroup of S_∞ and X a standard Borel G-space. Then E_G^X is Borel iff there is $\alpha < \omega_1$ such that $E_G^X \leq_B =^{\alpha+}$.

Exercise 12.2.1 Let $L = \{R_i\}_{i \in \omega}$, where each R_i is a unary relation symbol. Show that $\cong_L$ is Borel bireducible with $=^+$.

Exercise 12.2.2 Show that for any $\alpha < \omega_1 =^{\alpha+}$ is Borel bireducible with an S_∞-orbit equivalence relation on $\omega^{\epsilon(\alpha)}$.

Exercise 12.2.3 Show that for any $\alpha < \omega_1 \equiv_\alpha$ is continuously reducible to $=^{\alpha+}$.

12.3 The Scott rank and computable ordinals

In the preceding section we already saw that the canonicity of the objects in the Scott analysis gives far-reaching results on the isomorphism relation of invariant Borel classes. The complexity of the isomorphism relation is in some precise sense equivalent to the uniformity of the complexity of the models involved. In this section we focus on sharper characterizations of the complexity of countable models. This was first done by Nadel [127], whose idea was to analyze the lightface content of the objects of the Scott analysis. It will turn out that these characterizations do give us further results about Borelness of invariant Borel classes.

We now turn to Nadel's study of the Scott analysis. We will use the notation and results reviewed in Section 1.6. In particular, the boundedness principles for Σ_1^1 sets of computable well-orders and its relativizations will be used below.

Fix a countable relational language $L = \{R_i\}_{i \in I}$, where either $I \subseteq \omega$ is finite or $I = \omega$, and the arity function $i \mapsto n_i$ is computable. We say that such an L is **computably presented**. Then there is a computable bijection between the Baire space ω^ω and the space

$$X_L = \prod_{i \in I} 2^{\omega^{n_i}}.$$

If $M \in \mathrm{Mod}(L)$ we let $x_M \in X_L$ be correspondent with M, and write $\omega_1^{\mathrm{CK}(M)}$ for $\omega_1^{\mathrm{CK}(x_M)}$. Now the canonicity of the Scott formulas of M implies that the α-equivalence within M is hyperarithmetic in M for $\alpha < \omega_1^{\mathrm{CK}(M)}$.

Lemma 12.3.1
Let L be a countable relational language that is computably presented and $M \in \mathrm{Mod}(L)$. Then for any $\alpha < \omega_1^{\mathrm{CK}(M)}$ and $n \in \omega$, the set

$$\{(\vec{a}, \vec{b}) \in \omega^n \times \omega^n : (M, \vec{a}) \equiv_\alpha (M, \vec{b})\} \subseteq (\omega^n)^2$$

is $\Delta_1^1(M)$.

Proof. This is an effective version of Lemma 12.2.1. For any $i \in I$ and $n \geq n_i$, the set $\{\vec{a} \in \omega^n : M \models R_i[\vec{a}]\}$ is clearly computable in M. Since L is computably presented, it follows that for all $n \in \omega$ and $\vec{a}, \vec{b} \in \omega^n$,

$$(M, \vec{a}) \equiv_0 (M, \vec{b}) \iff \forall i \in I \ (n_i \leq n \rightarrow (M \models R_i[\vec{a}] \leftrightarrow M \models R_i[\vec{b}])).$$

Thus the set $\{(\vec{a}, \vec{b}) : (M, \vec{a}) \equiv_0 (M, \vec{b})\}$ is $\Pi_1^0(M)$. Now for any $\alpha < \omega_1^{\mathrm{CK}(M)}$ we may find $x \in \mathrm{WO}^M$ such that $\mathrm{ot}(<_x) = \alpha$. Now the inductive proof of Lemma 12.2.1 can be repeated up to α, giving that $\{(\vec{a}, \vec{b}) : (M, \vec{a}) \equiv_\alpha (M, \vec{b})\}$ is $\Pi_{1+2\cdot\alpha}^0(M)$. This shows that the sets are all $\Delta_1^1(M)$. $\qquad\square$

We now indicate how to eliminate the hypothesis that L is computably presented in the above lemma. Suppose L is an arbitrary countable relational language. Let $x \in \omega^\omega$ code its arity function. Then L is computable in x. There is a fixed bijection between the Baire space ω^ω and X_L that is also computable in x. Note that for every $M \in \mathrm{Mod}(L)$, x is computable in M. Thus by relativizing the above lemma, the assumption that L is presented computably can be removed.

We also need the following uniform version of Lemma 12.3.1.

Lemma 12.3.2
Let L be a countable relational language and $M \in \mathrm{Mod}(L)$. Then for any $n \in \omega$ both sets

$$\{(x, \vec{a}, \vec{b}) \in \omega^\omega \times (\omega^n)^2 : x \in \mathrm{WO}^M \to (M, \vec{a}) \equiv_{\mathrm{ot}(<_x)} (M, \vec{b})\}$$

and

$$\{(x, \vec{a}, \vec{b}) \in \omega^\omega \times (\omega^n)^2 : x \in \mathrm{WO}^M \to (M, \vec{a}) \not\equiv_{\mathrm{ot}(<_x)} (M, \vec{b})\}$$

are $\Sigma_1^1(M)$.

Proof. This clearly follows from the proof of Lemma 12.3.1 and the fact that WO^M is $\Pi_1^1(M)$. $\qquad\square$

Theorem 12.3.3 (Nadel)
Let L be a countable relational language and $M \in \mathrm{Mod}(L)$. Then $\mathrm{sr}(M) \le \omega_1^{\mathrm{CK}(M)}$.

Proof. Let $\lambda = \omega_1^{\mathrm{CK}(M)}$. By the definition of the Scott rank, and by symmetry, it suffices to show that if $n \in \omega$, $\vec{a}, \vec{b} \in \omega^n$ and $(M, \vec{a}) \equiv_\lambda (M, \vec{b})$, then for any $c \in \omega$ there is $d \in \omega$ such that $(M, \vec{a}^\frown c) \equiv_\lambda (M, \vec{b}^\frown d)$.

Assume not, and let $n \in \omega$, $\vec{a}, \vec{b} \in \omega^n$, $c \in \omega$ such that $(M, \vec{a}) \equiv_\lambda (M, \vec{b})$ but

$$\forall d \in \omega \; \exists \alpha < \lambda \; (M, \vec{a}^\frown c) \not\equiv_\alpha (M, \vec{b}^\frown d).$$

For each $d \in \omega$ we let $\alpha_d < \lambda$ be the least such ordinal. Then note that for any $\alpha \ge \alpha_d$, $(M, \vec{a}^\frown c) \not\equiv_\alpha (M, \vec{b}^\frown d)$. Define

$$S = \{(d, x) \in \omega \times \omega^\omega : x \in \mathrm{WO}^M \wedge \mathrm{ot}(<_x) = \alpha_d\}.$$

Then we have that $(d, x) \in S$ iff the following conditions hold:

(i) $x \in \mathrm{LO}$ and x is computable in M,

(ii) for all $y \in \Delta_1^1(M)$, if $y \in \mathrm{WO}^M$ then either

    (iia) $(M, \vec{a}\!^\frown c) \equiv_{\mathrm{ot}(<_y)} (M, \vec{b}\!^\frown d)$ and there is $j : \omega \to \omega$ an order-preserving injection from $<_y$ into an initial segment of $<_x$, or

    (iib) $(M, \vec{a}\!^\frown c) \not\equiv_{\mathrm{ot}(<_y)} (M, \vec{b}\!^\frown d)$ and there is $j : \omega \to \omega$ an order-preserving injection from $<_x$ into $<_y$.

The ($\Rightarrow$) direction of the equivalence is clearly true. To see ($\Leftarrow$), note that if (d, x) satisfies (i) and (ii) then $x \in \mathrm{WO}^M$. This is because, if $y \in \mathrm{WO}^M$ so that $\mathrm{ot}(<_y) \geq \alpha_d$, then (iia) fails and (iib) must hold, which implies that $<_x$ is a well-order. Now $\mathrm{ot}(<_x) = \alpha_d$ by (iia) and (iib). Now (iia) and (iib) are both $\Sigma_1^1(M)$ for parameters $x, y \in \mathrm{LO}$. It follows from Lemma 12.3.2 and Kleene's Theorem 1.7.5 that (ii) is $\Sigma_1^1(M)$ for x. Now it easily follows that S is $\Sigma_1^1(M)$.

Let $A = \{x \in \omega^\omega : \exists d \in \omega \ (d, x) \in S\}$. Then A is also $\Sigma_1^1(M)$. By definition $A \subseteq \mathrm{WO}^M$. By the boundedness principle Theorem 1.6.9 there is $\beta < \lambda = \omega_1^{\mathrm{CK}(M)}$ such that $A \subseteq \mathrm{WO}_\beta^M$. It thus follows that

$$\forall d \in \omega \ \exists \alpha \leq \beta \ (M, \vec{a}\!^\frown c) \not\equiv_\alpha (M, \vec{b}\!^\frown d).$$

However, since $(M, \vec{a}) \equiv_{\beta+1} (M, \vec{b})$, there is in fact some $d \in \omega$ such that $(M, \vec{a}\!^\frown c) \equiv_\beta (M, \vec{b}\!^\frown d)$, a contradiction.   □

The following theorem of Sacks is a useful tool in the study of isomorphism relations.

Theorem 12.3.4 (Sacks)
Let L be a countable relational language and φ an $L_{\omega_1\omega}$-sentence. Then $\cong_\varphi$ is Borel iff there is $\alpha < \omega_1$ such that for all $M \in \mathrm{Mod}(\varphi)$, if $\mathrm{sr}(M) > \alpha$, then $\mathrm{sr}(M) < \omega_1^{\mathrm{CK}(M)}$.

Proof. The direction ($\Rightarrow$) follows from Theorem 12.2.4. We use the boundedness principle to show ($\Leftarrow$). Fix $\alpha_0 < \omega_1$ such that for all $M \in \mathrm{Mod}(\varphi)$, either $\mathrm{sr}(M) \leq \alpha_0$ or $\mathrm{sr}(M) < \omega_1^{\mathrm{CK}(M)}$. Fix $x_0 \in \mathrm{WO}$ such that $\mathrm{ot}(<_{x_0}) = \alpha_0$. For all $x \in \mathrm{WO}$ let P_x be the set of all $M \in \mathrm{Mod}(\varphi)$ such that for all $n \in \omega$ and $\vec{a}, \vec{b} \in \omega^n$,

$$(M, \vec{a}) \equiv_{\mathrm{ot}(<_x)} (M, \vec{b}) \Rightarrow (M, \vec{a}) \equiv_{\mathrm{ot}(<_x)+1} (M, \vec{b}).$$

Then each P_x is Borel. Let

$$P = \{(M, x) \in \mathrm{Mod}(\varphi) \times \omega^\omega : x \in \mathrm{WO} \Rightarrow M \in P_x\}$$

and
$$Q = \{(M, x) \in \text{Mod}(\varphi) \times \omega^\omega : x \in \text{WO} \Rightarrow M \notin P_x\}.$$
Then both P and Q are Σ_1^1. Let
$$C = \{(M, x) \in \text{Mod}(\varphi) \times \omega^\omega : \text{ot}(<_x) = \text{sr}(M)\}.$$
Then by our assumption $(M, x) \in C$ iff either of the following conditions (i) and (ii) holds:

(i) there is $j : \omega \to \omega$ an order-preserving injection of $<_x$ into $<_{x_0}$, $(M, x) \in P$, and for all $n \in \omega$, there is $x' \in \text{LO}$ and $j' : \omega \to \omega$ an order-preserving bijection from $<_{x'}$ onto $<_x \restriction n$ such that $(M, x') \in Q$;

(ii) for all $y \in \Delta_1^1(M)$, if $y \in \text{WO}$ then either

(iia) $(M, y) \in Q$ and there is $j : \omega \to \omega$ an order-preserving injection from $<_y$ into $<_x$; or

(iib) $(M, y) \in P$ and there is $j : \omega \to \omega$ an order-preserving injection from $<_x$ into $<_y$.

Clause (i) corresponds to the case $\text{ot}(<_x) = \text{sr}(M) \leq \alpha_0$ and clause (ii) corresponds to the case $\text{ot}(<_x) = \text{sr}(M) < \omega_1^{\text{CK}(M)}$. It is straightforward to see that (i) is equivalent to $\text{ot}(<_x) = \text{sr}(M) \leq \alpha_0$. Assume either (i) or (ii) holds but (i) fails, then not only (ii) holds, but also $\text{sr}(M) > \alpha_0$, from which it follows that $\text{sr}(M) < \omega_1^{\text{CK}(M)}$. Hence if we take $y \in \text{WO}^M$ so that $\text{ot}(<_y) \geq \text{sr}(M)$, we have further that (iia) fails and therefore (iib) is true. It follows that $<_x$ is a well-order. Then $\text{ot}(<_x) = \text{sr}(M)$ by (iia) and (iib).

A computation shows that C is Σ_1^1. Let
$$A = \{x \in \text{LO} : \exists M \in \text{Mod}(\varphi) \ (M, x) \in C\}.$$
Then A is a Σ_1^1 subset of WO. By the boundedness principle there is $\alpha < \omega_1$ such that $A \subseteq \text{WO}_\alpha$. This implies that for all $M \in \text{Mod}(\varphi)$, $\text{sr}(M) \leq \alpha$. By Theorem 12.2.4 $\cong_\varphi$ is Borel. ☐

Exercise 12.3.1 Show that the set S in the proof of Nadel's Theorem 12.3.3 is actually $\Delta_1^1(M)$.

Exercise 12.3.2 Let $\mathcal{C}$ be an invariant Borel class such that for all $M \in \mathcal{C}$ there is $\alpha < \omega_1^{\text{CK}(M)}$ with $[M]_\cong \in \Pi_\alpha^0$. Show that $\cong \restriction \mathcal{C}$ is Borel.

12.4   A topological variation of the Scott analysis

In this section we follow an approach of Hjorth [70] to investigate the Scott analysis. This leads to general results about S_∞ actions, and much of the ideas presented in this section can be generalized to arbitrary Polish group actions.

Throughout this section, and for the most part of the next one, we fix a Polish S_∞-space X and a countable base $\{U_n\}_{n \in \omega}$ for X. For each $n \in \omega$ let $V_n = \{g \in S_\infty : \forall i < n \; g(i) = i\}$. Each V_n is a clopen subgroup, and $\{V_n\}_{n \in \omega}$ forms a nbhd base for the identity of S_∞. Note that the collection $\{V_n g : n \in \omega, \; g \in S_\infty\}$ is actually countable and is a base for S_∞. Similarly the collection $\{gV_n : g \in S_\infty, \; n \in \omega\}$ is also a countable base for S_∞.

Definition 12.4.1
For $x \in X$ and $\alpha < \omega_1$, the **canonical Hjorth–Scott type** of rank α for x with respect to V_n is inductively defined as follows:

$$\varphi_0(x, V_n) = \{l \in \omega : V_n \cdot x \cap U_l \neq \emptyset\},$$

$$\varphi_{\alpha+1}(x, V_n) = \{(\varphi_\alpha(\hat{x}, V_m))_{m \geq n} : \hat{x} \in V_n \cdot x\},$$

$$\varphi_\lambda(x, V_n) = (\varphi_\alpha(x, V_n))_{\alpha < \lambda}, \text{ if } \lambda \text{ is a limit.}$$

The definition is obviously motivated by that of the canonical Scott formulas; however, there are important differences in details. First, apparently $\varphi_0(x, V_n)$ is an element of 2^ω, but in general $\varphi_{\alpha+1}(x, V_n)$ is a set of countable sequences of types of rank α. We show below that $\varphi_{\alpha+1}$ is a countable set, and with this observation we may code $\varphi_\alpha(x, V_n)$ by elements of $2^{\epsilon(\alpha)}$.

Lemma 12.4.2
For all $\alpha < \omega_1$, $x, x_1, x_2 \in X$ and $n \in \omega$, the following hold:

(a) if $x_1 \in V_n \cdot x_2$ then $\varphi_\alpha(x_1, V_n) = \varphi_\alpha(x_2, V_n)$;

(b) $\varphi_{\alpha+1}(x, V_n)$ is countable.

Proof. We first show (a) by induction on $\alpha < \omega_1$. Assume $x_1 \in V_n \cdot x_2$. Note that V_n is a subgroup of S_∞, and therefore $V_n \cdot x_1 = V_n \cdot x_2$. This implies immediately that $\varphi_0(x_1, V_n) = \varphi_0(x_2, V_n)$. For the successor case

$$\varphi_{\alpha+1}(x_1, V_n) = \{(\varphi_\alpha(\hat{x}, V_m))_{m \geq n} : \hat{x} \in V_n \cdot x_1\}.$$

Since $V_n \cdot x_1 = V_n \cdot x_2$, this gives that

$$\varphi_{\alpha+1}(x_1, V_n) = \{(\varphi_\alpha(\hat{x}, V_m))_{m \geq n} : \hat{x} \in V_n \cdot x_2\} = \varphi_{\alpha+1}(x_2, V_n).$$

The limit case follows easily from the inductive hypothesis. This finishes the proof of (a).

Now for any $x_1, x_2 \in V_n \cdot x$ and $m \geq n$, it follows from (a) that if $x_1 \in V_m \cdot x_2$ then $\varphi_\alpha(x_1, V_m) = \varphi_\alpha(x_2, V_m)$. Since V_m is an open subgroup of V_n, there is a countable set $g_0, g_1, \cdots \in V_n$ such that $V_n = \bigcup_k V_m g_k$. It follows that

$$\varphi_{\alpha+1}(x, V_n) = \{(\varphi_\alpha(g_k \cdot x, V_m))_{m \geq n} : k \in \omega\},$$

hence is a countable set. ▯

In view of the above lemma we may define $\psi_\alpha(x, V_n) \in 2^{\epsilon(\alpha)}$ to code $\varphi_\alpha(x, V_n)$ so that

$$\psi_0(x, V_n) = \varphi_0(x, V_n),$$

$$\psi_{\alpha+1}(x, V_n) = (\psi_\alpha(g_k \cdot x, V_m))_{m \geq n, \; k \in \omega},$$

$$\psi_\lambda(x, V_n) = (\psi_\alpha(x, V_n))_{\alpha < \lambda}, \text{ if } \lambda \text{ is a limit.}$$

where $(g_k)_{k \in \omega}$ is a canonical countable sequence such that $V_n = \bigcup_k V_m g_k$. Then by an easy induction we can see that $\psi_\alpha(x, V_n)$ is an element of $2^{\epsilon(\alpha)}$. Moreover, $\varphi_\alpha(x_1, V_n) = \varphi_\alpha(x_2, V_n)$ iff $\psi_\alpha(x_1, V_n) =^{\alpha+} \psi_\alpha(x_2, V_n)$.

To summarize, the canonical Hjorth–Scott types are hereditarily countable sets describing the action of V_n on X. Its definition resembles, but is not literally equivalent to, the original Scott analysis. The main difference is in the definition of the types of rank 0. Despite the difference, the canonical Hjorth–Scott types give rise to equivalence relations approximating the orbit equivalence. These equivalence relations are Borel reducible and correspondent to the ones in the Friedman–Stanley tower. In the following we will recover all the crucial properties of elements of the original Scott analysis for the Hjorth–Scott types.

We remark that there are logical methods which can make the two approaches literally equivalent. However, we prefer to give the full details of the topological approach, first because it can be independently presented, and secondly because we are going to generalize it in the next section.

Lemma 12.4.3
For all $\alpha \leq \beta < \omega_1$, $x_1, x_2 \in X$ and $n \in \omega$, if $\varphi_\beta(x_1, V_n) = \varphi_\beta(x_2, V_n)$ then $\varphi_\alpha(x_1, V_n) = \varphi_\alpha(x_2, V_n)$.

Proof. We fix α and prove the lemma by induction on $\beta \geq \alpha$. The statement is a tautology when $\beta = \alpha$. For the successor case assume $\varphi_{\beta+1}(x_1, V_n) = \varphi_{\beta+1}(x_2, V_n)$. In particular, there is $\hat{x}_1 \in V_n \cdot x_1$ such that $\varphi_\beta(\hat{x}_1, V_m) = \varphi_\beta(x_2, V_m)$ for all $m \geq n$. By the inductive hypothesis $\varphi_\alpha(\hat{x}_1, V_n) = \varphi_\alpha(x_2, V_n)$. By Lemma 12.4.2 (a) $\varphi_\alpha(x_1, V_n) = \varphi_\alpha(\hat{x}_1, V_n)$. Hence $\varphi_\alpha(x_1, V_n) = \varphi_\alpha(x_2, V_n)$. Finally for the limit case assume β is a limit, $\beta > \alpha$, and $\varphi_\beta(x_1, V_n) = \varphi_\beta(x_2, V_n)$. Then by definition $\varphi_\alpha(x_1, V_n) = \varphi_\alpha(x_2, V_n)$. ▯

Thus as in the original Scott analysis we may define the equivalence relation $\equiv_\alpha$ on $X \times \omega$ by

$$(x, n) \equiv_\alpha (y, m) \iff n = m \wedge \varphi_\alpha(x, V_n) = \varphi(y, V_n),$$

and let $\equiv_\alpha$ on X be defined as

$$x \equiv_\alpha y \iff (x, 0) \equiv_\alpha (y, 0).$$

Lemma 12.4.4
$\equiv_\alpha$ *is a* $\mathbf{\Pi}^0_{2+2\cdot\alpha}$ *equivalence relation that is Borel reducible to* $=^{\alpha+}$.

Proof. $\equiv_0$ is obviously smooth and apparently G_δ. For general α we have that

$$(x, m) \equiv_\alpha (y, m) \iff n = m \wedge \psi(x, V_n) =^{\alpha+} \psi(y, V_n).$$

It is easy to verify that the mapping $\psi'(x, n) = \psi(x, V_n)$ is a Borel function from $X \times \omega$ to $2^{\epsilon(\alpha)}$. $\quad\square$

We need to show that $E^X_{S_\infty} = \bigcap_{\alpha < \omega_1} \equiv_\alpha$. First we show that a rank similar to the Scott rank exists for every $x \in X$.

Lemma 12.4.5
For all $x \in X$ *there is some* $\gamma(x) < \omega_1$ *such that for all* $\alpha < \omega_1$, $n \in \omega$, *and* $x_1, x_2 \in [x]_{S_\infty}$, *if* $\varphi_{\gamma(x)}(x_1, V_n) = \varphi_{\gamma(x)}(x_2, V_n)$ *then* $\varphi_\alpha(x_1, V_n) = \varphi_\alpha(x_2, V_n)$.

Proof. By Lemma 12.4.2 (a) for any $\hat{x}_1 \in V_n \cdot x_1$, $\varphi_\alpha(\hat{x}_1, V_n) = \varphi_\alpha(x_1, V_n)$ for all $\alpha < \omega_1$. Let $h_0, h_1, \ldots$ enumerate a countable set such that $G = V_0 = \bigcup_k V_n h_k$. Then for any $g \in G$ there is $k \in \omega$ such that $g \cdot x \in V_n \cdot (h_k \cdot x)$, and therefore $\varphi_\alpha(g \cdot x, V_n) = \varphi_\alpha(h_k \cdot x, V_n)$ for all $\alpha < \omega_1$. On the other hand, by Lemma 12.4.3, if $\varphi_\alpha(h_k \cdot x, V_n) \neq \varphi_\alpha(h_{k'} \cdot x, V_n)$ then for all $\beta \geq \alpha$, $\varphi_\beta(h_k \cdot x, V_n) \neq \varphi_\beta(h_{k'} \cdot x, V_n)$. Thus for each pair $k, k' \in \omega$ we may let $\gamma^n_{k,k'}(x)$ be the least α such that $\varphi_\alpha(g_k \cdot x, V_n) \neq \varphi_\alpha(g_{k'} \cdot x, V_n)$ if it exists, and 0 otherwise. Let $\gamma(x) = \sup\{\gamma^n_{k,k'}(x) : k, k', n \in \omega\}$. Then $\gamma(x) < \omega_1$ is as required. $\quad\square$

Definition 12.4.6
The **Hjorth–Scott rank** of $x \in X$, denoted $\gamma(x)$, is the least ordinal γ such that for all $n \in \omega$, $x_1, x_2 \in [x]_{S_\infty}$ and $\alpha < \omega_1$, if $\varphi_{\gamma(x)}(x_1, V_n) = \varphi_{\gamma(x)}(x_2, V_n)$ then $\varphi_\alpha(x_1, V_n) = \varphi_\alpha(x_2, V_n)$. The **Hjorth–Scott invariant** of $x \in X$, denoted $\varphi(x)$, is $\varphi_{\gamma(x)+2}(x, V_0)$.

By definition and Lemma 12.4.2(a) the Hjorth–Scott rank and invariant are both invariants of the orbit of x. The information for the Hjorth–Scott rank is contained in the Hjorth–Scott invariant, as the following lemma shows.

Lemma 12.4.7
If $\varphi_{\gamma(x)+2}(y, V_0) = \varphi(x)$ *then* $\gamma(y) = \gamma(x)$.

Proof.   Let $\gamma = \gamma(x)$. We first show $\gamma(y) \le \gamma(x)$. For this let $y_1, y_2 \in [y]_{S_\infty}$, $n \in \omega$, and assume $\varphi_\gamma(y_1, V_n) = \varphi_\gamma(y_2, V_n)$. We need to verify that $\varphi_\alpha(y_1, V_n) = \varphi_\alpha(y_2, V_n)$ for all $\alpha < \omega_1$. The case $\alpha \le \gamma$ follows from Lemma 12.4.3. We prove the case $\alpha \ge \gamma$ by induction on α. For the successor case we need to show that $\varphi_{\alpha+1}(y_1, V_n) = \varphi_{\alpha+1}(y_2, V_n)$, or by definition

$$\{(\varphi_\alpha(\hat{y}_1, V_m))_{m \ge n} : \hat{y}_1 \in V_n \cdot y_1\} = \{(\varphi_\alpha(\hat{y}_2, V_m))_{m \ge n} : \hat{y}_2 \in V_n \cdot y_2\}.$$

By our assumption that $\varphi_{\gamma+2}(y, V_0) = \varphi(x)$, we get that

$$\varphi_{\gamma+2}(y_1, V_0) = \varphi_{\gamma+2}(y_2, V_0) = \varphi_{\gamma+2}(y, V_0) = \varphi_{\gamma+2}(x, V_0)$$

by Lemma 12.4.2(a). Hence there are $x_1, x_2 \in V_0 \cdot x$ such that

$$\varphi_{\gamma+1}(y_1, V_n) = \varphi_{\gamma+1}(x_1, V_n), \quad \varphi_{\gamma+1}(y_2, V_n) = \varphi_{\gamma+1}(x_2, V_n).$$

By Lemma 12.4.5 we have

$$\varphi_\gamma(y_1, V_n) = \varphi_\gamma(x_1, V_n), \quad \varphi_\gamma(y_2, V_n) = \varphi_\gamma(x_2, V_n).$$

It follows that $\varphi_\gamma(x_1, V_n) = \varphi_\gamma(x_2, V_n)$ by our assumption that $\varphi_\gamma(y_1, V_n) = \varphi_\gamma(y_2, V_n)$. Now the definition of $\gamma = \gamma(x)$ implies that $\varphi_{\gamma+1}(x_1, V_n) = \varphi_{\gamma+1}(x_2, V_n)$. Thus it follows that $\varphi_{\gamma+1}(y_1, V_n) = \varphi_{\gamma+1}(y_2, V_n)$, or

$$\{(\varphi_\gamma(\hat{y}_1, V_m))_{m \ge n} : \hat{y}_1 \in V_n \cdot y_1\} = \{(\varphi_\gamma(\hat{y}_2, V_m))_{m \ge n} : \hat{y}_2 \in V_n \cdot y_2\}.$$

By the inductive hypotheses

$$\{(\varphi_\alpha(\hat{y}_1, V_m))_{m \ge n} : \hat{y}_1 \in V_n \cdot y_1\} = \{(\varphi_\alpha(\hat{y}_2, V_m))_{m \ge n} : \hat{y}_2 \in V_n \cdot y_2\},$$

and we are done with the successor case. The limit case is straightforward. This finishes the induction and the proof that $\gamma(y) \le \gamma(x)$.

Now since $\gamma(y) + 2 \le \gamma(x) + 2$ it follows that $\varphi(y) = \varphi_{\gamma(y)+2}(y, V_0) = \varphi_{\gamma(y)+2}(x, V_0)$ by Lemma 12.4.5. Thus by symmetry we have that $\gamma(x) \le \gamma(y)$. This shows that $\gamma(y) = \gamma(x)$.   □

Theorem 12.4.8 (Hjorth)
Let X be a Polish S_∞-space. Then $x E^X_{S_\infty} y$ iff $\varphi(x) = \varphi(y)$.

Proof.   It suffices to prove ($\Leftarrow$). So suppose $\varphi(x) = \varphi(y)$. Then by Lemma 12.4.7 $\gamma(x) = \gamma(y)$. Let γ be this common value. Let d be a compatible complete metric on X with $d < 1$. We construct sequences of elements

$$g_i, h_i \in S_\infty, \ x_i, y_i \in X, \ m_i, n_i \in \omega$$

for $i \in \omega$ so that the following hold:

(i) $g_0 = h_0 = 1_{S_\infty}$;

(ii) $x_i = g_i \cdot x$, $y_i = h_i \cdot y$;

(iii) $x_i, y_i \in U_{m_i}$, $\mathrm{diam}(U_{m_i}) \le 2^{-i}$, $U_{m_{i+1}} \subseteq U_{m_i}$;

(iv) $g_{i+1}g_i^{-1} \in V_{n_i}$, $h_{i+1}h_i^{-1} \in V_{n_i}$, $n_i < n_{i+1}$;

(v) $\varphi_\gamma(x_i, V_{n_i}) = \varphi_\gamma(y_i, V_{n_i})$;

(vi) if i is even then $k_i = \sup\{g_i(j), g_i^{-1}(j) : j < i\} < n_i$;

(vii) if i is odd then $l_i = \sup\{h_i(j), h_i^{-1}(j) : j < i\} < n_i$.

Granting the construction, (g_i) and (h_i) both have limits in S_∞. Let $g_i \to g_\infty$ and $h_i \to h_\infty$ as $i \to \infty$. Then by the continuity of the S_∞-action, (ii) and (iii) we have $g_\infty \cdot x = h_\infty \cdot y$. Thus $x E_{S_\infty}^X y$.

The construction is by induction on i. As (vi) and (vii) suggest the even and odd inductive steps form a back-and-forth argument. By symmetry it suffices to assume that i is even and that $g_i, h_i, x_i, y_i, m_i, n_i$ have been defined to satisfy (i) through (vi). We define $g_{i+1}, h_{i+1}, x_{i+1}, y_{i+1}, m_{i+1}, n_{i+1}$ to satisfy (ii) through (v) and (vii).

First we note that $\varphi_{\gamma+1}(x_i, V_{n_i}) = \varphi_{\gamma+1}(y_i, V_{n_i})$. This is because, by $\varphi(x) = \varphi(y)$, we have $\varphi_{\gamma+2}(x_i, V_0) = \varphi_{\gamma+2}(y_i, V_0)$, and therefore there is $\hat{y} \in V_0 \cdot y_i = V_0 \cdot y$ such that $\varphi_{\gamma+1}(x_i, V_{n_i}) = \varphi_{\gamma+1}(\hat{y}, V_{n_i})$. By Lemma 12.4.5 and (v), $\varphi_\gamma(y_i, V_{n_i}) = \varphi_\gamma(x_i, V_{n_i}) = \varphi_\gamma(\hat{y}, V_{n_i})$. Now by the definition of $\gamma = \gamma(y)$, $\varphi_{\gamma+1}(y_i, V_{n_i}) = \varphi_{\gamma+1}(\hat{y}, V_{n_i})$. By transitivity, $\varphi_{\gamma+1}(x_i, V_{n_i}) = \varphi_{\gamma+1}(y_i, V_{n_i})$ as claimed.

We define $h_{i+1} = h_i$. Then $y_{i+1} = y_i$ and $l_{i+1} = \sup\{h_i(j), h_i^{-1}(j) : j \le i\}$. Let $n_{i+1} > \max\{n_i, l_{i+1}\}$. Then (vii) is satisfied. By the above claim there is $\hat{x} \in V_{n_i} \cdot x_i$ such that $\varphi_\gamma(\hat{x}, V_{n_{i+1}}) = \varphi_\gamma(y_i, V_{n_{i+1}})$. By Lemma 12.4.5 in particular $\varphi_0(\hat{x}, V_{n_{i+1}}) = \varphi_0(y_i, V_{n_{i+1}})$. Let m_{i+1} be such that $y_i \in U_{m_{i+1}}$ and $\mathrm{diam}(U_{m_{i+1}}) \le 2^{-i-1}$. Then $V_{n_{i+1}} \cdot \hat{x} \cap U_{m_{i+1}} \ne \emptyset$. Let $x_{i+1} \in V_{n_{i+1}} \cdot \hat{x} \subseteq V_{n_{i+1}} V_{n_i} \cdot x_i = V_{n_i} \cdot x_i$. Let $\hat{g} \in V_{n_i}$ be such that $\hat{g} \cdot x_i = x_{i+1}$. Let $g_{i+1} = \hat{g}g_i$. Then (ii) through (iv) are satisfied.

It remains to check that $\varphi_\gamma(x_{i+1}, V_{n_{i+1}}) = \varphi_\gamma(y_{i+1}, V_{n_{i+1}})$. Since $x_{i+1} \in V_{n_{i+1}} \cdot \hat{x}$, by Lemma 12.4.2(a) we have $\varphi_\gamma(x_{i+1}, V_{n_{i+1}}) = \varphi_\gamma(\hat{x}, V_{n_{i+1}})$. By our construction $\varphi_\gamma(\hat{x}, V_{n_{i+1}}) = \varphi_{\gamma+1}(y_i, V_{n_{i+1}})$. Thus, since $y_{i+1} = y_i$, we have $\varphi_\gamma(x_{i+1}, V_{n_{i+1}}) = \varphi_\gamma(y_{i+1}, V_{n_{i+1}})$, as required. $\qquad\square$

The topological Scott analysis discussed in this section is the beginning of a train of powerful ideas that lead to remarkable results. In the next section we will give an example of how it can be used to obtain sharp results about S_∞ actions and orbit equivalence relations. Much of this can be generalized to arbitrary Polish group actions.

Exercise 12.4.1 Show that the function $\psi'(x, n) = \psi(x, V_n)$ from $X \times \omega$ to $2^{\epsilon(\alpha)}$ is Borel.

Exercise 12.4.2 Replace S_∞ by a closed subgroup G of S_∞ and give a topological Scott analysis for Polish G-spaces.

12.5   Sharp analysis of S_∞-orbits

In this section we give an application of the topological Scott analysis in obtaining sharp results about S_∞-orbit equivalence relations. The theorems presented here are special cases of more general results obtained by the same methods (see References [70], [101], [77], and [82]).

We first define a concept and prove some basic properties.

Definition 12.5.1
Let X be a Polish space, E an equivalence relation on X, and $\alpha < \omega_1$. We say that E is **potentially $\mathbf{\Pi}_\alpha^0$** (or **potentially $\mathbf{\Sigma}_\alpha^0$**) if there is a Polish topology τ on X finer than the original topology such that E is a $\mathbf{\Pi}_\alpha^0$ (or $\mathbf{\Sigma}_\alpha^0$, respectively) subset of (X^2, τ^2).

Note that if E is potentially $\mathbf{\Pi}_\alpha^0$ or potentially $\mathbf{\Sigma}_\alpha^0$ for $\alpha < \omega_1$ then E is Borel. Thus the concept makes sense only for Borel equivalence relations. The following lemmas are easy consequences of earlier theorems.

Lemma 12.5.2
Let X be a Polish space and E an equivalence relation on X. Then the following are equivalent:

(i) *E is potentially $\mathbf{\Sigma}_1^0$;*

(ii) *$E \leq_B \mathrm{id}(\omega)$;*

(iii) *E is a Borel equivalence relation with only countably many equivalence classes.*

Lemma 12.5.3
Let X be a Polish space and E an equivalence relation on X. Then the following are equivalent:

(i) *E is potentially $\mathbf{\Pi}_1^0$;*

(ii) *E is potentially $\mathbf{\Pi}_2^0$;*

(iii) *E is smooth.*

Proof.   This is Theorem 6.4.4 (i) through (iii).   ⬜

For orbit equivalence relations the potential classes are closely connected with the Borel reducibility.

Lemma 12.5.4
Let G be a Polish group, X a Polish G-space, and $\alpha < \omega_1$. Then E_G^X is potentially $\mathbf{\Pi}_\alpha^0$ (or potentially $\mathbf{\Sigma}_\alpha^0$) iff there is a Polish space Y and a $\mathbf{\Pi}_\alpha^0$ (or $\mathbf{\Sigma}_\alpha^0$, respectively) equivalence relation F on Y such that $E_G^X \leq_B F$.

Proof. If E_G^X is potentially $\mathbf{\Pi}_\alpha^0$, as witnessed by the topology τ on X, then let $Y = (X, \tau)$ and $F = E$, and certainly $E_G^X \leq_B F$. Conversely, suppose f is a reduction from X to Y witnessing that $E_G^X \leq_G F$. Let $\{U_n\}$ be a countable base for Y. Let τ be a finer topology on X such that every set in the countable collection $\{f(U_n)\}$ is clopen, and such that the action of G on (X, τ) is still continuous. Then f is now continuous from (X, τ) to Y, and hence E_G^X is $\mathbf{\Pi}_\alpha^0$ as a subset of (X^2, τ^2). This shows that E_G^X is potentially $\mathbf{\Pi}_\alpha^0$. The proof for potentially $\mathbf{\Sigma}_\alpha^0$ equivalence relations is similar. $\qquad\square$

The following theorem is an application of the topological Scott analysis.

Theorem 12.5.5 (Hjorth–Kechris–Louveau)
Let X be a Polish S_∞-space. Then E_G^X is potentially $\mathbf{\Pi}_3^0$ iff $E_G^X \leq_B =^+$.

Proof. Since $=^+$ is $\mathbf{\Pi}_3^0$ the implication ($\Leftarrow$) follows from Lemma 12.5.4. Now suppose E_G^X is potentially $\mathbf{\Pi}_3^0$. By the technique of change of topology we may assume without loss of generality that E_G^X is $\mathbf{\Pi}_3^0$.

Fix a countable base $\{U_l\}$ for X and clopen subgroups (V_n) of S_∞ as in the preceding section. Let τ be the original Polish topology on X. Let $\mathcal{B}$ be a countable base for S_∞. For $n, l \in \omega$ let $W_{n,l} = \{z \in X : U_l \cap V_n \cdot z = \emptyset\}$. Then each $W_{n,l}$ is τ-closed. It follows that $\{U_l, W_{n,l} : n, l \in \omega\}$ generates a Polish topology τ' on X in which $W_{n,l}$ is τ'-clopen.

Let (g_k) enumerate a countable dense subgroup of S_∞. By Lemma 12.4.2 for any $x \in X$ and $n \in \omega$,

$$\varphi_1(x, V_n) = \{(\varphi_0(g_k \cdot x, V_m))_{m \geq n} : k \in \omega\}.$$

Fix $x \in X$. For all $n, k \in \omega$ let

$$Q_{n,k} = \{z \in X : \varphi_0(z, V_n) = \varphi_0(g_k \cdot x, V_n)\}.$$

Then

$$z \in Q_{n,k} \iff \forall l \in \omega \, (g_k \cdot x \in W_{n,l} \leftrightarrow z \in W_{n,l}).$$

Since $W_{n,l}$ is τ'-clopen, $Q_{n,k}$ is τ'-closed. It follows that $\tau' \cup \{Q_{n,k} : n, k \in \omega\}$ generates again a Polish topology finer than τ', which we denote by $\tau(x)$.

It is easy to see that $\tau(x)$ is invariant, that is, if $x' \in [x]_{S_\infty}$ then $\tau(x') = \tau(x)$. Let $C(x)$ be the closure of $[x]_{S_\infty}$ in $(X, \tau(x))$. Then $(C(x), \tau(x))$ is again invariant.

Since $[x]_{S_\infty}$ is $\mathbf{\Pi}_3^0$, there are closed sets $F_{i,j} \subseteq (X, \tau)$ for $i, j \in \omega$ such that

$$z \in [x]_{S_\infty} \iff \forall i \in \omega \, \exists j \in \omega \, z \in F_{i,j}.$$

This implies the following description of the orbit of x:

$$z \in [x]_{S_\infty}$$

$$\Longleftrightarrow \forall^* g \in S_\infty \ g \cdot z \in [x]_{S_\infty}$$

$$\Longleftrightarrow \forall^* g \in S_\infty \ \forall i \in \omega \ \exists j \in \omega \ g \cdot z \in F_{i,j}$$

$$\Longleftrightarrow \forall i \in \omega \ \forall^* g \in S_\infty \ \exists j \in \omega \ g \cdot z \in F_{i,j}$$

$$\Longleftrightarrow \forall i \in \omega \ \forall V \in \mathcal{B} \ \exists^* g \in V \ \exists j \in \omega g \cdot z \in F_{i,j}$$

$$\Longleftrightarrow \forall i \in \omega \ \forall V \in \mathcal{B} \ \exists j \in \omega \ \exists V' \in \mathcal{B} \ (V' \subseteq V \wedge \forall^* g \in V' \ g \cdot z \in F_{i,j})$$

$$\Longleftrightarrow \forall i \in \omega \ \forall V \in \mathcal{B} \ \exists j \in \omega \ \exists V' \in \mathcal{B} \ (V' \subseteq V \wedge z \in F_{i,j}^{*V'})$$

$$\Longleftrightarrow \forall i \in \omega \ \forall V \in \mathcal{B} \ \exists j \in \omega \ \exists n \in \omega \ \exists g \in S_\infty \ (gV_n \subseteq V \wedge z \in F_{i,j}^{*gV_n}).$$

In view of this, consider for each $i \in \omega$ and $V \in \mathcal{B}$ the set

$$O(i, V) = \{\, y \in C(x) \, : \, \exists j \in \omega \ \exists n \in \omega \ \exists \hat{x} \in [x]_{S_\infty} \ \exists g \in S_\infty$$

$$gV_n \subseteq V \ \wedge \ \hat{x} \in F_{i,j}^{*gV_n} \ \wedge \ \varphi_0(y, V_n) = \varphi_0(\hat{x}, V_n) \,\}.$$

Then it follows from the above description that $[x]_{S_\infty} \subseteq O(i, V)$ for all $i \in \omega$ and $V \in \mathcal{B}$. Hence $O(i, , V) \subseteq C(x)$ is dense in $\tau(x)$. Also from the definition of $O(i, V)$ it is clear that it is $\tau(x)$-open. Thus $O(i, V)$ is comeager in $(C(x), \tau(x))$. Let $C_0 = \bigcap \{O(i, V) \, : \, i \in \omega, \ V \in \mathcal{B}\}$. Then $C_0 \subseteq (C(x), \tau(x))$ is comeager.

We note that if $\hat{x} \in F_{i,j}^{*gV_n}$ and $\varphi_0(y, V_n) = \varphi_0(\hat{x}, V_n)$, then $y \in F_{i,j}^{*gV_n}$. This is because $F_{i,j}^{*gV_n} = (g^{-1} \cdot F_{i,j})^{*V_n}$, where $g^{-1} \cdot F_{i,j}$ is τ-closed by the continuity of the action. If $y \notin F_{i,j}^{*gV_n}$, then there is $l \in \omega$ such that $U_l \cap g^{-1} \cdot F_{i,j} = \emptyset$ and $V_n \cdot y \cap U_l \neq \emptyset$. Since $\varphi_0(y, V_n) = \varphi_0(\hat{x}, V_n)$, it follows that for this l, $V_n \cdot \hat{x} \cap U_l \neq \emptyset$, hence $\hat{x} \notin (g^{-1} \cdot F_{i,j})^{*V_n}$, a contradiction. It follows immediate from the description of $[x]_{S_\infty}$ that $C_0 \subseteq [x]_{S_\infty}$.

We have shown that $(C(x), \tau(x))$ is a complete invariant of $[x]_{S_\infty}$. If $(C(x), \tau(x)) = (C(y), \tau(y))$, then from the above argument $[x]_{S_\infty}$ and $[y]_{S_\infty}$ are both comeager in this space, and hence have a nonempty intersection. This implies that $[x]_{S_\infty} = [y]_{S_\infty}$.

It is clear from the construction that $(C(x), \tau(x))$ depends only on $\varphi_1(x, V_0)$. In other words, if $\varphi_1(x, V_0) = \varphi_1(y, V_0)$ then $(C(x), \tau(x)) = (C(y), \tau(y))$. We have thus obtained that $x E_G^X y$ iff $\varphi_1(x, V_0) = \varphi_1(y, V_0)$. Since $\equiv_1 \leq_B =^+$, we have that $E_G^X \leq_B =^+$ as required. $\qquad \square$

The method illustrated in the proof of the theorem can be generalized. In general one can obtain a Polish topology $\tau_\alpha(x)$ for $\alpha < \omega_1$ to encode the canonical Hjorth–Scott type of rank α. This method can even be generalized to topological Scott analysis for arbitrary Polish group actions.

In the rest of this section, we investigate potentially Σ^0_3 S_∞-orbit equivalence relations and show that they are essentially countable. We need the following lemma of Kechris for essential countability.

Lemma 12.5.6 (Kechris)
Let X, Y be standard Borel spaces, E a Borel equivalence relation on X, and $f : X \to Y$ a Borel function. Suppose

(i) *for any $x \in X$, $\{f(y) : yEx\}$ is countable, and*

(ii) *for any $x, y \in X$, if $f(x) = f(y)$ then xEy.*

Then E is essentially countable.

Proof. We may assume without loss of generality that $X = Y = \omega^\omega$. Define

$$A = \{y \in Y : \exists x \in X \ f(x) = y)\} = f(X),$$

$$B = \{(y_1, y_2) \in Y^2 : \exists x_1, x_2 \in X \ (f(x_1) = y_1 \ \wedge \ f(x_2) = y_2 \ \wedge \ x_1 E x_2) \},$$

$$C = \{(y_1, y_2) \in Y^2 : \forall x_1, x_2 \in X \ [(f(x_1) = y_1 \wedge f(x_2) = y_2) \to x_1 E x_2] \}.$$

Then $B \in \Sigma^1_1$, $C \in \Pi^1_1$, and by assumption (ii) $B \cap A^2 = C \cap A^2$. By the Luzin separation theorem (Theorem 1.6.1) there is a Borel set D such that $B \subseteq D \subseteq C$, and hence $D \cap A^2 = B \cap A^2 = C \cap A^2$.

Note that the Σ^1_1 set A has the following properties:

(1) D is an equivalence relation on A, and

(2) for all $y_1 \in A$, the set $\{y_2 \in A : (y_1, y_2) \in D\}$ is countable.

We claim that there is a Borel set $A_0 \supseteq A$ such that (1) and (2) hold for A_0 instead of A. To prove this claim let $g : Y \to \mathrm{LO}$ be a Borel function with $f^{-1}(\mathrm{WO}) = Y - A$. For any $z \notin A$ let $\alpha_z = \mathrm{ot}(<_{g(z)})$. For $y, z \in Y$ let

$$R(z, y) \iff y \notin A \ \wedge \ [z \in A \ \vee \ (z \notin A \ \wedge \ \alpha_y < \alpha_z)].$$

Then

$$R(z, y) \iff y \notin A \ \wedge \ \text{there is no } j : \omega \to \omega \text{ order-preserving}$$
$$\text{from } <_{g(z)} \text{ into } <_{g(y)},$$

and hence $R \subseteq Y^2$ is Π^1_1. Let $R_z = \{y \in X : R(z, y)\}$ for all $z \in Y$. Then R_z is Π^1_1.

Now assume that claim fails, that is, there is no Borel set $A_0 \supseteq A$ such that (1) and (2) hold for A_0 instead of A. We make a subclaim that $z \in A$ iff (1) and (2) hold for $Y - R_z$ instead of A. First let $z \in A$. Then $R_z = Y - A$ and so (1) and (2) hold for $Y - R_z = A$. On the other hand, let $z \notin A$, then

$$y \in R_z \iff y \notin A \land \alpha_y < \alpha_z$$
$$\iff \text{there is } n \in \omega \text{ and } j : \omega \to \omega \text{ order-preserving}$$
$$\text{from } <_{g(y)} \text{ into } <_{g(z)} \restriction n.$$

Thus R_z is Σ_1^1, and moreover Borel. Clearly $R_x \cap A = \emptyset$. By our assumption, (1) or (2) does not hold for $X - R_x$ since it is now a Borel set containing A. This proves the subclaim.

The subclaim states that $z \in A$ iff

(1') D is an equivalence relation on $Y - R_z$, and

(2') for all $y_1 \in Y - R_z$, the set $\{y_2 \in Y - R_z : (y_1, y_2) \in D\}$ is countable.

We note that these are Π_1^1 conditions on z. Unraveling (1') we get

$$\forall y_1, y_2, y_3 \; [y_1 \in R_z \lor y_2 \in R_z \lor y_3 \in R_z \lor$$
$$((y_1, y_1) \in D \land (y_1, y_2) \in D \leftrightarrow (y_2, y_1) \in D \land$$
$$(y_1, y_2), (y_2, y_3) \in D \to (y_1, y_3) \in D)],$$

which is Π_1^1 for z. For (2') we use the perfect set theorem for analytic sets and note that (2') is equivalent to

(2") for all $y_1 \in Y - R_z$, there is no perfect subset of $\{y_2 \in Y - R_z : (y_1, y_2) \in D\}$.

A standard descriptive computation shows that this is a Π_1^1 condition (for more details, see Exercise 12.5.2).

Now we have a contradiction since on the one hand, we assume the contrary of the claim, and on the other hand we find A to be both Σ_1^1 and Π_1^1, hence A itself witnesses the claim.

Now let $A_0 \supseteq A$ be Borel such that (1) and (2) hold for A_0. Then D is an equivalence relation on A_0 and every D-class in A_0 is countable. This means that D is a countable equivalence relation on the standard Borel space A_0, and f is a Borel reduction from E to $D \restriction A_0$. ▯

The original proof of Kechris' lemma uses a theorem called Π_1^1 reflection and is much shorter. Here we have basically recovered the proof of Π_1^1 reflection in order to keep the argument self-contained.

Theorem 12.5.7 (Hjorth–Kechris–Louveau)
Let X be a Polish S_∞-space. Then the following are equivalent:

(i) E_G^X is potentially Σ_2^0;

(ii) E_G^X is potentially Σ_3^0;

(iii) E_G^X *is essentially countable.*

Proof. (iii)$\Rightarrow$(i) is immediate from Lemma 12.5.4. (i)$\Rightarrow$(ii) is obvious. It remains to show (ii)$\Rightarrow$(iii).

Assume E_G^X is potentially Σ_3^0. By the technique of change of topology we may assume without loss of generality that E_G^X is Σ_3^0. Let $\{U_l\}$ be a countable base for X. Let $\{W_m\}$ be a countable base for S_∞ such that for all $g \in S_\infty$ and $m \in \omega$, there is $m' \in \omega$ such that $W_m g = W_{m'}$. Note that the collection $\{V_n g : n \in \omega, g \in S_\infty\}$ is such a countable base for S_∞.

Fix $x \in X$. Since $[x]_{S_\infty}$ is Σ_3^0 there are G_δ sets F_n such that $[x]_{S_\infty} = \bigcap_n F_n$. We have that $\forall^* g \in S_\infty \ \exists n \in \omega \ g \cdot x \in F_n$. Thus in particular there is $n \in \omega$ and $m \in \omega$ such that $\forall^* g \in W_m \ g \cdot x \in F_n$. It follows that $F_n \cap \overline{W_m \cdot x}$ is a dense subset of $\overline{W_m \cdot x}$. Since it is also a G_δ subset of $\overline{W_m \cdot x}$ it must be comeager. We therefore conclude that $[x]_{S_\infty}$ is comeager in $\overline{W_m \cdot x}$ since $F_n \subseteq [x]_{S_\infty}$. We have shown that for any $x \in X$ there is $m \in \omega$ such that $[x]_{S_\infty}$ is comeager in $\overline{W_m \cdot x}$. Let $m(x)$ be the least such $m \in \omega$.

Now define a function θ from X into the Effros Borel space $F(X)$ by

$$f(x) = \overline{W_{m(x)} \cdot x}.$$

We verify that θ is Borel. For this let a basic open set U_l be given. Then

$$
\begin{aligned}
&f(x) \cap U_l \neq \emptyset \\
\Longleftrightarrow \ & \overline{W_{m(x)} \cdot x} \cap U_l \neq \emptyset \\
\Longleftrightarrow \ & \exists m \in \omega \ (\overline{W_m \cdot x} \cap U_l \neq \emptyset \ \wedge \ \forall^* y \in \overline{W_m \cdot x} \ y E_{S_\infty}^X x \\
& \wedge \ \forall m' < m \ \exists^* y \in \overline{W_m \cdot x} \ (x,y) \notin E_{S_\infty}^X).
\end{aligned}
$$

This is Borel by Kuratowski–Ulam (see Exercise 3.2.6).

We check that the function f satisfies the hypotheses of Kechris' lemma. Fix $x \in X$. Let $S = \{\overline{W_m \cdot x} : m \in \omega\}$. Then S is countable. We claim that for any $\hat{x} \in [x]_{S_\infty}$, $f(\hat{x}) \in S$. In fact $f(\hat{x}) = \overline{W_{m(\hat{x})} \cdot \hat{x}}$. Let $g \in S_\infty$ be such that $\hat{x} = g \cdot x$. Then by the invariance property of the base $\{W_m\}$, there is some $m \in \omega$ such that $W_{m(\hat{x})} g = W_m$. Then $f(\hat{x}) = \overline{W_{m(\hat{x})} g \cdot x} = \overline{W_m \cdot x} \in S$. This proves Lemma 12.5.6(i). To show Lemma 12.5.6(ii) let $x, y \in X$ and assume $f(x) = f(y)$. Then $[x]_{S_\infty}$ is comeager in $f(x)$ and $[y]_{S_\infty}$ is comeager in $f(y)$, and it follows that $[x]_{S_\infty}$ and $[y]_{S_\infty}$ have a nonempty intersection, and therefore they coincide.

Now by Lemma 12.5.6 E_G^X is essentially countable. $\Box$

The methods introduced in this section can be applied to investigate arbitrary potential classes of S_∞-orbit equivalence relations. In this manner Hjorth, Kechris, and Louveau have completely determined the possible potential classes of all Borel isomorphism relations.

Exercise 12.5.1 Prove Lemma 12.5.2.

Exercise 12.5.2 For any tree T on $\omega \times \omega$ let

$$p[T] = \{x \in \omega^\omega : \exists y \in \omega^\omega \ (x, y) \in [T]\}.$$

Note that T is an element of $X = 2^{(\omega \times \omega)^{<\omega}}$. Let P be the set of all trees T on $\omega \times \omega$ such that $p[T]$ is uncountable.

(1) Show that $p[T]$ is uncountable iff there is a subtree S of T such that the following property holds for S:

      for every $(s, u) \in S$ there are $(t, v), (r, w) \in S$ with $t \perp r$ and $(s, u) \subseteq (t, v), (r, w)$.

(2) Show that the set P is Σ_1^1.

Chapter 13

Natural Classes of Countable Models

In the previous two chapters we have developed the theory of countable models in some depth from the perspectives of both mathematical logic and descriptive set theory. For invariant Borel classes with Borel isomorphism relations we have obtained satisfactory characterizations of the models. These also led to powerful results about Borel S_∞-orbit equivalence relations. However, when natural classes of countable models are considered the isomorphism relations turn out to be mostly non-Borel, and often correspond to the universal S_∞-orbit equivalence relation. In this chapter we consider various natural classes of countable models and substantiate this observation. Historically these natural invariant Borel classes are not only examples to interpret the theoretical results, but also an integral part of the countable model theory and source of inspirations for further theoretical tools. In this chapter we will focus on countable graphs, trees, linear orderings, and groups. Similar work has been done for countable lattices, fields, and Boolean algebras.

13.1   Countable graphs

In previous discussions of countable model theory our general setup has always contained an arbitrary countable relational language L. There were few results that rely on assumptions of the composition of L. However, different languages L give rise to very different isomorphism relations $\cong_L$. In Section 3.6 we proved that $\mathrm{Mod}(L)$ is a universal Borel S_∞-space iff L contains relation symbols of arbitrarily high arity. On the other extreme, it is easy to see that if L contains only unary relation symbols then the isomorphism relation on $\mathrm{Mod}(L)$ is Borel (and in fact Borel reducible to $=^+$).

In this section we will show that if L contains at least one relation symbol of arity ≥ 2 then the isomorphism relation on $\mathrm{Mod}(L)$ is Borel bireducible with the universal S_∞-orbit equivalence relation. This completely classifies the languages L in terms of the complexity of the isomorphism relation $\cong_L$.

Throughout this section we fix a language $L^\Gamma = \{R\}$ with one binary relation symbol R. Let γ be the L^Γ-sentence

$$\forall x\, \forall y\, [\, \neg R(x,x) \,\wedge\, (\, R(x,y) \leftrightarrow R(y,x)\,)\,].$$

Then every model $M \in \mathrm{Mod}(\gamma)$ is essentially a countable graph (V, E) with the vertex set $V = \omega$ and the edge set $E = \{\{a, b\} : M \models R[a, b]\}$. Conversely every countable graph is obviously represented by an element of $\mathrm{Mod}(\gamma)$. In view of this we will refer to the elements of $\mathrm{Mod}(\gamma)$ as **countable graphs** and the isomorphism relation $\cong_\gamma$ the **graph isomorphism**. We will use the usual graph theoretic terminology when discussing the ingredients of a countable graph, such as vertex (or node), edge, degree, cycle, subgraph, and so on. If a, b are nodes in a graph, we denote the edge $\{a, b\}$ by ab or ba for brevity.

We will also use the following terminology.

Definition 13.1.1
Let C be an invariant Borel class.

(a) We say that C (or $\cong\restriction C$) is **Borel complete** if $\cong\restriction C$ is Borel bireducible with the universal S_∞-orbit equivalence relation.

(b) We say that C (or $\cong\restriction C$) is **faithfully Borel complete** if every S_∞-orbit equivalence relation is faithfully Borel reducible to $\cong\restriction C$.

(c) The **topological Vaught conjecture** for C, denoted $\mathrm{TVC}(C)$, is the statement $\mathrm{TVC}(C, \cong\restriction C)$.

Clearly faithful Borel completeness implies Borel completeness. By Theorem 9.5.5 if C is faithfully Borel complete then $\mathrm{TVC}(C)$ implies $\mathrm{TVC}(S_\infty)$, the $L_{\omega_1\omega}$-Vaught conjecture. Our objective is to show that the class of all countable graphs is faithfully Borel complete.

Theorem 13.1.2
The class of all countable graphs is faithfully Borel complete.

Proof. Let $L = \{R_n\}_{n\geq 2}$ be a countable relational language with each R_n an n-ary relation symbol. By Theorem 3.6.1 $\mathrm{Mod}(L)$ is a universal Borel S_∞-space. Hence by Proposition 5.2.2 $\mathrm{Mod}(L)$ is faithfully Borel complete. It suffices to construct a faithful Borel reduction from $\mathrm{Mod}(L)$ to the class of all countable graphs $\mathrm{Mod}(\gamma)$. So for each L-structure $M \in \mathrm{Mod}(L)$, we need to associate a countable graph $\Gamma(M)$.

For $n \geq 1$ an n-**tag** is a graph T_n with the vertex set

$$\{a_1, \ldots, a_n\} \cup \{b_{1,1}, b_{2,1}, b_{2,2}, \ldots, b_{n,1}, \ldots, b_{n,n}, c, d_1, d_2, d_3, f\} = A \cup B,$$

where the two displayed sets A and B are disjoint and the demonstrated elements of B are distinct, and with the following set of edges:

$$\{ a_1 b_{1,1}, b_{1,1} c,$$
$$a_2 b_{2,1}, b_{2,1} b_{2,2}, b_{2,2} c,$$
$$\ldots\ldots$$
$$a_n b_{n,1}, b_{n,1} b_{n,2}, \ldots, b_{n,n-1} b_{n,n}, b_{n,n} c,$$
$$c d_1, d_1 d_2, d_2 d_3, d_3 d_1, f d_2 \}.$$

Figure 13.1 illustrates an n-tag. It is important that the n-tags have no

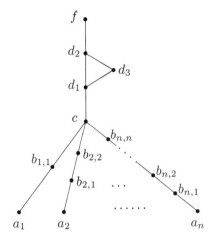

Figure 13.1 An n-tag $T_n(a_1, \ldots, a_n)$.

symmetry; each vertex in an n-tag is uniquely determined by its properties. To be specific, note that in such a graph f has degree 1, d_1, d_2, d_3 form a 3-cycle, c has degree $n + 1$, and each other vertex in B has degree 2. Such an n-tag will be used to code the tuple $(a_1, \ldots, a_n)$, and we denote such an n-tag $T_n(a_1, \ldots, a_n)$. We call the vertex c the **center** of the n-tag. If $a_1, \ldots, a_n$ are (not necessarily distinct) vertices in some graph Γ_0, then by adding $T_n(a_1, \ldots, a_n)$ to Γ_0 we mean to add fresh elements from B and the edges of T_n to form a new graph.

Now given an L-structure M we first let Γ_0 be the graph on $|M| = \omega$ with a 1-tag added for each $a \in |M|$. Then for each R_n, $n \geq 2$, and tuple $(a_1, \ldots, a_n)$, add an n-tag $T_n(a_1, \ldots, a_n)$ iff $M \models R_n[a_1, \ldots, a_n]$. The resulting graph is denoted $\Gamma(M)$.

Note that in $\Gamma(M)$ each 3-cycle is created only by the addition of an n-tag. From this the centers of n-tags can be identified because of their adjacency to the 3-cycles. Let $C(M)$ denote the set of centers of all n-tags for $n \geq 1$.

Use the abbreviation $\exists^{\neq} y_1 \ldots y_k \setminus x_1 \ldots x_l \, \varphi$ for the formula

$$\exists y_1 \ldots \exists y_k \left[\bigwedge_{\substack{1 \leq i < i' \leq k \\ 1 \leq j \leq l}} (\neg y_i = y_{i'} \wedge \neg y_i = x_j) \wedge \varphi \right].$$

Let $\theta(x)$ be the $L_{\omega\omega}^{\Gamma}$-formula

$$\exists^{\neq} y_1, y_2, y_3, z \setminus x \, [\, R(x, y_1) \wedge R(x, z) \wedge R(y_1, y_2) \wedge R(y_2, y_3) \wedge R(y_3, y_1)\,].$$

Then $\Gamma(M) \models \theta[c]$ iff $c \in C(M)$.

For $i \geq 1$, let $\eta_i(x)$ be the $L_{\omega\omega}^{\Gamma}$-formula

$$\exists^{\neq} y_1 \ldots y_i \bigwedge_{1 \leq j \leq i} R(x, y_j) \wedge \neg \exists^{\neq} y_1 \ldots y_{i+1} \bigwedge_{1 \leq j \leq i+1} R(x, y_j).$$

Then $\Gamma(M) \models \eta_i[c]$ iff c has degree i. Also for $l \geq 1$ let $\lambda_l(x, y)$ be the $L_{\omega\omega}^{\Gamma}$-formula

$$\exists^{\neq} z_1 \ldots z_l \setminus x, y \; [\neg x = y \wedge R(x, z_1) \wedge R(z_l, y) \wedge$$

$$\bigwedge_{1 \leq j \leq l} \eta_2(z_j) \wedge \bigwedge_{1 \leq j < l} R(z_j, z_{j+1})].$$

Then $\Gamma(M) \models \lambda_l[a, c]$ iff there is a path between a and c with l elements of degree 2 in between.

Now note that for each $c \in C(M)$, if the degree of c is 2 then the tag it is a center of is a 1-tag. It follows that a vertex a of $\Gamma(M)$ is in $|M|$ iff there is a path of length 2 from a to a center of a 1-tag. Let $\delta(x)$ be the $L_{\omega\omega}^{\Gamma}$-formula

$$\exists y \, (\lambda_1(x, y) \wedge \theta(y) \wedge \eta_2(y)).$$

Then $\Gamma(M) \models \delta[a]$ iff $a \in |M|$.

In general if $c \in C(M)$ then the degree of c is exactly one more than the arity of the tuple it codes. Thus if c has degree $n + 1$, then c codes a unique n-tuple $(a_1, \ldots, a_n)$. For $n \geq 2$, let $\rho_n(x_1, \ldots, x_n)$ be the $L_{\omega\omega}^{\Gamma}$-formula

$$\exists y \bigwedge_{1 \leq i \leq n} (\delta(x_i) \wedge \theta(y) \wedge \eta_{n+1}(y) \wedge \lambda_i(x_i, y)).$$

Then $\Gamma(M) \models \rho_n[a_1, \ldots, a_n]$ iff there is $c \in C(M)$ and c codes the tuple $(a_1, \ldots, a_n)$ in the sense that the unique n-tag $T_n(a_1, \ldots, a_n)$ has center c. It is now clear that $\Gamma(M) \models \rho_n[a_1, \ldots, a_n]$ iff $M \models R_n[a_1, \ldots, a_n]$.

We are ready to show that for $M, N \in \mathrm{Mod}(L)$, $M \cong N$ iff $\Gamma(M) \cong \Gamma(N)$. In fact, the invariance of the map Γ is obvious. We only check the converse. For this suppose $\pi : \Gamma(M) \cong \Gamma(N)$. Then for all $a \in \Gamma(M)$, $a \in |M|$ iff $\Gamma(M) \models \delta[a]$ iff $\Gamma(N) \models \delta[\pi(a)]$ iff $\pi(a) \in |N|$. Thus π restricted on $|M|$ is a bijection between $|M|$ and $|N|$. Now for any $n \geq 2$, $a_1, \ldots, a_n \in |M|$, $M \models R_n[a_1, \ldots, a_n]$ iff $\Gamma(M) \models \rho_n[a_1, \ldots, a_n]$ iff $\Gamma(N) \models \rho_n[\pi(a_1), \ldots, \pi(a_n)]$ iff $N \models R_n[\pi(a_1), \ldots, \pi(a_n)]$. Thus $\pi \upharpoonright |M|$ is an isomorphism between M and N.

Before proceeding to show that Γ is a faithful Borel reduction, we note the following further properties of the graphs $\Gamma(M)$:

(i) There is an $L_{\omega\omega}^{\Gamma}$-formula $\mu_1(x_1, y_{1,1}, z, u_1, u_2, u_3)$ such that for any countable graph Γ, $\Gamma \models \mu_1[a_1, b_{1,1}, c, d_1, d_2, d_3]$ iff the subgraph of Γ with the vertex set $\{a_1, b_{1,1}, c, d_1, d_2, d_3\}$ forms a 1-tag $T_1(a_1)$ as demonstrated in the definition of 1-tags;

(ii) More generally for each $n \geq 2$ there is a formula $\mu_n(x_1, \ldots, x_n, \ldots)$ such that for any countable graph Γ, $\Gamma \models \mu_n[a_1, \ldots, a_n, \ldots]$ iff the subgraph of Γ with the vertex set $\{a_1, \ldots, a_n, \ldots\}$ forms an n-tag as demonstrated in the definition of n-tags;

(iii) Furthermore, for each $n \geq 1$ and $1 \leq p \leq N(n) =_{\text{def}} (n^2 + 3n + 8)/2$, there is a formula $\xi_{n,p}(x)$ such that for any Γ, $\Gamma \models \xi_{n,p}[a]$ iff a is contained in an n-tag and is the p-th element in the demonstration in the definition of n-tags.

Let χ be the conjunction of the following $L_{\omega_1\omega}^{\Gamma}$-sentences:

$$\forall x \bigvee_{n\geq 1, p} \xi_{n,p}(x),$$

$$\forall x \bigwedge_{n,p>n} \left(\xi_{n,p}(x) \rightarrow \bigwedge_{(m,q)\neq(n,p)} \neg\xi_{m,q}(x) \right),$$

$$\forall x \bigvee_{n,p\leq n} \left(\xi_{n,p}(x) \rightarrow \xi_{1,1}(x) \right).$$

By now it is easy to verify that any countable graph satisfying χ is isomorphic to $\Gamma(M)$ for some $M \in \text{Mod}(L)$.

We are now ready to show that Γ is a faithful Borel reduction from $\text{Mod}(L)$ to $\text{Mod}(\gamma)$. In the above we have shown that $\Gamma : \text{Mod}(L) \rightarrow \text{Mod}(\gamma)$ is a reduction. With any reasonable coding of $\Gamma(M)$ by a genuine element of $\text{Mod}(\gamma)$ Γ can be easily seen to be a Borel map. We check finally that Γ is faithful. For this let $\mathcal{C}$ be an invariant Borel subset of $\text{Mod}(L)$. Then there is an $L_{\omega_1\omega}$-sentence ψ such that $\mathcal{C} = \text{Mod}(\psi)$. We can obtain an $L_{\omega_1\omega}^{\Gamma}$-sentence ψ^{Γ} by replacing, for all $n \geq 2$, all occurrences of R_n in ψ by ρ_n. It is clear that whenever $M \models \psi$ we have $\Gamma(M) \models \psi^{\Gamma}$. We claim that $\text{Mod}(\gamma \wedge \chi \wedge \psi^{\Gamma})$ is the smallest invariant Borel class containing $\Gamma(\mathcal{C})$. In fact, if $\Gamma \in \text{Mod}(\gamma \wedge \chi \wedge \psi^{\Gamma})$, then there is $M \in \text{Mod}(L)$ such that $\Gamma \cong \Gamma(M)$. Now it must be the case that $M \models \psi$, since otherwise $M \models \neg\psi$, and by our construction $\Gamma(M) \models \neg\psi^{\Gamma}$, a contradiction. $\qquad\square$

In model theory the reduction constructed in the above proof is called an interpretation because it has a much stronger property than we need for reducibility of the isomorphism relations. As shown in the proof the countable graph used to code the countable L-structure preserves all the $L_{\omega_1\omega}$-properties in a uniform way. Also, since the formulas ρ_n are first-order, the coding formulas stay in the same countable fragment generated by the original formulas. This implies that the structure $\Gamma(M)$ has approximately the same Scott rank as the original L-structure M. More examples of interpretations can be found in Reference [84] Chapter 5.

We remark that the $L_{\omega_1\omega}$-Vaught conjecture, or equivalently $\text{TVC}(S_\infty)$, is still a major open problem. By the above theorem it is equivalent to the topological Vaught conjecture for countable graphs.

We state the following corollary of the theorem and leave the details of the proof to the reader.

Corollary 13.1.3
Let L be a countable relational language with at least one relation symbol of arity ≥ 2. Then $\mathrm{Mod}(L)$ is faithfully Borel complete.

Exercise 13.1.1 Prove Corollary 13.1.3.

Exercise 13.1.2 Give the details of the formulas μ_n and $\xi_{n,p}$, $n \geq 1$ and $1 \leq p \leq N(n)$, in the proof of Theorem 13.1.2.

Exercise 13.1.3 Let Γ be the reduction defined in the proof of Theorem 13.1.2. Give a formal definition of $\Gamma(M)$ as an element of $\mathrm{Mod}(\gamma)$.

Exercise 13.1.4 Let Γ be the reduction defined in the proof of Theorem 13.1.2. Show that if L is a finite relational language then there is an $L_{\omega\omega}^{\Gamma}$-sentence χ such that any countable graph satisfying χ is isomorphic to $\Gamma(M)$ for some $M \in \mathrm{Mod}(L)$.

13.2 Countable trees

Recall that in descriptive set theory a tree T on ω is a subset of $\omega^{<\omega}$ closed under initial segments, that is, if $s \subseteq t$ and $t \in T$ then $s \in T$. By definition, every tree on ω contains the empty sequence $\emptyset$. For trees S, T on ω, an isomorphism is a bijection $\pi : S \to T$ preserving initial segments, that is, for all $s_1, s_2 \in S$, $s_1 \subseteq s_2$ iff $\pi(s_1) \subseteq \pi(s_2)$. For any $s \in S$, $\mathrm{lh}(\pi(s)) = \mathrm{lh}(s)$. In particular, $\pi(\emptyset) = \emptyset$. Every tree on ω is apparently an element of $2^{(\omega^{<\omega})}$. Let Tr be the set of all trees on ω. Then Tr is a closed subset of the Polish space $2^{(\omega^{<\omega})}$, hence is itself Polish.

In graph theory, a tree is an acyclic connected graph, that is, a graph with no cycles and in which there is a path between any pair of vertices. Here a tree on ω would be called a rooted tree, meaning a tree with a distinguished element known as the root. Conversely, a graph theoretic rooted tree can clearly be coded by a tree on ω. By this correspondence the class of all trees on ω becomes an invariant Borel class.

Theorem 13.2.1 (Friedman–Stanley)
The class of all countable trees on ω is Borel complete.

Proof. We define a Borel reduction from the class of all countable graphs to that of trees on ω. For each countable graph Γ we associate a tree $T(\Gamma)$ on ω.

For this fix a countable graph Γ with the underlying set ω and the edge relation R. Let T_0 be the full tree of nonrepeating finite sequences in $\omega^{<\omega}$. $T(\Gamma)$ will be obtained by adding at most one new, terminal, immediate successor to each node in T_0. For $m, n \in \omega$, let $\langle m, n \rangle = 2^m 3^n$. Then, for all $m, n > 0$, if

$$s = (x_1, \ldots, x_{\langle m,n \rangle}) \in T_0,$$

then add $s^\frown x_1$ to T_0 iff $R(x_m, x_n)$. The resulting tree is denoted $T(\Gamma)$.

Clearly the map $T : \Gamma \mapsto T(\Gamma)$ is continuous. We check that it is a reduction. First let $\pi : \Gamma \cong \Gamma'$. Let $\pi^* : \omega^{<\omega} \to \omega^{<\omega}$ be the automorphism induced by π:

$$\pi^*(x_1, \ldots, x_k) = (\pi(x_1), \ldots, \pi(x_k)).$$

Then $\pi^*(T_0) = T_0$. Now

$$(x_1, \ldots, x_k, x_1) \in T(\Gamma)$$
$$\Longleftrightarrow \exists m, n > 0 \ (k = \langle m, n \rangle \ \wedge \ R^\Gamma(x_m, x_n)$$
$$\Longleftrightarrow \exists m, n > 0 \ (k = \langle m, n \rangle \ \wedge \ R^{\Gamma'}(\pi(x_m), \pi(x_n))$$
$$\Longleftrightarrow (\pi(x_1), \ldots, \pi(x_k), \pi(x_1)) \in T(\Gamma').$$

Thus $\pi^*(T(\Gamma)) = T(\Gamma')$ and $T(\Gamma) \cong T(\Gamma')$.

Conversely, suppose $\sigma : T(\Gamma) \cong T(\Gamma')$. By a back-and-forth argument we find two permutations π and π' of ω such that for all $l \in \omega$,

$$\sigma(\pi(0), \ldots, \pi(l)) = (\pi'(0), \ldots, \pi'(l)).$$

This is done by induction on l. For the base case, let $\pi(0) = 0$, $s_0 = (\pi(0))$, and define $\pi'(0)$ so that $\sigma(s_0) = (\pi'(0))$. Next let $\pi'(1)$ be the least element of $\omega - \{\pi'(0)\}$, and $t_1 = (\pi'(0), \pi'(1))$. Define $\pi(1)$ so that $\sigma^{-1}(t_1) = (\pi(0), \pi(1))$. Since t_1 is not a terminal node of $T(\Gamma')$, neither is $\sigma^{-1}(t_1)$, and hence $\pi(1) \neq \pi(0)$. In general suppose distinct $\pi(0), \ldots, \pi(l)$ and distinct $\pi'(0), \ldots, \pi'(l)$ have been defined. Suppose l is odd. Let $\pi(l+1)$ be the least element of $\omega - \{\pi(0), \ldots, \pi(l)\}$ and $s_{l+1} = (\pi(0), \ldots, \pi(l+1))$. Then $\sigma(s_{l+1}) = (\pi'(0), \ldots, \pi'(l), y)$ for some $y \notin \{\pi'(0), \ldots, \pi'(l)\}$ since s_{l+1} and $\sigma(s_{l+1})$ are not terminal nodes. Define $\pi'(l+1) = y$ and continue the construction. If l is even then the definition is similar to the case $l = 0$.

Now we claim that $\pi' \circ \pi^{-1}$ is an isomorphism between Γ and Γ'. To see this suppose $R^\Gamma(a, b)$. Let $m = \pi^{-1}(a) - 1$, $n = \pi^{-1}(b) - 1$ and $k = \langle m, n \rangle$. Then the node $(\pi(0), \ldots, \pi(k-1), \pi(0))$ is a terminal node of $T(\Gamma)$. It follows that $\sigma(\pi(0), \ldots, \pi(k-1), \pi(0))$ is a terminal node of $T(\Gamma')$. Hence $\sigma(\pi(0), \ldots, \pi(k-1), \pi(0)) = (\pi'(0), \ldots, \pi'(k-1), \pi'(0)) \in T(\Gamma')$. This implies that $R^{\Gamma'}(\pi'(m), \pi'(n))$ or $R^{\Gamma'}(\pi' \circ \pi^{-1}(a), \pi' \circ \pi^{-1}(b))$. By symmetry we have $R^\Gamma(a, b)$ iff $R^{\Gamma'}(\pi' \circ \pi^{-1}(a), \pi' \circ \pi^{-1}(b))$ for any $a, b \in \omega$. $\Box$

There are some important differences between the above reduction and the one to countable graphs in Theorem 13.1.2. One cannot recover the original graph by the graph theoretic properties of the rooted tree constructed, and the reduction defined is not faithful. Gao [52] has shown that the class of trees on ω is not faithfully Borel complete, hence the topological Vaught conjecture for this class (which is known by results of Marcus [117], A. Miller [122], and Steel [150]) does not imply the $L_{\omega_1\omega}$-Vaught conjecture.

As we remarked above the theorem immediately implies the Borel completeness of all countable graph theoretic rooted trees. The proof of the theorem also has the following corollary about the other classes of countable graphs related to trees.

Corollary 13.2.2
The classes of all countable trees and of all countable acyclic graphs are both Borel complete.

Proof. In the proof of Theorem 13.2.1 the tree $T(\Gamma)$ constructed has the graph theoretic property that for every vertex v of $T(\Gamma)$ there is at most one vertex u of $T(\Gamma)$ of degree 1 and $uv \in T(\Gamma)$. Now let $S(\Gamma)$ be the graph theoretic tree obtained from $T(\Gamma)$ by adding two new vertices v_0, v_1 and edges between each of them and the root of $T(\Gamma)$. Then $\Gamma \cong \Gamma'$ iff $S(\Gamma) \cong S(\Gamma')$ since any isomorphism between $S(\Gamma)$ and $S(\Gamma')$ gives rise to an isomorphism between $T(\Gamma)$ and $T(\Gamma')$. ∎

The class of all finite splitting trees on ω is a Borel subset of Tr, hence is a standard Borel space. If we restrict our attention to finite splitting trees on ω then the isomorphism relation is much simpler, as the following theorem shows.

Theorem 13.2.3
The isomorphism relation for finite splitting trees on ω is smooth.

Proof. Let F denote the countable set of all finite trees on ω. Let $F_0 \subseteq F$ contain a tree of each isomorphism type. Let $\theta : F \to F_0$ be such that for any $T \in F$, $\theta(T) \cong T$. Let S be a finite splitting tree on ω. For each $n \in \omega$ let $S_n = \{s \in S : \text{lh}(s) \leq n\}$. Then define $f(S) = (\theta(S_n))_{n \in \omega}$.

Now f is a Borel function from the class of all finite splitting trees to F_0^ω. We claim that $S \cong S'$ iff $f(S) = f(S')$. The invariance of f is obvious, since every isomorphism between S and S' is also an isomorphism between S_n and S'_n for all $n \in \omega$.

For the converse, suppose $f(S) = f(S')$. Then $S_n \cong S'_n$ for all $n \in \omega$. Let σ_n be an isomorphism between S_n and S'_n. We note that for each $s \in S$, for all $n \geq \text{lh}(s)$, $\text{lh}(\sigma_n(s)) = \text{lh}(s)$. Since there are only finitely many nodes of length $\text{lh}(s)$ in S', there is a subset N_s of ω such that for all $n, m \in N_s$,

$\sigma_n(s) = \sigma_m(s)$, that is, $\sigma_n(s)$ is constant. The same observation applies to $s' \in S'$ and σ^{-1}.

Now using this observation, and by a back-and-forth construction, we may obtain a subset N of ω such that

(i) for all $s \in S$, for all but finitely many $n \in N$, the value $\sigma_n(s)$ is a constant; and

(ii) for all $s' \in S'$, for all but finitely many $n \in N$, the value $\sigma_n^{-1}(s)$ is constant.

Define $\sigma(s)$ to be the eventual constant value of $\sigma_n(s)$ for sufficiently large $n \in N$. By (ii) σ is a bijection between S and S'. We check that σ is an isomorphism between S and S'. To see this let $s, t \in S$ and assume $s \subseteq t$. Then for sufficiently large $n \in N$, $\sigma(s) = \sigma_n(s) \subseteq \sigma_n(t) = \sigma(t)$. By symmetry $s \subseteq t$ iff $\sigma(s) \subseteq \sigma(t)$. □

It is easy to see that there are uncountably many pairwise nonisomorphic finite splitting trees on ω. Thus the isomorphism relation for them is Borel bireducible with $\mathrm{id}(2^\omega)$.

A graph theoretic tree is locally finite if every vertex has finite degree. The following result shows a subtle difference between the descriptive set theoretic concept and the graph theoretic one. It also shows that there is no way to identify the root as we did in the proof of Corollary 13.2.2.

Theorem 13.2.4 (Jackson–Kechris–Louveau)
The isomorphism relation for countable locally finite trees is Borel bireducible with E_∞.

Proof. To see that this isomorphism relation is essentially countable, we use Kechris' Lemma 12.5.6. For each locally finite tree T and any vertex t in T, let S_t be the finite splitting tree on ω corresponding to the rooted tree with root t. Let F_0 and f be given by the proof of Theorem 13.2.3. Then for each $t \in T$, $f(S_t)$ is an element of F_0^ω.

Now given a locally finite tree T (with the underlying set a subset of ω) let $t \in T$ be a canonically selected element (for instance the least element of $T \subseteq \omega$). Then define $g(T) = f(S_t)$. We check that g satisfies the hypotheses of Lemma 12.5.6. Fix a locally finite tree T, let

$$A(T) = \{f(S_t) : t \in T\}.$$

Then $A(T)$ is countable. We note that $A(T) \supseteq \{g(T') : T' \cong T\}$. This is because, if $\pi : T' \cong T$ and $t' \in T'$ is chosen, then $f(S'_{t'}) = f(S_{\pi(t')})$ since $S'_{t'} \cong S_{\pi(t')}$. This shows that the range of g on every isomorphism class is countable. On the other hand, if $g(T') = g(T)$ then there is $t' \in T'$ and $t \in T$ such that $f(S'_{t'}) = f(S_t)$. This implies that $S'_{t'} \cong S_t$ and hence in particular

$T' \cong T$. Thus Lemma 12.5.6 is applicable, and it follows that the isomorphism of locally finite countable trees is essentially countable.

For the converse we code the shift equivalence of $2^{\mathbb{F}_2}$ by the isomorphism of locally finite trees. For each subset A of $\mathbb{F}_2$ we associate a locally finite tree $T(A)$ as follows. Let a, b be the generators of $\mathbb{F}_2$. Let K be a labeled directed tree with vertex set $\mathbb{F}_2$ and two edge relations R_a and R_b defined by

$$R_a(x, y) \iff xa = y, \text{ and } R_b(x, y) \iff xb = y.$$

If $R_a(x, y)$ we say that there is a directed edge xy with label a, and similarly for $R_b(x, y)$. Then we let T_0 be a locally finite tree obtained by the following operations illustrated by Figure 13.2:

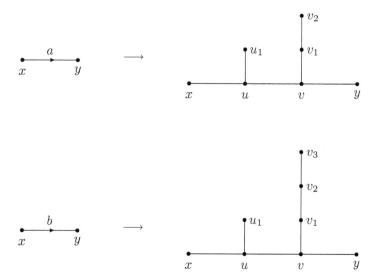

Figure 13.2   Coding $R_a(x, y)$ by $T_a(x, y)$ and $R_b(x, y)$ by $T_b(x, y)$.

(i) For each directed edge xy with label a in K, we replace the edge by a tree $T_a(x, y)$ with vertex set

$$\{x, y\} \cup \{u, u_1, v, v_1, v_2\}$$

and edge set

$$\{xu, uv, vy, uu_1, vv_1, v_1v_2\}.$$

(ii) For each directed edge xy with label b in K, we replace the edge by a tree $T_b(x, y)$ with vertex set

$$\{x, y\} \cup \{u, u_1, v, v_1, v_2, v_3\}$$

and edge set

$$\{xu, uv, vy, uu_1, vv_1, v_1v_2, v_2v_3\}.$$

Note that T_0 is a tree, and in T_0 every element of $\mathbb{F}_2$ has degree 4, whereas every other element of T_0 has degree at most 3. Now we obtain $T(A)$ by adding to T_0 a new vertex x^* for each element x of $\mathbb{F}_2$ and a new edge xx^*. In $T(A)$ a vertex v has degree ≥ 4 iff $v \in \mathbb{F}_2$, and v has degree 5 iff $v \in A$.

We claim that $T : A \mapsto T(A)$ is a Borel reduction from E_∞ to the isomorphism. First suppose $A, A' \subseteq \mathbb{F}_2$ and $g \in \mathbb{F}_2$ such that $gA = A'$. Consider $\sigma_g : \mathbb{F}_2 \to \mathbb{F}_2$ defined by $\sigma_g(x) = gx$. Then it is easy to see that σ_g is an automorphism of the labeled directed tree K. It follows that σ_g induces an automorphism σ_g^* of T_0. Moreover, σ_g^* can be extended to an isomorphism of $T(A)$ onto $T(A')$, since for each additional edge xx^* in $T(A)$, $x \in A$, and so $gx \in A'$ and $(gx)(gx)^* \in T(A')$; thus we may let $\sigma_g^*(x^*) = (gx)^*$.

Finally suppose $A, A' \subseteq \mathbb{F}_2$ and $\sigma : T(A) \cong T(A')$. Then by the degree properties we have that $\sigma(\mathbb{F}_2) = \mathbb{F}_2$ and $\sigma(A) = A'$. Now because of the difference between the constructions of (i) and (ii), as well as the asymmetry involved, we must have that σ induces an automorphism of K with $\sigma(A) = A'$. Let now $g = \sigma(1_{\mathbb{F}_2})$. Then for any $x \in \mathbb{F}_2$, we must have $\sigma(x) = gx$. This is because we can write x as $x_1^{\epsilon_1} \ldots x_n^{\epsilon_n}$, where $x_1, \ldots, x_n \in \{a, b\}$ and $\epsilon_1, \ldots, \epsilon_n \in \{+1, -1\}$. Then there is a directed path from $1_{\mathbb{F}_2}$ to x with labels given by the sequence of $x_i^{\epsilon_i}$. By isomorphism there is a directed path from $g = \sigma(1_{\mathbb{F}_2})$ to $\sigma(x)$ with the same sequence of labels. This implies that $\sigma(x) = gx$. We thus have that $gA = A'$ and the proof is complete. □

In the remainder of this section we consider well-founded trees on ω. Let WF be the class of all well-founded trees on ω. Then WF is a $\mathbf{\Pi}_1^1$, non-Borel subset of Tr. However, for each $\alpha < \omega_1$, if we let WF_α be the class of all well-founded trees on ω of rank $\leq \alpha$, then each WF_α is a Borel subset of Tr, and hence is a standard Borel space. For each $\alpha < \omega_1$, let $\cong_\alpha$ denote the isomorphism relation on WF_α. The following theorem shows that the collection of equivalence relations is essentially the same as the Friedman–Stanley tower. These isomorphism relations are in fact the original tower of equivalence relations considered by Friedman and Stanley.

Theorem 13.2.5
For each $\alpha < \omega_1$, $\cong_{3+\alpha}$ is Borel bireducible with $=^{\alpha+}$.

Proof. The proof is by induction on $\alpha < \omega_1$. For the base case we need to show that $\cong_3$ and $\text{id}(2^\omega)$ are Borel bireducible. For this note that a tree T on ω of rank ≤ 3 has depth ≤ 3, that is, for any $s \in T$, $\text{lh}(s) \leq 2$. Let T_1 be the set of elements in T of length 1. Then it is easy to see that the isomorphism type of T is determined by the cardinality of T_1 as well as the set of number of immediate successors for each element of T_1. Thus it is determined essentially by a subset of ω, or an element of 2^ω. Conversely, it is also easy to see that every subset of ω can be coded by a tree of depth ≤ 3 in the same sense.

For the successor inductive step, without loss of generality assume T has rank $3 + \alpha + 1$. Again let T_1 be the set of nodes in T of length 1. For each $t \in T_1$ let T_t be the tree on ω given by

$$s \in T_t \iff t^\frown s \in T.$$

If $\mathrm{lh}(t) = 1$ and $t \notin T_1$, let T_t be the empty tree. Then $T \cong T'$ iff there is a bijection π between T_1 and T_1' such that for all $t \in T_1$, $T_t \cong T'_{\pi(t)}$. Let f_α be a Borel reduction form $\cong_{3+\alpha}$ to $=^{\alpha+}$. Define $f_{\alpha+1}(T) = (f_\alpha(T_t))_{\mathrm{lh}(t)=1}$. Then $T \cong T'$ iff $f_{\alpha+1}(T) =^{(\alpha+1)+} f_{\alpha+1}(T')$. The converse reduction is similar.

Finally, if T has rank α, where α is a limit, then for any $t \in T_1$, T_t has rank $\beta(t) < \alpha$. By similar reductions as defined for the successor case, we obtain Borel bireducibility between $\cong_{3+\alpha}$ and $=^{\alpha+}$ from the inductive hypotheses. ☐

Exercise 13.2.1 Let $L^T = \{R_1, R_2\}$, where R_1 is unary and R_2 is binary. Give an $L_{\omega_1\omega}^T$-sentence φ such that for any $M \in \mathrm{Mod}(\varphi)$, R_1^M is a distinguished element of M and (M, R_2^M) is a tree.

Exercise 13.2.2 Show that the isomorphism relation for all pruned trees on ω is Borel complete. Deduce that the class of all countable trees with no vertices of degree 1 is Borel complete.

Exercise 13.2.3 Show that the isomorphism relation for countable locally finite acyclic graphs is Borel bireducible with $=^+$.

Exercise 13.2.4 Show that reduction defined in the proof of Theorem 13.2.4 is a faithful Borel reduction.

Exercise 13.2.5 Give an invariant Borel subclass of countable locally finite trees so that the isomorphism relation is Borel bireducible with E_0.

13.3   Countable linear orderings

In this section we prove the Borel completeness of the class of all countable linear orderings of ω.

Let $L = \{<\}$ be the language with one binary relation symbol. Let ρ be the $L_{\omega\omega}$ sentence that is the conjunction of the axioms of linear orders:

$$\forall x \ \neg(x < x),$$
$$\forall x \, \forall y \ (x < y \ \lor \ x = y \ \lor \ y < x),$$
$$\forall x \, \forall y \, \forall z \ [(x < y \ \land \ y < z) \to x < z].$$

Then every element of $\mathrm{Mod}(\rho)$ is a linear ordering of ω.

The proof that the class of all countable linear orderings is Borel complete will have some resemblance to the proof of Theorem 13.2.1, in the sense that some base linear order will be defined first and then the linear orders coding other structures will be obtained by a uniform operation on the base linear order. Before we start the proof of the theorem let us first describe the base linear order we will use.

We will be working with dense linear orders without endpoints. Of course, a standard back-and-forth argument shows that there is only one such countable order up to isomorphism. The natural linear order $(\mathbb{Q}, <)$ on the set of all rational numbers is such an order. If $\mathcal{P} = \{P_m : m \in \omega\}$ is a partition of $\mathbb{Q}$, we say that $\mathcal{P}$ is **mutually dense** if for any $p < q \in \mathbb{Q}$ and any $m \in \omega$, there is $r \in P_m$ with $p < r < q$. In particular, if $\mathcal{P}$ is a mutually dense partition of $\mathbb{Q}$, then every $(P_m, <)$ is a dense linear order without endpoints.

Lemma 13.3.1
There exists a mutually dense partition of $(\mathbb{Q}, <)$.

Proof. Enumerate $\mathbb{Q}$ as a nonrepeating sequence $q_0, q_1, \ldots$. We define a function $f : \mathbb{Q} \to \omega$ so that the partition $\mathcal{P} = \{f^{-1}(m) : m \in \omega\}$ is mutually dense. For any $i, j, m \in \omega$, let $\langle i, j, m \rangle = 2^i 3^j 5^m$. Then $\langle \cdot, \cdot, \cdot \rangle$ is an injection from ω^3 into ω. By induction on $n \in \omega$ we define a finite set D_n and $f(q)$ for each $q \in D_n$ so that the following properties hold:

(i) $q_n \in D_n$ and $D_n \subseteq D_{n+1}$ for all $n \in \omega$;

(ii) if $n = \langle i, j, m \rangle$ for some $i, j, m \in \omega$ with $i \neq j$, then there is $r \in D_n$ with $f(r) = m$ and either $q_i < r < q_j$ or $q_j < r < q_i$.

For the base step of the induction let $D_0 = \{q_0\}$ and $f(q_0) = 0$. For the inductive step $n > 0$ assume D_{n-1} has been defined and $f(q)$ have been defined for all $q \in D_{n-1}$. If $n = \langle i, j, m \rangle$ for $i, j, m \in \omega$ and $i \neq j$, we have either $q_i < q_j$ or $q_j < q_i$. In either case by the density of $\mathbb{Q}$ there is $r \notin D_{n-1}$ such that $q_i < r < q_j$ or $q_j < r < q_i$. Let k be the least so that q_k has this property. Let $D_n = D_{n-1} \cup \{q_n, q_k\}$. Let $f(q_k) = m$. If $q_n \in D_{n-1} \cup \{q_k\}$ then $f(q_n)$ is already defined, otherwise let $f(q_n) = 0$. We have that (i) and (ii) are satisfied in this case. Now if $n \notin \{\langle i, j, m \rangle : i \neq j, m \in \omega\}$, then let $D_n = D_{n-1} \cup \{q_n\}$. If $q_n \in D_{n-1}$ then $f(q_n)$ is already defined; otherwise let $f(q_n) = 0$. This finishes the inductive definition.

Now by (i) we have that $\bigcup_n D_n = \omega$, hence f is defined on the entire $\mathbb{Q}$. To see that f has the required property, let $p < q \in \mathbb{Q}$ and $m \in \omega$. For some unique i, j we have $p = q_i$ and $q = q_j$. Also $i \neq j$. Then for $n = \langle i, j, m \rangle$, by (ii) we have some $r \in D_n$ with $f(r) = m$ with $p = q_i < r < q_j = q$, as required. ☐

A standard back-and-forth argument shows that any two mutually dense partitions of $\mathbb{Q}$ are isomorphic, that is, if $\mathcal{P}_1 = \{P_{m,1} : m \in \omega\}$ and $\mathcal{P}_2 = \{P_{m,2} : m \in \omega\}$ are mutually dense partitions of $\mathbb{Q}$, then there is an order-preserving bijection $\pi : \mathbb{Q} \to \mathbb{Q}$ such that for any $m \in \omega$ and $q \in \mathbb{Q}$, $q \in P_{m,1}$ iff $\pi(q) \in P_{m,2}$.

Fix a mutually dense partition $\mathcal{P} = \{P_m : m \in \omega\}$ for $\mathbb{Q}$. We define a labeled linear order $Q_{<\omega}$ with labels in $\omega^{<\omega}$. $Q_{<\omega}$ will be the union of a sequence of inductively defined linear orders Q_n with labels in $\omega^{\leq n}$ so that $Q_n \subseteq Q_{n+1}$ for all $n \in \omega$. The labeling function will be denoted $\lambda : Q_{<\omega} \to \omega^{<\omega}$. We also define $l : Q_{<\omega} \to \omega$ by $l(x) = \mathrm{lh}(\lambda(x))$. Thus $l(x)$ represents the level of $\lambda(x)$, and $l(x) = n$ iff $\lambda(x) \in \omega^n$. We will say that $x \in Q_{<\omega}$ is of level n if $l(x) = n$.

To begin the inductive definition, let $Q_0 = (\mathbb{Q}, <)$, and for every $x \in Q_0$, define $\lambda(x) = \emptyset$ and $l(x) = 0$. Suppose Q_n has been defined and λ and l have been defined for elements of Q_n. Let Q_{n+1} be the linear order obtained from adding a copy of $(\mathbb{Q}, <)$ to the immediate right of each element of Q_n. Formally, $Q_{n+1} = Q_n \times (\{-\infty\} \cup \mathbb{Q})$ with the lexicographic order, where $(\{-\infty\} \cup \mathbb{Q}, <)$ is an extension of $(\mathbb{Q}, <)$ with $-\infty < q$ for all $q \in \mathbb{Q}$. In this formal definition we identify each $x \in Q_n$ with $(x, -\infty) \in Q_{n+1}$, thus maintaining $Q_n \subseteq Q_{n+1}$. Note that the new elements of Q_{n+1} form the product set $Q_n \times \mathbb{Q}$. Thus for each $x \in Q_n$ and $q \in \mathbb{Q}$, we define $l(x, q) = n+1$ and $\lambda(x, q) = \lambda(x)^{\frown} m$, where m is the unique number such that $q \in P_m \in \mathcal{P}$. Then λ on $Q_n \times \mathbb{Q}$ has the properties:

(i) for each $x \in Q_n$ and $q \in \mathbb{Q}$, $\lambda(x, q) \supseteq \lambda(x)$ and $\lambda(x, q) \in \omega^{n+1}$; and

(ii) for each $x \in Q_n$, the partition $\{\lambda^{-1}(\lambda(x)^{\frown} m) : m \in \omega\}$ is mutually dense in $\{x\} \times \mathbb{Q}$.

This finishes the inductive definition of Q_n, and also of $Q_{<\omega}$.

Theorem 13.3.2 (Friedman–Stanley)
The class of all countable linear orderings is Borel complete.

Proof. It suffices to define a Borel reduction from any Borel complete class to the class of countable linear orderings. For notational convenience we will use the class of all binary relations on ω. Note that it is Borel complete since the class of all countable graphs is a subclass.

Let $L^R = \{R\}$ where R is a binary relation symbol. Since the language is finite, for each $n \in \omega$ the set

$$\Phi_n = \{\varphi_0^{M,\bar{a}} : M \in \mathrm{Mod}(L^R), \bar{a} \in \omega^n\}$$

of formulas with n free variables $v_0, \dots, v_{n-1}$ is finite. Recall that $\varphi_0^{M,\bar{a}}$ is the atomic type of $\bar{a}$ over M defined in the Scott analysis in Section 12.1. Let $\Phi = \bigcup_n \Phi_n$. Fix a bijection $c : \Phi \to \omega$ so that for all $\varphi, \psi \in \Phi$ if $\varphi \in \Phi_n$,

$\psi \in \Phi_m$ and $n < m$, then $c(\varphi) < c(\psi)$. Thus the function c gives a coding of atomic types of tuples by natural numbers.

For each $n \in \omega$ define a linear order $(B_n, <)$ by

$$B_n = D_1 \cup F_n \cup D_2$$

where D_1 and D_2 are dense linear orders without endpoints, F_n contains $n+2$ elements, and for any $p \in D_1, r \in F_n$ and $q \in D_2$, $p < r < q$. Note that there is an $L_{\omega\omega}$-formula $\theta_n(u, v)$ for any $n \in \omega$ such that for any linear ordering $N \in \text{Mod}(\rho)$ and $a, b \in |N|$, $N \models \theta_n[a, b]$ iff $a < b$ and the linear order $\{x \in N : a < x < b\}$ is isomorphic to B_n. We let $\psi_n(u)$ be the formula

$$\exists v \, \theta_n(u, v) \lor \exists v \, \theta_n(v, u).$$

We are finally ready to define for each $M \in \text{Mod}(L^R)$ a countable linear order $Q(M)$. As before we describe $Q(M)$ informally and leave the exact details of defining $Q(M)$ as an element of $\text{Mod}(L)$ as an exercise. Note that each $x \in Q_{<\omega}$ gives rise to a tuple $\lambda(x)$, which in turn is coded by $c(\varphi_0^{M,\lambda(x)})$. In view of this define $c_x = c(\varphi_0^{M,\lambda(x)})$. Now $Q(M)$ is obtained from $Q_{<\omega}$ by replacing each element x of $Q_{<\omega}$ by a copy of the linear order B_{c_x}. This finishes the definition of the map $Q : M \mapsto Q(M)$. The proof for the Borelness of the map Q is routine.

We check that Q is a reduction. First suppose $\pi : M \to M'$, where $M, M' \in \text{Mod}(L^R)$. Since π is a bijection from ω onto ω, it induces an automorphism π^* of the tree $\omega^{<\omega}$, where $\pi^*(a_1, \ldots, a_n) = (\pi(a_1), \ldots, \pi(a_n))$. Furthermore, by an easy induction π^* induces an automorphism π_n^* of Q_n as labeled linear orders such that $\pi_{n+1}^* \upharpoonright Q_n = \pi_n^*$. Let $\pi_{<\omega}^* = \bigcup_n \pi_n^*$. Then $\pi_{<\omega}^*$ is an automorphism of $Q_{<\omega}$. Now if $x \in Q_{<\omega}$ then $\pi^*(\lambda(x)) = \lambda(\pi_{<\omega}^*(x))$ and $c_x = c_{\pi_{<\omega}^*(x)}$. Thus for any $x \in Q_{<\omega}$, the copy of B_{c_x} in $Q(M)$ replacing x is isomorphic to the copy of $B_{c_{\pi_{<\omega}^*(x)}}$ in $Q(M')$ replacing $\pi_{<\omega}^*(x)$. This shows that $Q(M) \cong Q(M')$.

Conversely, assume $M, M' \in \text{Mod}(L^R)$ and $\sigma : Q(M) \cong Q(M')$. Note that for any $a \in Q(M)$ there is some $n \in \omega$ such that $Q(M) \models \psi_n[a]$, which implies that $Q(M') \models \psi_n[\sigma(a)]$. It follows that σ induces an order-preserving bijection σ' from the copy of $Q_{<\omega}$ in the construction of $Q(M)$ to the copy of $Q_{<\omega}$ in the construction of $Q(M')$. Moreover, for each $x \in Q_{<\omega}$, $c_x = c_{\sigma'(x)}$, and hence $\varphi_0^{M,\lambda(x)} = \varphi_0^{M',\lambda(\sigma'(x))}$. Also by our construction, if $x, y \in Q_{<\omega}$ then $\lambda(x) \subseteq \lambda(y)$ iff $x < y$ and for all z with $x < z < y$ we have $l(x) < l(z)$. Since this property is preserved by σ', we have that for all $x, y \in Q_{<\omega}$, $\lambda(x) \subseteq \lambda(y)$ iff $\lambda(\sigma'(x)) \subseteq \lambda(\sigma'(y))$. By these and using a back-and-forth argument similar to the proof of Theorem 13.2.1 we can construct two permutations π and π' of ω such that for all $n \in \omega$,

$$\varphi_0^{M,(\pi(0),\ldots,\pi(n))} = \varphi_0^{M',(\pi'(0),\ldots,\pi'(n))}.$$

Then $\pi' \circ \pi^{-1}$ is an isomorphism from M to M' just as in the proof of Theorem 13.2.1. ∎

Gao [52] has shown that the class of countable linear orderings is not faithfully Borel complete. Rubin has proved the Vaught conjecture for countable linear orderings (see Reference [150]). Steel [150] has considered a concept of generalized trees defined as a partial order $<$ such that the initial segment below any element is linearly ordered, that is, satisfying the formula

$$\forall x \, \forall y \, \forall z [\, (y < x \wedge z < x) \rightarrow (y < z \vee y = z \vee z < y)\,].$$

This apparently defines a larger class than both that of countable trees and of countable linear orderings. He proved the topological Vaught conjecture for this class.

Exercise 13.3.1 Show that any two mutually dense partitions of $\mathbb{Q}$ are isomorphic.

Exercise 13.3.2 Let $(B_n, <)$ be the linear order defined in the proof of Theorem 13.3.2. Let $\mathcal{P} = \{P_m : m \in \omega\}$ be a mutually dense partition of $\mathbb{Q}$. For each $A \subseteq \omega$ let $Q(A)$ be obtained from $\mathbb{Q}$ by replacing each element of P_m by a copy of B_m iff $m \in A$. Show that $A = A'$ iff $Q(A) \cong Q(A')$.

Exercise 13.3.3 Give an alternative proof of Theorem 13.3.2 by defining a linear order $Q(T)$ for each countable tree T on ω so that $T \cong T'$ iff $Q(T) \cong Q(T')$.

13.4 Countable groups

Countable groups already played an important role in invariant descriptive set theory because of their connections to countable Borel equivalence relations (see Chapter 7). In this section we study the isomorphism relation for various classes of countable groups. Apparently there is a vast subfield of group theory which concentrates exclusively on the classification of countable groups. Moreover, there has already been a lot of interactions between model theory and group theory. Thus it is both unnecessary and impossible to develop the entire subject in this short section. Instead we will focus on several group theoretic results with interaction with descriptive set theory. In fact, the results we present here have become standard knowledge in countable model theory. Most of their proofs are lengthy; we will omit them since they can be found in standard group theory textbooks.

We need to interpret various classes of countable groups as invariant Borel classes for a relational language. Of course, the standard language of group

theory involves a multiplication operation (and very often an inverse operation as well). But as remarked before, the multiplication can be represented
by a 3-ary relation $P(x, y, z)$ with the intended meaning $x \cdot y = z$. Thus we
fix a language $L^G = \{P\}$ with a 3-ary relation symbol P. The axioms of
groups can then be expressed by an $L^G_{\omega\omega}$-sentence ξ. The class of countable
groups is therefore $\text{Mod}(\xi)$. In our discussions below we will freely use con-
ventional notation and terminology of group theory; for instance we refer to
multiplication and inverse without bothering with the relation P.

The following theorem of Mekler [120] confirms the common wisdom that
the isomorphism for countable groups is complicated.

Theorem 13.4.1 (Mekler)
The class of all countable groups is faithfully Borel complete.

The proof is complicated, and we omit it here. However, we remark that
in the proof Mekler constructed countable nilpotent groups of rank 2 to code
the isomorphism types of countable graphs. Moreover, the coding is in fact a
two-way interpretation in the sense that properties of the graphs correspond
to properties of the groups (see Reference [84] Sections 5.5 and A.3). As we
remarked before, this is stronger than the faithfulness of the Borel reduction
and has other model theoretic consequences.

Now recall that a group G is rank 2 nilpotent iff $G/Z(G)$ is abelian, where
$Z(G) = \{x \in G : \forall y \in G \ x \cdot y = y \cdot x\}$ is the center of G. Thus rank 2
nilpotent groups are characterized by the following sentence,

$$\forall g, h, x \, [\zeta(x) \rightarrow \exists y \, (\zeta(y) \, \wedge g \cdot h \cdot x = h \cdot g \cdot y)],$$

where $\zeta(x)$ is the formula characterizing the center, that is, $Z(G) = \{x \in G : \zeta(x)\}$. It follows that the class of all countable nilpotent groups of rank 2 is
invariant Borel.

Theorem 13.4.2 (Mekler)
*The class of all countable nilpotent groups of rank 2 is faithfully Borel com-
plete.*

The class of all countable abelian groups is obviously an invariant Borel
class. However, it is an open problem if it is Borel complete.

In the rest of this section we discuss countable abelian groups, and in doing
this we follow the convention of group theory to denote the group additive
rather than multiplicative.

Note that every countable abelian group G can be uniquely expressed as
the direct sum of a torsion abelian group (its torsion part) and a torsion-free
abelian group

$$G \cong G_T \oplus G_F.$$

Recall that an abelian group A is torsion-free if for all $g \in A$, $g \neq 0 = 1_A$,
and integer $n > 0$, $ng \neq 0$. For an abelian group G the torsion subgroup G_T

of G is defined by

$$G_T = \{g \in G : ng = 0 \text{ for some integer } n > 0\}.$$

An abelian group A is torsion if $A_T = A$. If p is a prime number, we may define further the p-torsion subgroup $G_{T,p}$ of G by

$$G_{T,p} = \{g \in G : p^n g = 0 \text{ for some integer } n > 0\}.$$

An abelian group A is p-torsion if $A = A_{T,p}$. We have in general

$$G_T = \bigoplus_p G_{T,p}.$$

The definitions make it clear that the subclasses of countable abelian groups consisting of p-torsion, torsion, or torsion-free groups respectively are all invariant Borel classes. Thus it makes sense to think of their isomorphism relations as S_∞-orbit equivalence relations.

Ulm gave a complete classification of countable torsion abelian groups up to isomorphism. His classification motivated the concept of Ulm classifiability, introduced in Section 9.2. We describe the original Ulm invariants below.

Let p be a prime and A a countable abelian group. For $\alpha < \omega_1$ define a subgroup $p^\alpha A$ by induction as follows:

$$p^0 A = A,$$

$$p^{\alpha+1} A = p(p^\alpha A) = \{pg : g \in p^\alpha A\},$$

$$p^\lambda A = \bigcap_{\alpha < \lambda} p^\alpha A, \text{ if } \lambda \text{ is a limit.}$$

It is easy to see by induction that each $p^\alpha A$ is a subgroup and $p^\alpha A \subseteq p^\beta A$ if $\alpha \geq \beta$. Since A is countable there exists $\alpha_0 < \omega_1$ such that $p^{\alpha_0} A = p^{\alpha_0+1} A$. The least such α_0 is called the p-**length** of A, and we denote it by $l_p(A)$.

Now suppose A is p-torsion. Let $A[p] = \{g \in A : pg = 0\}$. Then it is easy to see that

$$A[p] \cong \bigoplus_{i \in I} (\mathbb{Z}/p\mathbb{Z})$$

for an index set I that is at most countable. The cardinality of I is called the **dimension** of $A[p]$. Note that $(pA)[p] = A[p] \cap pA$ and there is $I' \subseteq I$ such that $(pA)[p] \cong \bigoplus_{i \in I'}(\mathbb{Z}/p\mathbb{Z})$. Thus it makes sense to speak of the dimensions for both $(pA)[p]$ and $A[p]/(pA)[p]$, to be respectively the cardinalities of I' and $I - I'$.

Now for each $\alpha < \omega_1$, let $f_\alpha(A)$ be the dimension of $(p^\alpha A)[p]/(p^{\alpha+1}A)[p]$; $f_\alpha(A)$ is called the α-**th Ulm invariant** of A. Note that $f_\alpha(A) = 0$ for all $\alpha \geq \alpha_0$. In general, $f_\alpha(A)$ is an element of $\omega \cup \{\infty\}$.

Theorem 13.4.3 (Ulm)
Let p be a prime and G, H be countable p-torsion abelian groups. Then $G \cong H$ iff for all $\alpha < \omega_1$, $f_\alpha(G) = f_\alpha(H)$.

For general torsion abelian groups A we obtain a transfinite sequence of Ulm invariants $f_\alpha^p(A)$ for each prime p, and the mixed sequence $(f_\alpha^p(A))_{p,\alpha}$ is a complete invariant for the group A.

In Section 9.2 we defined the abstract Ulm invariants to be elements of the space $2^{<\omega_1}$. This can be seen to be equivalent to the original Ulm invariants via a Δ_2^1 isomorphism. Moreover, the Ulm classification of countable p-torsion abelian groups provides a proof of Δ_2^1 bireducibility between the isomorphism relation and the equality of Ulm invariants. By metamathematical methods it can be shown that this isomorphism relation is nonsmooth but E_0 is not Borel reducible to it. Thus the Glimm–Effros dichotomy fails for it.

For the remainder of this section we turn to countable torsion-free abelian groups. Around the same time Ulm classified countable torsion abelian groups, Baer investigated torsion-free abelian groups in a somewhat similar manner.

Let A be a countable torsion-free abelian group and p a prime. For each $g \in A$, the p-**height** of g is defined by

$$h_p(g) = \begin{cases} \sup\{n \in \omega : g \in p^n A - p^{n+1}A\}, & \text{if } g \notin p^\omega A, \\ \infty, & \text{otherwise.} \end{cases}$$

Let P be the set of all prime numbers. Let $h(g) = (h_p(g))_{p \in P}$. Then $h : A \to (\omega \cup \{\infty\})^P$.

Call two elements $g, h \in A$ **linearly dependent** if there are integers $m, n \neq 0$ such that $mg + nh = 0$.

Lemma 13.4.4
Let A be a countable torsion-free abelian group and $g, h \in A$ are linearly dependent. Then

(i) *for all $p \in P$, $h_p(g) = \infty$ iff $h_p(h) = \infty$, and*

(ii) *for all but finitely many $p \in P$, $h_p(g) = h_p(h)$.*

Proof. Suppose $mg + nh = 0$ for integers $m, n \neq 0$. Since A is torsion-free we may assume $(m, n) = 1$. Suppose $h_p(g) = \infty$, that is, $g \in p^\omega A$. It follows that $mg \in p^\omega A$ and hence $nh \in p^\omega A$. Now assume $n = p^i q$ with $(q, p) = 1$. Then for any $k > 0$, we have that $nh \in p^{k+i}A$, that is, there is $x \in A$ with $p^i qh = p^{k+i}x$. Again by torsion-freeness we obtain that $qh = p^k x$. Since $(p^k, q) = 1$ there are integers r, s with $rp^k + qs = 1$. Let $y = sx + rh$. Then

$$p^k y = p^k sx + p^k rh = qsh + rp^k h = h.$$

This shows that $h \in p^k A$ for all integers $k > 0$. Hence $h \in p^\omega A$ and $h_p(h) = \infty$. For (ii) let p be a prime with $(p, m) = (p, n) = 1$. Suppose $h_p(g) = k$, that is, $g \in p^k A - p^{k+1} A$. Then for some $x \in A$, $g = p^k x$ and so $nh = p^k(-mx)$. If $k = 0$ then $h \in p^k A$. If $k > 0$ then a similar argument as above shows that $h \in p^k A$ as well. By symmetry we have both (i) and (ii). $\quad\square$

In view of the above lemma, define an equivalence relation $\sim$ on $(\omega \cup \{\infty\})^P$ by $f = (f_p)_{p \in P} \sim f' = (f'_p)_{p \in P}$ iff for all $p \in P$, $f_p = \infty$ iff $f'_p = \infty$, and for all but finitely many $p \in P$, $f_p = f'_p$.

Lemma 13.4.5
Let A be a countable torsion-free abelian group and $x \in A$. If $f \in (\omega \cup \{\infty\})^P$ is such that $f \sim h(x)$, then there is $y \in A$ such that $f = h(y)$.

Proof. Let $F = \{p \in P : f_p \neq h_p(x)\}$. Then F is finite and for all $p \in F$, $f_p, h_p(x) < \infty$. If $F = \emptyset$ there is nothing to prove. Otherwise enumerate F as $p_1 < \cdots < p_n$. Let $n_i = h_{p_i}(x)$ for $1 \leq i \leq n$. We define $x_1, \ldots, x_n \in A$ successively as follows. If $n_1 = 0$ then let $x_1 = x$. Otherwise, $x \in p^{n_1} A$, so there is x_1 with $x = p^{n_1} x_1$. By the argument in the preceding proof we have that $h_{p_1}(x_1) = 0$ and $h_p(x_1) = h_p(x)$ for all $p \neq p_1$. In a similar manner successively define $x_2, \ldots, x_n$ such that for each $2 \leq i \leq n$, $h_{p_i}(x_i) = 0$ and $h_{p_i}(x_i) = h_{p_i}(x_{i-1})$ for all $p \neq p_i$. It follows that $h_{p_i}(x_n) = 0$ for all $1 \leq i \leq n$ and $h_p(x_n) = h_p(x)$ for all $p \notin F$. Now let $y = p_1^{f_1} \ldots p_n^{f_n} x_n$. Then it is easy to check that $h_{p_i}(y) = f_i$ for all $1 \leq i \leq n$ and $h_p(y) = h_p(x)$ for all $p \notin F$. Thus y is as required. $\quad\square$

A countable torsion-free group A is of **rank 1** if every two nonzero elements of A are linearly dependent. It is easy to see that every subgroup of $(\mathbb{Q}, +)$ is a torsion-free group of rank 1. The following theorem of Baer completely classifies countable torsion-free groups of rank 1.

Theorem 13.4.6 (Baer)
If G, G' be countable torsion-free abelian groups of rank 1, then $G \cong G'$ iff there is $g \in G$ and $g' \in G'$ such that $h(g) \sim h(g')$.

Proof. The implication ($\Rightarrow$) is obvious. For ($\Leftarrow$) let $g \in G$ and $g' \in G'$ be such that $h(g) \sim h(g')$. Let $N = \{p \in P : h_p(g) = \infty\}$. Then $N = \{p \in P : h_p(g') = \infty\}$. We define an isomorphism π from G onto G' with $\pi(g) = g'$. For this consider an arbitrary element h of G. Let $(m, n) = 1$ be integers such that $mh + ng = 0$. Write $m = m_F m_N$ where $(m_F, m_N) = 1$, m_F contains no prime factor in N, and m_N contains only prime factors in N, unless $m_N = 1$. Similarly write $n = n_F n_N$. It is easy to see that $h_p(n_N g) = h_p(g)$ and $h_p(m_N h) = h_p(h)$ for all $p \in P$. By Lemma 13.4.4 $h(m_N h) \sim h(n_N g)$. We have $h(n_N g') = h(g') \sim h(m_N h)$. By the proof of Lemma 13.4.5 there is $x \in G'$ such that $m_F x + n_F(n_N g') = 0$. Such x is unique. Since $h_p(x) \sim h_p(h)$ for all $p \in P$, there is a unique h' such that $x = n_N h'$. We thus obtain a unique

h' such that $mh' + ng' = 0$. Define $\pi(h) = h'$. By symmetry π is a bijection. It is now easy to check that π is an isomorphism. □

Corollary 13.4.7
The isomorphism for countable torsion-free abelian groups of rank 1 is Borel bireducible with E_0.

Proof. Given an arbitrary countable torsion-free abelian group G we let $g_0 \in G$ be a canonically distinguished element (for instance when G has underlying set ω we may let $g_0 = 0$). The let $N_G = \{p \in P : h_p(g) = \infty\}$ and $F_G = P - N_G$. Define $h_G : F_G \to \omega$ by letting $h_G(p) = h_p(g_0)$ for all $p \in F_G$. By Theorem 13.4.6 and Lemmas 13.4.4 and 13.4.5, we have that $G \cong G'$ iff $N_G = N_{G'}$ and for all but finitely many $p \in F_G$, $h_G(p) = h_{G'}(p)$. The function $G \mapsto (N_G, h_G)$ is easily seen to be Borel, and is a reduction from the isomorphism relation to $E = \mathrm{id}(2^\omega) \times E_0$. But E is hyperfinite, hence $E \leq_B E_0$.

Conversely we show that E_0 on ω^ω is Borel reducible to the isomorphism of subgroups of $\mathbb{Q}$. Let P be enumerated as $p_0, p_1, \ldots$. For $x \in \omega^\omega$, let G_x be the subgroup of $\mathbb{Q}$ generated by

$$\{1, p_i^{-x(i)} : i \in \omega\}.$$

Then G_x is torsion-free and $h_{p_i}(1) = x(i)$ by our construction. Now for $x, y \in \omega$, $x E_0 y$ iff $h_{G_x}(1) \sim h_{G_y}(1)$. It follows from Theorem 13.4.6 and Lemmas 13.4.4 and 13.4.5 that $x E_0 y$ iff $G_x \cong G_y$. □

In general elements $g_1, \ldots, g_n$ in a torsion-free abelian group G are said to be **linearly independent** if for any integers $m_1, \ldots, m_n$, if $m_1 g_1 + \cdots + m_n g_n = 0$ then $m_1 = \cdots = m_n = 0$. A torsion-free abelian group G is of **rank** n if there is a linearly independent set with n elements but there are no linearly independent sets with $n + 1$ elements. If G is of rank n for some $1 \leq n < \infty$, then G is said to be of **finite rank**. It is not hard to see that any torsion-free abelian group of rank $\leq n$ is isomorphic to a subgroup of $\mathbb{Q}^n$. Thus the standard Borel space $S(\mathbb{Q}^n)$ of all subgroups of $\mathbb{Q}^n$, as a Borel subspace of the Polish space $2^{\mathbb{Q}^n}$ is another representation of the invariant Borel class of countable torsion-free abelian groups of rank $\leq n$.

Lemma 13.4.8
For each integer $n \geq 1$, the isomorphism relation for countable torsion-free abelian groups of rank $\leq n$ is essentially countable.

Proof. It suffices to show that the isomorphism relation on $S(\mathbb{Q}^n)$ is countable. For this fix a countable subgroup G of $\mathbb{Q}^n$. We verify that there are only countably many subgroups of $\mathbb{Q}^n$ isomorphic to G. Let $g_1, \ldots, g_m$ be a linearly independent set in G, where $m \leq n$ is the rank of G. Then every element of G is uniquely expressed as a rational linear combination of $g_1, \ldots, g_m$. It follows that if $G' \in S(\mathbb{Q}^n)$ is isomorphic to G, then there are linearly independent $g_1', \ldots, g_m' \in G'$ such that for all rationals $q_1, \ldots, q_m \in \mathbb{Q}$,

$$q_1 g_1 + \cdots + q_m g_m \in G \iff q_1 g_1' + \cdots + q_m g_m' \in G'.$$

In this sense the group G' is completely determined by the set of linearly independent elements $g_1', \ldots, g_m'$. Since there are only countably many such sets, the number of subgroups of $\mathbb{Q}^n$ isomorphic to G is countable. □

Using deep results in Zimmer's superrigidity theory, Thomas [153] has shown that the isomorphism relation for countable torsion-free abelian groups of rank $\leq n$ form a strictly increasing chain of essentially countable equivalence relation in the Borel reducibility hierarchy. Such results are significant not only for group theory but also for descriptive set theory of equivalence relations.

Countable torsion-free abelian groups that are not of finite rank are said to have **infinite rank**. It is an open problem whether the class of all countable torsion-free abelian groups of infinite rank is Borel complete. Hjorth [71] has shown that all equivalence relations in the Friedman–Stanley tower are Borel reducible to the isomorphism relation of this class.

Exercise 13.4.1 Show that for any $n \geq 2$ the class of all countable nilpotent groups of rank n is an invariant Borel class. Deduce that the class of all countable nilpotent groups is an invariant Borel class.

Exercise 13.4.2 Show that for any $n \geq 2$ the class of all countable solvable groups of rank n is an invariant Borel class. Deduce that the class of all countable solvable groups is an invariant Borel class.

Exercise 13.4.3 Show that each countable torsion-free abelian group of rank n is isomorphic to a subgroup of $\mathbb{Q}^n$.

Exercise 13.4.4 Let $GL(\mathbb{Q}^n)$ be the group of all $n \times n$ matrices with nonzero determinant. Show that two subgroups G, H of $\mathbb{Q}^n$ iff there is $\mu \in GL(\mathbb{Q}^n)$ such that $\mu G = H$.

Part IV

Applications to Classification Problems

Chapter 14

Classification by Example: Polish Metric Spaces

In this final part of the book we focus on applications of the theory of equivalence relations to classification problems in mathematics. In these applications classification problems in mathematics are first identified as equivalence relations on standard Borel spaces. Then the exact complexity of each equivalence relation is determined by a comparison to some benchmark equivalence relation in the Borel reducibility hierarchy.

In fact in Chapter 13 we already discussed isomorphic classification problems for invariant Borel classes of countable structures. Precise measurement of the complexity of some classification problems has been achieved, but in the process knowledge is needed from both descriptive set theory and the particular field (combinatorics and algebra, to be more specific).

In this chapter we continue to consider classification problems for uncountable mathematical structures, thus expanding the applicability of invariant descriptive set theory to other mathematical fields (such as topology, geometry, and analysis). As such classification/nonclassification results are numerous and are still mushrooming as we speak, it is impossible to give a comprehensive account of applications in many directions.

To make this book more useful, we will instead focus on the fundamentals, or the framework of these applications. In this chapter we will concentrate on one subject, the classification of Polish metric spaces, and illustrate how results about equivalence relations are relevant to the classification problems. When the reader attempts to consider classification problems for structures of a different nature, he will find the framework flexible enough to be used again.

14.1 Standard Borel structures on hyperspaces

When considering isomorphic classification of a family of countable structures, we have been setting up the problem by identifying the family as an invariant Borel class. Thus the isomorphism relation for the class of structures becomes an equivalence relation on a standard Borel space. And then it becomes mathematically sound to apply the Borel reducibility notion and results about the Borel reducibility hierarchy.

If the structures considered are uncountable, then in theory the above method is no longer valid and the applicability of invariant descriptive set theory is limited. However, in practice, a large number of classes of mathematical structures can be turned into standard Borel spaces. Examples of such classes are ubiquitous in topology, geometry, and analysis.

In this section we take only the example of Polish metric spaces, that is, separable complete metric spaces. Let

$$\mathcal{X} = \text{the hyperspace of all Polish metric spaces.}$$

We show that, even if in general Polish metric spaces are uncountable, $\mathcal{X}$ can be turned into a standard Borel space by a suitable coding. In fact we give two coding methods to do this, and prove that they are equivalent.

The first method, used first by Vershik [163], is based on the following simple observation. Every Polish metric space contains a countable dense subset, and the metric structure on the whole space is completely determined by and can be fully recovered from the metric structure on any dense subset by the operation of completion. Thus a Polish metric space can be coded by a canonical enumeration of a canonical countable dense subset. In practice, when a space is presented to us a canonical countable dense subset together with a canonical enumeration is often immediately recognizable. For instance, if the space $\mathbb{R}$ is presented to us we immediately recognize a canonical countable dense subset $\mathbb{Q}$ and a canonical enumeration of its elements.

In view of this we define $\mathbb{X}$ to be the subspace of $\mathbb{R}^{\omega \times \omega}$ consisting of elements $(r_{i,j})_{i,j \in \omega}$ such that

(1) for all $i, j \in \omega$, $r_{i,j} \geq 0$, and $r_{i,j} = 0$ iff $i = j$;

(2) for all $i, j \in \omega$, $r_{i,j} = r_{j,i}$;

(3) for all $i, j, k \in \omega$, $r_{i,j} \leq r_{i,k} + r_{k,j}$.

Since $\mathbb{R}^{\omega \times \omega}$ is a Polish space with the product topology, and $\mathbb{X}$ is a G_δ subset of $\mathbb{R}^{\omega \times \omega}$, $\mathbb{X}$ is Polish.

Now if a Polish metric space (X, d) is given, as we required, together with an enumeration $(x_i)_{i \in \omega}$ of a canonical countable dense set D_X, we let

$$r_X = (r_{i,j})_{i,j \in \omega} = (d(i,j))_{i,j \in \omega}.$$

Note that r_X depends also on the sequence (x_i), but as our assumption goes it is a part of the presentation of the space X, and hence we suppress mentioning (x_i) for notational simplicity. Conversely, for any $r \in \mathbb{X}$ we define metric spaces $D_r = (\omega, d_\omega)$ with $d_\omega(i,j) = r_{i,j}$ and X_r to be the completion of D_r. It is clear that r_X contains the full information of the metric structure on X, since the map $\varphi : \omega \to D$ with $\varphi(i) = x_i$ is an isometry and is uniquely extended to an isometry φ^* from X_r onto X.

In this sense we have established a correspondence between elements of $\mathcal{X}$ with those of $\mathbb{X}$, and this correspondence induces a standard Borel structure (in fact a Polish topological structure) on $\mathcal{X}$.

We now turn to the second method of endowing $\mathcal{X}$ with a standard Borel structure following the approach of Gao and Kechris [58]. This method is based on the following more sophisticated observations. There is a universal Polish metric space, for instance, the universal Urysohn space $\mathbb{U}$, in which every Polish metric space is isometrically embedded into (see Section 1.2); the Effros Borel space $F(\mathbb{U})$ is standard Borel (see Section 1.4). Thus in an informal sense we may let elements of $F(\mathbb{U})$ represent the elements of $\mathcal{X}$. However, to pin down the standard Borel structure on $\mathcal{X}$ exactly, and to allow a proof that this Borel structure is equivalent to the one defined via $\mathbb{X}$, we need to well define the correspondence between $\mathcal{X}$ and $F(\mathbb{U})$.

This turns out to be highly nontrivial, as the following discussion shows. In the following we need the reader to be familiar with Section 1.2.

Let a Polish metric space (X, d) be given, together with an enumeration $(x_i)_{i \in \omega}$ of a canonical countable dense set D_X. We need to associate with X a particular closed subset F_X of $\mathbb{U}$. Recall from Section 1.2 that for each separable metric space (Y, δ), a sequence $(Y_n, \delta_n)_{n \leq \omega}$ of consecutive extensions of (Y, δ) is defined:

$$(Y_0, \delta_0) = (Y, \delta),$$

$$(Y_{n+1}, \delta_{n+1}) = E(Y_n, \omega),$$

$$(Y_\omega, \delta_\omega) = \bigcup_{n \in \omega}(Y_n, \delta_n),$$

with the completion of $(Y_\omega, \delta_\omega)$ isometric to $\mathbb{U}$. In particular, $\mathbb{U}$ is the completion of $(\mathbb{R}_\omega, d_\omega)$ where d is the usual metric on $\mathbb{R}$. Thus given X above we may follow the constructions of $(X_n)_{n \leq \omega}$ to obtain an isometric embedding from X into $\mathbb{U}$.

We claim that every step of the construction can be carried out canonically given the enumeration of the countable dense set D_X. To see this, we need to verify that a canonical countable dense subset can be obtained for each step of the construction, and that a canonical isometry between X_ω and $\mathbb{U}$ can be obtained with a canonical countable dense subset of X_ω.

For the first part, it suffices to note that if a separable metric space (Y, δ) and a countable dense subset D are given, then the set

$$E(Y, D, \omega) =_{\text{def}} \{f \in E(Y) : \text{ the support of } f \text{ is a finite subset of } D\}$$

is a countable dense subset of $E(Y, \omega)$. Moreover under the canonical isometric embedding $x \mapsto f_x$ from Y to $E(Y, \omega)$, D is a subset of $E(Y, D, \omega)$. Thus $E(Y, D, \omega)$ is a canonical countable dense subset of $E(Y, \omega)$. Since every element of $E(Y, D, \omega)$ is completely determined by its support, we also obtain a canonical enumeration of $E(Y, D, \omega)$ by canonically enumerating all finite sequences in D. Applying this observation for all X_n, we obtain a canonical enumeration of a canonical countable dense subset of X_ω, which is also dense in the completion of X_ω. In the same fashion, we also obtain a canonical enumeration of a countable dense subset of $\mathbb{R}_\omega$ that is also dense in $\mathbb{U}$.

For the second part of the claim, that a canonical isometry between the completion of X_ω and $\mathbb{U}$ can be obtained from canonical countable dense subsets of X_ω and $\mathbb{R}_\omega$ respectively, simply note that the proof of Theorem 1.2.5 relies only on countable dense subsets of two complete spaces with the Urysohn property.

Therefore we have associated a closed subset of $\mathbb{U}$ to every element of $\mathcal{X}$ presented to us. We note two properties of this associated map, that it is one-to-one, and that it is *not* onto $F(\mathbb{U})$. To see that it is one-to-one, note that if two distinct elements X and X' of $\mathcal{X}$ are given, then the two canonical enumerations (x_i) and (x_i') are distinct, and there are $i, j \in \omega$ such that $d(x_i, x_j) \neq d(x_i', x_j')$. By our construction above, the canonical countable dense subsets of X_ω and X_ω' contain the given ones respectively, hence X_ω and X_ω' continue to be distinct elements of $\mathcal{X}$. Finally the proof of Theorem 1.2.5 is carried out such that if the two countable dense subsets of X_ω and X_ω' respectively are distinct then so are the resulting isometries between the respective spaces and $\mathbb{U}$. To see that the associated map is not onto, note that X_ω contains X as a proper subset, and therefore $\mathbb{U}$ itself is not in the image.

At this moment we are still short of a satisfactory correspondence between $\mathcal{X}$ and $F(\mathbb{U})$. But for the record we have given all the essential ingredients for a proof of the following theorem.

Theorem 14.1.1
There is a Borel embedding J from $\mathbb{X}$ into $F(\mathbb{U})$ such that for any $r \in \mathbb{X}$, X_r is isometric with $J(r)$.

Proof. From the above discussions we obtain the following two Borel functions. Corresponding to the process of obtaining X_ω from X we have a function $E : \mathbb{X} \to \mathbb{X}$ such that for any $r \in \mathbb{X}$, $D_{E(r)}$ is the canonical countable dense subset of $(X_r)_\omega$. In fact there is a fixed injection $e : \omega \to \omega$ such that for any $r \in \mathbb{X}$ and $i, j \in \omega$, $r_{i,j} = E(r)_{e(i),e(j)}$. This extension function E is clearly Borel since each of its entries is defined from the entries of r using finitely many supremum operations (as it is a value of some d_n in X_n).

Corresponding to the next step of the construction, namely obtaining an isometry between the completion of X_ω and $\mathbb{U}$ by the proof of Theorem 1.2.5, we have a function $I : \mathbb{X} \to \mathbb{U}^\omega$ such that for any $r \in \mathbb{X}$, if D_r has the Urysohn property, then $I(r)$ enumerates a countable dense subset of $\mathbb{R}_\omega$. Here $I(r)(n)$ is in fact the element of $\mathbb{U}$ corresponding to the n-th element in D_r under the resulting isometry. Since this function is obtained by following the back-and-forth argument in the proof of Theorem 1.2.5, we have that for any n, $I(r)(n)$ is the limit of a sequence (x_m), where each x_m is an element of the canonical countable dense subset of $\mathbb{R}_\omega$. Furthermore the requirement that each x_m satisfies is a Borel condition involving the entries of r. It follows that each $I(r)(n)$ is a Borel function from X to $\mathbb{U}$, and hence I is Borel.

Now the embedding $J : X \to F(\mathbb{U})$ is defined, for each $r \in X$, by

$$J(r) = \text{the closure of } \{IE(r)(n) : n \in e(\omega)\} \text{ in } \mathbb{U}.$$

It only remains to check that J is Borel. For this let $U \subseteq \mathbb{U}$ be an open set. We have

$$r \in J^{-1}(\{F \in F(\mathbb{U}) : F \cap U \neq \emptyset\})$$

$$\Longleftrightarrow J(r) \cap U \neq \emptyset$$

$$\Longleftrightarrow \exists n \in e(\omega) \; IE(r)(n) \in U,$$

which is Borel since both I and E are Borel. □

It follows from the above theorem that $J(X)$ is a Borel subset of $F(\mathbb{U})$ and therefore a standard Borel space. Also from the definition of J and earlier discussions we may correspond the elements of $\mathcal{X}$ with those of $J(X)$ and thus obtain a standard Borel structure on $\mathcal{X}$. This standard Borel structure is certainly equivalent to the one induced by X by the above theorem.

Now the following theorems finally establish a satisfactory correspondence between $\mathcal{X}$ and $F(\mathbb{U})$.

Theorem 14.1.2
There is a Borel embedding j from $F(\mathbb{U})$ into $J(X)$ such that for any $F \in F(\mathbb{U})$, F is isometric with $j(F)$.

Proof. We let $X = \mathbb{U}$ be given with the canonical countable dense subset of $\mathbb{R}_\omega$, and obtain $r \in X$ to code X. By the constructions of Theorem 14.1.1, we obtain the extension X_ω and an isometry between the completion of X_ω with $\mathbb{U}$ again. Let $j(\mathbb{U})$ be the image of X in $\mathbb{U}$ following the construction. In symbols, $j(\mathbb{U}) = J(r)$. For each $F \in F(\mathbb{U})$, the construction yields a closed subset of $J(r)$ in a natural sense, and we let $j(F)$ be this closed subset of $J(r)$. It is clear that the map j is a Borel embedding. By definition, $j(F)$ is isometric with F for any $F \in F(\mathbb{U})$. □

Theorem 14.1.3
There is a Borel isomorphism Θ between $J(X)$ and $F(\mathbb{U})$ such that for any $r \in X$, X_r is isometric with $\Theta(r)$.

Proof. Let j be the Borel embedding in Theorem 14.1.2. The identity map i from $J(X)$ into $F(\mathbb{U})$ obviously satisfies that $i(F)$ is isomorphic to F for all $F \in J(X)$. By the usual proof of the Cantor–Bernstein Theorem 1.3.3, a

Borel isomorphism Θ can be obtained from the Borel embeddings i and j so that the isometry types of the elements are preserved. ∎

Now the correspondence between $\mathcal{X}$ and $J(\mathbb{X})$, composed with the Borel isomorphism Θ, finally gives a correspondence between $\mathcal{X}$ and $F(\mathbb{U})$.

Of course it follows immediately from Theorem 14.1.3 that $\Theta \circ J$ is a Borel isomorphism between $\mathbb{X}$ and $F(\mathbb{U})$ preserving the isometry types of elements. The existence of such an isomorphism means that the two approaches to equip $\mathbb{X}$ with a standard Borel structure are equivalent.

Note that it is mathematically nontrivial to prove this equivalence. However, in practice, whenever we have different approaches to equip standard Borel structures on hyperspaces such as $\mathcal{X}$, they always end up to be equivalent despite the difficulty of the proof. We summarize this phenomenon and generalize it philosophically in the following statement:

> For any hyperspace $\mathcal{H}$ of mathematical structures, if $\mathcal{B}_1$ and $\mathcal{B}_2$ are two natural standard Borel structures on $\mathcal{H}$, then there is a Borel isomorphism Ψ between $(\mathcal{H}, \mathcal{B}_1)$ and $(\mathcal{H}, \mathcal{B}_2)$ such that for any space X in $\mathcal{H}$, $\Psi(X)$ and X are isomorphic as mathematical structures.

This is apparently a philosophical statement rather than a mathematical one, since there is no mathematical way to define naturalness of a standard Borel structure. For readers familiar with computability theory, the statement is similar to the Church–Turing Thesis, which states that any two natural ways to define computability are equivalent. Such statements can be refuted mathematically, that is, if two different definitions are given and are proven inequivalent, and it is accepted that both definitions are natural. While there is no way to confirm such a statement, there might be mathematical theorems which can attest to its plausibility. One reason that the Church–Turing Thesis is widely accepted now is the practical need to reduce the work done and presented to establish its concrete instances again and again. To a large extent such work is mathematically insignificant and irrelevant to the understanding of the main problem. For our purpose the objective is to understand the complexity of various classification problems for, say, Polish metric spaces. While the theorems in this section are nontrivial and their proofs contain important ideas that can be used later to tackle the classification problems, the statements of the theorems themselves are not helping in our understanding of the complexity of the classification problems.

Exercise 14.1.1 Let $\mathbb{X}_{\mathrm{cpt}}$ be the set of all $r \in \mathbb{X}$ such that X_r is compact. Show that $\mathbb{X}_{\mathrm{cpt}}$ is a Borel subset of $\mathbb{X}$, and therefore a standard Borel space.

Exercise 14.1.2 Let $K(\mathbb{U})$ be the set of all $F \in F(\mathbb{U})$ such that F is compact. Show that $K(\mathbb{U})$ is a Borel subset of $F(\mathbb{U})$, and therefore a standard Borel space.

The next three exercise problems discuss a natural map from $F(\mathbb{U})$ into $\mathbb{X}$ preserving isometry types.

Exercise 14.1.3 Let X be a Polish space and $\{U_n\}_{n\in\omega}$ a countable base for X. Show that for each $n \in \omega$, the map $F \mapsto F \cap \overline{U_n}$ from $F(X)$ to $F(X)$ is Borel. (Note that in general the intersection operation on $F(X)$ is not Borel.)

Exercise 14.1.4 Let $\{U_n\}_{n\in\omega}$ be a countable base for $\mathbb{U}$. Let $s : F^*(\mathbb{U}) \to \mathbb{U}$ be a selection function given by Theorem 1.4.6, where $F^*(\mathbb{U}) = \{F \in F(\mathbb{U}) : F \neq \emptyset\}$. For each $n \in \omega$ define $p_n : F(\mathbb{U}) \to \mathbb{U}$ by

$$p_n(F) = \begin{cases} s(F \cap \overline{U_n}), & \text{if } F \cap U_n \neq \emptyset, \\ s(F), & \text{otherwise.} \end{cases}$$

Show that each p_n is a Borel function and for any $F \in F^*(\mathbb{U})$, $\{p_n(F)\}_{n\in\omega}$ is dense in F.

Exercise 14.1.5 Let $F^\infty(\mathbb{U}) = \{F \in F(\mathbb{U}) : F \text{ is infinite}\}$.

(a) Show that $F^\infty(\mathbb{U})$ is a Borel subset of $F(\mathbb{U})$.

(b) Using Exercise 14.1.4 define a Borel map $k : F^\infty(\mathbb{U}) \to \mathbb{X}$ such that for any $F \in F^\infty(\mathbb{U})$, $X_{k(F)}$ is isometric with F.

(c) Show that k is not an embedding.

14.2   Classification versus nonclassification

In mathematics, a classification problem is associated with a hyperspace of mathematical structures and a notion of equivalence. As we did in the preceding section, if the hyperspace can be turned into a standard Borel space X, then the equivalence notion becomes an equivalence relation E on the standard Borel space X. In our example, the space $\mathcal{X}$ of all Polish metric spaces have been equipped with a standard Borel structure. The isometric classification of Polish metric spaces is thus an equivalence relation on $\mathcal{X}$, which we denote by $\cong_i$. In this section we discuss classification and nonclassification results in mathematics.

We first give an example of a complete and satisfactory solution to a classification problem. The problem is the isometric classification of compact metric spaces. Let $\mathcal{X}_{\text{cpt}}$ be the hyperspace of all compact metric spaces. Since $\mathcal{X}_{\text{cpt}}$ is a subspace of $\mathcal{X}$, to see that it is a standard Borel space it suffices to show that it is a Borel subset of $\mathcal{X}$, which was done in Exercises 14.1.1 and 14.1.2 for both approaches to equip standard Borel structures on $\mathcal{X}$. In particular, recall that for any Polish space X, the space $K(X)$ of all compact subsets of X form a standard Borel space with the Effros Borel structure. We will use this fact in the proof of the following theorem.

Theorem 14.2.1 (Gromov)
The isometric classification problem for compact metric spaces is smooth.

Proof. Given a compact metric space X, for any $n \in \omega$ we define a map $M_n : X^{n+1} \to \mathbb{R}^{(n+1)^2}$ by

$$M_n(x_0, \ldots, x_n) = (d(x_i, x_j))_{0 \le i,j \le n}.$$

Then M_n is a continuous function from X^{n+1} into $\mathbb{R}^{(n+1)^2}$. Since X is compact, so is X^{n+1} and its image $D_n(X^{n+1})$ as a subset of $\mathbb{R}^{(n+1)^2}$. We let $M(X)$ be the sequence $(M_n(X^{n+1}))_{n \in \omega}$. Then M is a map from $\mathcal{X}_{\mathrm{cpt}}$ into the product space

$$Y = \prod_{n \in \omega} K(\mathbb{R}^{(n+1)^2}).$$

We note that M is a Borel map. To see this, we use the correspondence between $\mathcal{X}$ and $\mathbb{X}$. For any $X \in \mathcal{X}_{\mathrm{cpt}}$ let D_X, a canonical countable dense subset of X, and an enumeration (x_i) of D_X be given. Then for any $n \in \omega$, D_X^{n+1} is dense in X^{n+1}, and therefore $M_n(X^{n+1})$ is the closure of $M_n(D_X^{n+1})$ in $\mathbb{R}^{(n+1)^2}$. Now for any open $U \subseteq \mathbb{R}^{(n+1)^2}$,

$$M_n(X^{n+1}) \cap U \ne \emptyset \iff M_n(D_X^{n+1}) \cap U \ne \emptyset$$

$$\iff \exists i_0, \ldots, i_n \in \omega \; M_n(x_{i_0}, \ldots, x_{i_n}) \in U.$$

This shows that M is Borel as a function from $\mathcal{X}_{\mathrm{cpt}}$ into Y.

To prove the theorem we show that for any $X, X' \in \mathcal{X}_{\mathrm{cpt}}$, $X \cong_i X'$ iff $M(X) = M(X')$. By the definition of M it is clear that $X \cong_i X'$ implies $M(X) = M(X')$. For the converse, assume $(X, d), (X', d')$ are compact metric spaces with $M(X) = M(X')$.

We first define an isometric embedding φ from X into X'. Let $D \subseteq X$ be countable dense and $(x_i)_{i \in \omega}$ be an enumeration of D. For each $n \in \omega$, $M_n(x_0, \ldots, x_n) \in M_n(X^{n+1}) = M_n(X'^{n+1})$, hence there are $y_0^n, \ldots, y_n^n \in X'$ such that $M_n(x_0, \ldots, x_n) = M_n(y_0^n, \ldots, y_n^n)$. Now consider the sequence $(y_0^n)_{n \in \omega}$. Since X' is compact there is a convergent subsequence of (y_0^n); this means that there is a subset $A_0 \subseteq \omega$ and $x_0' \in X'$ such that $\lim_{n \in A_0} y_0^n = x_0'$. Without loss of generality we may assume $0 \notin A_0$. By similar arguments we obtain an infinite sequence

$$A_0 \supsetneq A_1 \supsetneq A_2 \supsetneq \cdots$$

of subsets of ω and $x_0', x_1', \ldots$ of elements of X' such that $\inf A_m \notin A_{m+1}$ and $\lim_{n \in A_m} y_m^n = x_m'$ for all $m \in \omega$. Let $A = \{\inf A_m : m \in \omega\}$. Then $A - A_m$ is finite for each $m \in \omega$, and so $\lim_{n \in A} y_n^n = \lim_{n \in A_m} y_m^n = x_m'$. We claim that $d'(x_i', x_j') = d(x_i, x_j)$ for all $i, j \in \omega$. For this fix $i, j \in \omega$ and let $m = \max\{i, j\}$. Note that for any $n \geq m$, by $M_n(x_0, \ldots, x_n) = M_n(y_0^n, \ldots, y_n^n)$, we have that $d(x_i, x_j) = d'(y_i^n, y_j^n)$. By continuity we get that $d(x_i, x_j) = \lim_{n \in A} d'(y_i^n, y_j^n) = d'(x_i', x_j')$. Thus we may define $\varphi(x_i) = x_i'$. Then φ is an isometric embedding from D into X', which uniquely extends to an isometric embedding from X into X', still denoted by φ.

By symmetry we can also define an isometric embedding $\psi : X' \to X$. Thus $\psi \circ \varphi : X \to X$ is an isometric embedding of X into itself. It follows that $\psi \circ \varphi$ must be onto. To see this, let $\theta = \psi \circ \varphi$ and assume $x \in X - \theta(X)$. Let $r = d(x, \theta(X))$. Since $\theta(X)$ is compact, $r > 0$. Now consider the sequence

$$x, \ \theta(x), \ \theta^2(x), \ \ldots.$$

Note that $d(x, \theta^n(x)) \geq r$ for all $n \geq 1$, and by isometry $d(\theta^m(x), \theta^n(x)) \geq r$ for all $n > m$. This implies that the sequence does not contain any Cauchy subsequences. But being a sequence in a compact space it contains a convergent subsequence, and therefore a Cauchy one, a contradiction. We thus have that $\psi \circ \varphi$ is onto, which implies that both ψ and φ are onto. Hence φ is an isometry from X onto X', as required. □

It is trivial to see that there are perfectly many pairwise nonisometric compact metric spaces (even with two points). Thus as an equivalence relation $\cong_i$ for compact metric spaces is Borel bireducible with $\mathrm{id}(2^\omega)$. In the intuitive sense a complete classification only requires the determination of complete invariants for the objects in question, and therefore corresponds to a Borel reduction to a known equivalence relation. A satisfactory classification, on the other hand, should leave no room for significant improvement, and therefore can be interpreted as a natural equivalence relation that is Borel bireducible with the classification problem. When both these are achieved, we refer to the result as a **classification theorem**.

Next we turn to the full isometric classification problem for all Polish metric spaces. Before we obtain a classification theorem for this problem we first show that it is essentially more complex than the isometric classification of compact metric spaces. Thus not only the Gromov invariants for compact metric spaces are no longer complete invariants for general Polish metric spaces, but also there is no way to expand the Gromov invariants to such complete invariants. It has to be that the *nature* of the complete invariants for general Polish metric spaces is different than that of the Gromov invariants. We therefore refer to a result such as the following one as a **nonclassification theorem**.

Theorem 14.2.2
The graph isomorphism is Borel reducible to $\cong_i$. In particular, the isometric classification of Polish metric spaces is not smooth.

Proof. By Theorem 13.2.1 the graph isomorphism is Borel bireducible with the isomorphism relation for countable connected graphs. Thus it suffices to define a Borel reduction from the latter to $\cong_i$.

For any countable connected graph Γ, we define a metric space $S(\Gamma)$ on Γ with the metric given by the geodesic distance:

$$d_S(x,y) = \text{ the length of the shortest path from } x \text{ to } y.$$

$S(\Gamma)$ is countable and discrete, hence it is Polish and d_S is complete. Γ can be recovered from $S(\Gamma)$ since $xy \in \Gamma$ iff $d_S(x, y) = 1$. It is easy to see that $\Gamma \mapsto S(\Gamma)$ is a Borel reduction from $\cong$ to $\cong_i$. □

With the background of theory of equivalence relations, this simple theorem now has far-reaching consequences. For instance, it implies that $\cong_i$ is non-Borel and is above the Friedman–Stanley tower. It now becomes a simple abstract argument to see that, for instance, a countable set of compact metric spaces can not be coded by a simple compact metric space, but can be coded by a Polish metric space.

Invariant descriptive set theory makes it possible to prove nonclassification theorems by proving *reductions*. In the usual practice of mathematics, non-classification is usually suggested but seldom proved; here, as we saw above, nonclassification theorems are precise mathematical theorems.

In the rest of this section we give a complete classification for Polish metric spaces up to isometry. The equivalence relation we use as complete invariants is an Iso($\mathbb{U}$)-orbit equivalence relation. More precisely, we consider the natural Iso($\mathbb{U}$) action on the standard Borel space $F(\mathbb{U})$, and denote the induced orbit equivalence relation by E_I.

Theorem 14.2.3 (Gao–Kechris)
$\cong_i \leq_B E_I$. That is, the isometric classification of Polish metric spaces is Borel reducible to the Iso($\mathbb{U}$)-orbit equivalence relation on $F(\mathbb{U})$.

Proof. The reduction function is the map J defined in Theorem 14.1.1. We showed that it is a Borel embedding from $\mathcal{X}$ into $F(\mathbb{U})$. Here we check that for Polish metric spaces X and X', $X \cong_i X'$ iff $J(r_X) E_I J(r_{X'})$. First assume $J(r_X) E_I J(r_{X'})$, that is, for some $\varphi \in \text{Iso}(\mathbb{U})$, $\varphi(J(r_X)) = J(r_{X'})$. Then in particular, $J(r_X) \cong_i J(r_{X'})$. By Theorem 14.1.1, $X \cong_i J(r_X)$ and $X' \cong_i J(r_{X'})$. Thus by transitivity we have $X \cong_i X'$. For the other direction, assume $X \cong_i X'$. Then from the definition of J, we have isometries ψ between the completion of X_ω and $\mathbb{U}$ and ψ' between the completion of X'_ω and $\mathbb{U}$. Now if π is an isometry between X and X', by the construction of X_ω and X'_ω there is an extension π^* between X_ω and X'_ω such that $\pi^* \restriction X = \pi$. π^* can be uniquely extended to an isometry between the completions of X_ω and X'_ω, which we continue to denote by π^*. Then $\psi' \circ \pi^* \circ \psi^{-1}$ is an element of

Iso($\mathbb{U}$) so that

$$\psi' \circ \pi^* \circ \psi^{-1}(J(r_{X'})) = \psi' \circ \pi^*(X) = \psi'(X') = J(r_{X'}).$$

This shows that $J(r_X)E_I J(r_{X'})$. ☐

In the next section we will show in a precise sense that there is no room to improve this classification significantly. Thus, regardless of the reader's familiarity with the equivalence relation used as complete invariants, the theorem cannot be materially improved with the use of another kind of invariant.

Exercise 14.2.1 Show that the isometric biembeddability between compact metric spaces is a smooth equivalence relation.

Exercise 14.2.2 Give a direct coding of countable sets of compact metric spaces by single Polish metric spaces, that is, associate with each countable set $\{X_n\}_{n\in\omega}$ of pairwise nonisometric compact metric spaces a Polish metric space X so that $X \cong_i X'$ iff there is a permutation π of ω with $X_n \cong_i X'_{\pi(n)}$ for all $n \in \omega$.

The following three problems give an alternative proof of Gromov's Theorem 14.2.1.

Exercise 14.2.3 Let $(X, d_X), (Y, d_Y)$ be compact metric spaces. Suppose that for any $\epsilon > 0$, $n \in \omega$, and $x_0, \ldots, x_n \in X$, there are $y_0, \ldots, y_n \in Y$ such that $|d(x_i, x_j) - d(y_i, y_j)| < \epsilon$. Show that there is an isometric embedding from X into Y.

Exercise 14.2.4 Let $K(\mathbb{U})$ be the space of all compact subsets of $\mathbb{U}$ with the Vietoris topology (see Section 1.1 and Exercise 1.4.4). Show that every $\cong_i$-equivalence class is closed.

Exercise 14.2.5 For compact metric spaces X, Y, the **Gromov–Hausdorff metric** $d_G(X, Y)$ is defined as the infimum of $d_Z(\varphi(X), \psi(Y))$, where Z is a metric space, $\varphi : X \to Z$ and $\psi : Y \to Z$ are isometric embeddings, and d_Z is the Hausdorff metric on $K(Z)$. Show that for $X, Y \in K(\mathbb{U})$,

$$d_G(X, Y) = \inf\{d_H(X', Y') : X' \cong_i X, Y' \cong_i Y\},$$

where d_H is the Hausdorff metric on $K(\mathbb{U})$.

14.3   Measurement of complexity

At first sight the equivalence relation E_I considered in Theorem 14.2.3 is far from canonical as complete invariants for Polish metric spaces. However, in this section we will show that E_I is universal for all orbit equivalence relations induced by Borel actions of Polish groups. Moreover, we will show that the classification problem has the same complexity as that of E_I, that is, that they are Borel bireducible to each other. Putting these results together, we obtain a complete classification for Polish metric spaces up to isometry that cannot be improved; and the complete invariants we use have a characteristic property making them in a sense canonical.

We now start to present a proof that any orbit equivalence relation is Borel reducible to $\cong_i$. This recent theorem was proved by Gao and Kechris, and also independently by Clemens (see References [22] and [58]). We present Clemens' proof here since it is more elementary.

Let G be a Polish group and X a Borel G-space. We will associate with each element $x \in X$ a Polish metric space M_x so that $x E_G^X y$ iff $M_x \cong_i M_y$. First, in view of Exercises 3.3.6 and 3.1.9 we may assume X is a compact Polish G-space. We will need a special compatible metric on G given by the following lemma.

Lemma 14.3.1
Let G be a Polish group and X a compact Polish G-space. Let $d_X \leq 1$ be a compatible metric on X. Then there is a left-invariant compatible metric $d_G \leq 1$ on G such that, for any $x \in X$ and $g, h \in G$,

$$d_G(g, h) \geq \frac{1}{2} d_X(g^{-1} \cdot x, h^{-1} \cdot x).$$

Proof.   Let $d_0 \leq 1$ be a compatible left-invariant metric on G. Define

$$d_G(g, h) = \frac{1}{2} d_0(g, h) + \frac{1}{2} \sup\{d_X(g^{-1} \cdot x, h^{-1} \cdot x) : x \in X\}.$$

It is easy to see that $d_G \leq 1$ is a left-invariant metric on G satisfying $d_G(g, h) \geq \frac{1}{2} d_X(g^{-1} \cdot x, h^{-1} \cdot x)$ for all $x \in X$. It only remains to check that d_G is a compatible metric on G. Since $d_G(g, h) \geq \frac{1}{2} d_0(g, h)$ for all $g, h \in G$, the topology induced by d_G is finer than that induced by d_0. For the converse let $\epsilon > 0$. By left-invariance of d_0 and d_G it suffices to find $\delta > 0$ such that, whenever $d_0(g, 1_G) < \delta$, we have $d_G(g, 1_G) < \epsilon$.

Since the action is continuous as a function from $G \times X$ to X, we have that for any $x \in X$, there is $0 < \delta < \epsilon/2$ such that for all $g \in G$ and $y \in X$ with $d_0(g, 1_G) < \delta_x$ and $d_X(y, x) < \delta_x$, we have $d_X(g^{-1} \cdot y, x) < \epsilon/2$. It follows that for such g and y,

$$d_X(g^{-1} \cdot y, y) \leq d_X(g^{-1} \cdot y, x) + d_X(y, x) < \frac{\epsilon}{2} + \frac{\epsilon}{2} = \epsilon.$$

By compactness of X there are finitely many points $x_0, \ldots, x_n \in X$ such that for all $y \in X$, there is $i \leq n$ with $d_X(y, x_i) < \delta_{x_i}$. Let $\delta = \inf\{\delta_{x_0}, \ldots, \delta_{x_n}\}$. Then for all $g \in G$ with $d_0(g, 1_G) < \delta$ and all $y \in X$, $d_X(g^{-1} \cdot y, y) < \epsilon$; hence

$$d_G(g, 1_G) = \frac{1}{2} d_0(g, 1_G) + \frac{1}{2} \sup\{d_X(g^{-1} \cdot y, y) : y \in X\} < \frac{1}{2}\epsilon + \frac{1}{2}\epsilon = \epsilon.$$

$\square$

Now fix a compatible metric $d_X \leq 1$ on X and a compatible left-invariant metric $d_G \leq 1$ on G given by Lemma 14.3.1. Without loss of generality we may assume that there are elements $x, y \in X$ with $d_X(x, y) = 1$. Also we may assume that $\sup\{d_G(g, h) : g, h \in G\} = 1$. By left-invariance this implies that for any $g \in G$, $\sup\{d_G(g, h) : h \in G\} = 1$. Fix a countable dense subset D of X, and let $(x_n)_{n \in \mathbb{Z}}$ be an enumeration of D. We are now ready to define the Polish metric spaces M_x for each $x \in X$.

Figure 14.1 The space $H = G \times \mathbb{Z} \times \{0, 1\}$.

Let $H = G \times \mathbb{Z} \times \{0, 1\}$. Let $\pi : \mathbb{Z} \to \omega$ be the bijection defined as

$$\pi(n) = \begin{cases} 2n, & \text{if } n \geq 0, \\ -2n - 1, & \text{otherwise.} \end{cases}$$

Define a metric d_x on H by letting, for $(g_1, n_1, i_1), (g_2, n_2, i_2) \in H$,

$d_x((g_1, n_1, i_1), (g_2, n_2, i_2))$

$$= \begin{cases} d_G(g_1, g_2), & \text{if } n_1 = n_2 \text{ and } i_1 = i_2, \\[2mm] \dfrac{3}{2} + 4^{-|n_1 - n_2|}(1 + d_G(g_1, g_2)), & \text{if } n_1 \neq n_2 \text{ and } i_1 = i_2, \\[2mm] 1 + 4^{-\pi(n_1 - n_2) - 1}(1 + d_X(x_{n_1 - n_2}, g_2^{-1} \cdot x)), & \text{if } i_1 = 0 \text{ and } i_2 = 1, \\[2mm] 1 + 4^{-\pi(n_2 - n_1) - 1}(1 + d_X(x_{n_2 - n_1}, g_1^{-1} \cdot x)), & \text{if } i_1 = 1 \text{ and } i_2 = 0. \end{cases}$$

It is straightforward to check that d_x is a metric (see Exercise 14.3.1). With this definition each H can be viewed as the union of countably many copies of G indexed by elements of $\mathbb{Z} \times \{0,1\}$, as illustrated in Figure 14.1. Note that for any two elements in different copies of G, their distance is greater than 1, while the metric d_x on each copy of G coincides with $d_G \leq 1$.

Let $\hat{G}$ be the completion of G with d_G, and continue to denote the metric by d_G. Define M_x to be the completion of H with d_x, and we continue to denote the metric on M_x by d_x. Then M_x has the underlying set

$$\hat{H} = \hat{G} \times \mathbb{Z} \times \{0,1\},$$

which is a union of countably many copies of $\hat{G}$ indexed by elements of $\mathbb{Z} \times \{0,1\}$, while the completed metric d_x on each copy of $\hat{G}$ coincides with the completed metric d_G.

We note that the map $x \mapsto M_x$ is a Borel function from X into $\mathcal{X}$. To see this just note that a canonical countable dense subset D_x of M_x can be obtained from any canonical countable dense subset of G, and the distance between two points in D_x is explicitly defined in the definition of d_x, which is clearly Borel.

Lemma 14.3.2
If $x E_G^X y$ then $M_x \cong_i M_y$.

Proof. Suppose $y = h \cdot x$. We check that the map $(g, n, i) \mapsto (hg, n, i)$ is an isometry from (H, d_x) onto (H, d_y). Since this isometry extends uniquely to an isometry from M_x onto M_y, we have $M_x \cong_i M_y$. The map is obviously a permutation of H. We only need to verify that it preserves the metric. Among the four cases of the definition of d_x, the first two cases follow easily by left-invariance of d_G. By the symmetry of the remaining cases, we only show one of them. So consider two points $(g_1, n_1, i_1), (g_2, n_2, i_2) \in H$ with $i_1 = 0$ and $i_2 = 1$. Then

$$d_x((g_1, n_1, i_1), (g_2, n_2, i_2))$$

$$= 1 + 4^{-\pi(n_1 - n_2) - 1}(1 + d_X(x_{n_1 - n_2}, g_2^{-1} \cdot x))$$

$$= 1 + 4^{-\pi(n_1 - n_2) - 1}(1 + d_X(x_{n_1 - n_2}, (hg_2)^{-1} \cdot y))$$

$$= d_y((hg_1, n_1, i_1), (hg_2, n_2, i_2)).$$

▯

Lemma 14.3.3
If $M_x \cong_i M_y$ then $x E_G^X y$.

Proof. Suppose $\varphi : M_x \cong_i M_y$. Since M_x contains countably many copies of $\hat{G}$, with the diameter of each copy at most 1 and distances between points from different copies greater than 1, it follows that φ must send each copy of $\hat{G}$ in M_x onto a copy of $\hat{G}$ in M_y. This means that φ induces a bijection $f : \mathbb{Z} \times \{0,1\} \to \mathbb{Z} \times \{0,1\}$ such that for all $(h,n,i) \in \hat{H}$, $\varphi(h,n,i) \in \hat{G} \times \{f(n,i)\}$.

For convenience we will use the following terminology to address subsets of $\hat{H}$. Call $\hat{G} \times \{(n,i)\}$ the (n,i)-copy of $\hat{G}$, the collection $\hat{G} \times \mathbb{Z} \times \{0\}$ the 0-chain of copies of $\hat{G}$, and $\hat{G} \times \mathbb{Z} \times \{1\}$ the 1-chain of copies of $\hat{G}$.

Now note further that points from different copies of $\hat{G}$ in the 0-chain have distance greater than $3/2$, whereas points from different chains have distance at most $3/2$. It follows that φ sends copies of $\hat{G}$ in the same chain in M_x to those in the same chain in M_y. Furthermore, for $i \in \{0,1\}$ and $n_1 \neq n_2 \in \mathbb{Z}$,

$$\sup\{d_x((h_1,n_1,i),(h_2,n_2,i)) \; : \; h_1, h_2 \in \hat{G}\} = \frac{3}{2} + 2 \cdot 4^{-|n_1 - n_2|},$$

since $\sup\{d_G(h_1,h_2) \; : \; h_1, h_2 \in \hat{G}\} = 1$. This implies the following rigidity property of φ: if $f(0,i) = (n_0, j)$, then either $f(m,i) = (n_0 + m, j)$ for all $m \in \mathbb{Z}$ or $f(m,i) = (n_0 - m, j)$ for all $m \in \mathbb{Z}$.

We now consider two cases. Case 1: $f(0,0) = (n_0, 0)$ for some $n_0 \in \mathbb{Z}$. It follows that $f(0,1) = (m_0, 1)$ for some $m_0 \in \mathbb{Z}$. However, if $n_0 \neq m_0$, then for $g_1, g_2 \in G$,

$$d_x((g_1,0,0),(g_2,0,1)) = 1 + 4^{-1}(1 + d_X(x_0, g_2^{-1} \cdot x)) \geq \frac{5}{4}$$

but

$$d_y((g_1,n_0,0),(g_2,m_0,1)) = 1 + 4^{-\pi(n_0-m_0)-1}(1 + d_X(x_{n_0-m_0}, g_2^{-1} \cdot y))$$

$$\leq 1 + 2 \cdot 4^{-\pi(n_0-m_0)-1} \leq \frac{9}{8}.$$

Thus we must have $m_0 = n_0$. Also for $g_1, g_2 \in G$ and any $m \in \mathbb{Z}$,

$$d_x((g_1,m,0),(g_2,0,1)) \in [1 + 4^{-\pi(m)-1}, 1 + 2 \cdot 4^{-\pi(m)-1}]$$

$$d_y((g_1,n_0+m,0),(g_2,n_0,1)) \in [1 + 4^{-\pi(m)-1}, 1 + 2 \cdot 4^{-\pi(m)-1}]$$

$$d_y((g_1,n_0-m,0),(g_2,n_0,1)) \in [1 + 4^{-\pi(-m)-1}, 1 + 2 \cdot 4^{-\pi(-m)-1}]$$

The first two intervals being the same and disjoint from the third, it follows that $f(m,0) = (n_0 + m, 0)$ for all $m \in \omega$. Since G is comeager in $\hat{G}$, it follows that there is a comeager subset C of G such that for all $g \in C$, and for all $m \in \omega$, $\varphi(g,m,0) \in H$ and $\varphi(g,0,1) \in H$. Let $g_1, g_2 \in C$. We have for all $m \in \omega$,

$$d_x((g_1,m,0),(g_2,0,1)) = 1 + 4^{-\pi(m)-1}(1 + d_X(x_m, g_2^{-1} \cdot x)).$$

Let $h_{1,m} \in G$ be such that $\varphi(g_1, m, 0) = (h_{1,m}, n_0 + m, 0)$, and let $h_2 \in G$ be such that $\varphi(g_2, 0, 1) = (h_2, n_0, 1)$. Then

$$d_x((g_1, m, 0), (g_2, 0, 1)) = d_y(\varphi(g_1, m, 0), \varphi(g_2, 0, 1))$$

$$= d_y((h_{1,m}, n_0 + m, 0), (h_2, n_0, 1)) = 1 + 4^{-\pi(m)-1}(1 + d_X(x_m, h_2^{-1} \cdot y)).$$

It follows that for all $m \in \omega$, $d_X(x_m, g_2^{-1} \cdot x) = d_X(x_m, h_2^{-1} \cdot y)$, and hence $g_2^{-1} \cdot x = h_2^{-1} \cdot y$. Thus $y = h_2 g_2^{-1} \cdot x$ and $y E_G^X x$.

Case 2: $f(0,0) = (n_0, 1)$ for some $n_0 \in \mathbb{Z}$. Then a similar argument shows that $f(0,1) = (n_0, 0)$ and $f(m,1) = (n_0 - m, 0)$ for all $m \in \mathbb{Z}$. Again there are $g_1, h_{1,m}, g_2, h_2 \in G$ such that $\varphi(g_1, m, 1) = (h_{1,m}, n_0 - m, 0)$ and $\varphi(g_2, 0, 1) = (h_2, n_0, 0)$. Thus

$$d_x((g_1, m, 1), (g_2, 0, 0)) = d_y(\varphi(g_1, m,), \varphi(g_2, 0, 0))$$

$$= d_y((h_{1,m}, n_0 - m, 0), (h_2, n_0, 1)) = 1 + 4^{-\pi(m)-1}(1 + d_X(x_m, h_2^{-1} \cdot y)),$$

from which it follows as before that $y E_G^X x$. □

We have thus proved the following theorem.

Theorem 14.3.4 (Clemens–Gao–Kechris)
Let G be a Polish group and X a Borel G-space. Then $E_G^X \leq_B \cong_i$.

Recall that a universal orbit equivalence relation is one of the form E_G^X, where G is a Polish group and X a Borel G-space, such that for all Polish groups H and Borel H-space Y, $E_H^Y \leq_B E_G^X$. The following corollary is now immediate.

Corollary 14.3.5 (Gao–Kechris)
E_I is a universal orbit equivalence relation.

Proof. Note that E_I is an orbit equivalence relation. By Theorems 14.2.3 and 14.3.4, every orbit equivalence relation is Borel reducible to E_I. □

Corollary 14.3.6 (Gao–Kechris)
The isometric classification for Polish metric spaces is Borel bireducible to the universal orbit equivalence relation.

Proof. It follows from Theorem 14.3.4 that $E_I \leq_B \cong_i$, and thus $\cong_i \sim_B E_I$. □

Our result is therefore a proper classification theorem for the isometric problem of Polish metric spaces, and the universality property of the equivalence relation E_I makes it a benchmark equivalence relation.

Exercise 14.3.1 Show that d_x defined on the set H is a metric.

Exercise 14.3.2 Let G be a cli Polish group, that is, a Polish group with a compatible complete left-invariant metric and X a Polish G-space. Show that E_G^X is Borel reducible to the isometric classification of Polish metric groups.

Exercise 14.3.3 Let G be an abelian Polish group and X a Polish G-space. Show that E_G^X is Borel reducible to the isometric classification of abelian Polish metric groups.

Exercise 14.3.4 Let G be Iso($\mathbb{U}$) with a compatible left-invariant metric $d \leq 1$. Show that the G-orbit equivalence relation on $L(G, d)$ is a universal orbit equivalence relation.

14.4   Classification notions

In this section we discuss some different classification notions for Polish metric spaces.

The isometric classification is the finest notion of equivalence for metric spaces. In mathematics we are often interested in other classification notions for the same class of mathematical structures. In the case of metric spaces, we recall the following list of notion of equivalence, all considered natural and helpful in our understanding of the properties of Polish metric spaces.

Definition 14.4.1
Let (X, d_X) and (Y, d_Y) be metric spaces.

(1) We say that X and Y are **homeomorphic** if there is a bijection $f : X \to Y$ such that both f and f^{-1} are continuous.

(2) We say that X and Y are **uniformly homeomorphic** if there is a bijection $f : X \to Y$ such that both f and f^{-1} are uniformly continuous.

(3) We say that X and Y are **Lipschitz isomorphic** if there is a bijection $f : X \to Y$ such that both f and f^{-1} are Lipschitz.

(4) We say that X and Y are **isometrically biembeddable** if there are isometric embeddings $f : X \to Y$ and $g : Y \to X$.

This definition does not exhaust all natural notions of equivalence for metric spaces. Note that uniform homeomorphism, Lipschitz isomorphism, and isometric biembeddability are still metric notions, but homeomorphism is a topological notion.

Each of the above classification notions corresponds to an equivalence relation on the standard Borel space $\mathcal{X}$ of all Polish metric spaces. One can check that the metric notions are Σ_1^1 (see Exercise 14.4.1). Here we show that the homeomorphism problem is Σ_2^1, as noted by Ferenczi, Louveau, and Rosendal [42]. Similar claims have been shown in Reference [21]. The proof below is due to Kechris.

Proposition 14.4.2
The homeomorphic classification for all Polish metric spaces is Σ_2^1.

Proof. We consider the homeomorphism relation $\approx$ on $F(\mathbb{U})$. We claim that, for closed subsets X, Y of $\mathbb{U}$, $X \approx Y$ iff

> there are countable dense subsets D_X, D_Y of X, Y, respectively, and a homeomorphism f between D_X and D_Y, such that for any sequence $(x_k)_{k \in \omega}$ in D_X, (x_k) is Cauchy iff $(f(x_k))$ is Cauchy.

To see this, suppose $X \approx Y$ and $\varphi : X \to Y$ is a homeomorphism. Let D_X be any countable dense subset of X. Then $\varphi(D_X)$ is dense in Y. Thus if we let $D_Y = \varphi(D_X)$ and $f = \varphi \restriction D_X$, φ is an extension of f to the completion of D_X, which is X, as required. Conversely, suppose the displayed property holds for $X, Y \in F(\mathbb{U})$. Then define $\varphi : X \to Y$ as follows. For any $x \in X$ let (x_k) be a Cauchy sequence in D_X with $\lim_k x_k = x$ and define $\varphi(x) = \lim_k f(x_k)$. The definition of φ does not depend on the choice of the Cauchy sequence (x_k), since if (x_k') is another Cauchy sequence in D_X converging to x, we may define a new Cauchy sequence (x_k'') by letting $x_{2k}'' = x_k$ and $x_{2k+1}'' = x_k'$ for all $k \in \omega$; then by the assumption $(f(x_k''))$ is Cauchy and hence $\lim_k f(x_k) = \lim_k f(x_k')$. A similar argument also shows that φ is a bijection between X and Y, and that φ is a homeomorphism. This finishes the proof of the claim. We next show that the displayed property is Σ_2^1.

Fix a countable base $\{U_n\}_{n \in \omega}$ for $\mathbb{U}$. Note that the relation $x \in X$ for $x \in \mathbb{U}$ and $X \in F(\mathbb{U})$ is Borel since

$$x \in X \iff \forall n \, (x \in U_n \to X \cap U_n \neq \emptyset).$$

Moreover, if $(x_k)_{k \in \omega} \in \mathbb{U}^\omega$ and $X \in F(\mathbb{U})$, then $\{x_k : k \in \omega\}$ is a dense subset of X iff

$$\forall k \in \omega \, x_k \in X \, \wedge \, \forall n \in \omega \, \exists k \in \omega \, (X \cap U_n \neq \emptyset \to x_k \in U_n),$$

and hence is Borel as well.

Let d be the metric on $\mathbb{U}$. For $(x_k)_{k \in \omega} \in \mathbb{U}^\omega$,

$$(x_k) \text{ is } d\text{-Cauchy} \iff \forall n \in \omega \, \exists m \in \omega \, \forall p, q \geq m \, d(x_p, x_q) < 2^{-n}$$

is a Borel relation. If $(x_k)_{k \in \omega}, (y_k)_{k \in \omega} \in \mathbb{U}^\omega$, define $f : \{x_k : k \in \omega\} \to \{y_k : k \in \omega\}$ by letting $f(x_k) = y_k$; then f is continuous iff

$$\forall k, n \in \omega \, \exists m \in \omega \, \forall l \in \omega \, (d(x_l, x_k) < 2^{-m} \to d(y_l, y_k) < 2^{-n}).$$

It follows that the relation that $x_k \mapsto y_k$ is a homeomorphism between $\{x_k : k \in \omega\}$ and $\{y_k : k \in \omega\}$ is Borel.

Finally for $X, Y \in F(\mathbb{U})$, we have that $X \approx Y$ iff

$$\exists (x_k)_{k\in\omega}, (y_k)_{k\in\omega} \in \mathbb{U}^\omega$$

$$[\{x_k : k \in \omega\} \text{ is a dense subset of } X \wedge \{y_k : k \in \omega\} \text{ is a subset of } Y$$

$$\wedge \ x_k \mapsto y_k \text{ is a homeomorphism between } \{x_k : k \in \omega\} \text{ and } \{y_k : k \in \omega\}$$

$$\wedge \ \forall z \in \omega^\omega \ ((x_{z(k)})_{k\in\omega} \text{ is } d\text{-Cauchy} \leftrightarrow (y_{z(k)})_{k\in\omega} \text{ is } d\text{-Cauchy })]$$

This shows that $\approx$ is $\mathbf{\Sigma}^1_2$. ⬜

It is not known that the statement of the proposition is optimal, but this computation of descriptive complexity suggests that the homeomorphism relation is much more complicated than all the metric notions.

For the rest of this section we consider the homeomorphic classification of compact metric spaces. We denote this homeomorphism equivalence relation by $\approx^c$. As a contrast to Proposition 14.4.2 we note that $\approx^c$ is $\mathbf{\Sigma}^1_1$.

Proposition 14.4.3
The homeomorphic classification for compact metric spaces is $\mathbf{\Sigma}^1_1$.

Proof. The main difference here is that continuous functions between compact metric spaces have much nicer properties. First, they are closed maps, which implies that any continuous bijection is a homeomorphism. Second, they are uniformly continuous, and therefore completely determined by restrictions on countable dense subsets.

To be more specific, let X and Y be compact metric spaces and D a countable dense subset of X. Then a function $f : X \to Y$ is continuous iff $f \restriction D$ is a uniformly continuous function from D into Y, and for a continuous function $f : X \to Y$, f is onto iff $f(D)$ is dense in Y. It follows that f is a homeomorphism from X onto Y iff $f \restriction D$ is one-to-one, $f(D)$ is dense, and $f \restriction D$ is uniformly continuous. These are now Borel conditions on $\mathcal{X}_{\mathrm{cpt}}$. ⬜

In fact, $\approx^c$ is Borel reducible to the universal orbit equivalence relation. But we will not prove it here.

In contrast to the Gromov Theorem 14.2.1 we show that $\approx^c$ is nonsmooth.

Proposition 14.4.4
The equivalence relation $=^+$ is Borel reducible to $\approx^c$. In particular, the homeomorphic classification of compact metric spaces is not smooth.

Proof.   First we define a reduction of certain finite splitting trees to compact metric spaces. Let $\mathcal{T}_0$ be the set of all finite splitting trees on ω such that the following conditions hold:

(i) $\emptyset$ has exactly three immediate successors,

(ii) every node in T has at least three immediate successors.

For each tree T in $\mathcal{T}_0$ we define a compact subset S_T of $\mathbb{R}^2$ as follows. First let $(0,1) \in S_T$. This point corresponds to $\emptyset \in T$. Then corresponding to the three nodes t_1, t_2, t_3 in T of length 1, we let $p_1 = (-2^{-1}, 2^{-1}), p_2 = (0, 2^{-1}), p_3 = (2^{-1}, 2^{-1}) \in S_T$ together with the line segments linking each of them with $(0,1)$. Next if t_1 has n immediate successors $s_1, \ldots, s_n$ then we find n points $q_1, \ldots, q_n$ with y-coordinate 2^{-2} and x-coordinates in $[-5/8, -3/8]$, add $q_1, \ldots, q_n$ to S_T, together with the line segment linking each of them with p_1. The construction up to this point is illustrated in Figure 14.2. Repeat

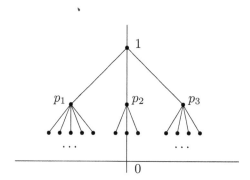

Figure 14.2   The compact metric space S_T.

this process for the rest of the nodes in T, and we obtain a subset of $\mathbb{R}^2$. The construction should be carried out that the set contains all of its accumulation points with nonzero y-coordinates. Finally let S_T be the closure of the resulting set, that is, by adding to the resulting set all of its accumulation points (and note that they all have zero y-coordinates). It is clear that S_T is path-connected.

In a natural sense we may regard T as a subset of S_T. Then the degree of nodes in T can be topologically characterized: each node of degree $d \geq 3$ becomes a point x in S_T with the property that $S_T - \{x\}$ has d many path-connected components. It is now easy to see that $T \cong T'$ iff $S_T \approx^c S_{T'}$.

By Theorem 13.2.3 the isomorphism on $\mathcal{T}_0$ is Borel bireducible to $\mathrm{id}(2^\omega)$. This implies that there is a Borel map $x \mapsto S_x$ from 2^ω into $\mathcal{X}_{\mathrm{cpt}}$ such that $S_x \not\approx^c S_y$ for $x \neq y$. Now if A is a countable subset of 2^ω, then we let S_A be the countable union of a copy of S_x for each $x \in A$. For each $x \in A$, S_x is homeomorphic to a connected component of S_A. It then follows easily that $A = A'$ iff $S_A \approx S_{A'}$. □

In fact Proposition 14.4.3 can be proved by many different constructions. We present this particular proof here to give the flavor of the subject and to show the power of invariant descriptive set theory in gaining new insight. The homeomorphic classification problem for compact metric spaces has been studied extensively in topology and descriptive set theory. However, its exact complexity is still unknown.

Exercise 14.4.1 Show that uniform homeomorphism, Lipschitz isomorphism, and isometric biembeddability are all Σ_1^1 equivalence relations on $\mathcal{X}$.

Exercise 14.4.2 Show that $=^+$ is Borel reducible to the homeomorphism classification of connected compact metric spaces.

Exercise 14.4.3 Give a Borel reduction from $\mathrm{id}(2^\omega)$ to the homeomorphism classification of 0-dimensional compact metric spaces.

Chapter 15

Summary of Benchmark Equivalence Relations

We have given in this book an introduction of definable equivalence relations. In decreasing order of comprehensiveness, we have treated the topics of orbit equivalence relations, general Borel equivalence relations, general $\boldsymbol{\Sigma}_1^1$ equivalence relations, and general $\boldsymbol{\Pi}_1^1$ equivalence relations. While these classes do not exhaust all definable equivalence relations, they are most relevant to many other areas of mathematics and results about them often do not go beyond the usual axioms of mathematics and set theory. Many equivalence relations we have considered have characteristic properties which give them distinct places in the hierarchy of Borel reducibility. This kind of canonicity makes them benchmark equivalence relations suitable for use in gauging the complexity of other equivalence relations and classification problems arising in mathematics. In this chapter we summarize these equivalence relations and mention classification problems with identical complexity.

15.1 Classification problems up to essential countability

On a very crude scale there are four benchmark equivalence relations up to the universal countable Borel equivalence relation, and they are

$$\mathrm{id}(\omega), \ \mathrm{id}(2^\omega), \ E_0, \ E_\infty.$$

We recall the basic facts we proved about the Borel reducibility hierarchy related to these equivalence relations.

Below $\mathrm{id}(\omega)$ are equivalence relations with finitely many equivalence classes, which are obviously determined by the cardinality of the quotient space. By the Silver dichotomy (Theorem 5.3.5) every $\boldsymbol{\Pi}_1^1$ equivalence relation is either at most $\mathrm{id}(\omega)$ or at least $\mathrm{id}(2^\omega)$ in the Borel reducibility quasiorder. This is no longer true for $\boldsymbol{\Sigma}_1^1$ equivalence relations (Section 9.1). The statement that it holds for orbit equivalence relations is the topological Vaught conjecture and is one of the major open problems of the entire field. The Glimm–Effros dichotomy proved by Harrington–Kechris–Louveau (Theorem 6.3.1) states that every Borel equivalence relation is either at most $\mathrm{id}(2^\omega)$ or else at least E_0.

Smooth equivalence relations are the ones up to $\mathrm{id}(2^\omega)$ in the Borel reducibility hierarchy. Up to Borel bireducibility they can be listed as

$$\mathrm{id}(1),\ \mathrm{id}(2),\ \ldots,\ \mathrm{id}(\omega),\ \mathrm{id}(2^\omega).$$

Essentially hyperfinite equivalence relations are either smooth or Borel bireducible with E_0, thus up to Borel bireducibility they are listed as

$$\mathrm{id}(1),\ \mathrm{id}(2),\ \ldots,\ \mathrm{id}(\omega),\ \mathrm{id}(2^\omega),\ E_0.$$

Equivalence relations beyond E_0 cannot be listed sequentially any more.

E_∞ is a universal countable Borel equivalence relation. It is known that the Borel reducibility hierarchy between E_0 and E_∞ is very complicated, but we did not get into the details in this book, and interested readers can find further results in References [1] and [86].

Many classification problems in mathematics have been identified to have the same complexity as one of the four equivalence relations. The following are some examples.

Example 15.1.1
In topology, a compact orientable surface is completely classified up to homeomorphism by its genus. Since the genus is a natural number, and its computation involves only finite triangulations and is obviously Borel, we obtain that the homeomorphic classification problem for closed orientable surfaces is Borel bireducible with $\mathrm{id}(\omega)$.

Example 15.1.2
In algebra, a finitely generated abelian group can be uniquely expressed, up to isomorphism, as a direct sum

$$\mathbb{Z}/p_1^{r_1}\mathbb{Z} \oplus \cdots \oplus \mathbb{Z}/p_n^{r_n}\mathbb{Z} \oplus \underbrace{\mathbb{Z} \oplus \cdots \oplus \mathbb{Z}}_{m}$$

where $m, n \in \omega$, $p_1, \ldots, p_n$ are prime numbers (not necessarily distinct) and $r_1, \ldots, r_n \geq 1$. It follows that the isomorphic classification problem for finitely generated abelian groups is Borel bireducible with $\mathrm{id}(\omega)$.

Example 15.1.3
In linear algebra, the similarity type of a complex square matrix is completely determined by its Jordan normal form. Recall that two $n \times n$ complex matrixes A and B are **similar** if there is an invertible matrix P such that $P^{-1}AP = B$. Every square matrix is similar to one in Jordan normal form, which is determined by the eigenvalues of the matrix as well as their multiplicities. These finitely many eigenvalues, together with their multiplicities, can be coded by a single real number. It follows that the similarity classification problem for complex square matrices is smooth.

Example 15.1.4

In dynamical systems, a **Bernoulli shift** is a triple (X, μ, T), where for some $n \geq 1$, $X = \{1, 2, \ldots, n\}^{\mathbb{Z}}$ with the standard Borel structure given by the product topology, μ is a product measure on X given by nonnegative real numbers $p_1, \ldots, p_n$ with $\sum_{i=1}^{n} p_i = 1$, and T is the shift: for any $x = (x_n)_{n \in \mathbb{Z}} \in X$, $(Tx)_{n+1} = x_n$ for all $n \in \mathbb{Z}$.

Given a Bernoulli shift as above, its **entropy** is defined as

$$E = - \sum_{i=1}^{n} p_i \log p_i.$$

Two Bernoulli shifts (X_1, μ_1, T_1) and (X_2, μ_2, T_2) are **isomorphic** if there is a map $\varphi : (X_1, \mu_1) \to (X_2, \mu_2)$ that is measure-preserving, that is, $\mu_2(\varphi(A)) = \mu_1(A)$ for any Borel set $A \subseteq X_1$, and such that for all $x \in X_1$, $\varphi(T_1 x) = T_2 \varphi(x)$.

The following celebrated theorem of Ornstein classifies Bernoulli shifts up to isomorphism by their entropies. Since the entropy is a real number, it essentially shows that the isomorphic classification problem for Bernoulli shifts is smooth.

Theorem 15.1.5 (Ornstein)

Two Bernoulli shifts are isomorphic iff they have the same entropy.

Somewhat related is the shift action of $\mathbb{Z}$ on $2^{\mathbb{Z}}$. The induced orbit equivalence relation is Borel bireducible with E_0. This is also useful in the study of dynamical systems.

As another example in algebra, we have seen that E_0 comes up in Baer's classification of torsion-free abelian groups of rank 1.

Example 15.1.6

A metric space (X, d) is **Heine–Borel** if every closed bounded subset is compact. A Heine–Borel metric space is separable and complete. Consider the isometric classification problem for all Heine–Borel ultrametric spaces. It is a theorem of Gao and Kechris [58] that it is Borel bireducible with E_0. The proof is left as exercise.

The universal countable Borel equivalence relation E_∞ is less well known. But Kechris' theorem (Corollary 7.5.3) that any locally compact Polish group action induces essentially countable orbit equivalence relations gives a sweeping classification for a wide variety of objects considered in algebra, dynamical systems, geometry, and analysis. We mention two results that employ E_∞ as the exact complexity of classification problems (from References [58] and [79] respectively).

Theorem 15.1.7 (Hjorth)

The isometric classification of all Heine–Borel Polish metric spaces is Borel bireducible with E_∞.

Theorem 15.1.8 (Hjorth–Kechris)
The conformal equivalence for Riemann surfaces is Borel bireducible with E_∞.

There are many essentially countable equivalence relations strictly in between E_0 and E_∞ (see References [1] and [153]). Many results involving them require the deep theory of superrigidity developed by Zimmer and others in dynamical systems. However, so far none has been recognized as a benchmark as in the cases of E_0 and E_∞.

Exercise 15.1.1 Show that any Heine–Borel metric space is separable and complete.

Exercise 15.1.2 Let (X, d) be a Heine–Borel ultrametric space. For any $x \in X$ and $r > 0$, let $B_r^c(x) = \{y \in X : d(x, y) \leq r\}$.

(a) Show that for any $x, y \in X$ and $r > d(x, y)$, $B_r^c(x) = B_r^c(y)$.

(b) Show that two Heine–Borel ultrametric spaces (X, d_X) and (Y, d_Y) are isometric iff for any $x \in X$ and $y \in Y$ there is $n \in \omega$ such that for all $m \geq n$, $B_m^c(x) \cong_i B_m^c(y)$.

(c) Deduce from (b) that the isometric classification of Heine–Borel ultrametric spaces is Borel reducible to E_0.

(d) Give a Borel reduction from E_0 to the isometry of countable Heine–Borel ultrametric spaces.

15.2 A roadmap of Borel equivalence relations

In this section we identify benchmark Borel equivalence relations. Recall that any Borel equivalence relation E is strictly below its Friedman–Stanley jump E^+ (Theorem 8.3.6). Also we may define finite or countable products of Borel equivalence relations; in particular, for any Borel equivalence relation E we have $E \leq_B E^\omega$. These operations allow us to generate a complicated list of equivalence relations even starting from a relatively short list of basic equivalence relations. We will first focus on giving a list of $\mathbf{\Pi}_3^0$ equivalence relations.

In Chapter 8 we have introduced the following $\mathbf{\Pi}_3^0$ equivalence relations:

$$E_0, \ E_\infty, \ \ell_p(1 \leq p < \infty), \ E_1, \ \ell_\infty, \ E_0^\omega, \ c_0, \ =^+, \ \equiv_m .$$

Dougherty and Hjorth showed that for $1 \leq p, q < \infty$, $\ell_p < \ell_q$ iff $p < q$ (Theorem 8.4.3). It is a theorem of Kechris and Louveau [102] that E_1 is the unique nonhyperfinite hypersmooth equivalence relation up to Borel bireducibility.

They also showed that E_1 is not below any orbit equivalence relation (Theorem 10.6.1). Rosendal's Theorem 8.4.2 identifies ℓ_∞ as a universal K_σ equivalence relation; this positions it above every equivalence relation that comes before it in the above list. It is known that E_0^ω is below c_0 and $=^+$, and that ℓ_2 and $=^+$ are below the measure equivalence $\equiv_m$ (Section 8.5). In fact $=^+$ is the universal $\mathbf{\Pi}_3^0$ S_∞-orbit equivalence relation (Theorem 12.5.5). And by the theory of turbulence, neither ℓ_1 nor c_0 is below $=^+$ (Section 10.5). This implies that none of ℓ_p $(1 \le p \le \infty)$, c_0, or $\equiv_m$ is below $=^+$. We summarize these reductions in the following figure.

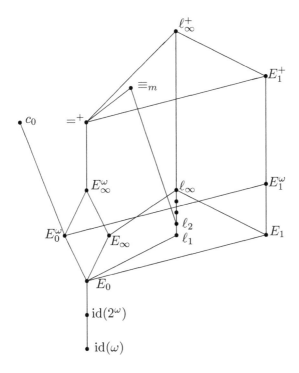

Figure 15.1 Borel reductions among $\mathbf{\Pi}_3^0$ equivalence relations.

In Figure 15.1 each line represents a known Borel reduction from the horizontally lower equivalence relation to the higher one. Other Borel reductions can be obtained by compositions. For pairs of equivalence relations for which no Borel reduction is implied by the figure, some nonreducibility results are known, such as the ones mentioned in the preceding paragraph, but more often the question is an open problem. For instance, it is not known if $\ell_p \le_B \equiv_m$ iff $p \le 2$, or whether $c_0 \le_B \equiv_m$.

Among $\mathbf{\Pi}_3^0$ equivalence relations ℓ_∞ and $\equiv_m$ show up the most often in classification problems. The measure equivalence $\equiv_m$ is known to be Borel bireducible with the unitary equivalence of bounded normal operators (or unitary operators, or bounded self-adjoint operators) by the Spectral Theory of functional analysis (see, for example, Reference [104]). By Rosendal's theorem ℓ_∞ is known to be bireducible with a number of classification problems involving biembeddability. For instance, the isometric biembeddability problem for compact metric spaces is known to have this complexity [132]. This is in contrast with Gromov's theorem that the isometric classification of compact metric spaces is smooth (Theorem 14.2.1).

There are very few benchmarks of Borel equivalence relations beyond $\mathbf{\Pi}_3^0$. The Friedman–Stanley tower is a cofinal ω_1-sequence of Borel S_∞-orbit equivalence relations (Section 12.2) (we postpone their further discussion to the next section). Rosendal [132] recently constructed a cofinal ω_1-sequence of Borel equivalence relations, but their connection with classification problems has not been well studied.

Exercise 15.2.1 Show that $E^\omega \leq_B (E \times \mathrm{id}(2^\omega))^+$. Thus if E is a Borel equivalence relation such that $E \times \mathrm{id}(2^\omega) \leq_B E$, then $E^\omega \leq_B E^+$.

Exercise 15.2.2 Verify all the reducibility claims in Figure 15.1.

15.3   Orbit equivalence relations

In Figure 15.1 most equivalence relations are induced by Polish group actions. The exceptions are all the equivalence relations from E_1 and above. It is an open problem if nonreducibility from E_1 characterizes orbit equivalence relations.

Among the orbit equivalence relations the best understood are S_∞-orbit equivalence relations. In Figure 15.1 these include all the equivalence relations from $=^+$ and below. The rest of the Friedman–Stanley tower include $=^{\alpha+}$ for all $2 \leq \alpha < \omega_1$, which form a strictly increasing ω_1-sequence of Borel equivalence relations. By the Scott analysis the Friedman–Stanley tower is cofinal in all Borel S_∞-orbit equivalence relations. There are two distinguished non-Borel isomorphism relations which are frequently used as benchmarks to gauge the complexity of other equivalence relations and classification problems. The first is the isomorphism relation for countable torsion abelian groups. Because of its classification by the Ulm invariants (Sections 9.2 and 13.4), we denote this equivalence relation $\mathrm{id}(2^{<\omega_1})$. One can form a tower similar to the relativized Friedman–Stanley tower above $\mathrm{id}(2^{<\omega_1})$, but it has not been very useful for classification problems (mostly isomorphism problems for countable structures) considered so far. The other distinguished isomorphism relation is the universal S_∞-orbit equivalence relation, sometimes denoted by $E^\infty_{S_\infty}$. By our results in Chapter 13 it has many realizations, but we use the graph isomorphism most often. Figure 15.2 summarizes S_∞-orbit equivalence relations.

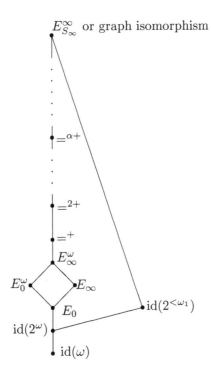

$E^\infty_{S_\infty}$ or graph isomorphism

Figure 15.2 S_∞-orbit equivalence relations.

Many classification problems allow countable structures as complete invariants, and therefore regardless of the nature of the objects their complexity becomes that of one of the S_∞-orbit equivalence relations. For instance, Camerlo and Gao [17] showed that the isomorphism relation of commutative almost finite-dimensional C^*-algebras is Borel complete, that is, Borel bireducible with the graph isomorphism. Gao and Kechris [58] showed that the isometric classification of Polish ultrametric spaces is also Borel complete.

Among other orbit equivalence relations the one with an obvious importance is the universal orbit equivalence relation. As we proved in the preceding chapter, it can be realized as a universal Iso($\mathbb{U}$)-orbit equivalence relation, which we denote by $E^\infty_{\text{Iso}(\mathbb{U})}$, or as the isometric classification problem for all Polish metric spaces, which is denoted $\cong_i$.

It has been shown by Melleray [121] and Weaver [165] that the linear isometric classification for separable Banach spaces is Borel bireducible with $\cong_i$.

Exercise 15.3.1 For any Polish group G let E_G^∞ be a universal G-orbit equivalence relation. Give a diagram of Borel reducibility among E_G^∞ where G varies over the following list:

$$\mathbb{Z}, \ \mathbb{F}_2, \ \ell_2, \ S_\infty, \ \mathrm{Iso}(\mathbb{U}).$$

15.4 General $\mathbf{\Sigma}_1^1$ equivalence relations

By general $\mathbf{\Sigma}_1^1$ equivalence relations we mean $\mathbf{\Sigma}_1^1$ equivalence relations that are non-Borel and nonorbit equivalence relations. As we discussed in Chapter 9 the definable reducibility notion suitable for general $\mathbf{\Sigma}_1^1$ equivalence relations seems to be that of $\mathbf{\Delta}_2^1$ reductions. There are more absolute notions such as C-measurable reducibility and absolutely $\mathbf{\Delta}_2^1$ reducibility. For equivalence relations and classification problems arising naturally, very often one can show Borel reductions or even continuous reductions.

There are two kinds of general $\mathbf{\Sigma}_1^1$ equivalence relations that come up often in practice. One kind includes equivalence relations with ω_1 many equivalence classes. We saw in Chapter 9 some examples of such equivalence relations. One can show that they are $\mathbf{\Delta}_2^1$ bireducible with one another, but probably not Borel bireducible. To be consistent with the notation we have been using, we denote any representative of this kind of equivalence relations by $\mathrm{id}(\omega_1)$. The Burgess trichotomy theorem (Theorem 9.1.5) is an analog of the Silver dichotomy for $\mathbf{\Sigma}_1^1$ equivalence relations. It states that any $\mathbf{\Sigma}_1^1$ equivalence relation has either countably many, ω_1 many, or perfectly many equivalence classes.

$\mathrm{id}(\omega_1)$ is above $\mathrm{id}(\omega)$ but $\mathbf{\Delta}_2^1$ reducible to $\mathrm{id}(2^{<\omega_1})$. By the Silver dichotomy no $\mathbf{\Pi}_1^1$ equivalence relation is $\mathbf{\Delta}_2^1$ bireducible with $\mathrm{id}(\omega_1)$. But whether it is $\mathbf{\Delta}_2^1$ bireducible with an orbit equivalence relation is equivalent to the topological Vaught conjecture and is therefore open. No natural classification problems have been identified with this equivalence relation.

Hjorth and Kechris [76] have proved an analog of the Glimm–Effros dichotomy for $\mathbf{\Sigma}_1^1$ equivalence relations.

At the top is the universal $\mathbf{\Sigma}_1^1$ equivalence relation, which we denote by $E_{\mathbf{\Sigma}_1^1}$. Louveau and Rosendal [111] have recently obtained combinatorial realizations of this equivalence relation. They showed that the biembeddability of trees on ω are universal $\mathbf{\Sigma}_1^1$. More recent theorems of Ferenczi, Louveau, and Rosendal [42] related this equivalence relation with classification problems. Among the classification problems proved to be universal $\mathbf{\Sigma}_1^1$ are the isomorphism problem of separable Banach spaces and the uniform homeomorphic classification of Polish metric spaces.

15.5   Beyond analyticity

Silver dichotomy is formulated for $\mathbf{\Pi}_1^1$ equivalence relations. Hjorth [67] has shown that there is a universal $\mathbf{\Pi}_1^1$ equivalence relation. However, in practice true $\mathbf{\Pi}_1^1$ equivalence relations appear far less frequently than $\mathbf{\Sigma}_1^1$ equivalence relations.

There are examples of classification problems on even higher levels of the projective hierarchy. We have seen in the preceding chapter that the homeomorphism problem for general Polish metric spaces is $\mathbf{\Sigma}_2^1$. There are no known classification problems beyond analytic ones whose exact complexity has been determined.

However, we should probably close by remarking that the objective of invariant descriptive set theory is not to reach equivalence relations ever higher in the reducibility hierarchy, but to obtain ever deeper understandings of the ones that are most relevant to mathematics and its applications.

Appendix A

Proofs about the Gandy–Harrington Topology

In this appendix we give proofs for the theorems mentioned in Section 1.8. These theorems are harder to find in the literature and not always included in standard textbooks. However, the techniques and some of the statements can be found in References [126] and [133].

A.1 The Gandy basis theorem

In this section we give a proof of the Gandy basis theorem and deduce some consequences toward a characterization of low elements. We will use Kleene's Theorem 1.7.5 in our proofs. This is not always necessary, but it allows us to reach the proofs faster.

Lemma A.1.1
The set $\{(x, y) : \omega_1^{CK(x)} \leq \omega_1^{CK(y)}\}$ is Σ_1^1.

Proof. By Spector's Theorem 1.6.12 the condition $\omega_1^{CK(x)} \leq \omega_1^{CK(y)}$ is equivalent to

$$\forall u \in \Delta_1^1(x) \ [u \in WO \to (\exists z)(\exists f)(\ z \text{ is computable in } y \text{ and}$$
$$f \text{ is an order-preserving bijection of } (\omega, <_u) \text{ onto } (\omega, <_z))],$$

which is Σ_1^1 by Kleene's Theorem 1.7.5. $\qquad\square$

On the flip side we have the following immediate consequence.

Lemma A.1.2
The set $\{(x, y) : \omega_1^{CK(x)} < \omega_1^{CK(y)}\}$ is Π_1^1.

A basic fact about Σ_1^1 sets is that for every Σ_1^1 set $A \subseteq \omega^\omega$ there is a computable tree T on $\omega \times \omega$ such that

$$x \in A \iff \exists y \in \omega^\omega \ \forall n \in \omega \ (x{\upharpoonright}n, y{\upharpoonright}n) \in T.$$

Recall that for any tree T on ω the Kleene–Brouwer ordering of T is defined by

$$s <^{T}_{\text{KB}} t \iff t \subsetneq s \ \lor \ \exists i \in \omega \ (\forall j < i \ s(j) = t(j) \ \land \ s(i) < t(i)).$$

It is easy to see that $<^{T}_{\text{KB}}$ is a linear order, and that it is a well-order iff T is well-founded. It is clear from the definition that $<^{T}_{\text{KB}}$ is computable in T. Via the standard coding we may similarly define the Kleene–Brouwer ordering for any tree T on $\omega \times \omega$. It follows easily that if T is computable, then so is $<^{T}_{\text{KB}}$, and that T is well-founded iff $<^{T}_{\text{KB}}$ is a well-order.

We need the following notation. Kleene's O is the set

$$O = \{e \in \omega : \{e\} \in \text{WO}\}.$$

Thus Kleene's O is just the set of indices of computable well-orders. If T is a computable tree on ω, then there is $e \in \omega$ so that $<^{T}_{\text{KB}}$ is computed by $\{e\}$, and moreover the map $T \mapsto e$ is computable. It follows that T is well-founded iff $e \in O$, and the relation "T is well-founded" is computable in Kleene's O. The same argument applies to computable trees on $\omega \times \omega$.

We are now ready to prove a basis theorem of Kleene, which will be used in the proof of the Gandy basis theorem.

Theorem A.1.3 (Kleene)
The set $\{x : x$ is computable in $O\}$ is a basis for Σ^1_1 sets. That is, any nonempty Σ^1_1 set contains an element x computable in Kleene's O.

Proof. Let $A \in \Sigma^1_1$ be nonempty and $T \subset \omega^{<\omega} \times \omega^{<\omega}$ be a computable tree such that

$$x \in A \leftrightarrow (\exists y)(\forall n) \ (x \restriction n, y \restriction n) \in T.$$

For each $(s,t) \in T$, let $T_{(s,t)} = \{(s',t') \in T : s \subseteq s', \ t \subseteq t'\}$. Then each $T_{(s,t)}$ is computable. For any $(s,t) \in T$, let

$$z(s,t) = 1 \iff T_{(s,t)} \text{ is well-founded} \iff [T_{(s,t)}] = \emptyset.$$

Then by the properties of the Kleene–Brouwer ordering we have that z is computable in Kleene's O. Define $T' = \{(s,t) \in T : z(s,t) \neq 1\}$. Then T' is computable in O and is a pruned tree. Let (x,y) be its leftmost branch in the lexicographic order. Then (x,y) is computable in T' and therefore computable in O. It follows that x is computable in O and $x \in A$. $\quad\square$

The Gandy basis theorem follows.

Theorem A.1.4 (Gandy)
The set $\{x : \omega^{\text{CK}(x)}_1 = \omega^{\text{CK}}_1\}$ is a basis for Σ^1_1 sets. That is, any nonempty Σ^1_1 set contains an element x such that $\omega^{\text{CK}(x)}_1 = \omega^{\text{CK}}_1$.

Proof. Now suppose $A \in \Sigma_1^1$. Consider the set

$$\{(x, y) : x \in A \text{ and } y \notin \Delta_1^1(x)\}.$$

It follows from Kleene's Theorem 1.7.5 that the set is Σ_1^1. If $A \neq \emptyset$ then the above set is nonempty, since there are only countably many reals in $\Delta_1^1(x)$. By the Kleene basis theorem above there are x, y such that (x, y) is computable in O, $x \in A$, and $y \notin \Delta_1^1(x)$. Thus $x \in \Delta_1^1(O)$ and $y \in \Delta_1^1(O)$ but $y \notin \Delta_1^1(x)$, and thus $O \notin \Delta_1^1(x)$.

We verify that $w_1^{CK(x)} = w_1^{CK}$. Toward a contradiction assume this is not the case. Then $w_1^{CK} < w_1^{CK(x)}$, and it follows that there is a u computable in x such that $\mathrm{ot}(<_u) = w_1^{CK}$. Now for any $e \in \omega$, $e \in O$ iff $\{e\} \in \mathrm{WO}$ iff there is an order-preserving injection f from $(\omega, \leq_{\{e\}})$ into $(\omega, <_u)$. This shows that O is $\Sigma_1^1(x)$. But since O is Π_1^1, it is also $\Pi_1^1(x)$, hence is $\Delta_1^1(x)$, a contradiction. $\square$

Corollary A.1.5
$\{x : w_1^{CK(x)} = w_1^{CK}\}$ is not Δ_1^1.

Proof. If it were Π_1^1 then its complement, which is nonempty, is also Σ_1^1. This contradicts the Gandy basis theorem. $\square$

We can now give a proof of Theorem 1.8.3.

Theorem A.1.6
The following are equivalent:

(i) $w_1^{CK(x)} = w_1^{CK}$.

(ii) $\forall A \in \Sigma_1^1 \ [x \notin A \rightarrow \exists B \in \Sigma_1^1 \ (x \in B \ \wedge \ A \cap B = \emptyset)]$.

(iii) $\forall A \in \Sigma_1^1 \ [x \notin A \rightarrow \exists B \in \Delta_1^1 \ (x \in B \ \wedge \ A \cap B = \emptyset)]$.

Proof. That (iii)$\Rightarrow$(ii) is obvious. The implication (ii)$\Rightarrow$(i) is an immediate corollary of the Gandy basis theorem. In fact, assume that it fails for x. Let $A = \{y : w_1^{CK(y)} = w_1^{CK}\}$, which is Σ_1^1 by Lemma A.1.1. Then $x \notin A$, and by the assumed property there is a Σ_1^1 set B with $x \in B$ and $A \cap B = \emptyset$. Since B contains x, it is nonempty, hence must meet A by the Gandy basis theorem. This contradicts $B \cap A = \emptyset$.

It remains to show (i)$\Rightarrow$(iii). For this let $w_1^{CK(x)} = w_1^{CK}$. Let $A \in \Sigma_1^1$ with $x \notin A$ and T be a computable tree such that

$$y \in A \leftrightarrow \exists z \, \forall n \, (y \restriction n, z \restriction n) \in T.$$

Now the tree $T_x = \{s : (x \restriction \mathrm{lh}(s), s) \in T\}$ is computable in x and is well-founded because $x \notin A$. Let γ be the order type of the Kleene–Brouwer ordering $<^{T_x}_{\text{KB}}$. Then $\gamma < \omega_1^{\text{CK}} = \omega_1^{\text{CK}(x)}$ since T_x is computable in x. Consider the set

$$B = \{y \;:\; \text{the order type of the Kleene–Brouwer ordering } <^{T_y}_{\text{KB}} \text{ is } \leq \gamma \}.$$

Since WO^0_γ is Δ^1_1, so is B. Now we have $x \in B$. But for any $y \in A$, T_y is ill-founded, and hence $y \notin B$. So $A \cap B = \emptyset$. □

A.2   The Gandy–Harrington topology on X_{low}

Let
$$X_{\text{low}} = \{x \;:\; \omega_1^{\text{CK}(x)} = \omega_1^{\text{CK}}\}.$$

Elements of X_{low} are called low.

We essentially showed in the preceding section that X_{low} is Σ^1_1 but not Π^1_1, and it meets every nonempty Σ^1_1 set. However, for the record we note the following fact.

Lemma A.2.1
X_{low} is Borel.

Proof.   Note that if $x \notin X_{\text{low}}$ then $\omega_1^{\text{CK}(x)} > \omega_1^{\text{CK}}$, and thus there is $e \in \omega$ such that $\{e\}^x \in \text{WO}$ and $\text{ot}(<_{\{e\}^x}) = \omega_1^{\text{CK}}$. We thus have that

$$x \notin X_{\text{low}} \iff \exists e \in \omega \; \{e\}^x \in \text{WO}_{\omega_1^{\text{CK}}} - \bigcup_{\alpha < \omega_1^{\text{CK}}} \text{WO}_\alpha.$$

Since WO_α is Borel for all $\alpha < \omega_1$, it follows that X_{low} is a countable intersection of Borel sets, and hence is Borel. □

Recall that the Gandy–Harrington topology on ω^ω is the topology generated by all Σ^1_1 sets. Being a Σ^1_1 set, X_{low} is basic open in the Gandy–Harrington topology. We consider the Gandy–Harrington topology restricted on X_{low}.

Theorem A.1.6 implies that every Σ^1_1 set restricted to X_{low} is clopen. Thus the Gandy–Harrington topology on X_{low} is 0-dimensional, and in particular regular. Since it is also second countable, it thus is metrizable by the Urysohn metrization theorem. We next show that it is completely metrizable, and hence is Polish.

Theorem A.2.2
X_{low} with the Gandy–Harrington topology is Polish.

Proof. We give an indirect proof using strong Choquet spaces. By Theorem 4.1.5 the Baire space ω^ω with the Gandy–Harrington topology is strong Choquet. Since X_{low} is an open subset of ω^ω with the Gandy–Harrington topology, by Theorem 4.1.2 (a) the Gandy–Harrington topology on X_{low} is strong Choquet. Now note that the Gandy–Harrington topology on X_{low} is second countable and regular. It is finer than the usual topology, and hence is Hausdorff, and furthermore T_3. Hence this topology is second countable, T_3, and strong Choquet. By Choquet's Theorem 4.1.4 it is Polish. $\square$

It is possible to prove directly that the Gandy–Harrington topology on X_{low} is completely metrizable by constructing a homeomorphism between this space and a G_δ subset of the Baire space ω^ω with the usual topology.

References

[1] S. Adams and A. S. Kechris, Linear algebraic groups and countable Borel equivalence relations, *Journal of the American Mathematical Society* 13 (2000), no. 4, 909–943.

[2] R. Baer, Abelian groups without elements of finite order, *Duke Mathematical Journal* 3 (1937), no. 1, 68–122.

[3] J. Barwise, *Admissible Sets and Structures: An Approach to Definability Theory.* Perspectives in Mathematical Logic. Springer-Verlag, Berlin, New York, 1975.

[4] H. Becker, The topological Vaught's conjecture and minimal counterexamples, *Journal of Symbolic Logic* 59 (1994), no. 3, 757–784.

[5] H. Becker, Polish group actions: dichotomies and generalized elementary embeddings, *Journal of the American Mathematical Society* 11 (1998), no. 2, 397–449.

[6] H. Becker, Topics in invariant descriptive set theory, *Annals of Pure and Applied Logic* 111 (2001), no. 3, 145–184.

[7] H. Becker and A. S. Kechris, Borel actions of Polish groups, *Bulletin of the American Mathematical Society* 28 (1993), no. 2, 334–341.

[8] H. Becker and A. S. Kechris, *The Descriptive Set Theory of Polish Group Actions.* London Mathematical Society Lecture Notes Series 232. Cambridge University Press, Cambridge, 1996.

[9] M. Benda, Remarks on countable models, *Fundamenta Mathematicae* 81 (1974), 107–119.

[10] *Logic and Its Applications.* Papers from the Conference on Logic and its Applications in Algebra and Geometry held at the University of Michigan, Ann Arbor, MI, April 11–13, 2003 and the Workshop on Combinatorial Set Theory, Excellent Classes and Schanuel's Conjecture held at the University of Michigan, Ann Arbor, MI, April 14–15, 2003. Edited by A. Blass and Y. Zhang. Contemporary Mathematics 380. American Mathematical Society, Providence, RI, 2005.

[11] G. S. Boolos and R. C. Jeffrey, *Computability and Logic.* Third edition. Cambridge University Press, Cambridge, 1989.

[12] D. S. Bridges, *Computability: A Mathematical Sketchbook.* Graduate Texts in Mathematics 146. Springer-Verlag, New York, 1994.

[13] J. P. Burgess, Equivalences generated by families of Borel sets, *Proceedings of the American Mathematical Society* 69 (1978), no. 2, 323–326.

[14] J. P. Burgess, A reflection phenomenon in descriptive set theory, *Fundamenta Mathematicae* 104 (1979), no. 2, 127–139.

[15] J. P. Burgess, A measurable selection theorem, *Fundamenta Mathematicae* 110 (1980), no. 2, 91–100.

[16] J. P. Burgess, A selection theorem for group actions, *Pacific Journal of Mathematics* 80 (1980), no. 2, 333–336.

[17] R. Camerlo and S. Gao, The completeness of the isomorphism relation for countable Boolean algebras, *Transactions of the American Mathematical Society* 353 (2001), no. 2, 491–518.

[18] *Proceedings of the International Congress of Mathematicians*, vol. 1, 2. Held in Zürich, August 3–11, 1994. Edited by S. D. Chatterji. Birkhäuser Verlag, Basel, 1995.

[19] *Computability Theory and Its Applications; Current Trends and Open Problems.* Proceedings of the AMS-IMS-SIAM Joint Summer Research Conference held at the University of Colorado, Boulder, CO, June 13–17, 1999. Edited by P. A. Cholak, S. Lempp, M. Lerman, and R. A. Shore. Contemporary Mathematics 257. American Mathematical Society, Providence, RI, 2000.

[20] *The Notre Dame Lectures.* Edited by P. Cholak. Lecture Notes in Logic 18. Association for Symbolic Logic, Urbana, IL, 2005.

[21] J. D. Clemens, Classifying Borel automorphisms, *Journal of Symbolic Logic* 72 (2007), no. 4, 1081–1092.

[22] J. D. Clemens, S. Gao, and A. S. Kechris, Polish metric spaces: their classification and isometry groups, *Bulletin of Symbolic Logic* 7 (2001), no. 3, 361–375.

[23] *Sets and Proofs.* Invited papers from Logic Colloquium '97 (European Meeting of the Association for Symbolic Logic) held in Leeds, July 6–13, 1997. Edited by S. B. Cooper and J. K. Truss. London Mathematical Society Lecture Note Series 258. Cambridge University Press, Cambridge, 1999.

[24] N. Cutland, *Computability: An Introduction to Recursive Function Theory.* Cambridge University Press, Cambridge-New York, 1980.

[25] L. Ding and S. Gao, Diagonal actions and Borel equivalence relations, *Journal of Symbolic Logic* 71 (2006), no. 4, 1081–1096.

[26] L. Ding and S. Gao, On generalizations of Lavrentieff's theorem for Polish group actions, *Transactions of the American Mathematical Society* 359 (2007), no. 1, 417–426.

[27] L. Ding and S. Gao, New metrics on free groups, *Topology and Its Applications* 154 (2007), 410–420.

[28] L. Ding and S. Gao, Graev metric groups and Polishable subgroups, *Advances in Mathematics* 213 (2007), 887–901.

[29] R. Dougherty and G. Hjorth, Reducibility and nonreducibility between ℓ_p equivalence relations, *Transactions of the American Mathematical Society* 351 (1999), no. 5, 1835–1844.

[30] R. Dougherty, S. Jackson, and A. S. Kechris, The structure of hyperfinite Borel equivalence relations, *Transactions of the American Mathematical Society* 341 (1994), no. 1, 193–225.

[31] R. Dougherty and A. S. Kechris, How many Turing degrees are there? in *Computability Theory and Its Applications* (Boulder, CO, 1999), 83–94. Contemporary Mathematics 257. American Mathematical Society, Providence, RI, 2000.

[32] E. G. Effros, Transformation groups and C^*-algebras, *Annals of Mathematics (2)* 81 (1965), no. 1, 38–55.

[33] I. Farah, Ideals induced by Tsirelson submeasures, *Fundamenta Mathematicae* 159 (1999), no. 3, 243–258.

[34] I. Farah, Basis problem for turbulent actions, I, Tsirelson submeasures, *Annals of Pure and Applied Logic* 108 (2001), no. 1-3, 189–203.

[35] I. Farah, Basis problem for turbulent actions, II, c_0-equalities, *Proceedings of the London Mathematical Society (3)* 82 (2001), no. 1, 1–30.

[36] I. Farah and S. Solecki, Two $F_{\sigma\delta}$ ideals, *Proceedings of the American Mathematical Society* 131 (2003), no. 6, 1971–1975.

[37] I. Farah and S. Solecki, Borel subgroups of Polish groups, *Advances in Mathematics* 199 (2006), no. 2, 499–541.

[38] J. Feldman, Borel structures and invariants for measurable transformations, *Proceedings of the American Mathematical Society* 46 (1974), 383–394.

[39] J. Feldman, P. Hahn, and C. C. Moore, Orbit structure and countable sections for actions of continuous groups, *Advances in Mathematics* 28 (1978), no. 3, 186–230.

[40] J. Feldman and C. C. Moore, Ergodic equivalence relations, cohomology, and von Neumann algebras, I, *Transactions of the American Mathematical Society* 234 (1977), no. 2, 289–324.

[41] J. Feldman and C. C. Moore, Ergodic equivalence relations, cohomology, and von Neumann algebras, I, *Transactions of the American Mathematical Society* 234 (1977), no. 2, 325–359.

[42] V. Ferenczi, A. Louveau, and C. Rosendal, The complexity of classifying separable Banach spaces up to isomorphism, preprint, 2006.

[43] *Analysis and Logic.* Lectures from the mini-courses offered at the International Conference held at the University of Mons-Hainaut, Mons, August 25–29, 1997. Edited by C. Finet and C. Michaux. London Mathematical Society Lecture Note Series 262. Cambridge University Press, Cambridge, 2002.

[44] *Descriptive Set Theory and Dynamical Systems.* Papers from the International Workshop held in Marseille-Luminy, July 1–5, 1996. Edited by M. Foreman, A. S. Kechris, A. Louveau, and B. Weiss. London Mathematical Society Lecture Note Series 277. Cambridge University Press, Cambridge, 2000.

[45] H. Friedman, On the necessary use of abstract set theory, *Advances in Mathematics* 41 (1981), 209–280.

[46] H. Friedman and L. Stanley, A Borel reducibility theory for classes of countable structures, *Journal of Symbolic Logic* 54 (1989), no. 3, 894–914.

[47] S. D. Friedman and B. Veličković, Nonstandard models and analytic equivalence relations, *Proceedings of the American Mathematical Society* 125 (1997), no. 6, 1807–1809.

[48] *General Topology and Its Relations to Modern Analysis and Algebra, VI.* Proceedings of the Sixth Prague Topological Symposium held in Prague, August 25–29, 1986. Edited by Z. Frolík. Research and Exposition in Mathematics 16. Heldermann Verlag, Berlin, 1988.

[49] S. Gao, On automorphism groups of countable structures, *Journal of Symbolic Logic* 63 (1998), no. 3, 891–896.

[50] S. Gao, A dichotomy theorem for mono-unary algebras, *Fundamenta Mathematicae* 163 (2000), no. 1, 25–37.

[51] S. Gao, A remark on Martin's conjecture, *Journal of Symbolic Logic* 66 (2001), no. 1, 401–406.

[52] S. Gao, Some dichotomy theorems for isomorphism relations of countable models, *Journal of Symbolic Logic* 66 (2001), no. 2, 902–922.

[53] S. Gao, Some applications of the Adams–Kechris technique, *Proceedings of the American Mathematical Society* 130 (2001), no. 3, 863–874.

[54] S. Gao, The homeomorphism problem for countable topological spaces, *Topology and Its Applications* 139 (2004), no. 1-3, 97–112.

[55] S. Gao, Unitary group actions and Hilbertian Polish metric spaces, in *Logic and Its Applications*, 53–72. Contemporary Mathematics 380. American Mathematical Society, Providence, RI, 2005.

[56] S. Gao, Equivalence relations and classical Banach spaces, in *Mathematical Logic in Asia* (Novosibirsk, Russia, 2005), 70–89. World Scientific, Hackensack, NJ, 2006.

[57] S. Gao, Complexity ranks of countable models, *Notre Dame Journal of Formal Logic* 48 (2007), no. 1, 33–48.

[58] S. Gao and A. S. Kechris, On the classification of Polish metric spaces up to isometry, *Memoirs of the American Mathematical Society* 161 (2003), no. 766, viii+78 pp.

[59] S. Gao and V. Pestov, On a universality property of some abelian Polish groups, *Fundamenta Mathematicae* 179 (2003), no. 1, 1–15.

[60] J. Glimm, Locally compact transformation groups, *Transactions of the American Mathematical Society* 101 (1961), 124–138.

[61] *Mathematical Logic in Asia*. Proceedings of the 9th Asian Logic Conference held in Novosibirsk, August 16–19, 2005. Edited by S. S. Goncharov, R. Downey, and H. Ono. World Scientific, Hackensack, NJ, 2006.

[62] M.I. Graev, Free topological groups, *American Mathematical Society Translation* 1951 (35) (1951), 61 pp.

[63] M. Gromov, *Metric Structures for Riemannian and Non-Riemannian Spaces*. Progress in Mathematics 152. Birkhäuser, Boston, 1999.

[64] L. A. Harrington, A. S. Kechris, and A. Louveau, A Glimm–Effros dichotomy for Borel equivalence relations, *Journal of the American Mathematical Society* 3 (1990), no. 4, 903–928.

[65] *Harvey Friedman's Research on the Foundations of Mathematics*. Edited by L. A. Harrington, M. D. Morley, A. Ščedrov, and S. G. Simpson. Studies in Logic and the Foundations of Mathematics 117. North-Holland, Amsterdam, 1985.

[66] G. Hjorth, A dichotomy for the definable universe, *Journal of Symbolic Logic* 60 (1995), no. 4, 1199–1207.

[67] G. Hjorth, Universal co-analytic sets, *Proceedings of the American Mathematical Society* 124 (1996), no. 12, 3867–3873.

[68] G. Hjorth, Sharper changes in topologies, *Proceedings of the American Mathematical Society* 127 (1999), no. 1, 271–278.

[69] G. Hjorth, Actions by the classical Banach spaces, *Journal of Symbolic Logic* 65 (2000), no. 1, 392–420.

[70] G. Hjorth, *Classification and Orbit Equivalence Relations.* Mathematics Surveys and Monographs 75. American Mathematical Society, Providence, RI, 2000.

[71] G. Hjorth, The isomorphism relation on countable torsion free abelian groups, *Fundamenta Mathematicae* 175 (2002), no. 3, 241–257.

[72] G. Hjorth, Countable models and the theory of Borel equivalence relations, in *The Notre Dame Lectures*, 1–43. Lecture Notes in Logic 18. Association for Symbolic Logic, Urbana, IL, 2005.

[73] G. Hjorth, A dichotomy theorem for being essentially countable, in *Logic and Its Applications*, 109–127. Contemporary Mathematics 380. American Mathematical Society, Providence, RI, 2005.

[74] G. Hjorth, Subgroups of abelian Polish groups, in *Set Theory*, 297–308. Trends in Mathematics. Birkäuser, Basel, 2006.

[75] G. Hjorth, A note on counterexamples to the Vaught conjecture, *Notre Dame Journal of Formal Logic* 48 (2007), no. 1, 49–51.

[76] G. Hjorth and A. S. Kechris, Analytic equivalence relations and Ulm-type classifications, *Journal of Symbolic Logic* 60 (1995), no. 4, 1273–1300.

[77] G. Hjorth and A. S. Kechris, Borel equivalence relations and classifications of countable models, *Annals of Pure and Applied Logic* 82 (1996), no. 3, 221–272.

[78] G. Hjorth and A. S. Kechris, New dichotomies for Borel equivalence relations, *Bulletin of Symbolic Logic* 3 (1997), no. 3, 329–346.

[79] G. Hjorth and A. S. Kechris, The complexity of the classification of Riemann surfaces and complex manifolds, *Illinois Journal of Mathematics* 44 (2000), no. 1, 104–137.

[80] G. Hjorth and A. S. Kechris, Recent developments in the theory of Borel reducibility, *Fundamenta Mathematicae* 170 (2001), no. 1-2, 21–52.

[81] G. Hjorth and A. S. Kechris, Rigidity theorems for actions of product groups and countable Borel equivalence relations, *Memoirs of the American Mathematical Society* 177 (2005), no. 833, viii+109 pp.

[82] G. Hjorth, A. S. Kechris, and A. Louveau, Borel equivalence relations induced by actions of the symmetric group, *Annals of Pure and Applied Logic* 92 (1998), 63–112.

[83] G. Hjorth and S. Solecki, Vaught's conjecture and the Glimm–Effros property for Polish transformation groups, *Transactions of the American Mathematical Society* 351 (1999), no. 7, 2623–2641.

[84] W. Hodges, *Model Theory*. Encyclopedia of Mathematics and Its Applications 42. Cambridge University Press, Cambridge, 1993.

[85] *Recent Progress in General Topology, II*. Edited by M. Hušek and J. van Mill. North-Holland, Amsterdam, 2002.

[86] S. Jackson, A. S. Kechris, and A. Louveau, Countable Borel equivalence relations, *Journal of Mathematical Logic* 2 (2002), no. 1, 1–80.

[87] *Set Theory of the Continuum*. Papers from the workshop held in Berkeley, California, October 16–20, 1989. Edited by H. Judah, W. Just, and W. H. Woodin. Mathematical Sciences Research Institute Publications 26. Springer-Verlag, New York, 1992.

[88] M. Katětov, On universal metric spaces, in *General Topology and Its Relations to Modern Analysis and Algebra, VI* (Prague, 1986), 323–330. Research and Exposition in Mathematics 16. Heldermann, Berlin, 1988.

[89] *Transformation Groups*. Proceedings of the International Conference held at Osaka University, Osaka, December 16–21, 1987. Edited by K. Kawakubo. Lecture Notes in Mathematics 1375. Springer-Verlag, Berlin, 1989.

[90] A. S. Kechris, Measure and category in effective descriptive set theory, *Annals of Mathematical Logic* 5 (1973), 337–384.

[91] A. S. Kechris, Amenable equivalence relations and Turing degrees, *Journal of Symbolic Logic* 56 (1991), no. 1, 182–194.

[92] A. S. Kechris, Countable sections for locally compact group actions, *Ergodic Theory and Dynamical Systems* 12 (1992), no. 2, 283–295.

[93] A. S. Kechris, The structure of Borel equivalence relations in Polish spaces, in *Set Theory of the Continuum* (Berkeley, CA, 1989), 89–102. Mathematical Sciences Research Institute Publications 26. Springer, New York, 1992.

[94] A. S. Kechris, Amenable versus hyperfinite Borel equivalence relations, *Journal of Symbolic Logic* 58 (1993), no. 3, 894–907.

[95] A. S. Kechris, Countable sections for locally compact group actions, II, *Proceedings of the American Mathematical Society* 120 (1994), no. 1, 241–247.

[96] A. S. Kechris, Topology and descriptive set theory, *Topology and Its Applications* 58 (1994), no. 3, 195–222.

[97] A. S. Kechris, *Classical Descriptive Set Theory*. Graduate Texts in Mathematics 156. Springer-Verlag, New York, 1995.

[98] A. S. Kechris, New directions in descriptive set theory, *Bulletin of Symbolic Logic* 5 (1999), no. 2, 161–173.

[99] A. S. Kechris, On the classification problem of rank 2 torsion-free abelian groups, *Journal of the London Mathematical Society* (2) 62 (2000), 437–450.

[100] A. S. Kechris, Descriptive dynamics, in *Descriptive Set Theory and Dynamical Systems* (Marseille-Luminy, 1996), 231–258. London Mathematical Society Lecture Note Series 277. Cambridge University Press, Cambridge, 2000.

[101] A. S. Kechris, Actions of Polish groups and classification problems, in *Analysis and Logic* (Mons, 1997), 115–187. London Mathematical Society Lecture Note Series 262. Cambridge University Press, Cambridge, 2002.

[102] A. S. Kechris and A. Louveau, The classification of hypersmooth Borel equivalence relations, *Journal of the American Mathematical Society* 10 (1997), no. 1, 215–242.

[103] A. S. Kechris and B. D. Miller, *Topics in Orbit Equivalence*. Lecture Notes in Mathematics 1852. Springer-Verlag, Berlin, 2004.

[104] A. S. Kechris and N. E. Sofronidis, A strong generic ergodicity property of unitary and self-adjoint operators, *Ergodic Theory and Dynamical Systems* 21 (2001), no. 5, 1459–1479.

[105] H. J. Keisler, *Model Theory for Infinitary Logic; Logic with Countable Conjunctions and Finite Quantifiers*. Studies in Logic and the Foundations of Mathematics 62. North-Holland, Amsterdam-London, 1971.

[106] J. L. Kelly, *General Topology*. Van Nostrand, New York, 1955.

[107] A. Louveau, Two results on Borel orders, *Journal of Symbolic Logic* 54 (1989), no. 3, 865–874.

[108] A. Louveau, Classifying Borel structures, in *Set Theory of the Continuum* (Berkeley, CA, 1989), 103–112. Mathematical Sciences Research Institute Publications 26. Springer, New York, 1992.

[109] A. Louveau, On the reducibility order between Borel equivalence relations, in *Logic, Methodology and Philosophy of Science, IX* (Uppsala, 1991), 151–155. Studies in Logic and the Foundations of Mathematics 134. North-Holland, Amsterdam, 1994.

[110] A. Louveau, On the size of quotients by definable equivalence relations, in *Proceedings of the International Congress of Mathematicians*, vol. 1, 2 (Zürich, 1994), 269–276. Birkhäuser, Basel, 1995.

[111] A. Louveau and C. Rosendal, Complete analytic equivalence relations, *Transactions of the American Mathematical Society* 357 (2005), no. 12, 4839-4866.

[112] A. Louveau and B. Veličković, A note on Borel equivalence relations, *Proceedings of the American Mathematical Society* 120 (1994), no. 1, 255–259.

[113] A. Louveau and B. Veličković, Analytic ideals and cofinal types, *Annals of Pure and Applied Logic* 99 (1999), no. 1-3, 171–195.

[114] G. W. Mackey, Infinite-dimensional group representations, *Bulletin of the American Mathematical Society* 69 (1963), 628–686.

[115] A. Manoussos and P. Strantzalos, On the group of isometries on a locally compact metric space, *Journal of Lie Theory* 13 (2003), no. 1, 7–12.

[116] R. Mansfield and G. Weitkamp, *Recursive Aspects of Descriptive Set Theory*. With a chapter by S. Simpson. Oxford University Press, Oxford; Clarendon Press, New York, 1985.

[117] L. Marcus, The number of countable models of a theory of one unary function, *Fundamenta Mathematicae* 108 (1980), no. 3, 171–181.

[118] D. Marker, An analytic equivalence relation not arising from a Polish group action, *Fundamenta Mathematicae* 130 (1988), 225–228.

[119] A.A. Markov, Three papers on topological groups: I. On the existence of periodic connected topological groups. II. On free topological groups. III. On unconditionally closed sets, *American Mathematical Society Translation* 1950 (30) (1950), 120 pp.

[120] A. H. Mekler, Stability of nilpotent groups of class 2 and prime exponent, *Journal of Symbolic Logic* 46 (1981), no. 4, 781–788.

[121] J. Melleray, Computing the complexity of the relation of isometry between separable Banach spaces, *Mathematical Logic Quarterly* 53 (2007), no. 2, 128–131.

[122] A. W. Miller, Vaught's conjecture for theories of one unary operation, *Fundamenta Mathematicae* 111 (1981), no. 2, 135–141.

[123] D. E. Miller, On the measurability of orbits in Borel actions, *Proceedings of the American Mathematical Society* 63 (1977), no. 1, 165–170.

[124] D. E. Miller, The invariant $\mathbf{\Pi}^0_\alpha$ separation principle, *Transactions of the American Mathematical Society* 242 (1978), 185–204.

[125] D. A. Martin and A. S. Kechris, Infinite games and effective descriptive set theory, in *Analytic Sets*, 403–470. Academic Press, London, 1980.

[126] Y. N. Moschovakis, *Descriptive Set Theory*. Studies in Logic and Foundations of Mathematics 100. North-Holland, Amsterdam, New York, 1980.

[127] M. Nadel, Scott sentences and admissible sets, *Annals of Mathematical Logic* 7 (1974), 267–294.

[128] M. Nadel, On models $\equiv_{\infty\omega}$ to an uncountable model, *Proceedings of the American Mathematical Society* 54 (1976), 307–310.

[129] D. Ornstein, Two Bernoulli shifts with infinite entropy are isomorphic, *Advances in Mathematics* 5 (1970), 339–348.

[130] *Logic, Methodology and Philosophy of Science, IX.* Proceedings of the Ninth International Congress held in Uppsala, August 7–14, 1991. Edited by D. Prawitz, B. Skyrms, and D. Westerstahl. Studies in Logic and the Foundations of Mathematics 134. North-Holland, Amsterdam, 1994.

[131] *Analytic Sets.* Edited by C. A. Rogers et al. Academic Press, London, 1980

[132] C. Rosendal, Cofinal families of Borel equivalence relations, *Journal of Symbolic Logic* 70 (2005), no. 4, 1325–1340.

[133] G. E. Sacks, *Higher Recursion Theory.* Perspectives in Mathematical Logic. Springer-Verlag, Berlin, Heidelberg, 1990.

[134] R. L. Sami, Polish group actions and the Vaught conjecture, *Transactions of the American Mathematical Society* 341 (1994), no. 1, 335–353.

[135] D. Shakhmatov, J. Pelant, and S. Watson, A universal complete metric abelian group of a given weight, in *Topology with Applications, Szekszárd, 1993*, 431–439. Bolyai Mathematical Studies 4. János Bolyai Mathematical Society, Budapest, 1995.

[136] S. A. Shkarin, On universal abelian topological groups, *Sbornik Mathematics* 190 (1999), no. 7-8, 1059–1076.

[137] J. R. Shoenfield, *Mathematical Logic.* Reprint of the 1973 second printing. Association for Symbolic Logic, Urbana, IL; A K Peters, Natick, MA, 2001.

[138] W. Sierpiński, *General Topology.* Mathematical Expositions 7. University of Toronto Press, Toronto, 1952.

[139] S. Solecki, Measurability properties of sets of Vitali's type, *Proceedings of the American Mathematical Society* 119 (1993), no. 3, 897–902.

[140] S. Solecki, Equivalence relations induced by actions of Polish groups, *Transactions of the American Mathematical Society* 347 (1995), no. 12, 4765–4777.

[141] S. Solecki, Analytic ideals, *Bulletin of Symbolic Logic* 2 (1996), no. 3, 339–348.

[142] S. Solecki, Analytic ideals and their applications, *Annals of Pure and Applied Logic* 99 (1999), no. 1-3, 51–72.

[143] S. Solecki, Polish group topologies, in *Sets and Proofs* (Leeds, 1997), 339–364. London Mathematical Society Lecture Note Series 258. Cambridge University Press, Cambridge, 1999.

[144] S. Solecki, Actions of non-compact and non-locally compact Polish groups, *Journal of Symbolic Logic* 65 (2000), no. 4, 1881–1894.

[145] S. Solecki, Descriptive set theory in topology, in *Recent Progress in General Topology, II*, 485–514. North-Holland, Amsterdam, 2002.

[146] S. Solecki, Translation invariant ideals, *Israel Journal of Mathematics* 135 (2003), 93–110.

[147] S. Solecki and S. M. Srivastava, Automatic continuity of group operations, *Topology and Its Applications* 77 (1997), no. 1, 65–75.

[148] S. M. Srivastava, *A Course on Borel Sets*. Graduate Texts in Mathematics 180. Springer, New York, 1998.

[149] L. J. Stanley, Borel diagonalization and abstract set theory: recent results of Harvey Friedman, in *Harvey Friedman's Research on the Foundations of Mathematics*, 11–86. Studies in Logic and Foundations of Mathematics 117. North-Holland, Amsterdam, 1985.

[150] J. R. Steel, On Vaught's conjecture, in *Cabal Seminar 76-77*, 193–208. Lecture Notes in Mathematics 689. Springer-Verlag, Berlin, 1978.

[151] P. Strantzalos, Action by isometries, in *Transformation Groups* (Osaka 1987), 319–325. Lecture Notes in Mathematics 1375. Springer, Berlin, 1989.

[152] Y. Suzuki, Orbits of denumerable models of complete theories, *Fundamenta Mathematicae* 67 (1970), 89–95.

[153] S. Thomas, The classification problem for torsion-free abelian groups of finite rank, *Journal of the American Mathematical Society* 16 (2003), no. 1, 233–258.

[154] S. Thomas and B. Veličković, On the complexity of the isomorphism relation for finitely generated groups, *Journal of Algebra* 217 (1999), no. 1, 352–373.

[155] S. Thomas and B. Veličković, On the complexity of the isomorphism relation for fields of finite transcendence degree, *Journal of Pure and Applied Algebra* 159 (2001), no. 2-3, 347–363.

[156] F. Topsøe and J. Hoffmann-Jorgensen, Analytic spaces and their application, in *Analytic Sets*, 317–403. Academic Press, London, 1980.

[157] P. Urysohn, Sur en espace métrique universel, *Bulletin des Sciences Mathématiques* 51 (1927), 43–64, 74–90.

[158] V. V. Uspenskij, A universal topological group with a countable basis, *Functional Analysis and Its Applications* 20 (1986), 86–87.

[159] V. V. Uspenskij, On the group of isometries of the Urysohn universal metric space, *Commentationes Mathematicae Universitatis Carolinae* 31 (1990), no. 1, 181–182.

[160] C. E. Uzcátegui A., Smooth sets for a Borel equivalence relation, *Transactions of the American Mathematical Society* 347 (1995), no. 6, 2025–2039.

[161] R. Vaught, Invariant sets in topology and logic, *Fundamenta Mathematicae* 82 (1974/75), 269–294.

[162] B. Veličković, A note on Tsirelson type ideals, *Fundamenta Mathematicae* 159 (1999), no. 3, 259–268.

[163] A. M. Vershik, The universal Urysohn space, Gromov's metric triples, and random metrics on the series of natural numbers, *Russian Mathematical Surveys* 53 (1998), no. 5, 921–928.

[164] S. Wagon, *The Banach–Tarski Paradox*. Cambridge University Press, Cambridge, 1985.

[165] N. Weaver, *Lipschitz Algebras*. World Scientific, River Edge, NJ, 1999.

[166] B. Weiss, The isomorphism problem in ergodic theory, *Bulletin of the American Mathematical Society* 78 (1972), 668–684.

[167] S. Willard, *General Topology*. Addison-Wesley, Reading, MA, 1970.

Index

Milton Keynes UK
Ingram Content Group UK Ltd.
UKHW021823071024
449327UK00021B/1406